Characterization of Biomaterials

Characterization of Biomaterials

Edited by

Amit Bandyopadhyay
School of Mechanical and Materials Engineering
Washington State University
Pullman, WA, USA

Susmita Bose
School of Mechanical and Materials Engineering
Washington State University
Pullman, WA, USA

ELSEVIER

AMSTERDAM • WALTHAM • HEIDELBERG • LONDON
NEW YORK • OXFORD • PARIS • SAN DIEGO
SAN FRANCISCO • SYDNEY • TOKYO

Elsevier
225 Wyman Street, Waltham, MA 02451, USA
The Boulevard, Langford Lane, Kidlington, Oxford OX5 1GB, UK

Library of Congress Cataloging-in-Publication Data
Characterization of biomaterials / edited by Amit Bandyopadhyay, School of Mechanical and Materials Engineering, Fellow AAAS, ASM International, AIMBE and ACerS, Washington State University, Pullman, WA, USA, Susmita Bose, School of Mechanical and Materials Engineering, Fellow AIMBE, Washington State University, Pullman, WA, USA.
pages cm
Summary: "Characterization of Biomaterials will serve as a comprehensive resource for biomaterials researchers requiring detailed information on physical, chemical, mechanical, surface, *in vitro* or *in vivo* characterization. The book is designed for materials scientists, bioengineers, biologists, clinicians and biomedical device researchers seeking input towards planning on how to test their novel materials or structures or biomedical devices towards a specific application. Chapters are developed considering the need for both industrial researchers as well as academics"– Provided by publisher.
Includes bibliographical references and index.
ISBN 978-1-4933-0137-9 (paperback)
1. Biomedical materials. I. Bandyopadhyay, Amit, editor of compilation. II. Bose, Susmita, editor of compilation.
R857.M3C47 2013
660.6'3–dc23
2012051185

British Library Cataloguing in Publication Data
A catalogue record for this book is available from the British Library

ISBN: 978-1-4933-0137-9

For information on all Elsevier publications
visit our website at www.store.elsevier.com

Printed and bound in USA

13 14 15 16 17 10 9 8 7 6 5 4 3 2 1

Cover Credits:
1. Taxus Liberté drug-eluting stent - Courtesy Boston Scientific Corporation.
2. Bone cell-materials interaction on ceramic scaffold surface- Courtesy Mr. Gary Fielding, WSU, USA.
3. Radiographic image of 3D printed ceramic scaffold in rabbit femur - Courtesy Dr. Samit Nandi, WBUAFS, India.

This book has been manufactured using Print on Demand technology. Each copy is produced to order and is limited to black ink. The online version of this book will show color figures where appropriate.

Contents

Preface

The field of biomaterials is inherently multidisciplinary involving materials science, physical, engineering, biological and clinical sciences. Therefore, it is important for biomaterials researchers to understand what to and how to characterise different biomaterials. In recent years, there are quite a few good books published on introduction to biomaterials topic, however, no specific book is available towards how to characterise different biomaterials. The aim to develop this book was to focus primarily on biomaterials characterisation, something that can help graduate students and biomaterials professionals to learn specifically what techniques are good for testing certain biomaterial properties based on applications and how to do it. Moreover, what properties are good to test during materials development versus which properties should be tested during device development? These questions may be simple to some experienced biomaterials scientists, but many times hard to find answers for others. Considering these issues, this book is developed with specific emphasis towards characterisation of biomaterials and biomedical devices for their physical, mechanical, surface, *in vitro* and *in vivo* biological properties. Special attention was given towards device level characterisation for orthopaedic and cardiovascular implants. Chapters dealing with physical, mechanical and surface properties of biomaterials are developed by researchers from physical and engineering sciences. These chapters are focused on techniques that can reveal basic properties such as atomic structures, bonding, chemical interactions, phase identification and transformation, strength and toughness measurements, and property measurements at the surface level. Most of these techniques are followed by biomaterials researchers during early stage of materials development towards certain applications. The following chapters include *in vitro* and *in vivo* testing of biomaterials including microbial interactions and biofilm characterisation. These levels of testing are needed before a biomaterial can be considered for clinical trials or in actual device manufacturing. The final part of the book deals with device level characterisation with special emphasis on orthopaedic and cardiovascular devices. Researchers with industrial product development background contributed towards these chapters.

Apart from all the contributing authors, we also like to thank many of our students for their support towards developing this book particularly Mr Solaiman Tarafder, Mr Gary Fielding, Mr Himanshu Sahasrabudhe and Ms Sahar Vahabzadeh. We are also grateful to our boys, Shohom and Aditya, without their cooperation we could not have completed this work. We hope that better understanding of biomaterials and biomedical device will not only save time but also will make them safer to use, a dream that resonates with every biomaterials researcher.

Susmita Bose and Amit Bandyopadhyay
Pullman, WA, USA
February 2013.

List of Contributors

Nehal I. Abu-Lail, Gene and Linda Voiland School of Chemical Engineering and Bioengineering, Washington State University, Pullman, WA, USA

Amit Bandyopadhyay, W. M. Keck Biomedical Materials Research Lab, School of Mechanical and Materials Engineering, Washington State University, Pullman, WA, USA

Haluk Beyenal, Gene and Linda Voiland School of Chemical Engineering and Bioengineering, Washington State University, Pullman, WA, USA

Subhasish Biswas, Department of Livestock Products Technology, West Bengal University of Animal & Fishery Sciences, Kolkata, India

Aldo R. Boccaccini, Institute of Biomaterials, University of Erlangen-Nuremberg, Erlangen, Germany

Susmita Bose, W. M. Keck Biomedical Materials Research Lab, School of Mechanical and Materials Engineering, Washington State University, Pullman, WA, USA

Hengchu Cao, Edwards Lifesciences LLC, One Edwards Way Irvine, CA, USA

Paul K. Chu, Department of Physics and Materials Science, City University of Hong Kong, Kowloon, Hong Kong, China

Rainer Detsch, Institute of Biomaterials, University of Erlangen-Nuremberg, Erlangen, Germany

Gautam Gupta, Biomet Inc, Warsaw, IN, USA

Imran Khan, Biomet Inc, Warsaw, IN, USA

T.S. Sampath Kumar, Department of Metallurgical and Materials Engineering, Indian Institute of Technology Madras, Chennai, India

Samit K. Nandi, Department of Veterinary Surgery and Radiology, West Bengal University of Animal & Fishery Sciences, Kolkata, India

Malcolm Naylor, Biomet Inc, Warsaw, IN, USA

Ryan K. Roeder, Department of Aerospace and Mechanical Engineering, Bioengineering Graduate Program, University of Notre Dame, Notre Dame, IN, USA

Mangal Roy, W. M. Keck Biomedical Materials Research Lab, School of Mechanical and Materials Engineering, Washington State University, Pullman, WA, USA

Chandra P. Sharma, Division of Biosurface Technology, Biomedical Technology Wing, Sree Chitra Tirunal Institute for Medical Sciences and Technology, Trivandrum, Kerala, India

Y.M. Thasneem, Division of Biosurface Technology, Biomedical Technology Wing, Sree Chitra Tirunal Institute for Medical Sciences and Technology, Trivandrum, Kerala, India

Huaiyu Wang, Department of Physics and Materials Science, City University of Hong Kong, Kowloon, Hong Kong, China

Julia Will, Institute of Biomaterials, University of Erlangen-Nuremberg, Erlangen, Germany

Ming H. Wu, Edwards Lifesciences LLC, One Edwards Way Irvine, CA, USA

Chapter 1

Introduction to Biomaterials

Susmita Bose and Amit Bandyopadhyay
W. M. Keck Biomedical Materials Research Lab, School of Mechanical and Materials Engineering, Washington State University, Pullman, WA, USA

Chapter Outline

1.1. INTRODUCTION

With the evolution of human civilization, the field of biomaterials evolved involving different materials at multiple length scales from nano- to micro- to macrolevel with a simple focus to extend human life and improve the quality of life. Over 1000 years back, silver in different forms was used as an antimicrobial agent to prevent infection. Different types of surgical procedures can also be found during early stages of civilization. However, probably the most significant developments took place in the field of biomaterials over the years 1901–2000. Artificial joints improved the quality of life for millions of people over the past 60 years, resorbable sutures simplified surgical procedures, and different cardiovascular devices saved millions of lives, just to name a few. The advent of tissue engineering and organ regeneration is pushing the frontiers of science today to make the years 2001–2100 more exciting in the field of biomaterials. However, to appreciate the benefits, it is not just the design of biomaterials that is important, but sound engineering design and appropriate materials and device characterization are also needed. Moreover, for a biomedical device to see the commercialization light, it is also important to carry out testing following appropriate standards to get regulatory approval. Overall, benefits of biomaterials research can only be appreciated when these

Characterization of Biomaterials. http://dx.doi.org/10.1016/B978-0-12-415800-9.00001-2

materials are characterized well at both the materials level and the device level following regulatory guidelines. Considering the multidisciplinary nature of the field, it is also not easy to carry out large variety of experiments using different techniques. Realizing this problem in biomaterials characterization, we have developed this book to offer an insight on various characterization tools focusing on biomaterials and biomedical devices.

1.2. TYPES OF MATERIALS

Materials can be classified into different groups based on their crystal structure, bonding, and macrostructures. Each subgroup of materials shows somewhat similar properties and then those materials can be clubbed together to study their performance for different applications. If we look at types of bonding, materials can be classified into three broad categories—metals, ceramics and polymers. Materials that are bonded via metallic bonds are called *metals*. Due to abundance of free electrons in metallic bonds, metals are both thermally and electrically conductive, and show malleability in terms of their mechanical properties. Materials that are primarily ionic and/or covalently bonded are called *ceramics*. Since ionic and covalent bonds do not offer any free electrons, ceramics are generally nonconducting materials both thermally and electrically. However, due to the movements of defects, some ceramics show conductivity at higher temperature. Materials that are based on long carbon chain and covalently bonded with some secondary bonding are polymers, where "mers" or units are connected in the long range. Due to covalent bonding, most polymers are nonconducting. When any three of these main materials are mixed together without losing its inherent characteristics, then we form a new class of materials, called *composites*. Some examples of natural composites are wood and bone.

When we look at the unit cell, the basic building block of any material, we can classify materials into three groups—crystalline, semicrystalline and amorphous. If the unit cell is repeated in all three directions and maintains a long-range order, then those materials are called a crystalline material such as iron, titanium, chromium or polycrystalline ceramics. The basic unit cell, defined by the three-dimensional shape and atom positions, can be for example body centered cubic or face centered cubic or hexagonal close packed (hcp), where "cubic" or "hexagonal" are crystal systems defined by the shape of the unit cell and "body centered" or "face centered" are specific atom positions that defines the Bravais lattice of the unit cell. These kinds of simple structures are common for most metallic systems, which are mainly crystalline in nature. For many materials, unit cell does not repeat itself in three dimension for long range, but only shows short-range repetition. Those materials with short-range ordering of unit cells are called amorphous materials or glass. Due to lack of ordering, glassy materials show unique properties such as glass transition temperature (T_g), where a liquid phase transforms to a solid rubbery phase. There are also a group of materials that are partially glassy and partially crystalline. Those materials are called semicrystalline materials. Among others, many polymeric materials show semicrystalline nature.

Materials can also be classified as natural or synthetic. Natural materials are those which are available in nature such as wood, rocks, corals and bones. Most natural materials are ceramics, polymers and their composites. Mother Nature designed these materials for a variety of purposes such as sensors, reservoirs, structural support or energy converters. Most of these materials have complex chemistry and structure and the mankind is still exploring those to enrich their learning. Synthetic materials are man-made materials designed for specific functionality. These materials include metallic materials such as steels for fracture management devices, to titanium and its alloys for implant, to polymers for ocular lenses and to ceramics for bone-tissue engineering. Synthetic materials are tailored for specific chemistry and structure for properties that can be utilized to improve our everyday life. Once processed, their physical, chemical, mechanical and sometimes biological property determination is necessary based on application need.

Materials can also be classified based on their macrostructures such as dense or porous. Most natural materials such as rocks, tissues, wood are porous materials. Porosity in these materials can serve various purposes. Most ceramic materials have residual porosity. Porosity can be nonuniform and vary in size and distribution. Porosity in materials can vary from 1% to 10% such as in cortical bone to as high as >70% in some cancellous bones. Because of porosity, these materials are lighter weight, and many times show nonuniform properties at different directions. However, it is very difficult to mimic such natural materials in terms of composition, structure and properties. When materials do not show any porosity, those are called dense materials. Most metallic materials are dense in nature with residual porosity <1%. Dense materials are typically isotropic in nature and can be shaped easily by various forming techniques.

Figure 1.1 schematically shows different types of materials. Any one material can also fall into many of these categories. For example, bone is a natural material, that is porous, and a ceramic–polymer composite.

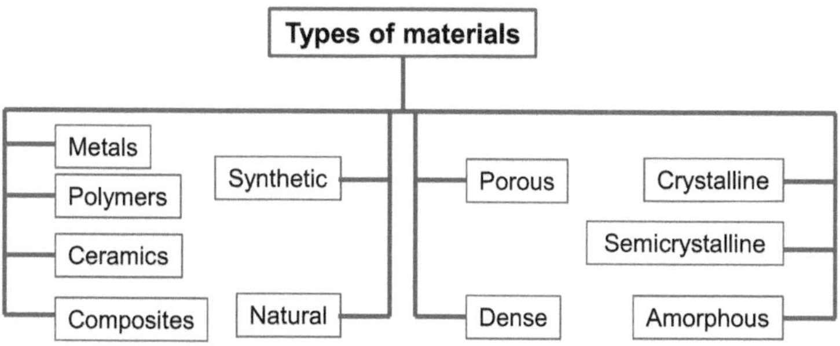

FIGURE 1.1 Different types of materials.

1.3. BIOMATERIALS AND BIOCOMPATIBILITY

A biomaterial is a material, synthetic or natural, that can be used in medical applications to perform a body function or replace a body part or tissue. A biomaterial is intended to interact at the interface of biological systems. It may also be used as a delivery system for drug or biological factor. Biomaterials are designed based on application needs. Reaching back to the beginnings of civilization, the Romans, Aztecs and Chinese used gold in dental applications. The Mayans were found to have fashioned dental implants out of sea shells with results indicating actual bone integration. A biomaterial must be biocompatible, i.e., it should be friendly to biological system and not do any harm to the system, whether at the cellular level or at the system level. A biocompatible material should elicit appropriate host response or able to perform its intended function in a specific application without the presence of adverse reactions. This is an emerging paradigm that requires and pushes unique multidisciplinary boundaries based on understanding and integration of concepts from various broad fields, but not limited to, chemistry, biology, materials science, mechanical, chemical and electrical engineering as well as medicine.

Biomaterials are used to augment, repair or replace any tissue, organ, or function of the body that has been lost through trauma, disease or injury. Recent practice in medicine often times uses tissue reconstruction using autograft, where tissue graft or organ transplant from one point to another of the same individual takes place. However, limited availability, donor site morbidity and above all, the need for a second surgery restrict their application. On the other hand, potential alternative is the use of allograft, i.e. tissue graft or organ transplant from a donor of the same species as the recipient. The other alternative could be xenograft, which is tissue graft or organ transplant from a donor of a different species from the recipient. Both allograft and xenograft use are somewhat restricted due to the immunogenic response and they may impose adverse biocompatibility in patients' body. These reasons draw our attention to the biomaterials that are available from other sources, which could be synthetic or natural. It was not until the turn of the twentieth century, modern biomaterials and their potential applications began to take shape. These first-generation modern biomaterials are mainly focused on basic functionality and biocompatibility.

Since early 1900s, metal plates were used to stabilize long bone fractures with the hopes of faster and more functional healing. By the 1930s, full joint replacement surgeries were being performed with varying degrees of success. The invention and widespread use of synthetic plastics created a boom in the biomaterials industry resulting in devices such as synthetic heart valve (metal cage surrounding a silicone elastomer ball) and total hip arthroplasty systems (ultrahigh molecular weight polyethylene). Some other important inventions in this era include the intraocular lenses (poly(methyl methacrylate)—PMMA) and the desk-sized pacemaker.

Further understanding of the complexities of biology ushered in the next generation of biomaterials beginning in the 1960s. Scientists and engineers began to design materials specifically for use in biomedical applications. The focus began to shift from materials that were simply biocompatible to technologies that took into consideration the specific needs of the specific biology. Many bioresorbable materials, those biomaterials resorbed in biological system, were also created during this time. Some of the major advances in materials came in the form of synthetic polymers such as Teflon, which is still widely used today in various vascular grafting and surgical applications. Hydrogels lead to the invention of the soft contact lenses. Poly(lactic–glycolic acid) (PLGA) was developed for resorbable sutures. Plastic materials were not the only advances made during this time period. Ceramics such as calcium phosphate synthetic bone analogs were also developed. Today, these synthetic materials are used more often than autografts (bone taken from another source on the patient's body) or allografts (processed cadaver bones). Titanium alloys with better biocompatibility were developed as a lighter alternative to traditional stainless steel orthopedic implants.

1.4. TYPES OF BIOMATERIALS

Like any other materials, biomaterials can be grouped into four major categories: (1) polymers, (2) metals, (3) ceramics, and (4) composites. Polymers can be used in both soft and hard tissue applications, and comprise the largest class of biomaterials. Polymers are also widely used in drug delivery applications. Polymers could be natural (examples: collagen, sodium alginate, and cellulose) or synthetic (examples: silicone rubber, PMMA, poly(vinyl chloride), and co-PLGA). Metals are mostly used for dental and orthopedic applications. Most commonly used metals are Ti and its alloy, stainless steels and Co–Cr alloy. Ceramics are mainly used in hard tissue repair, regeneration and augmentation, especially in nonload-bearing applications or as coatings on metal implants. Most widely used ceramic biomaterials are calcium phosphates (CaP), alumina (Al_2O_3), and bioglass. Polymer–ceramic composite represents the major part of composite biomaterials.

Biomaterials–tissue responses can be divided into four different types. (1) Toxic: toxic materials cause death to surrounding tissue. (2) Bioinert: this type of response is caused by nontoxic but biologically inactive materials. Fibrous tissue encapsulation of the bioinert material is caused *in vivo*, which leads to the loosening and ultimate failure of the implant. This is most commonly seen in metallic implants. A bioactive material coating on metallic implant may prevent fibrous tissue encapsulation. (3) Bioactive: this type of response is seen if the material is nontoxic and biologically active. The term biologically active means that an interfacial bond forms between the material and host tissue. Bioactive glasses, many types of polymers and most calcium phosphate ceramics fall in this category. (4) Bioresorbable: this type of response is seen when a nontoxic material dissolves *in vivo* such as calcium sulfate (plaster of paris), tricalcium phosphate, bioactive glasses and PLGA. As a result, the surrounding host tissue can replace the synthetic material.

1.5. PROPERTIES OF BIOMATERIALS

There are many good books written on properties of materials. The purpose of this brief introduction on materials properties is only to make the reader familiar with the subject matter. Among various materials properties, the most important ones related to biomaterials are chemical, physical, mechanical, and biological in relation to their surface and bulk properties. Chemical properties deal with the chemistry of the materials such as composition, bonding and atomic structures. Physical properties deal with microstructures, phases, density and different types of porosity. Mechanical properties deal with strength and toughness of the materials along with hardness and different failure mechanisms. Biological properties deal with how materials behave in a biological environment. If the biological environment is created in a Petri dish and properties are measured, then they are called *in vitro* properties; if the properties are measured inside an animal or a human body, then they are called *in vivo* properties. Surface properties deal with properties of materials at the outer layer where all bonds are not satisfied. Due to dangling bonds, surface properties are often quite different from the bulk properties of materials. All these material properties are linked with chemistry, structure and processing of materials. For example, carbon is a material with very different properties depending on its bonding and structure. If carbon is closely packed and fully covalently bonded, then it forms diamond, which is the hardest material known to mankind. However, if carbon forms an hcp structure with covalent bonding along with some secondary bonding, then it forms graphite. Graphite is a soft material. Therefore, just chemistry alone does not reflect properties of materials. Similarly, even if the chemistry and the structure are the same, materials processing will also have a significant impact on properties and performance of a material. A ceramic body with high-purity calcium phosphate can have strength values varied significantly depending on the residual porosity in the part, which will be dependent directly on the processing condition. Therefore, understanding properties of materials requires an in-depth knowledge of structure as well as processing of those materials.

In this context, it is also important to review advantages and disadvantages of different materials systems. Such generic properties only guide materials selection toward any specific application. For metallic system, the main advantage lies with ductility. Field of metallurgy is very rich due to long tradition. Apart from their high strength and toughness, metallic materials can be processed reliably due to thousands of years of experience. These materials are inherently conductive and available worldwide. Naturally, most large industrial structures are made of metallic materials. Metals are widely used in joint replacement and bone fracture fixation. Metallic materials provide high strength and resistance to fracture, and are designed to resist corrosion. In comparison to metals, ceramics are generally hard and offer low wear rate *in vivo*. Ceramics are resistant to microbial attack and strong in compression. Aluminum oxide, zirconium oxide, calcium phosphates and bioglasses are some of the most commonly used ceramic biomaterials. The first two compositions are widely used in load-bearing implants, and the last two are used as bioactive and bioresorbable ceramics in bone-tissue engineering and as coatings on metal implants to enhance bioactivity. Different forms of carbon are also used in devices such as heart valves as coatings or in dialysis chamber. A large variety of polymers are also used as biomaterials. Polymers can be flexible or rigid, with low or high strength, and can be biodegradable or bio-inert. Synthetic polymers can be functionalized with different biomolecules, and processed via different methods to achieve a variety of structures. Among others, polymers are used in artificial joints, artificial skin, sutures, facial implants and drug delivery devices.

Selection of materials for different biomedical devices is a complicated process, which depends on many factors including but not limited to mechanical loading requirements, chemical and structural properties, and biological properties. Biomaterials are also used in tissue engineering and drug delivery. Different types of biomolecules and drugs can be carried to desired organ using nanostructures and then released in a controlled manner. In addition, multi-functional systems can be achieved by engineering biomaterials. Magnetic nanoparticles are examples of multifunctional biomaterials. Different drugs can be loaded on magnetic nanoparticles and led to specific organ using external magnetic field. Some of these particles can also be used to enhance imaging contrast in magnetic resonance imaging and computed tomography scans.

Contemporary materials have made yet another paradigm shift. Not only are materials carefully designed to accommodate various physiological factors, but also they are moving into the realm of bioactive responses. In short, these materials go leaps and bounds beyond the traditional sense of biocompatible by incorporating factors that elicit a positive biological interaction. These materials focus on issues such as biofouling, regenerative medicine, controlled release of therapeutic substances and tissue/organ engineering. Although these materials have come a long way since the first steel plates used for fracture fixation, there are still many challenges to overcome. Biological requirements should be met for each application, and materials should be characterized to fulfill the needs. Human biology is very complex and we are just beginning to understand the intricate processes and interactions between synthetic materials and our body. Figure 1.2 shows generic applications of different types of biomaterials.

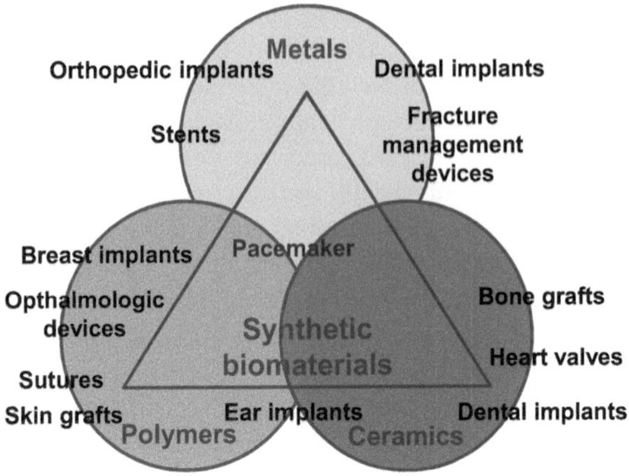

FIGURE 1.2 Applications of biomaterials in different devices.

1.6. BIOMATERIALS CHARACTERIZATION AND OUTLINE OF THIS BOOK

From the previous discussion, it is clear that the field of biomaterials is inherently multidisciplinary. Researchers from physical sciences, engineering sciences, biological sciences and medicine often flock together to work on biomaterials toward improving human health. Therefore, it is very important for all to understand what to and how to characterize different biomaterials, which is the main theme of this book. It is also important to understand that many times, characterization of just biomaterials may not be enough, and further systems-level testing may be necessary to validate actual performance of the device. Considering all those issues, this book is developed with specific emphasis toward characterization of biomaterials and biomedical devices for their physical, mechanical, surface, *in vitro* and *in vivo* biological properties. Special attention was given toward device-level characterization of orthopedic and cardiovascular implants.

Chapters dealing with physical, mechanical and surface properties of biomaterials are developed by researchers from physical and engineering sciences. These chapters are focused on techniques that can reveal basic properties such as atomic structures, bonding, chemical interactions, phase identification and transformation to strength and toughness measurements to understanding modifications at the surface level and their influence on material properties. Most of these techniques are followed by biomaterials researchers during early stage of materials development toward certain applications. The following chapters include *in vitro* and *in vivo* testing of biomaterials including microbial interactions and biofilm characterization. These levels of testing will be needed before a biomaterial can be considered

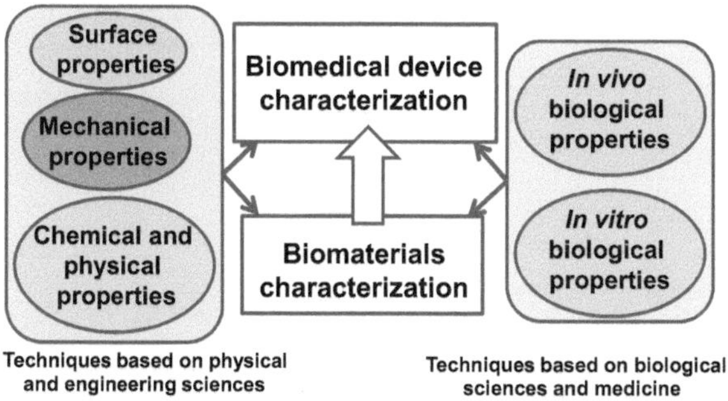

FIGURE 1.3 Schematic outline shows different properties related to biomaterials.

for clinical trials or in actual device manufacturing. The final part of this book deals with device-level characterization with special emphasis on orthopedic and cardiovascular devices. Researchers with industrial product development background contributed toward these chapters. Figure 1.3 outlines how different chapters are linked to understand properties of biomaterials and biomedical devices.

1.7. SUMMARY

This brief introductory chapter provides a broad overview of materials, biomaterials and the need to understand different techniques to characterize biomaterials. From this chapter, the reader can gain a perspective on how rest of the topics in different chapters are divided to fully comprehend this inherently multidisciplinary field. Application of appropriate characterization tools can not only save time to fully evaluate different biomaterials, it can also make commercial biomedical devices safer. In the long run, safer biomedical devices can only reduce the pain and suffering of mankind, a dream that resonates with every biomedical researcher.

SUGGESTED FURTHER READING

[1] Ratner BD, Hoffman AS, Schoen FJ, Lemons JE. Biomaterials science – an introduction to materials in medicine. Academic Press; 2012.

[2] Hollinger Jeffrey O, editor. An introduction to biomaterials. 2nd ed. CRC Press; 2011.

[3] Robert Lanza, Robert Langer, Joseph Vacanti, editors. Principles of tissue engineering. Associated Press.

[4] Burdick Jason A, Mauck Robert L, editors. Biomaterials for tissue engineering applications. Springer; 2011.

Chapter 2

Physical and Chemical Characterization of Biomaterials

T.S. Sampath Kumar
Department of Metallurgical and Materials Engineering, Indian Institute of Technology Madras, Chennai, India

Chapter Outline

2.1. MICROSTRUCTURAL CHARACTERIZATION

The physical, chemical and often mechanical properties of materials are interrelated to its internal microstructural features. The microstructure of the material involves specifying the crystallography, morphology and chemical composition of the material. Crystallographic analysis includes identifying the different phases which are present in the structure of the material and the nature of the atomic packing within the phases. Most phases are crystalline phases with highly ordered and regularly repeated arrangement of atoms. But some phases are amorphous or glassy, which do not posses any long-range ordered arrangement of atoms. The morphological analysis corresponds to the characterization of the size, shape and spatial distribution of the phases or particles. As stated above, the compositional analysis involves identification of the chemical constituents of the material and its relative abundance. In all the cases

Characterization of Biomaterials. http://dx.doi.org/10.1016/B978-0-12-415800-9.00002-4

of microstructural characterization, a carefully prepared specimen sample is interacted with some form of probe and the scattered or excited signal from the sample is collected for analysis. The visible light, X-rays and energetic electrons are the most commonly used probes for the characterization of materials. All the probes can interact with the sample through elastic or inelastic processes. The information that can be obtained for biomaterials from each of the microstructural characterization methods are discussed in the subsequent sections.

2.1.1. Optical Microscopy

'Seeing is believing' and hence, microscopy in its various forms still remains as the preeminent tool at the forefront for the characterization of biomaterials. Both light and electron microscopes are used to fabricate and characterize new materials, coatings, and devices. The microscopy techniques used for the biomaterials include scanning electron microscope (SEM), transmission electron microscope (TEM), scanning tunnelling microscope (STM) and atomic force microscope (AFM). As the interest is often focused on the biologic responses at the cell–biomaterial interface, a variety of light microscopes such as the recently developed near-field scanning optical microscope (NSOM) and confocal laser scanning microscopy are daily relied for tasks ranging from routine materials analysis to detailed study on the behavior of cells and tissues at the biomaterials interface.

The optical microscopy is the oldest and simplest of the microscopes. But its resolution, defined as the ability of an objective to separate clearly two points or details lying close together in the specimen, is poor. As the resolution is about half the wavelength of illumination used in the measurement, the resolving powder of an optical microscope is limited by the wavelength of the visible light (around 400–700 nm) and by aberrations to be around 200 nm. The magnification of the light microscope refers to the ratio of the dimension in the viewed image to the corresponding dimension in the original sample and is given as the product of the lenses and the distance over which the image is projected. So, magnification is more of a scaling phenomenon and is a function of the number of lenses of the microscope while resolution is a function of the ability of a lens to gather light. In light microscopy, the range of magnification reaches from 5× to about 1000× (a 100× objective lens and a 10× ocular). Nowadays, a digital camera is often employed to acquire an image as shown in Fig. 2.1, which uses a combination of transfer optics, intermediate and camera coupling lenses, etc. Hence, to convey the level of magnification, a scale bar is embedded in the images to have a direct feel for the size of the features in the image.

The contrast implies the ability to differentiate the light intensity between the sample and the adjacent background relative to the overall background intensity. The contrast in a microscope image is generated due to the interaction

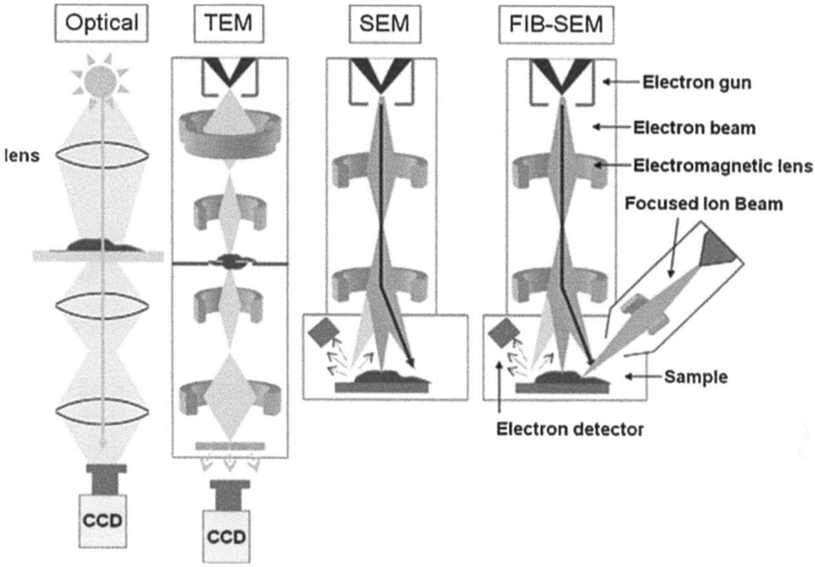

FIGURE 2.1 Principles of imaging by optical microscope, transmission electron microscope (TEM), scanning electron microscope (SEM) and FIB-SEM. *Source: Trends in Analytical Chemistry, Vol. 30, No. 1, 2011, P 29 Figure 1.*

of light with matter. As light can be characterized by its amplitude, wavelength, phase, and polarization, these many interactions of light and matter can be detected in the image. But human eyes and most of the electronic detectors are sensitive to amplitude and wavelength variations only. The various ways in which contrast is generated by converting these interactions of light and matter into differences in amplitude or wavelength have lead to the many different modalities of light microscopy such as phase-contrast microscopy, differential interference-contrast microscopy, fluorescence microscopy, confocal microscopy, etc. The contrast and resolution are related as one is often gained at the expense of the other. Apart from magnification, resolution, and contrast, the depth of focus (vertical resolution) is also an important parameter in microscopy and it is the ability to produce a sharp image from a nonflat surface. In light microscopes, the sample can be imaged by passing illuminating light through the sample (transmission microscopy) or by reflecting the illumination off the sample (reflected microscopy) to the eye/detector. For transmitted light microscopy (diascopic illumination), the sample must be thin enough to allow the light to pass through the sample. For opaque samples, reflected light techniques (episcopic illumination) such as compound microscopes are commonly used.

Titanium and titanium alloys have been the biomaterial of choice for hard tissue implants such as joint replacements, due to its biocompatibility, high

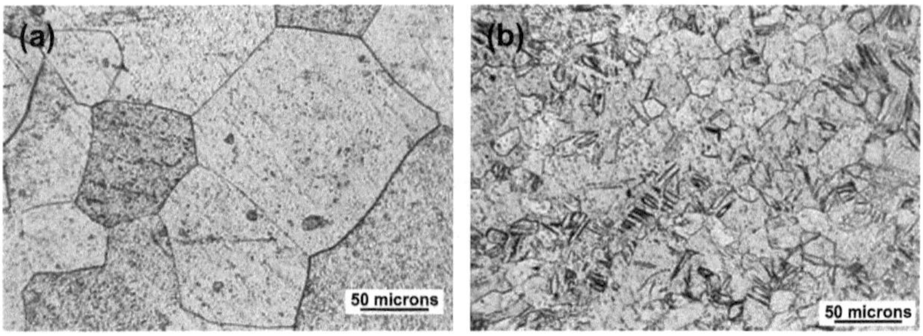

FIGURE 2.2 Optical micrographs of annealed (a) and SPD processed (b) cpTi samples.

corrosion resistance, high strength-to-weight ratio and good bioadhesion. Typical optical micrograph of annealed commercially pure titanium (cpTi) is shown in Fig. 2.2a, which indicates the initial grain size of the material as approximately 50 μm. However, the bioactivity of titanium, i.e. ability to form a hydroxyapatite (HA) layer, similar to the mineral phase of the bone on its surface in biological environment is poor. As the surface properties of the material are one of the key factors that determine the biological responses and integration of the implants, there is a great interest in the formation of submicron/nanosize grain materials over conventional biomaterials with micron-size grains. The optical micrograph of grain refined cpTi, which was obtained by severe plastic deformation (SPD) process, is also shown in Fig. 2.2b. The microstructure was significantly finer than in the annealed condition. The yield strength (stress that marks the onset of plastic flow) of a material is a structure-sensitive property and so, fine-grained cpTi has significantly higher yield strength than coarse grain material.

2.1.2. Electron Microscopy

The (de Broglie) wavelength of an electron is much smaller than that of a visible light photon. Using electron beams to illuminate materials, higher magnifications of around 1–2 million times can be obtained in electron microscopes. There are two types of electron microscopes such as transmission electron microscope (TEM) and scanning electron microscope (SEM). The TEM is one of the most powerful techniques to provide direct important information on the size, shape and morphology of biomaterials. In TEM, electrons generated from an electron gun are focused using electrostatic lenses on a thin specimen of the sample or particles held on a thin amorphous carbon or polymer films. A comparison of the TEM with the optical microscope is shown schematically in Fig. 2.1. Typical accelerating voltage is around 100 kV–1 MeV, for which the electrons have mean free paths of the order of a few tens of nanometre and so, thin specimens of thickness <50 nm is typically used. After penetrating through the sample, the transmitted electrons are either deflected or undeflected depending upon the scattering processes experienced by electrons during the passage. Inelastic scattering with the sample electrons at heterogeneities such as dislocations, defects, density variations, grain boundaries, second-phase particles, etc. causes complex absorption and scattering effects, leading to intensity variation of the undeflected transmitted electron beam. The deflection of the transmitted electron beam is due to elastic scattering with the sample, which gives rise to a diffraction pattern as shown schematically in Fig. 2.3.

The imaging mode can be selected by the use of an aperture. If most of the undeflected electron (direct beam) is allowed through, focused and magnified using electrostatic lenses to form images usually on a fluorescent screen, the resulting image is called a *bright-field image*. The image results from the weakening of the direct beam by the thickness/mass contrast (thicker regions of

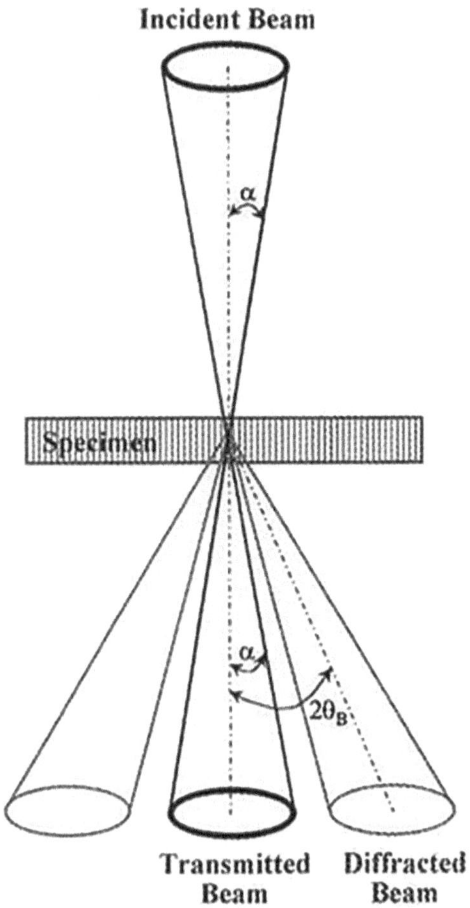

FIGURE 2.3 Schematic diagrams of transmitted and diffracted beams in TEM.

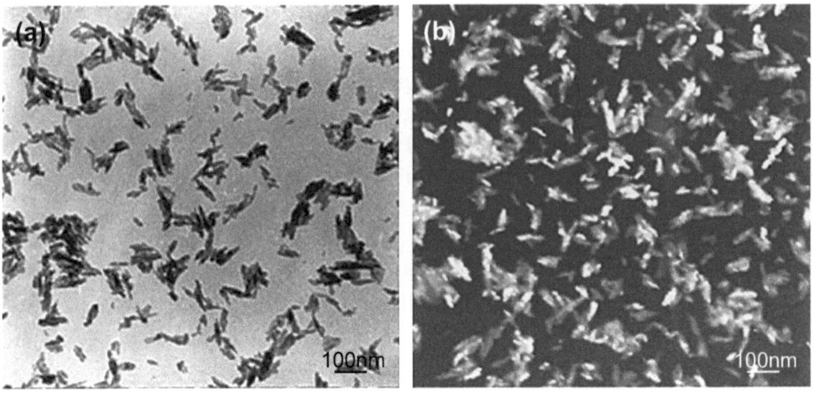

FIGURE 2.4 TEM bright-field image (a) and dark-field image (b) of hydroxyapatite nanoparticles.

the sample will appear dark), atomic number/Z-contrast (regions with a higher *Z* will appear dark) and diffraction contrast (Bragg diffraction of crystalline region removes intensity of primary beam). If the direct beam is blocked by the aperture and one or more of the deflected beams (diffracted beam) are selected to form images, the image is known as a *dark-field image*. The image formation is based on the dark contrast appearing on areas which are thicker, crystalline and with heavier atoms. This image gives useful information on particle size, defects and stacking faults. Using a larger aperture, the direct beam and the diffracted beams can be allowed to pass through to form image. The image formed by the interference of the direct beam with the diffracted beams is called *lattice image* or *phase-contrast image* due to atomic resolution achieved and the TEM is called high-resolution TEM (HRTEM). Typical bright and dark images of HA nanoparticles, a major mineral component of the calcified tissues (i.e. bones and teeth) are shown in Fig. 2.4. The morphological features are commonly described as *equiaxed* if the structure appears similar in all the directions otherwise described as *acicular* or *plate-like*. The apatite particles were of nanoplate-like morphology with 15–20 nm width and 60–80 nm length.

The electron beam can be focused to any region of interest on the specimen and so the diffraction pattern formed by the scattered electron in TEM is called selected-area diffraction (SAED). Three types of SAED patterns as shown in Fig. 2.5 can be distinguished as follows:

- A pattern of spots created by single crystalline materials from which the type of crystal structure, lattice parameters and crystal orientation can be determined.
- A pattern of concentric rings by polycrystalline material based on the superposition of many single-crystal patterns, from which the type of crystal structure and lattice parameters can be determined but not the crystal orientation.

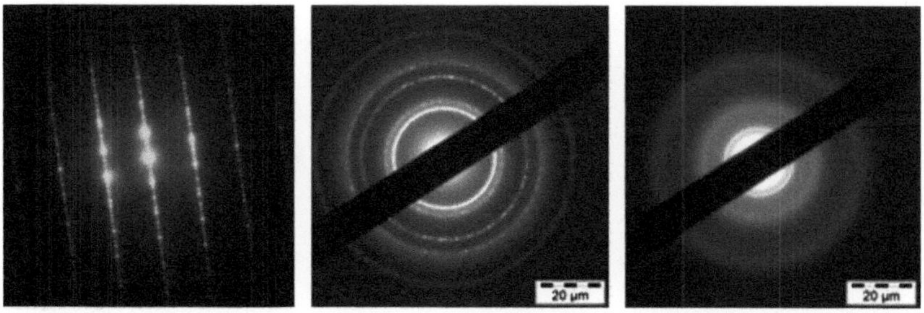

FIGURE 2.5 Typical SAED patterns from single crystals, polycrystalline and amorphous materials. *Source: Dr Michael Krüger, Freiburg Materials Research Centre (FMF) Nanomaterials Characterization of Nanomaterials – High resolution methods for materials characterization, IMTEK Uni. Freiburg.*

- A pattern of fuzzy rings by amorphous materials due to disordered lattice from which average nearest neighbour distance in the material can be related.

The greatest advantage of TEM is the higher resolution of the order of less than 1 nm due to the small effective electron wavelength. As the wavelength of 100 keV electrons is 0.37 nm, higher the accelerating voltage better will be the resolution of materials. Thus, by using TEM, either imaging or structural analysis of individual nanoparticles, nanorods and -wires, and different regions of the nanostructured materials can be carried out at very high resolution. However, to have realistic representation of the material, enough particles or regions should be examined.

Shape memory effect (SME) is a phenomenon, in which a material recovers to its original size and shape when heated above a certain characteristic transformation temperature. Nitinol, an alloy consisting of equal composition of nickel and titanium, exhibits SME. This NiTi alloy has been extensively tried for various medical applications such as orthodontic wires, stents, joints, blood clot filters, etc. due to the ability to return to its predeformation shape at body temperature. The SME has been related to the occurrence of the martensitic-phase transformation, which is a diffusionless phase transformation with a cooperative movement of atoms and often by a shear-like mechanism. The parent phase (high-temperature phase) in NiTi is cubic, and the martensite (low-temperature phase) is monoclinic. Typical TEM micrographs of austenite phase and transformed martensite phase of NiTi wire are shown in Fig. 2.6. As the wire was cold drawn, the austenite phase shows elongated grains. The SAED pattern shown along with the micrographs clearly indicates the different crystal structure of the respective phases.

The incident electron beam interacts with the atoms of the sample and releases element specific X-rays as well as suffers energy loss due to element-specific absorption. Chemical information of the nanomaterials can be carried out by analysing the energy of the X-rays emitted (EDAX) or by measuring the

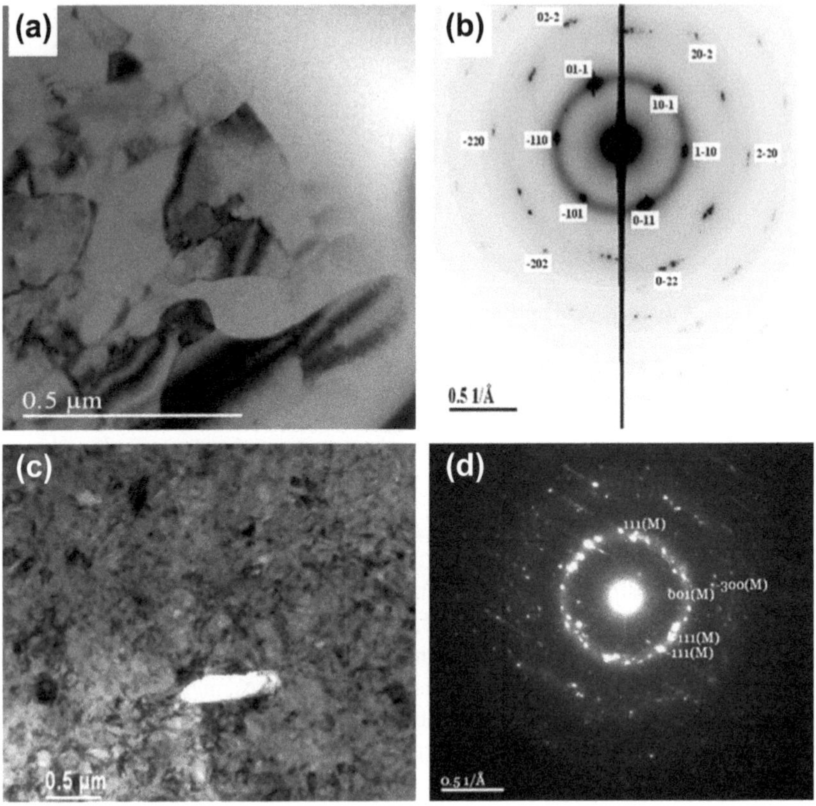

FIGURE 2.6 TEM micrograph and SAED pattern of austenite (a,b) and martensite (c,d) phases of NiTi alloy. *Source: Madhavi Tiyyagura, M.S. thesis, University of Central Florida, Orlando, Florida, 2005.*

energy loss of the primary beam (EELS). Direct probing of both the qualitative and quantitative analysis of biomaterials locally at regions <2 nm is possible. The TEM, thus, provides direct information on the morphology, structural and composition of biomaterials at very high resolution. However, enough particles/regions should be examined to have realistic representation of the sample. The major limitations of TEM is its limited depth of resolution leading to only two-dimensional images, sample damage due to energetic electron beam and low contrast for low atomic number materials.

The SEM is one of the multipurpose methods for characterization of biomaterials. In SEM, an electron beam of relatively low energy from few hundred eV to 50 keV but with a very fine spot size of ~5 nm is scanned across the sample surface instead of transmitted through the sample as in TEM. A variety of interactions occur during the penetration and passage through the sample which results in the emission of electrons and photons from the sample

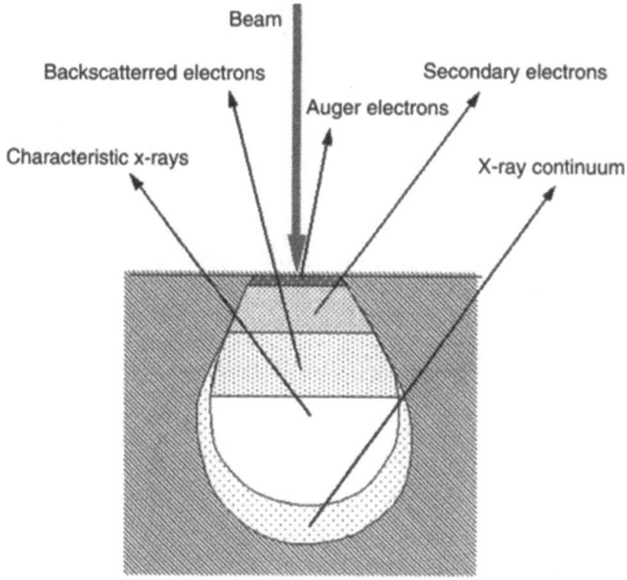

FIGURE 2.7 Schematic diagram of electron-material interactions. *Source: Dr Michael Krüger, Freiburg Materials Research Centre (FMF) Nanomaterials Characterization of Nanomaterials – High resolution methods for materials characterization, IMTEK Uni. Freiburg.*

as shown in Fig. 2.7. The reflected type electrons and photons from the sample are used for imaging and characterization. Some of the SEM modes widely used to characterize biomaterials include secondary and backscattered electrons, X-ray mapping, electron-beam-induced current, electron channelling, electron backscattered diffraction and Auger electron microscopy.

The secondary electrons with energies <50 eV (average 3–5 eV) are produced due to inelastic scattering of the incident electron beam and it is routinely used for surface and topography information of the sample. Basically, the secondary electrons are collected by an electron detector with a positive potential to have maximum numbers and the detector output is used to modulate the intensity of a display system such as cathode ray tube. The image is formed pixel by pixel as the beam is scanned across the sample. So, the resolution of the SEM is limited not by the wavelength of electron beam but by the size of the electron beam spot being scanned and is much less than that of a TEM. Biomaterials can be imaged at resolutions approaching 0.3 nm at 30 keV using a high-resolution SEM. Typical SEM micrograph of compacted magnesium metal powder is shown in Fig. 2.8.

The backscattered electron image is formed by detecting the elastically scattered high-energy primary electrons. To prevent the secondary electrons entering the detector, a slightly negative charge is applied to the Faraday cage fixed to the detector and is located at an angle or in column to the primary beam

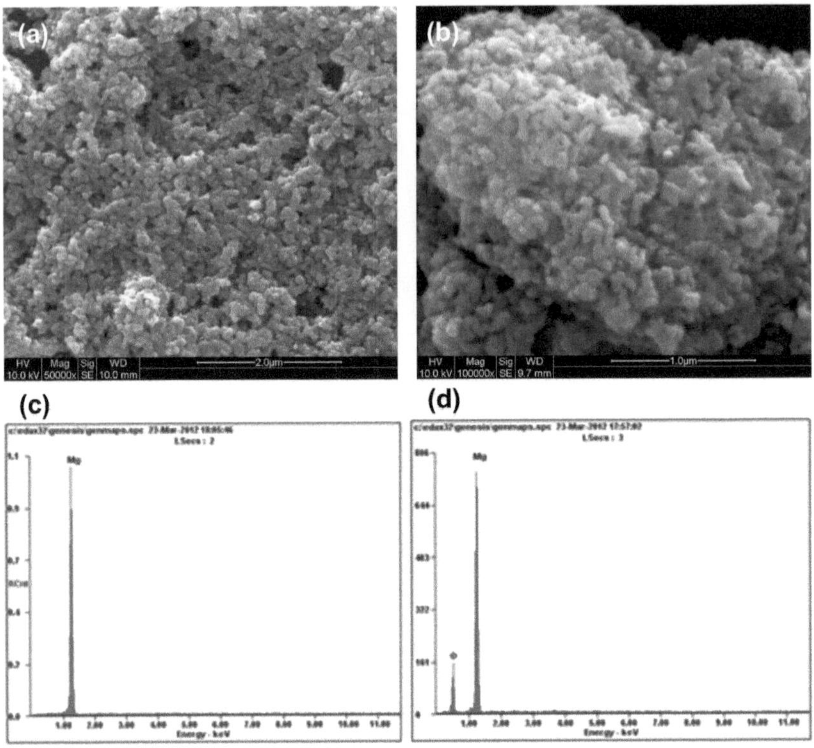

FIGURE 2.8 SEM micrographs and EDAX analysis of sintered pure magnesium pellets: (a) secondary image, (b) backscattered image, (c) grain composition and (d) grain boundary composition.

column. The backscattered electron image compliments the secondary electron image as well as gives additional information on elemental composition (*Z* imaging) as shown in Fig. 2.8. The *Z* contrast due to the presence of oxygen clearly distinguishes the particle boundary. Similar to TEM, chemical information of the biomaterials can also be evaluated using the characteristic X-rays or Auger electrons from the sample under electron beam. SEM in combination with a focused ion beam (FIB) source can be used as a nanofabrication tool for imaging and generation of nanostructures down to 10 nm as shown in Fig. 2.1.

2.2. SCANNING PROBE MICROSCOPY

Scanning probe microscope (SPM) is a group of techniques that provide three-dimensional real-space images directly as well as allow measurement of properties of biomaterials locally at atomic resolution. In all the SPM techniques, a sharp tip of atomic dimensions (typically of 3–50 nm size with one or few atoms tip) is scanned at few Angstroms above the surface of the sample and the interaction between the tip and the sample is measured as shown in Fig. 2.9.

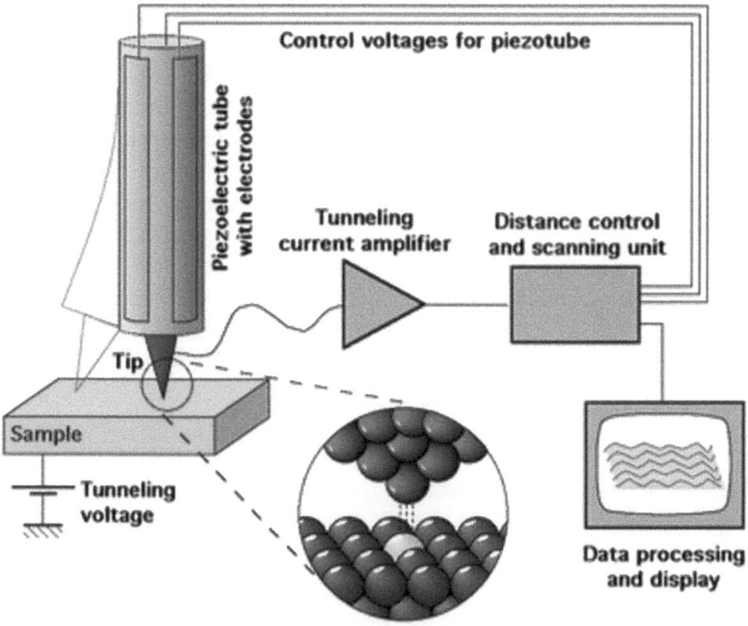

FIGURE 2.9 Schematic diagram of the operation of STM. *Source: Dr Michael Krüger, Freiburg Materials Research Centre (FMF) Nanomaterials Characterization of Nanomaterials – High resolution methods for materials characterization, IMTEK Uni. Freiburg.*

The various interactions between the probing tip and the sample surface results in different type of SPM techniques and is no longer restrained by the wavelength of electron or light. The sample surface is alternatively scanned under the stationary probe and also, the tip is mounted on a cantilever. Typically, scan may cover a distance of around 100 μm in the *x* and *y* directions and about 4 μm in the *z* direction.

The scanning tunnelling microscope (STM) forms the basis for the family of scanning probe tools and was one of the earliest tools to provide a three–dimensional profile of nanomaterials with atomic resolution. In STM, an atomically 'sharp' tip (typically pyramidal in shape of about 3 mm long and literally with one or two atoms at the tip) is moved very close to a sample's surface without hitting the surface but at a few Angstroms from making contact and the small electrical current (tunnelling current, typically nanoamperes), which crosses the gap between the tip and the sample, is measured. The tip mounted on a piezoelectric transducer is then scanned in steps of 0.1 nm across the surface in a regular pattern (much as one would move a lawn rover) and the tunnelling current is monitored. A topographical map of the sample's surface with atomic resolution (0.1 nm lateral resolution and 0.01 nm depth resolution) is generated by keeping the height of the tip constant (constant height mode) and measuring the changing electrical current (more when close to the sample, less when farther away). Alternatively, in the constant current mode, the deflection (rise and lowering) of the tip is measured by keeping the current constant to form a graded height map of the sample surface as similar to an atlas map. For the STM operation, it is necessary that a tunnel current flow or the sample must be a conductor or semiconductor. A thin layer of conductive material, such as gold, is coated to scan a surface which is not electrically conductive. The STM is still a fundamental tool in nanotechnology to obtain atomic-scale profile of surface roughness, observing surface defects and characterizing the size and conformation of molecules and aggregates on metallic and semiconducting surfaces. Interestingly, STM can be used as a manipulation tool to create nanostructures on a surface by moving individual atoms as shown in Fig. 2.10.

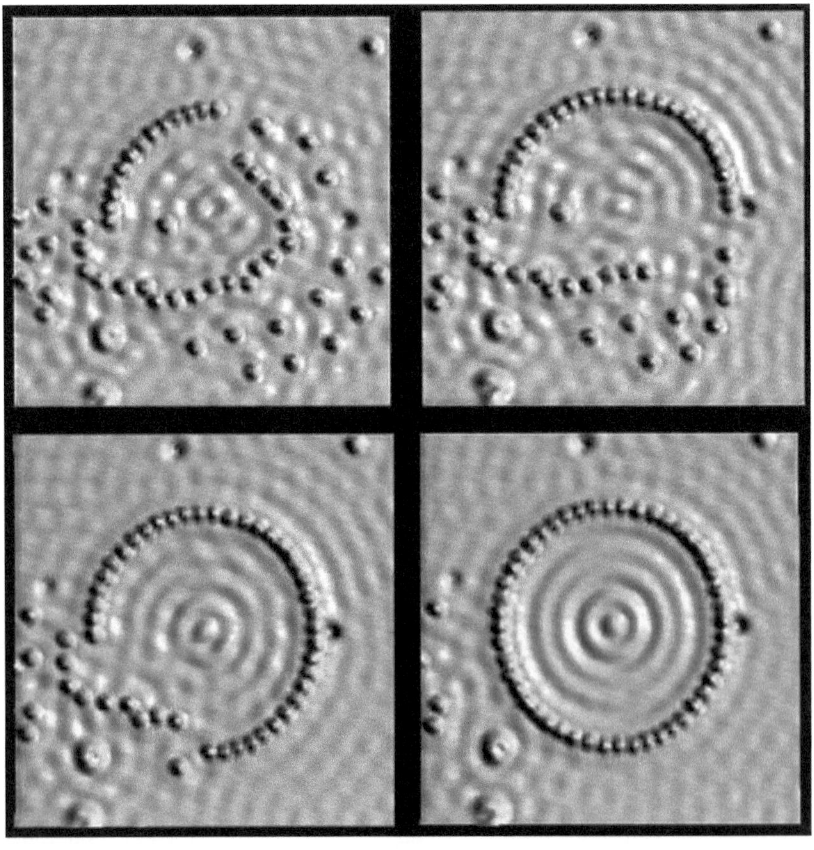

FIGURE 2.10 Building up an atomic configuration (Fe atom rings). *Source: Image originally created by IBM (http://www.almaden.ibm.com/vis/stm/gallery.html).*

The atomic force microscope (AFM) is the most widely used member of the SPM family and was developed to specifically to overcome the limitations of STM, to image nonconducting surfaces as well as to operate in liquid for imaging biological samples. AFM also uses an atomically sharp tip as similar to the STM, but instead of moving it just very close to the sample, the AFM actually drags its tip along the sample's surface. Effectively, the inter-atomic forces prevent atoms from 'dragging', they are simply pushed into very close proximity. So, the tip is mounted on a very flexible cantilever and the overall forces between the tip and the sample surface deform the cantilever elastically. The interatomic forces exerted can be monitored by measuring the deflection of the cantilever using an optical system that detects the reflection of a laser beam bouncing from the back of the cantilever. Similar to STM, the tip is scanned on the surface and the vertical movement of the tip is recorded from the cantilever fluctuations to form the topographic image of the surface. However, the applied force is kept constant by continuously changing the height of the tip through a feedback mechanism. The tip–surface interactions as shown by the force–distance curve in Fig. 2.11 can be attractive or repulsive and so, AFM can be operated in various modes depending on the interaction distance.

1) *Contact mode*: The tip experiences a strong repulsive force as it moves over the sample surface and the deflection of the cantilever tip is the measure of the surface topography of the sample. The tip always maintains a force on the sample in constant force mode.
2) *Noncontact mode*: The tip mounted on a stiff cantilever is not touching the sample but is quite close to the sample and experiences a weak attractive

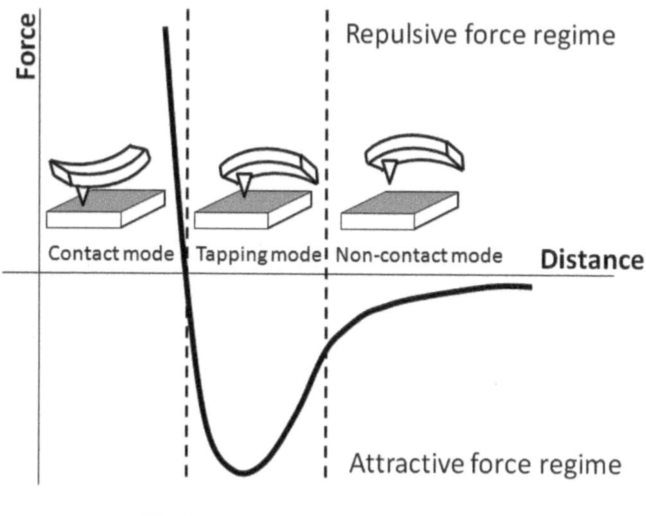

FIGURE 2.11 Force-probe distance curve.

force. The forces involved are as low as picoNewton and soft samples can be imaged without damaging it by probe penetration.

3) *Dynamic force mode*: The tip is oscillated vertically near its resonant frequency closer to the sample than in noncontact mode, such that part of oscillations that extends into repulsive region causes intermittent contact (touches/taps) with the surface. Hence, it is also commonly referred to as 'tapping mode' or intermittent-contact mode.

The intermittent tapping significantly reduces the damage that can be induced by the tip penetrating the sample and degrading the spatial resolution. At high-oscillation frequencies, the interaction time between the tip and the surface is reduced and so, it allows imaging of soft samples like biomolecules and cells with improved lateral resolution of the resulting micrograph. As the tip touches the surface very gently, the local elastic properties of the compressible samples can be characterized from the slope of the force–distance curve. Additionally, the phase shift of the oscillating tip relative to the driving signal can be used to measure the local variation in properties such as adhesion, friction, visco-elasticity, etc. along with the morphology.

The interatomic forces exerted can also be calculated by measuring the deflection of the cantilever and using the spring constant of the cantilever. In scanning force microscopy (SFM), interaction forces such as van der Waals forces, capillary forces, and frictional forces can be measured. The SFM does not record any morphological features but measures the interaction forces as the tip is moved towards the surface of the sample. The tip can also be modified to reveal the various local properties of the nanomaterials, for example, the local magnetic properties can be determined by coating the tip with a magnetic material, in the magnetic force microscopy (MFM). Similarly, in electrostatic force microscopy (EFM), the local electrostatic properties and by using scanning thermal microscopy, the local thermal distribution can be revealed. By measuring the twist of the cantilever rather than deflection, the local variation in the frictional properties of the surface can be qualitatively characterized in the lateral force microscopy. Similar to STM, AFM also can be used to create nanostructures on any material surface.

A favourable interaction between the surface properties of the material and the biological responses is one of the key factors that determine the integration of the implants, particularly longevity of biological adhesive joints. Hence, various surface treatments were employed to modify titanium surface not only for improving its bioactivity but also induces growth and fixation of the bone tissue. Surface modification by chemical treatments like acid treatment, alkali treatment, etc. is the simplest method to strengthen the titanium bonding to the hard tissue. The alkali treatment basically alters the surface chemistry by the formation of sodium titanate hydrogel layer on its surface when exposed to NaOH aqueous solution, which transforms to bone-like apatite in simulated body fluids. Nitric acid treatment ensures passivation on Ti surfaces, which,

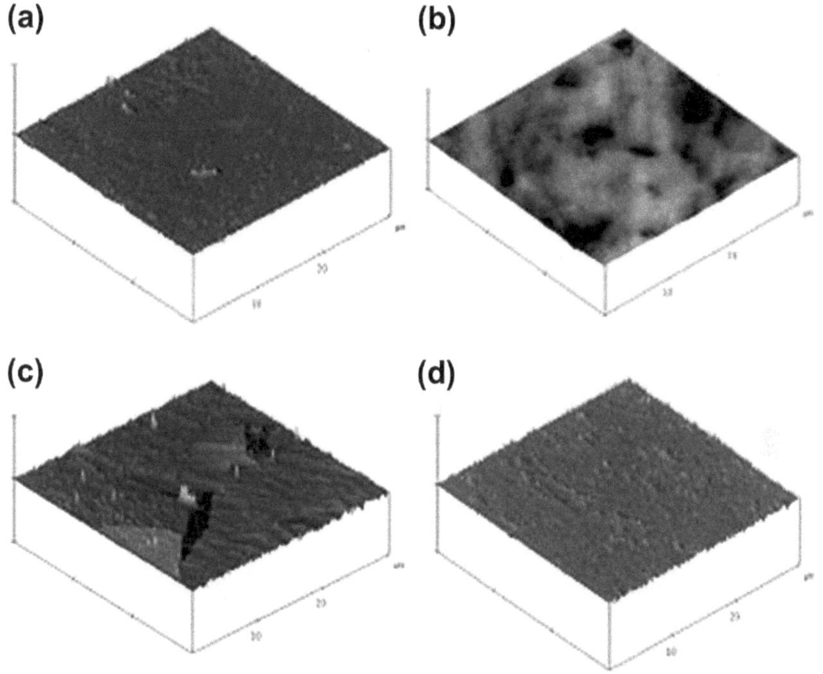

FIGURE 2.12 Atomic force microscope images of various surface-treated Ti samples: untreated, (b) HNO_3 treated, (c) dual acid treated, and (d) NaOH treated.

under simulated physiological conditions, induce biomimetic calcium phosphate (CaP) coatings. But, acid treatments using H_2SO_4 or HCl result in uniform microroughness on titanium surface, which may enhance the adhesion of sodium titanate gel. The AFM images by contact mode of untreated, HNO_3-treated, H_2SO_4 and HCl dual-acid-treated, and NaOH alkali-treated titanium are shown in Fig. 2.12. Passivation by HNO_3 treatment was clearly visible with the presence of oxide in the AFM morphology (Fig. 2.12b). The dual acid treatment reveals the grain boundaries of titanium (Fig. 2.12c). The roughness value (R_a) calculated from the AFM images was found to be higher (~87 nm) for acid-treated samples compared to the untreated sample (5.3 nm). The alkali-treated sample (Fig. 2.12d) showed few sodium titanate phases.

The AFM tip can be used to transmit light from an external light source with a diameter much smaller than the wavelength of light and scanned very close to the sample, much closer than the wavelength of the light. This region of illumination is 'near-field or evanescent field optics' and hence the technique is called NSOM or as scanning near-field optical microscopy (SNOM). Since both the dimension of the tip with a subwavelength diameter aperture and its distance from the surface are much less than the optical wavelength as shown schematically in Fig. 2.13. NSOM gives images at

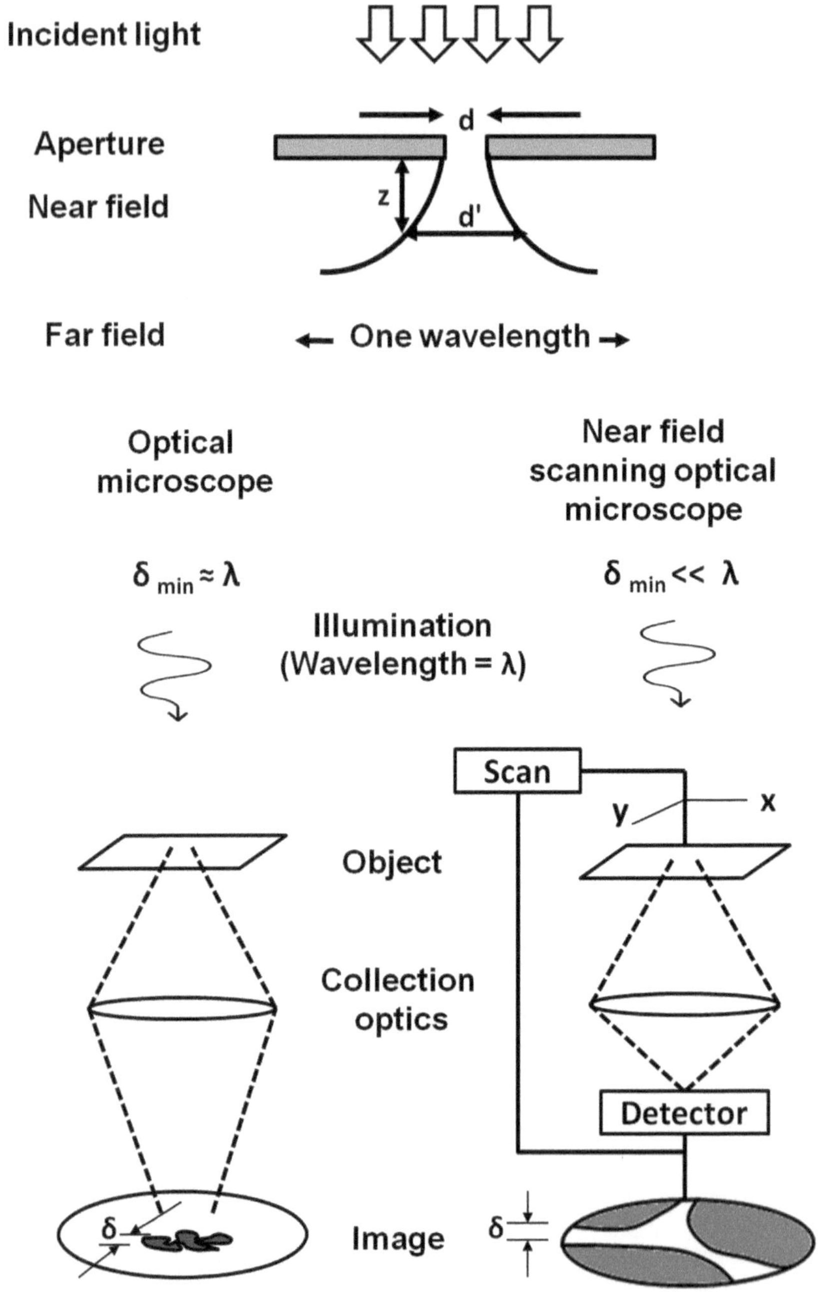

FIGURE 2.13 Schematic diagrams of near- and far-field regions in optical microscope comparison with NSOM.

very high-resolution approaching <50 nm beyond the diffraction limit ($\lambda/2$) of optical microscope, which has a best resolution of about 200 nm. Apart from improved spatial resolution, the other advantages of the near-field interactions are simultaneous topographic image, surface enhancement, rapidly varying electric fields, and the presence of an electric field normal to the surface. Typically, in NSOM, the aperture can be a tapered optical fibre coated with a metal (such as Al) or the apex of a hollow pyramid on the microfabricated AFM cantilever probe itself and laser light is fed to the aperture via an optical fibre.

Several modes of imaging for NSOM offer flexibility in the amount of light incident on the sample, microscope geometry, sample characteristics (opaque, transparent, etc.):

- Transmission-mode imaging – light passing through the transparent sample is collected and detected.
- Reflection-mode imaging – low-intense light reflected from typically opaque sample surface is collected and detected.
- Collection-mode imaging – sample is illuminated with a macroscopic light source from the top or bottom, and the probe collects the reflected light from the sample surface.
- Illumination/collection-mode imaging – probe both illuminates the sample and collects the reflected light.

Usually, the changes in the light intensity are used to create the image in NSOM. However, it is possible to use the various properties of the sample that can give contrast in the image such as changes in polarization, stress at certain points of the sample that changes its optical properties, magnetic properties, which can change the optical properties, fluorescent molecules, molecules excited through a Raman shift, SHG, or other effects.

Hence, SPM techniques can be used to characterize the surface of any nanomaterials from conductors to insulators, hard materials to soft materials, including biological samples, as well as transparent to opaque materials. It can also be used in any environments such as air, high vacuum, high pressures or even in liquids. Additionally, SPM has minimal sample preparation requirements compared to other techniques.

2.3. X-RAY DIFFRACTION AND SCATTERING METHODS

Knowledge of the structure and composition over small distances is a key requirement for understanding the properties of biomaterials. The wavelength of X-rays is on the atomic scale and, hence, provides a variety of primary tools such as X-ray diffraction (XRD) and small-angle X-ray scattering (SAXS) to reveal structural information of nanomaterials over a macroscopic sample volume. The X-ray techniques are also virtually nondestructive.

The XRD method provides a wealth of information – from phase identification to crystallite size, from lattice strain to crystallographic orientation of nanoparticles, nanowires and tubes, and nanostructured materials. In this method, a monochromatic X-ray beam is incident on the crystalline material and the intensity of the elastically scattered (diffracted) beam due to the periodic arrangement of atoms in the sample is measured as a function of the diffracted angle '2θ'. Then using the Bragg's law, the interplanar spacing *d* between atomic planes in the material is calculated as

$$2d \sin\theta = \lambda$$

where λ is the wavelength of the X-rays used and the structural parameters like unit cell dimensions, cell volume, etc. are subsequently obtained from *d*.

Calcium phosphate (CaP) ceramics have been used extensively for bone replacement and augmentation due to their similarity with the mineral component of bone. The most common types of CaPs are HA ($Ca_{10}(PO_4)_6(OH)_2$) and tricalcium phosphate (TCP, $Ca_3(PO_4)_2$), which have different characteristics *in vivo*, although both forms have Ca/P ratios within the range known to promote bone in growth (1.50–1.67). The HA is known to bond with bone directly and thus, can be used as a bone replacing material. The TCP is known to be a bone substituting material because as it dissolves gradually, new bone will be formed where it is resorbed. The tetracalcium phosphate (TTCP, $Ca_4P_2O_9$) is the other member of the CaP family, which hydrolyses slowly to form calcium hydroxide and a thin insoluble apatite layer that prevents further reaction and used clinically as a component of a number of CaP cements for bone defect repairs. The HA in bone and teeth are nanostructured crystals of dimension 20 nm in diameter and 50 nm long. From bionics viewpoint, synthetic apatites to be used for repairing damaged hard tissues are expected to have characteristics closer to those of biological apatite in both composition and structure. Nanocrystalline HA has proved to be of greater biological efficacy in terms of osteoblast adhesion, proliferation, osseointegration and formation of new bone on its surface.

A typical XRD pattern of the oven-dried HA nanoparticles prepared by microwave processing is shown in Fig. 2.14. As the peaks are not well resolved owing to the low temperature involved, a small portion of the apatite sample was heated to 650 °C for 2 h and the XRD pattern of this sample was also recorded. The diffraction peaks of the heat-treated HA are well resolved as shown in Fig. 2.14. The XRD pattern of the apatite can be indexed on the basis of hexagonal structure and the lattice parameters obtained by refinement of the diffraction lines were compared with reported values (JCPDS 9-432). The *a*-axis expands from 9.398 to 9.411 A° on heating from 70 to 1000 °C, whereas the change along *c*-axis is negligible (6.883–6.881 A°). The XRD patterns of HA, TCP and TTCP are shown in Fig. 2.15 for comparison. The thermal stability of CaPs increases from HA to TTCP and the higher intensity peak

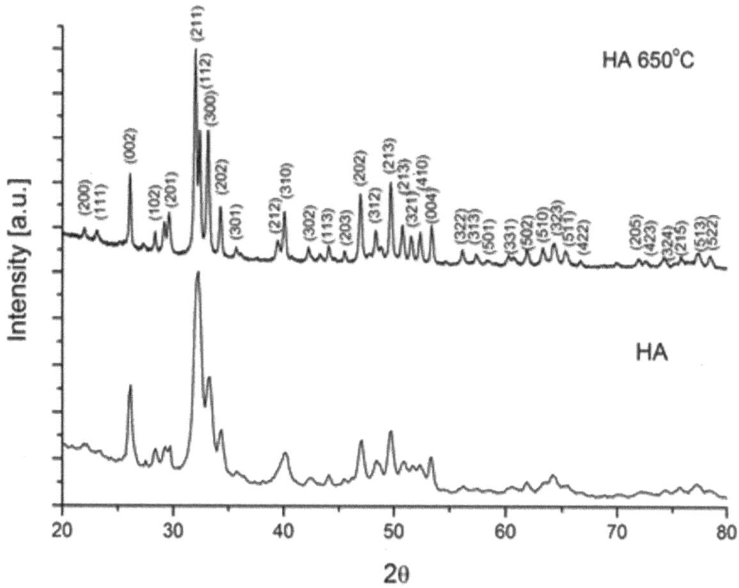

FIGURE 2.14 X-ray diffraction patterns of HA nanoparticles.

shifts from 30° to lower values for HA, TCP, TTCP, respectively. The fraction of crystalline phase (X_c) in HA powder can be calculated by the following equation:

$$X_c \approx 1 - (V_{112/300})/I_{300})$$

Where I_{300} is the intensity of the (300) diffraction peak and $V_{112/300}$ is the intensity of the hollow between (112) and (300) diffraction peaks. The X_c increases from 8% to 97% for the oven-dried and 1000 °C-heated HA powders, respectively.

Interestingly, broadening of the diffraction peak occurs due to deviations from ideal crystallinity, such as decrease in crystallite size for nanomaterials. By analysing this broadening, the nanocrystallite size $t_{(hkl)}$ perpendicular to a crystallographic plane (*hkl*) can be evaluated according to the Scherrer formula,

$$t_{(hkl)} = K\lambda/\mathrm{B}\cos\theta_{(hkl)},$$

where t is the crystallite size (nm); K is the shape factor ($K = 0.9$); λ is the wavelength of the X-rays ($\lambda = 1.54056$ Å for Cu K α radiation); B is the full width at half maximum ((FWHM) in radians) height of the diffraction peak, and $\theta_{(hkl)}$ is the Bragg's diffraction angle (°). Furthermore, the size that is measured by XRD is a measure of the 'fundamental' size of the coherently diffracting individual crystalline domains, rather than the size of particles

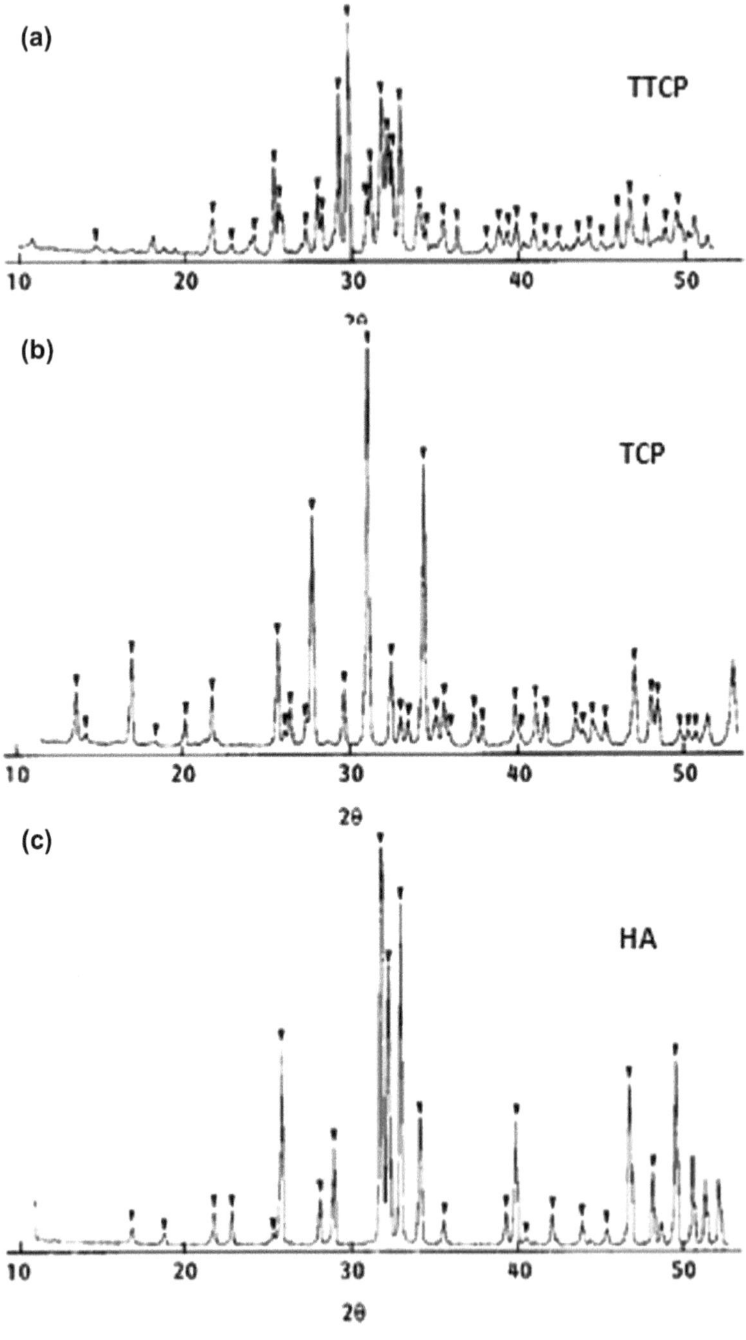

FIGURE 2.15 X-ray diffraction patterns of (a) hydroxyapatite; (b) β-tricalcium phosphate; and (c) tetracalcium phosphate.

formed by the agglomeration of crystals as important properties of nanomaterials is dependent on the crystallite sizes, but not the particle size. Additionally, the size of an individual phase within a mixture, and the dimensions in a specific crystallographic direction can be determined by XRD. Also, by measuring crystallite size in several different crystallographic directions, the crystallite shape can be determined.

The broad diffraction peaks of HA indicate nanocrystalline nature of the synthesized powder. The peak broadening of the XRD reflection was utilized to estimate the average crystallite size using the Scherrer formula. The diffraction peak at 25.9° was chosen for calculation of the crystallite size as it is sharper and isolated from others. This peak assigns to (0 0 2) Miller's plane family and shows the crystal growth along the *c*-axis of the HA crystalline structure. The estimated crystallite sizes were 28 and 56 nm for the oven-dried and 1000 °C-heated samples, respectively. The broadening of diffraction lines is also caused by inhomogeneous strains due to residual stress in the crystallite and from instrumental effects. Hence, careful analysis of the peak shapes should be done to discern the contribution due to crystallite size (diffraction-order-independent) broadening than from (diffraction-order-dependent) broadening of the peak width. Thus, XRD is a very simple technique that can measure millions of crystals in short period of time to accurately characterize the size distribution of nanomaterials as compared to commonly used imaging techniques, such as AFM, SEM, TEM, etc., which are very laborious and expensive.

SAXS evaluates the X-ray scattering pattern of nanosized particles and domains (size range: 1–100 nm) at small angles to get information about their particle structure, i.e. size, shape and internal structure. As shown schematically in Fig. 2.16, a highly collimated X-ray beam is passed through the sample and the scattered X-rays at small angles within the range of 0°–5° are measured in SAXS [Severgun 2003]. The SAXS phenomenon is caused by difference in the electron density within a 100 nm due to the inhomogeneities in atomic density in the interior of the grains and the interface region. In contrast to SAXS, the conventional XRD called wide-angle X-ray scattering (WAXS) examines smaller structures (<1 nm) such as atoms and interatomic distances, which scatter towards large angles and provide information on the phase state, crystal symmetry and the molecular structure. In SAXS, the scattering angle is known as momentum transfer q instead of the scattering angle 2θ as normally used in XRD to obtain results independent of the wavelength. The q is calculated as

$$q = (4\pi/\lambda)\sin\theta.$$

SAXS is used to determine not only the size, size distribution and shape of nanoparticles but also its internal structure as well as surface per volume (porosity) of nanostructured samples. However, as all phase information is lost in the scattering and only scattering intensity can be measured, the exact

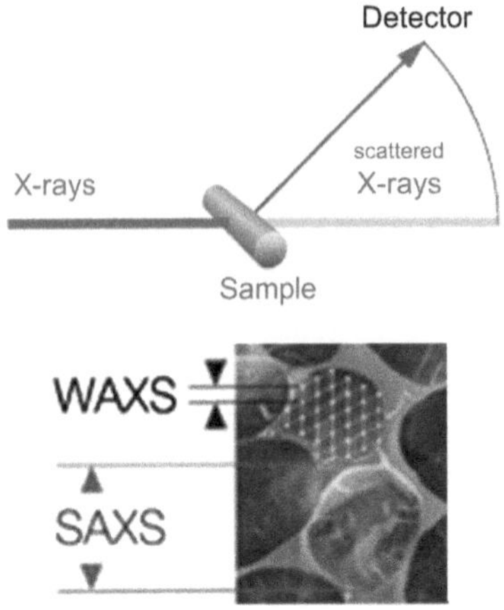

FIGURE 2.16 Schematic diagram of principle of SAXS and the difference between WAXS and SAXS. *Source: Anton Paar.*

microstructure may be difficult to reconstruct from the SAXS results. Since, SAXS pattern is essentially the Fourier transform of the electron density, it must also be deduced in order to resolve the morphology as shown in Fig. 2.16.

The SAXS profiles of the HA (coded as HAP)-reinforced polycaprolactone (PCL) bioactive composites are shown in Fig. 2.17. The scattering intensity gradually decreases with an increase in HA content suggesting good dispersion

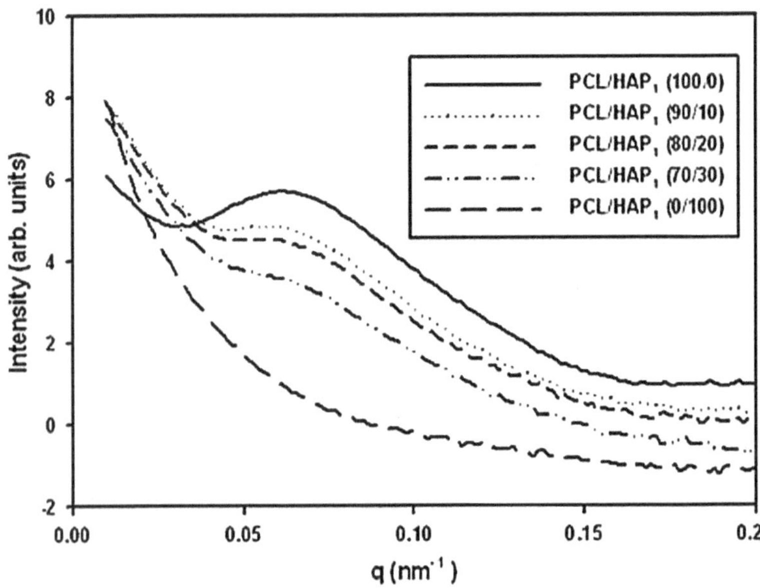

FIGURE 2.17 SAXS patterns of HAP-filled PCL with varying HAP concentration. *Source: Baji et al. J Biomed Mater Res Part B: Appl Biomater 81B: 343–350, 2007.*

of filler in the polymer matrix. The crystalline domains in the PCL exhibits an orthorhombic structure with two distinct diffraction peaks corresponding to (110) and (200) planes by conventional XRD analysis. The presence of HA filler reduces the crystalline domains in the PCL phase and creates smaller crystallites in the polymer blends [2].

2.4. FT-IR SPECTROSCOPY

IR spectroscopy is a fast, relatively inexpensive and widely used analytical technique for the characterization of biomaterials. It is a form of vibrational spectroscopy based on the interaction of IR radiation and natural vibrations of the chemical bonds among atoms that compose the material. In IR spectroscopy, the sample is irradiated with IR radiation and the changes in the absorption of this radiation by the sample are measured. The sample absorbs radiation in the IR region only whenever there is a coincidence (resonance) among the frequencies of the IR radiation with the molecular vibration and the natural vibration causes a change in the dipole moment during vibration. Molecular vibrations are two types: *stretching* (that changes the bond length) and *bending* (that changes the bond angle). These changes in vibrational motion give rise to absorption bands in the vibrational spectrum. For the IR region, the position of absorption bands in the spectra are presented not as wavelength (λ) but as wave number (ν), using the reciprocal centimetre as its unit (cm^{-1}), because it is directly proportional to energy (E) and frequency (ν) of radiation. The IR region is subdivided into three spectral regions, i.e. the near IR (NIR – from 4000 to approximately 14,000 cm^{-1}), mid IR (MIR – from 400 to 4000 cm^{-1}) and far IR (FIR – from approximately 25–400 cm^{-1}). The MIR is the most common and widely employed region for the characterization of materials as it depicts the primary molecular vibrations. The NIR and FIR are not frequently employed because only skeletal and secondary vibrations (overtones) occur in these regions producing spectra that are difficult to analysis.

The IR spectroscopy was greatly improved with the use of interferometer and by the application of fast Fourier transform algorithm, which allowed for the simultaneous detection of all the transmitted energy. The components of a Fourier transform IR (FT-IR) spectrometer are IR source such as Ever-Glo ceramic, Michelson interferometer and IR detector like the triglycine sulphate (TGS) pyroelectric bolometer. The spectra can be obtained by transmission or reflection method. However, the transmission method is the oldest and most commonly used arrangement wherein, the detector measures the intensity of the IR radiation passing through the sample. Various sampling techniques can be used for the analysis by the transmission method. For example, most of the solid biomaterials are too opaque in their normal forms for direct transmission analysis in the MIR region. As potassium bromide (KBr) is completely transparent in the MIR region, a common method for handling solid samples is mixing a small amount of sample with a dry KBr powder and then pressing into transparent disc, known as the compressed alkali metal halide pellet method (KBr pellet or disk method). Depending on the nature of the sample, spectroscopy measurements can be performed in solvents like chloroform, mineral oils, etc.

The FT-IR spectra of the oven-dried and 1000 °C-heated HA nanoparticles obtained by KBr pellet method is shown in Fig. 2.18. Both the spectra exhibit the characteristic bands for HA: 900–1200 cm^{-1} for phosphate (PO_4^{3-} ions) bending and stretching, 602 cm^{-1} for phosphate bending, 632 cm^{-1} and 3571 cm^{-1} for hydroxyl (OH^- ions) structural vibrations. Information about the HA crystallinity can be inferred from the intensities of both the hydroxyl and the phosphate bands (962 cm^{-1}). The intensities of these bands increase with heating indicating an increase in HA crystallinity. But the intensity of the broad bands observed at 1630 and 3400 cm^{-1} drastically decreases with the heating as it corresponds to adsorbed water in HA powder. The effect of nano size of the HA particles were indicated by the characteristic bands for phosphate bending near 1040 and 1090 cm^{-1}, which were broadened and blue shifted for the oven-dried sample compared to 1000 °C-heated sample. The absorption bands at 873, 1421 and 1455 cm^{-1} of oven-dried samples were due to carbonate (CO_3^{2-}) ions present in the HA sample. The carbonate ion may be incorporated during the synthesis due to the reaction between carbon dioxide

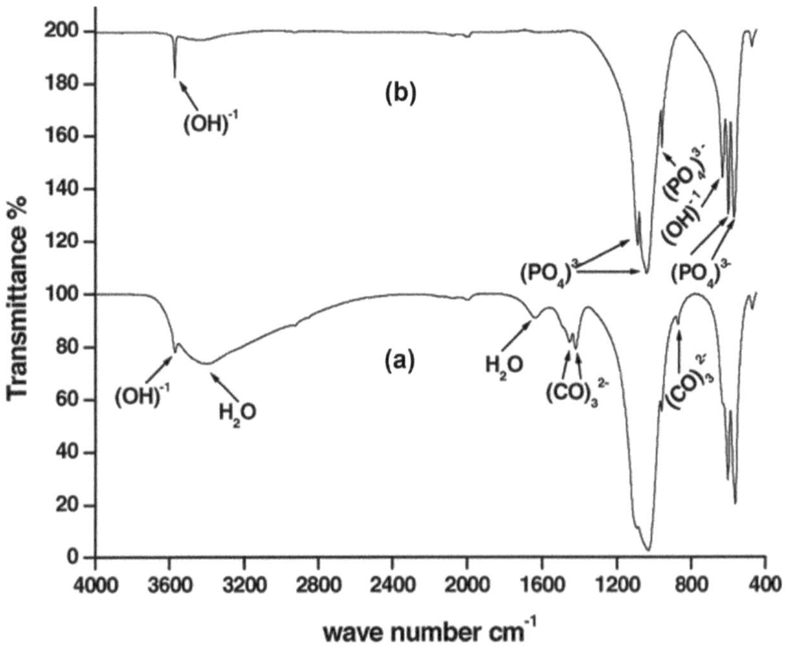

FIGURE 2.18 FT-IR spectra of oven-dried (a) and 1000 °C-heated (b) HA nanoparticles.

and high solution pH. But the heated sample do not show carbonate bands due to the loss of carbonate ions during heating.

Samples that cannot transmit light such as opaque films or surfaces of solid samples, various reflectance methods, such as the attenuated total reflectance (ATR), diffuse and specular reflectance methods can be used. These are generally nondestructive methods, and require only a minimum amount of sample preparation. The ATR is the most common method for obtaining IR spectra of the thin films/plates from near-surface region. In ATR attachment, the IR radiation passes through the optically denser crystal and reflects at the surface of the sample. According to Maxwell's theory, a standing wave forms perpendicular to the total reflecting surface, which interacts with the sample and (its energy or frequency) becomes attenuated. The sample absorbs a fraction of this radiation giving rise to a reflection spectra corresponding to a sampling depth of ~5 μm. The ATR spectrum of HA, calcium deficient hydroxyapatite (CDHA) and TCP is shown in Fig. 2.19. Although ATR method lend to rapid and reproducible sampling, it has an impact on the spectrum, and does not provide a true spectrum of the sample free from some form of distortion or aberration as evident from Fig. 2.19. The specular reflectance and

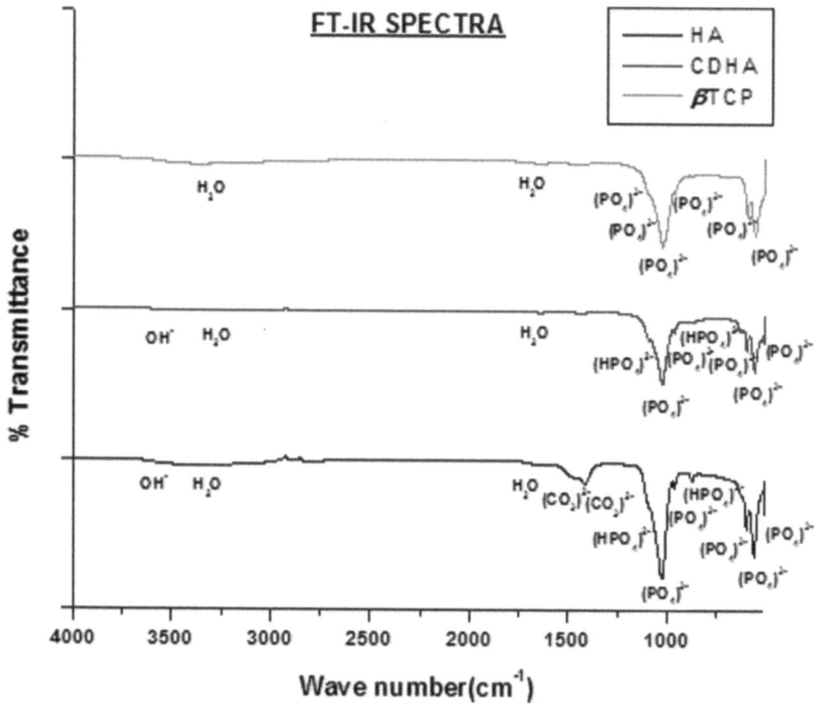

FIGURE 2.19 ATR-IR spectra of HA, calcium deficient hydroxyapatite (CDHA) and TCP nanoparticles.

transflectance methods are greatly used with microscope-based methods of spectral measurement.

2.5. DLS TECHNIQUES

The nanoparticles also scatter (part of the radiation in all directions) visible light and occur everywhere as a natural phenomenon, e.g. the light scattering by atmospheric particles is the reason for the blue color of the sky and brilliant colors seen at sunrise and sunset. It is basically due to the discrete variation in the refractive index within the medium of travel such as solid particles in a liquid. The scattered light may be analyzed either in terms of its intensity or in terms of its fluctuations. The former is called static light scattering (SLS), also as multiangle laser scattering, Rayleigh or classical scattering technique, where-in a laser is used to illuminate the particles suspended in a liquid and the time-averaged intensity of the scattered light by the particles is measured. A complex function of maxima and minima with respect to angle is observed if the size of the particles is similar to the wavelength of the laser used and by modelling the scattering profile with optical model such as Mie theory, size distribution is obtained. However, if the particles are small compared to the wavelength of the illuminating light, typically less than wavelength/10 (around 60 nm for a He–Ne laser), then as per Rayleigh approximation, the scattering will be essentially isotropic, i.e. equal in all directions and also of low intensity, i.e. smaller particles generate low-intensity signals at wide angles, whereas larger particles generate high-intensity signals at small scattering angles. Additionally, the small (nano)particles in suspension undergo Brownian motion (random thermal motion), which causes rapid fluctuations in the intensity of light scattered due to interference, as the distance between the particles is varying. The fluctuation contains the information about the particles' size as the frequency of fluctuation relates to the size of the particles as schematically shown in Fig. 2.20.

DLS, which is also called quasi-elastic light scattering (QELS), is a method for studying fluctuation frequency of the intensity of the scattered light, which can be used for measuring *in situ* the sizes, size distributions, and (in some cases) the shapes of nanoparticles in liquids [3]. Basically due to Doppler broadening, the frequency of the scattered light by the moving particle is shifted and as smaller particles due to higher average velocity produces a greater shift in frequency than larger particles, this difference in frequency of the scattered light is used to determine the particle size distribution. Typically, two successive images are captured to get the time dependence of the scattered intensity and the degree of correlation between the images is measured. Hence, this method is also called photon correlation spectroscopy (PCS).

The optical configuration of a typical DLS system comprises a laser light source to illuminate the stable, well-dispersed particles in dilute concentrations contained in a sample cell and a detector to measure the scattered light at an

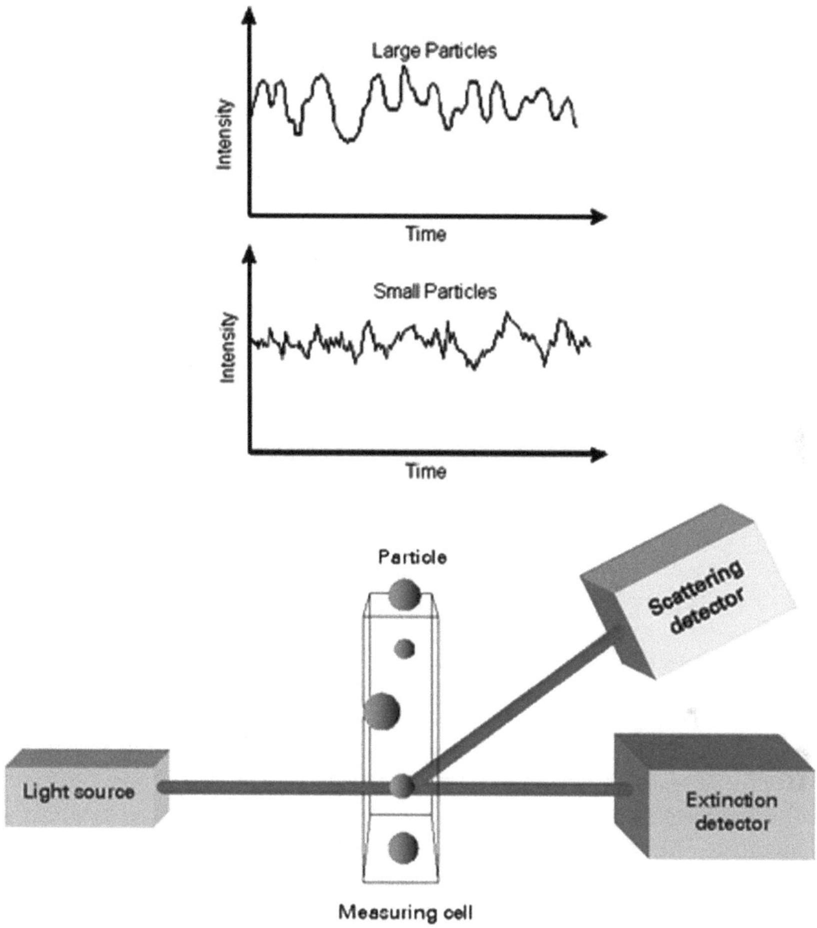

FIGURE 2.20 Typical scattering intensity fluctuations for large and small particles. Schematic diagram of the DLS system. *Source: Malvern DLS technical note, Figure 4.*

angle to the direct beam (Fig. 2.20). Although DLS measurements are usually made at a single angle, data obtained at several angles can be useful. The velocity of the Brownian motion, called translational diffusion coefficient (D_t) is calculated from the intensity fluctuations measured by DLS and is converted into hydrodynamic diameter (D_h particle size) using the Stokes–Einstein equation:

$$D_h = \frac{k_B T}{3\pi\eta D_t}$$

where k_B is the Boltzmann's constant, T is the thermodynamic temperature and η is the dynamic viscosity. It should be noted that the size measured by

DLS is called a *hydrodynamic diameter* as it refers to the size of a sphere that diffuses the way as a particle diffuses within a fluid and the diameter of a sphere corresponds to the same translational diffusion coefficient as the particle.

The diffusion speed of the particles are affected by the ionic strength of the medium (by changing the thickness of the electric double layer), surface structure (by changing the apparent size) and shape of the particle (by changing the hydrodynamic size). DLS is one of the few ensemble techniques (i.e. measures many particles rather than a single particle measurement) that can be used for determination of average particle size between 3 and 3000 nm and with a main advantage that the size information can be obtained in the order of a few seconds to a few minutes over its rival techniques. This technique is also completely noninvasive as the particle motion continues whether or not it is being probed by DLS. Additionally, DLS method has the advantage of being used for *in vitro* study due to the possibility of an *in situ* size determination of nanoparticles in the cell culture medium.

2.6. CONTACT ANGLE MEASUREMENTS

The surface properties of biomaterial have profound effect on the biomaterial–tissue interactions with subsequent consequences on the host responses. This part is specifically addressed in Chapter 4 of this book. Hence, it is very essential to understand the unique properties of the surfaces and the methods to characterize them in striving towards more 'biocompatible materials' which either elicits a specific cellular response or evokes a minimal inflammatory response. For example, the surface composition is different from its bulk composition due to energy minimization, which drives some of the elements of the bulk material to its surface. Similarly, the absence of periodic arrangement of the outermost atomic layers of the surface causes relaxation of the atoms from their regular periodic position leading to rearrangement of atoms with different lattice spacing with same periodicity or reconstruction to new periodic arrangement of atoms at the surface. The surfaces are also always covered with few contaminated layers as hydrocarbons, adsorbed layers of oxygen and carbon, etc. due to the exposure to the atmosphere. So, many techniques are needed for a complete description of the surface. However, the surface energy of the biomaterial has been found to be strongly correlated with the biological interactions such as cell adhesion. The surface energy can be quantitatively characterized from the contact angle of a liquid drop (θ) with the biomaterial surface and geometrically θ is defined as the angle formed by the liquid at the phase boundary of liquid, gas and solid interacting phases. Experimentally, θ is measured by drawing a tangent at the contact where the liquid and solid phases intersect.

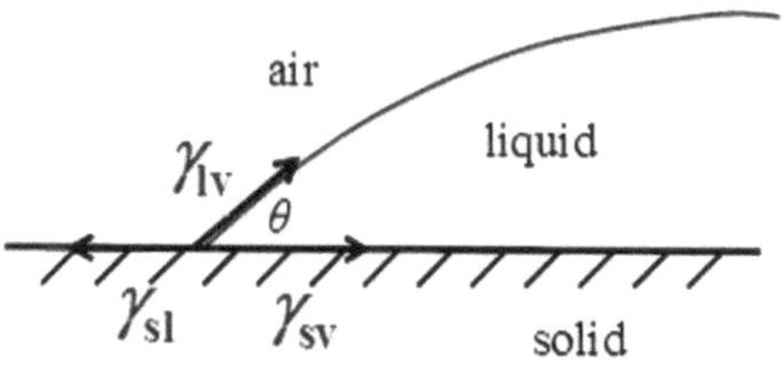

FIGURE 2.21 Contact angle between liquid and solid surface in air medium.

The surface tension can be thought as the energy required for creating a unit area of an interface, the surface energy can be calculated from the Young's equation by measuring the contact angle and the surface tension of the liquid:

$$\gamma_s = \gamma_{sl} + \gamma_l \cdot \cos\theta$$

where γ_{sl} represents the interfacial tension between the two phases while γ_s and γ_l describe the surface tension components of the solid and liquid phases, respectively. As shown in Fig. 2.21, θ corresponds to the angle between vectors γ_l and γ_{sl} and hence, γ_s provides the surface energy of the biomaterial. However, to determine the surface energy of solids from contact angle measurements alone, the value of γ_{sl} must be known. In liquid homologue method (also called Zissman method), the contact angle of a series of homologous nonpolar liquids is measured on the specific biomaterial surface and a plot is drawn between the cosine of the angle and the liquid tension of the liquids. The critical surface tension of the solid is obtained by extrapolation of the surface tension at which the contact angle is zero.

The solid surfaces can be divided into high-energy surfaces and low-energy surfaces based on the surface energy values. The high-surface-energy materials typically have surface tension around 200–500 dyn/cm and are mainly composed of metallic and inorganic materials. Low-energy materials have surface tension <100 dyn/cm, which includes organic compounds and polymeric materials. Often, to reduce the surface free energy of the system, low-energy materials such as oils contaminate high-energy solid surfaces. Hence, it is very essential to have a proper surface cleaning procedure or to tailor the surfaces of implant materials. So, for adhesion of proteins or cells, the aim of many treatments is to bring the surface to an intermediate state between water-repellent (hydrophobic) and completely wettable (hydrophilic). Thus, contact angle measurement provides a simple way to establish the intermediate values [4].

Contact angle also signifies physical manifestation of more fundamental concepts of surface tension and surface energy like the chemical bonding of the uppermost surface layers of the biomaterial. The contact angle values gives an easy-to-measure indication of this bonding, which determine the wettability, adhesion, surface cleanliness, effect of surface treatments, and properties of coatings. The wettability or wetting is an important characteristic property of the surface for biomaterial applications as it refers to the actual process

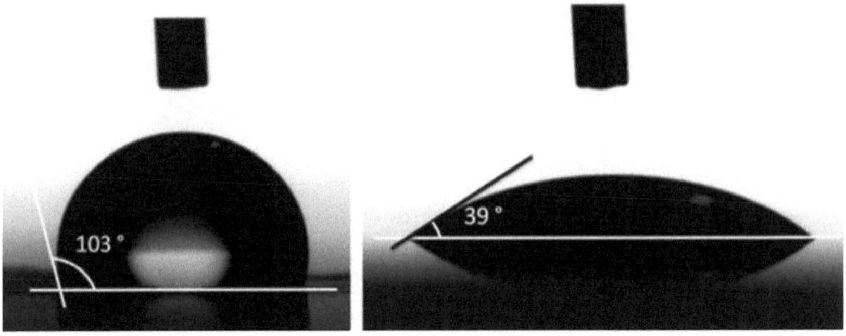

FIGURE 2.22 Photograph of water droplet on the surface of Teflon and glass surfaces.

of spreading of a liquid on the solid surface. As shown in Fig. 2.22, contact angle is a quantitative measure of the wetting of a solid by a liquid, i.e. for low contact angles, the liquid spreads or wets well, meaning the liquid has a strong affinity for the solid (total wetting), and high contact angles indicate partial/poor wetting conditions. To distinguish between the total and partial wetting, a spreading parameter S is also used, which measures the difference between the surface energy of the sample under wet and dry conditions as

$$S = \gamma_s - (\gamma_l + \gamma_{sl})$$

For $S < 0$, the liquid does not spread but forms a spherical drop resting on the surface of the solid with a contact angle (partial wetting), but if $S > 0$, the liquid spreads completely with zero θ to lower its surface energy like ethanol with low value of γ_l spreading on a high-value γ_s solid-like glass (total wetting). In partial wetting condition ($S < 0$), a liquid is said to be mostly wetting a solid if $\theta < 90°$ and if $\theta > 90°$, the liquid is said to be mostly nonwetting. Thus, contact angle gives a direct evaluation of interactions occurring between liquid, gas (or a second immiscible liquid) and solid phases.

The sessile drop method is the most common and convenient method used to measure the contact angle. This method involves placing a drop of liquid of about 5 μL on the solid surface and measuring the angle between the baseline of the drop and the tangent at the drop boundary using optics. Hence, it is called static contact angle method and is ideally suited for solids which have different properties at various surfaces as well as for curved surfaces and can measure average θ as a function of time or temperature. In the dynamic contact angle method, the volume of the drop is increased in steps while the θ is measured until a plateau called 'advancing contact angle' (θ_A) is reached. Similarly, the 'receding contact angle' (θ_R) is determined immediately by measuring the θ for successive removal of equivalent volume of the liquid from the droplet just before the wetting is receded. The contact angle hysteresis is the difference in $\theta_A - \theta_R$ as the contact angle for an advancing liquid across a surface is significantly higher than that of one receding from the surface (i.e. $\theta_A > \theta_R$). Due to contact angle hysteresis, a liquid drop can remain struck on a tilted plane, despite gravity, as θ is larger at the front than at the rear of the drop and the capillary force, which exists due to the difference between the θ_A and θ_R, opposes the gravity force. The θ_A and θ_R are experimentally measured in the 'tilting plate method' using an inclined surface until the rolling of the drop starts.

Overall, contact angle is one of the simple, sensitive, reliable and inexpensive methods for the characterization of the surface as well as physical properties of biomaterials to prepare the biomaterials for specific applications.

2.7. MERCURY INTRUSION POROSIMETRY

Porosity, defined as the percentage of void space in a solid, is an essential requirement of a biomaterial to allow migration and proliferation of cells, mass-transport requirements for cell nutrition as well as vascularization for tissue formation. In addition, a porous surface provides greater mechanical stability by improving the mechanical interlocking between the implant biomaterial and the surrounding natural tissue. Recently, porous biomaterials have been intensively studied and widely used as scaffolds in tissue engineering and cell therapy and as drug delivery system [5]. Although the minimum pore size required to allow the in-growth of mineralized tissue seems to be in the order of 50 μm, larger pore sizes seem to accelerate the depth of penetration of mineralized tissues into the biomaterial, but weakens the mechanical properties. Smaller pore results in larger surface area and seems to contribute to higher adsorption of cell-inducing proteins. As the optimal pore size is dependent on the biomaterial used and its application, it is very essential to characterize the porosity and pore size of biomaterials.

Porous materials might have two important types of pores: open and closed pores as shown in Fig. 2.23. The closed pores are inside the material and not accessible to external fluids. The open pore is a cavity or channel that is connected to the material surface and hence accessible to fluids. The open pores are further divided into dead-end/blind pores, which terminate inside the material, and interconnected/through pores, which make possible for the complete passage of fluids. Based on the pore diameter, the International Union of Pure and Applied Chemistry (IUPAC) has divided pores into three groups as micropores ($D < 2$ nm), mesopores (2 nm $< D <$ 50 nm) and macropores ($D >$ 50 nm). For implant applications, interconnected macropores are of particular interest as it has the potential to facilitate tissue in-growth.

Mercury intrusion porosimetry (MIP) is a widely used method to measure both porosity and pore sizes in biomaterials as it enables detection from macropores down to larger mesopores. MIP is based on the principle that the pressure (P) required to force a nonwetting liquid to fill the pores of a sample is inversely proportional to the diameter (D) of the pores (assuming the pore is cylindrical)

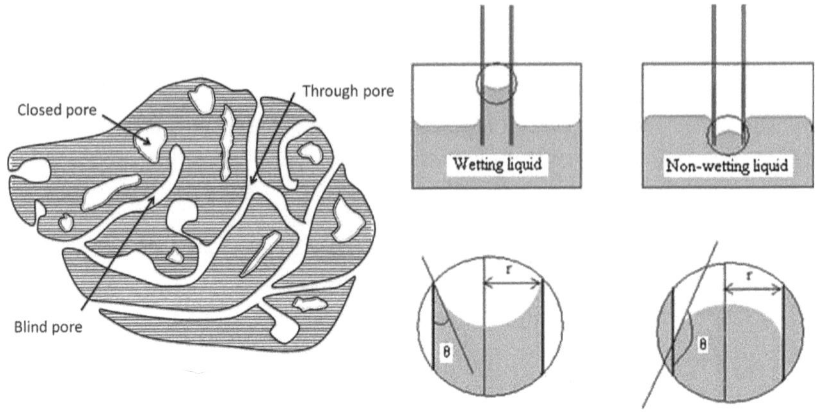

FIGURE 2.23 Schematic diagram of various pore types and capillary action of a wetting and nonwetting liquid. *Source: John I. D'Souzae and H. N. More, Mercury Intrusion Porosimetry, Pharmaceutical Reviews Volume: 6; Issue: 2; 2008; http://www.pharmainfo.net/reviews/mercury-intrusion-porosimetry-tool-pharmaceutical-particle-characterization*

and directly proportional to the surface tension of the liquid (γ) and the angle of contact (θ) with the solid surface as shown by the Washburn equation:

$$P = -4\gamma\cos\theta/D$$

Mercury is the liquid of choice for intrusion porosimetry as it is nonwetting to most solid materials and will not spontaneously penetrate pores by capillary action. For mercury-solid system (θ is usually taken as 140° and γ of the mercury is $\approx$0.48 N/m), the numerator is constant, thus the applied pressure is inversely proportional to the size of the pores, i.e. only slight pressure being required to intrude mercury into large macropores, whereas much greater pressures are required to force the required equilibrated mercury into small pores. It also indicates that under given external pressure P, mercury can resist entry into pores smaller than D, but cannot resist entry into pores of sizes larger than D. So, for any pressure, it can be determined which pore sizes have been invaded with mercury and which sizes have not. Monitoring the progressive intrusion of mercury into a porous structure under stringently controlled pressures permits the generation of pore size/volume distributions within the porous biomaterial.

Typical MIP test involves placing the biomaterial sample into a container and evacuating the container to remove contaminant gases and vapours (usually water). Then, while still evacuated, mercury is allowed to fill the container and afterwards, the pressure is increased towards ambient while monitoring the volume of mercury entering larger openings in the sample. This corresponds to pores of diameters down to about 12 mm have been filled when ambient pressure has reached. For the remainder of the test, the sample container is then placed in a pressure vessel and volume of mercury that intrudes into the sample due to an increase in pressure is determined by substituting pressure values into Washburn's equation. To have a greater volume sensitivity down to <1 μL, a capillary tube is attached to the sample cup which serves both as the mercury reservoir during analysis and as an element of the mercury volume transducer since only a small volume of mercury is required to produce a long 'string' of mercury in a small capillary. The combination of sample cup and capillary stem is referred to as penetrometer (Fig. 2.24). The main source of mercury is removed after filling and the pressure applied to the mercury in the capillary is transmitted from the far end of the capillary to the mercury surrounding the sample in the sample cup. The outer surface of the glass capillary stem is plated with metal (an electrical conductor) to form a co-axial capacitor along with the mercury column in the stem. The length of the mercury column is the only variable in the value of the capacitance and a small volume of mercury entering or leaving a small capillary causes a large change in length (and area) of the mercury column significantly. Thus, capacitance measurements of the stem

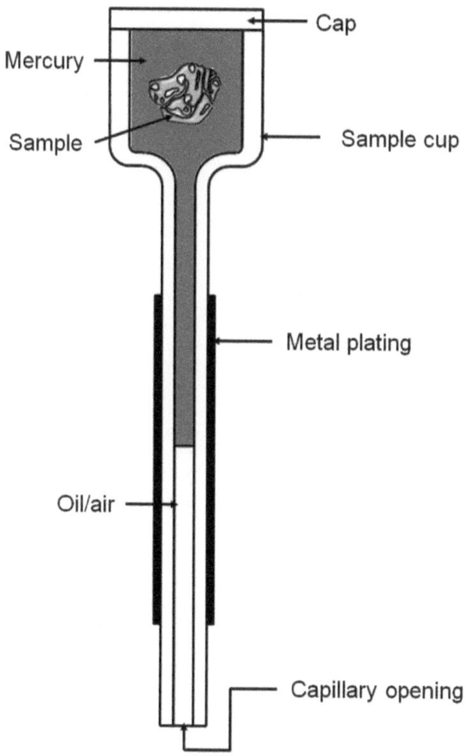

FIGURE 2.24 Cross-sectional view of a typical mercury penetrometer. *Source: Mercury Intrusion Porosimetry Theory, Presented by Micromeritics Instrument Corporation, www.micromeritics.com.*

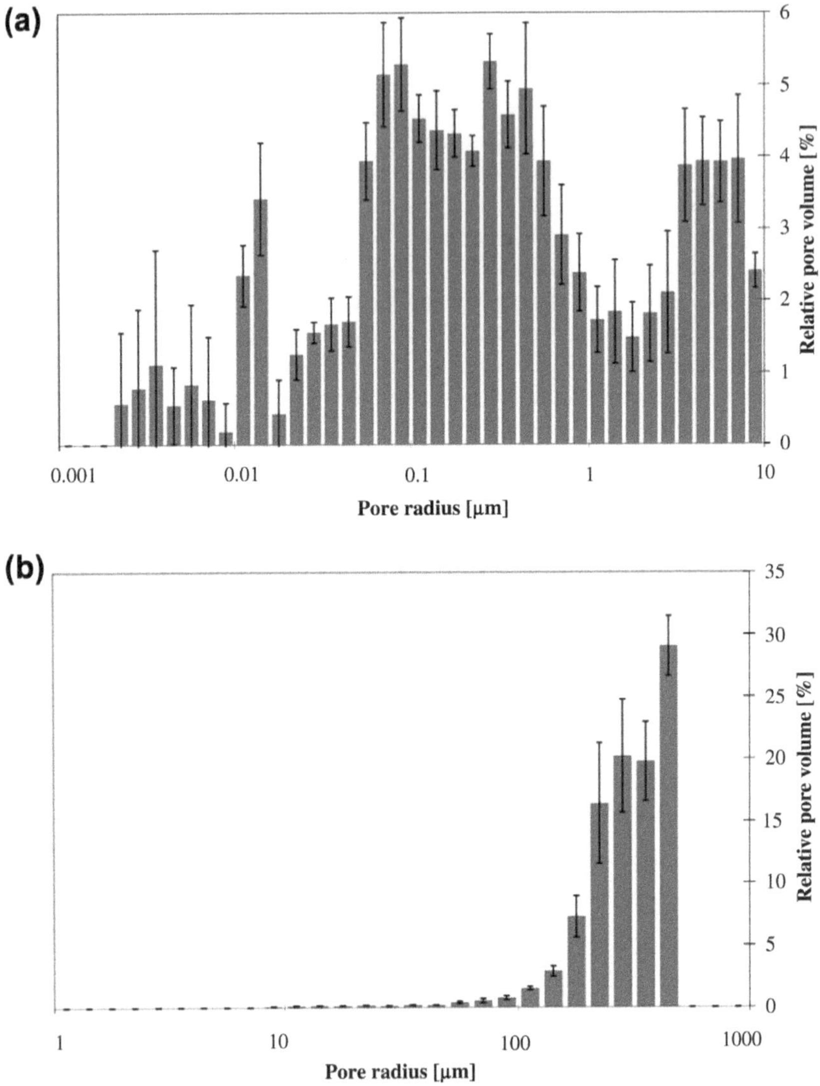

FIGURE 2.25 Relative pore volume of pores determined by mercury intrusion porosimetry (mean ± SD for $n = 5$). (a) 2.3 nm–9 µm; (b) 9–500 µm. *Source: Joschek et al. (2000) Chemical and physicochemical characterization of porous hydroxyapatite ceramics made of natural bone.* Biomaterials ***21*** *1645–1658.*

provide a very high measuring sensitivity and resolution for the mercury volume passing into or out of the sample cup for changes in the external pressure. A plot of the applied pressures and the cumulative volumes of mercury intruded at each pressure is called the intrusion curve. Similarly, as pressure is reduced, mercury leaves the pores, or extrudes, and a plot of this process gives the extrusion curve. The intrusion and extrusion curves do not follow the same path due to the shape of the pores and other physical phenomena. However, both the intrusion and extrusion curves give information about the pore network in the biomaterial.

The porosity of the Endobont® (Merck, Darmstadt, Germany), a porous HA ceramic derived from bovine bone by sintering was determined by MIP. The pore size distributions calculated from the Washburn equation show (Fig. 2.25) two types of pores: macropores with an average size of ~300 μm and micropores with an average size of ~1.3 μm [6].

The porosity and pore sizes can also be characterized by SEM, TEM, micro-computed tomography (μCT), AFM and ultrasound (broadband ultrasound attenuation) methods. However, MIP measurements provide a border pore size distribution more accurately and with greater speed than other methods. The extensive characterization of the biomaterial includes details about total pore volume, incremental volume, differential volume, log-differential volume, percent total volume, total pore surface area, incremental area, median and mean pore diameter, pore-size distributions, bulk and skeletal densities, percent porosity, tortuosity, compressibility, etc.

2.8. GAS ADSORPTION MEASUREMENTS

As particles become smaller, the proportion of atoms found at the surface increases relative to the proportion inside its bulk and so, characterization of surfaces and interfaces are very important for biomaterials. The surface area, particle size and porous structure of biomaterials can be characterized by the measurement of the amount of adsorbed inert gas such as nitrogen, argon, krypton or carbon dioxide on solid surfaces. The gas adsorption can be classified as either physisorption or chemisorption. The physical adsorption of a gas involves weak molecular forces, such as van der Waals forces while formation of a chemical bond provides the driving force in chemisorption. The physically adsorbed gases can be easily removed by reducing the partial pressure, but the chemisorbed gases are usually difficult to remove from the solid surface. The physisorption is used not only to measure surface area but also to probe the entire surface including irregularities and pore interiors as shown schematically in Fig. 2.23. The equilibrium relation between the amounts of adsorbed gas with the partial pressure of the gas at constant temperature is called gas adsorption isotherm. To measure the surface area and for pore analysis, the BET 'Brunauer, Emmett, Teller' adsorption isotherm is frequently used, which involves adsorption of a monolayer of nitrogen gas molecules on the surface of the powder by cooling and then heating to vapourize (desorption) the monolayer to measure the amount of nitrogen adsorbed. By assuming spherical shape and unimodal distribution of the nanoparticles, the surface area is determined from adsorption measurements and the surface area can then be easily

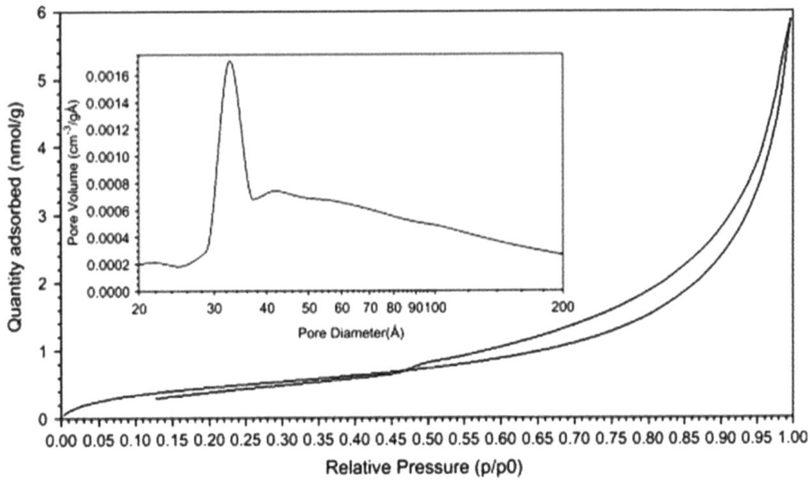

FIGURE 2.26 N_2 adsorption/desorption isotherm of HA nanorods and pore size distribution plot (inset) of the sample. *Source: Self-organization of hydroxyapatite nanorods through oriented attachment, Chen et al.,* Biomat. *28 (2007) 2275–2280.*

converted into particle size. Prior to the measurements, the sample is pretreated at elevated temperature in vacuum or flowing gas to remove any contaminants. The nitrogen adsorption–desorption isotherm of oriented attachment of HA nanorods processed by hydrothermal treatment is shown in Fig. 2.26. The sample exhibits type II isotherm behavior with a large hysteresis loop at high relative pressure. The pore size distributions indicate that the nanorods are self-organized with around 3.5 nm pores (inset) [7]. The BET surface area of the sample is 42 $m^2\ g^{-1}$.

2.9. SUMMARY

The future for biomaterials is exciting and the research for newer biomaterials is truly becoming more diverse, bringing together knowledge from a variety of disciplines. The emphasis of this chapter was concerned with the determination of the physical characteristics, specifically microstructural features, porosity, and surface properties. These parameters were evaluated using optical, electron and AFM images, mercury porosimetry, gas adsorption and contact angle measurements complemented with XRD analysis and Fourier transform IR spectroscopy. The capabilities of various characterization methods for the microstructural analysis of biomaterials are summarized in Table 2.1. To completely characterize the biomaterial, it is not only necessary to know a multitude of chemical and physical parameters but also essential to correlate the effect of materials processing with the properties of the materials. In addition to looking at a variety of characterization techniques and their

TABLE 2.1 List of Characterization Methods and Their Capabilities for Microstructural Analysis of Biomaterials

Characterization Methods	Crystallography	Morphology	Chemical
Optical microscopy		√	
TEM	√	√	√
SEM	X1	√	√
STM		√	
AFM		√	X2
XRD	√	X3	√
FT-IR	X4		√

X1 – grain orientations, texture analysis.
X2 – magnetic probes.
X3 – nano-size analysis.
X4 – allowed vibrations for the symmetry groups.

application to biomaterials, it is also important to understand the interactions occurring on the nano-, micro- and macrolength scales at the material–biological host interface for the development of more efficient biomaterial, fit for purpose.

REFERENCES

[1] Figueiredo Margarida, Henriques Jose, Martins Gabriela, Guerra Fernando, Judas Fernando, Figueiredo Helena. Physicochemical characterization of biomaterials commonly used in dentistry as bone substitutes—comparison with human bone. J Biomed Mater Res Part B: Appl Biomater 2010;92B:409–19.

[2] Baji A, Wong SC, Liu T, Li T, Srivatsan TS. Morphological and X-ray diffraction studies of crystalline hydroxyapatite-reinforced polycaprolactone, J Biomed Mater Res Part B: Appl Biomater 2007:343–50.

[3] Pecora R. Dynamic light scattering measurement of nanometer particles in liquids. J Nanoparticle Res 2000;2:123–31.

[4] Vogler Erwin A. Protein adsorption in three dimensions. Biomaterials 2012;33:1201–37.

[5] Karageorgiou Vassilis, Kaplan David. Porosity of 3D biomaterial scaffolds and osteogenesis. Biomaterials 2005;26:5474–91.

[6] Joschek S, Nies B, Krotz R, GoKpferich A. Chemical and physicochemical characterization of porous hydroxyapatite ceramics made of natural bone. Biomaterials 2000;21:1645–58.

[7] Chen JD, Wang YJ, Wei K, Zhang SH, Shi XT. Self-organization of hydroxyapatite nanorods through oriented attachment. Biomaterials 2007;28:2275–80.

Chapter 3

Mechanical Characterization of Biomaterials

Ryan K. Roeder
Department of Aerospace and Mechanical Engineering, Bioengineering Graduate Program, University of Notre Dame, Notre Dame, IN, USA

Chapter Outline

Characterization of Biomaterials. http://dx.doi.org/10.1016/B978-0-12-415800-9.00003-6

3.1. INTRODUCTION

Mechanical behavior is crucial to the performance of biomaterials used in most biomedical implants and devices. In many cases, the importance is obvious and the connection direct. The femoral stem of a total hip replacement must bear static and cyclic loads imposed by body weight and muscle forces without failure for the lifetime of the implant. In other cases, the importance of mechanical behavior is less obvious and the connection is indirect. If the aforementioned femoral stem is too stiff, cells in the adjacent bone tissue are shielded from mechanical stimuli by the implant, which can ultimately lead to implant failure via osteolysis, bone resorption, and loosening of the stem in the femur [1]. Thus, in almost all cases, including the femoral stem, *optimum device performance is predicated by optimum mechanical properties, which involves design tradeoffs*.

In other examples, middle-ear implants may preferably utilize biomaterials rigid enough to readily transmit sound waves, but ductile enough to permit intraoperative shaping [2]. A contact lens is designed to be "soft" for user comfort and is minimally loaded in use, but should not tear during repeated handling and placement [3]. Biomaterial mechanical properties may also involve spatiotemporal variations. Sutures and fixation screws used, for example, in arthroscopic reconstruction of a torn anterior cruciate ligament, may have mechanical properties that intentionally degrade over a desirable time scale after implantation [4]. Thus, *the design of biomedical devices almost always involves some form of mechanical characterization of biomaterials.*

Therefore, *the objective of this chapter is to provide a broad overview of experimental methods and important considerations for the mechanical characterization of biomaterials*. Note that the purpose here is *not* to cover the mechanical behavior of biomaterials, including structure-property relationships and deformation mechanisms. For this, the reader is referred to a welcomed new textbook focused on the mechanical behavior of biomaterials [5], and other excellent textbooks on the mechanical behavior of engineering materials [6–8]. Instead, this chapter is intended to address the practical needs of engineers and scientists who encounter a need to characterize the mechanical properties of a biomaterial, but may not know where to begin or what the key considerations should be. Of course, individuals and companies without expertise in mechanical characterization are wise to collaborate with colleagues or hire consultants who possess this expertise. However, delegation does not abdicate responsibility. The author has encountered multiple examples where a lack of basic knowledge on the part of the primary stakeholder allowed a consultant or collaborator to provide misleading or even incorrect mechanical characterization. Therefore, this chapter is also intended to be useful for those who may not perform mechanical characterization themselves.

Finally, many details are necessarily omitted from this broad overview, but references are provided to direct the reader seeking greater technical depth on a particular topic. Of particular note, common methodologies are standardized and published by the American Society for Testing Materials (ASTM) and International Standards Organization, which provide a wealth of mechanical testing procedures and guidelines. Any laboratory or institution is wise to adopt or develop internal standard operating procedures in order to ensure the repeatability and reliability of test results, while demonstrating transparency and a mechanism for corrective measures.

3.2. FUNDAMENTAL CONCEPTS

3.2.1. Stress and Strain versus Force and Displacement

Mechanical properties of biomaterials are almost always characterized by the stresses and strains in a material that result from applied loads and displacements. Consider the schematic depiction of two different interbody spinal fusion cage designs shown in Fig. 3.1. Interbody spinal fusion cages are designed to alleviate back pain caused by a degenerated or herniated intervertebral disc by restoring the intervertebral height and facilitating the formation of bony fusion in the disc space between two vertebral bodies [9–11]. Bone formation is guided through a center cavity shown in each device.

Suppose device (1) is designed with a larger cavity in the center and a smaller cross-sectional area (A) compared to device (2), but both are composed of the same material (Fig. 3.1). If the same load (P, e.g., body weight) is applied to each device, the stress ($\sigma = P/A$) acting on the material in device (1) will exceed that in device (2), inversely proportional to the ratio of

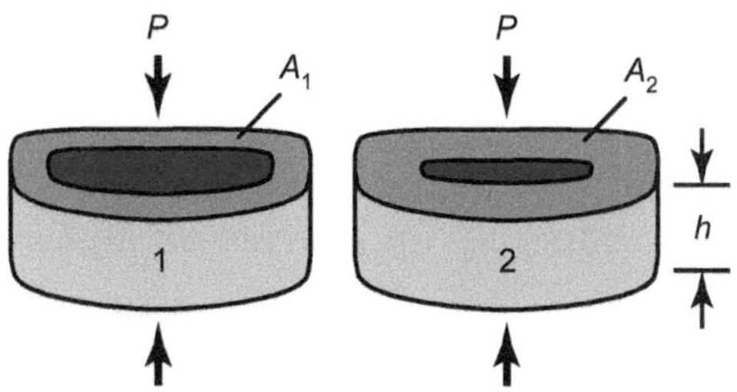

$$\left(\sigma_1 = \frac{P}{A_1}\right) > \left(\sigma_2 = \frac{P}{A_2}\right)$$

$$\left(\Delta h_1 = \frac{P}{k_1} = \frac{P.h}{E_1.A_1}\right) > \left(\Delta h_2 = \frac{P}{k_2} = \frac{P.h}{E_2.A_2}\right)$$

FIGURE 3.1 Differences between stress (σ) and strain (ε) versus force (P) and displacement (Δh) are introduced using the example of two interbody spinal fusion cages, where device (1) is designed with a larger cavity in the center and a smaller cross-sectional area (A) compared to device (2). For the examples described, each device has the same height (h) and the same applied load (P, e.g., body weight). The mechanics of the device, including the device stiffness (k), is shown to be governed by both the device design (e.g., A, h) and material properties (e.g., E).

cross-sectional areas (Fig. 3.1). Therefore, while it may be desirable to increase the size of the cavity for a greater fusion volume, the engineer must also consider whether greater stress in the device could result in implant failure before a fusion is achieved, or whether greater contact stress between the device with reduced cross-sectional area and adjacent bone tissue could result in subsidence of the implant into the adjacent bone tissue with loss of vertebral height [10]. On the other hand, the potentially deleterious effect of the reduced cross-sectional area on device failure can be mitigated by utilizing a material in device (1) that exhibits a greater failure strength (or stress) compared to the material used in device (2). In summary, internal stresses are related to applied loads by the inverse of the cross-sectional area (Fig. 3.1).

Deformations are also related to the applied load and cross-sectional area, as well as the material's stiffness. The amount of elastic (recoverable) deformation (Δh) due to the applied load will also be greater in device (1) compared to (2) if composed of the same material because device (1) will exhibit a lower stiffness (k, analogous to a spring) than (2) due to the smaller cross-sectional area (A) (Fig. 3.1). Internal material strains (ε) are calculated by normalizing deformations by an initial dimension (e.g., $\Delta h/h$). Therefore, a device with reduced stiffness may be advantageous for transmitting greater osteogenic strains to the bone tissue forming within the center cavity. However, notice that a reduced stiffness could also be achieved while maintaining the same cross-sectional area in device (2) by utilizing a different material to achieve a reduced stiffness (elastic modulus, E, explained below). Thus, the mechanics of a device are governed by both the device design (e.g., A, h) and material properties (e.g., E). Moreover, from a different perspective, the above example also demonstrated that *the mechanical properties of materials should always be characterized using stress and stain, instead of force and displacement, in order to avoid the influence of specimen size and facilitate interstudy comparison.*

3.2.2. Mechanical Properties on a Stress–Strain Curve

The most fundamental mechanical properties of materials are measured from a stress–strain curve (Fig. 3.2), which is calculated from the measured load and deformation during the course of a quasistatic (time-independent) test as described above. For example, consider a simple and reproducible specimen geometry, such as a right circular cylinder or a rectangular prism, loaded in quasistatic uniaxial tension. The stress–strain curve shows that this material (e.g., a metal) exhibited three regimes of deformation (Fig. 3.2a). Elastic deformation is fully recoverable, meaning that upon unloading the specimen returns to its original size and shape (Fig. 3.2b, $\Delta\varepsilon_e$). At the yield point (y), deformation is no longer fully recoverable but plastic, including permanent or inelastic deformation. Unloading at a point in the plastic range of the stress–strain curve will result in a return to zero stress but a permanent strain

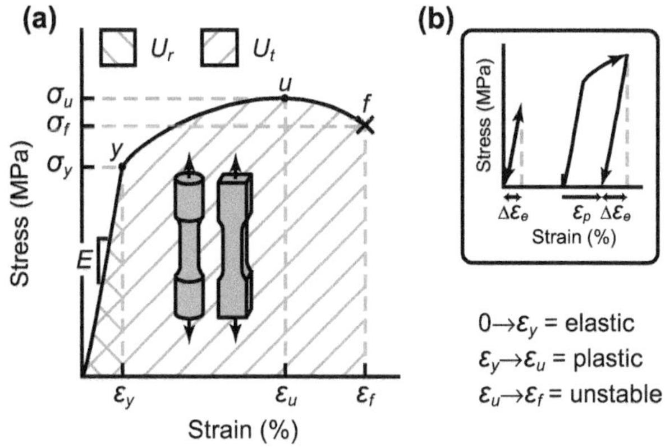

FIGURE 3.2 Schematic stress–strain (σ–ε) curves for a right circular cylinder or a rectangular prism of a material (e.g., a metal) loaded in quasistatic uniaxial tension showing (a) three regimes of deformation (elastic, plastic, and unstable) separated by critical points (y = yield, u = ultimate and f = fracture), fundamental material properties, and (b) the difference between recoverable elastic deformation ($\Delta\varepsilon_e$) and plastic or permanent deformation (ε_p). Fundamental material properties include the elastic modulus (E), yield strength and strain (σ_y, ε_y), ultimate strength and strain (σ_u, ε_u), fracture strength (σ_f), strain-to-failure (ε_f), modulus of resilience (U_r), and modulus of toughness (U_t).

(Fig. 3.2b, ε_p), typically following the slope of the elastic region. After reaching the ultimate point (u), deformation becomes unstable due to a localized reduction in cross-sectional area (necking) and fracture (f) is imminent.

The slope of the linear elastic region of the stress–strain curve ($\Delta\sigma/\Delta\varepsilon$) is termed the *elastic modulus* (E), or Young's modulus, and reflects the material's stiffness (Fig. 3.2a). Elastic moduli should be reported in units of Pa, typically GPa for most materials, although highly compliant materials such as foams and hydrogels are more conveniently reported in MPa or kPa. In polymers or ductile metals where a linear region is not readily apparent, a tangent or secant modulus may also be measured [12]. Note also that in order to facilitate accurate and repeatable measurement of strain-based properties, the presence of a "toe" region (observed as nonlinear region of increasing slope) at low stress and strain if often removed by extrapolating the measured elastic modulus to zero stress and setting strain equal to zero at that point.

The stress at the point of yielding (y) is termed the *yield strength* (σ_y or Y), and is typically reported in MPa or kPa (Fig. 3.2). Identification of the yield point is often not obvious due to a gradual transition from elastic to plastic deformation. Therefore, the yield stress is often more accurately and reproducibly measured using a line parallel to the measured elastic modulus but offset by a set amount of strain, typically 0.2% and thus termed the *0.2%-offset yield strength*. Notice that due to a phenomenon termed *strain hardening* the yield strength may be increased upon reloading a specimen previously

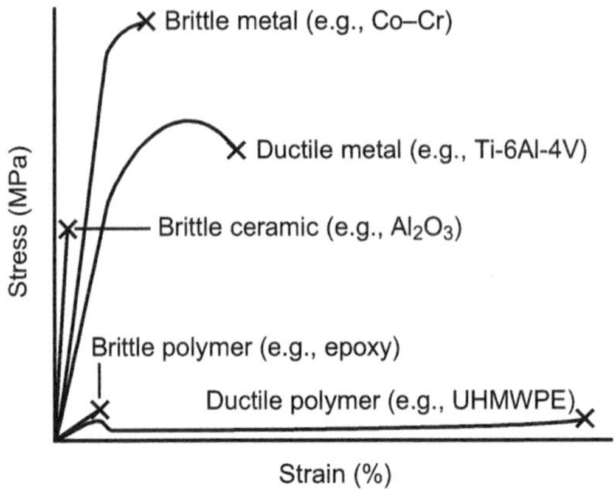

FIGURE 3.3 Schematic stress–strain (σ–ε) curves showing general differences in behavior between ceramics, metals, and polymers under quasistatic tensile loading.

plastically deformed (Fig. 3.2b). The stress at the ultimate point (u) is termed the *ultimate strength* or ultimate tensile strength (σ_u or *UTS*), and is typically reported in MPa or kPa (Fig. 3.2a). Similarly, the stress at the point of fracture (f) is termed the *fracture strength* (σ_f), and is typically reported in MPa or kPa. Note that brittle materials may not exhibit a yield or ultimate strength, but only fracture strength (Fig. 3.3).

The corresponding strains measured at the yield strength, ultimate strength, and fracture strength are termed the *yield strain* (ε_y), *ultimate strain* (ε_u), and *strain-to-failure* (ε_f), respectively (Fig. 3.2), with the latter being most commonly reported as measure of ductility. Strains are reported either in nondimensional units of the change in length divided by the original length (mm/mm), as a percentage (mm/mm $\times$ 100), or in microstrain ($\mu\varepsilon$ = mm/mm $\times$ 1000). *Ductility* is also commonly reported by the *percent elongation* (*%EL*) in tensile tests, which is the same as the percent strain-to-failure, and/or the *percent area reduction* (*%AR*), which is the percent change in cross-sectional area divided by the original cross-sectional area. *Poisson's ratio* (ν) is the negative of the ratio of a strain normal to the applied load relative to the strain in the direction of applied load. Poisson's ratio can range from $-1.0 < \nu < 0.5$ for an isotropic, linear elastic material, but is typically near 0.3 for most materials.

The area under a force–displacement curve represents the amount of work (w) to deform the material and is typically reported in N-m or N-mm; the area under the stress–strain curve represents the total *strain energy* per unit volume (U), or strain energy density, of the material, and is typically reported in J/m^3. The total strain energy up to the yield point and fracture is termed the *modulus of resilience* (U_r) and *modulus of toughness* (U_t), respectively (Fig. 3.2). While these properties provide important and useful measures of material toughness or energy absorption, they should not be mistaken or substituted as a measure of fracture toughness, which is introduced in Section 3.2.6.

3.2.3. Material Classes

The main classes of engineering materials with respect to mechanical behavior are *ceramics*, *metals*, *polymers*, and *composites* thereof. Schematic tensile stress–strain curves for examples from each class show general differences in behavior (Fig. 3.3). Under ambient and physiological conditions, ceramics typically exhibit the highest elastic modulus and high strength, but are brittle, fracturing during linear elastic deformation. Note, however, that ceramics exhibit even greater strength, by up to an order of magnitude, in compressive loading rather than tensile loading. Therefore, ceramics are typically only used in applications designed for primarily compressive stresses. Metal alloys typically exhibit a relatively high elastic modulus, the highest strength in tension, and ductility ranging from relatively brittle to highly ductile. Note that pure metals (not shown) exhibit significantly lower strength and greater ductility than metal alloys. Finally, polymers typically exhibit the lowest elastic modulus and strength. Thermoplastics (e.g., ultrahigh molecular weight polyethylene (UHMWPE)) may exhibit the greatest ductility, while thermosets (e.g., epoxy) may be stiffer and stronger, but more brittle. Composite materials exhibit properties characterized by a rule of mixtures for two or more constituent phases [13,14]. Note that all the above generalities can have exceptions. For example, ceramics can exhibit ductility at high temperature, and polymer fibers (e.g., Kevlar® or spider silk) can be prepared to exhibit extremely high strength. Nonetheless, the general *differences in mechanical properties between material classes have profound influences on the methods and analyses used for mechanical characterization*, which will become evident throughout the remainder of this chapter.

The reasons for these general differences in mechanical properties are many and a primary subject of textbooks on mechanical behavior [5–8], but a few general points are worth pointing out here. Differences in the elastic modulus and strength reflect differences in atomic bonding. In fact, mechanical properties like the elastic modulus and fracture strength can be predicted from first principles through atomistic/molecular modeling incorporating bond strengths and atomic/molecular structure [15]. The large difference in electronegativity between metal and nonmetal atoms in ceramics results in strong covalent or ionic bonding, due to electrons being shared or donated, respectively, to fill the outer shell. A small difference in electronegativity between metal atoms and excess electrons characterizes metallic bonding. However, the ductility of metals is mainly derived from crystalline defects called dislocations [5–8]. Polymers have strong covalent bonds between atoms within a molecule, but weak van der Waals bonds between molecules, which result in lower elastic modulus and strength. However, the influence of weak van der Waals bonding can be decreased and strength increased by cross-linking (thermosets) or alignment of large molecules (e.g., Kevlar® fibers or spider silk).

3.2.4. Types of Stress–Strain Curves and Constitutive Models

Stress–strain curves can take on a number of different forms including and beyond those described above (Fig. 3.4). Constitutive models of stress–strain curves are useful for characterizing the mechanical properties of materials under quasistatic loading. Constitutive models are empirical or theoretical mathematical descriptions of physical behavior. No one model is suitable for the diverse behavior exhibited by different materials under different loading conditions, but a number of common examples for quasistatic loading are shown in Fig. 3.4.

Elastic deformation is most commonly linear or approximated as linear in most metals, ceramics, polymers, and composites (Fig. 3.4a). However, elastomers and soft biological tissues like tendon may exhibit nonlinear elasticity or

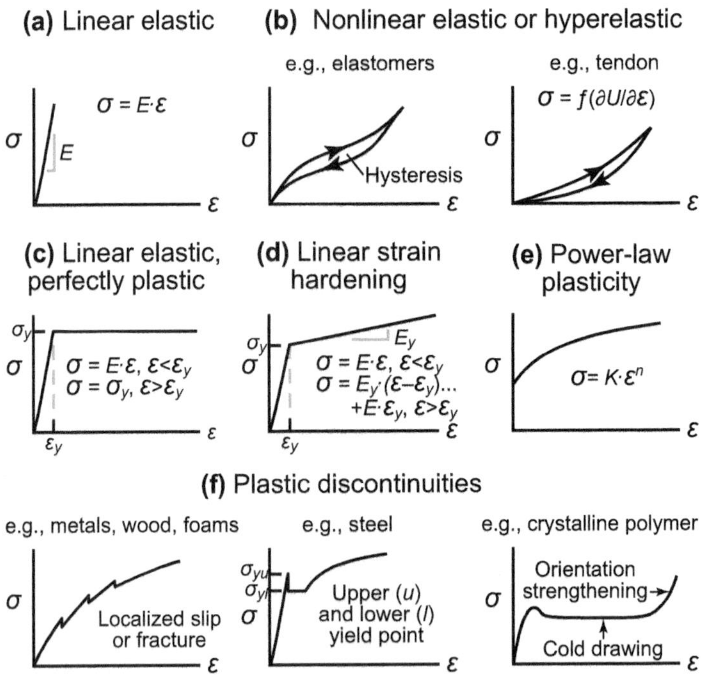

FIGURE 3.4 Common forms of stress–strain ($\sigma-\varepsilon$) curves for quasistatic loading and corresponding constitutive models for (a) linear and (b) nonlinear elastic deformation, (c) perfectly plastic deformation, plastic deformation with (d) linear or (e) power-law strain hardening, and (f) discontinuities during plastic deformation, where E is the elastic modulus, U is strain energy density, σ_y is the yield stress, ε_y is the yield strain, E_y is a "modulus" for linear strain hardening, K is the strength coefficient, n is the strain hardening exponent, σ_{yu} is the upper yield stress, and σ_{yl} is the lower yield stress. Note that all stress–strain curves are schematic and not to scale relative to each other.

hyperelasticity (Fig. 3.4b). Linear elastic deformation is readily described by Hooke's Law ($\sigma = E \cdot \varepsilon$) and its generalized form for three-dimensional deformation [6,7], while nonlinear elastic constitutive models are based on functions of the change in internal strain-energy density per change in strains [16,17].

The use of true stress and strain is an important consideration for plastic deformation that requires introduction before discussing additional constitutive models. All preceding discussions used *engineering stress and strain*, which normalized force and deformation by the *original* specimen cross-sectional area ($\sigma = F/A_0$) and length ($\varepsilon = \Delta l/l_0$), respectively. However, as noted in Section 3.2.2, plastic deformation results in permanent changes in specimen dimensions. Therefore, *true stress and strain* normalize force and deformations by the *instantaneous* specimen cross-sectional area ($\sigma = F/A_i$) and length ($\varepsilon = \ln[l/l_0]$), respectively. There is no difference between engineering and true stress/strain during elastic deformation, but during plastic deformation, and especially at large strains, true stress and strain should be used for experimental data and constitutive models. Under an assumption of volume conservation, true stress can be related to engineering stress as $\sigma_{true} = \sigma_{eng} \cdot (1+\varepsilon_{eng})$, and true strain can be related to engineering strain as $\varepsilon_{true} = \ln(1-\varepsilon_{eng})$. The assumption of volume conservation is valid for incompressible solids, which includes all relatively dense materials, but not porous materials (e.g., foams, scaffolds, and particulate compacts) where the apparent volume is not conserved during plastic deformation.

Plastic deformation may occur at a constant flow stress (Fig. 3.4c), such as during hot working of metals, or this may be used as a simplifying assumption. A strain hardening material may be modeled as linear (Fig. 3.4d), including a "modulus" for plastic deformation, or more commonly using a power-law (Fig. 3.4e). While experimental data are empirically fit by a power-law, the constants K and n are material properties with physical meaning. The strength coefficient (K) is the true stress at a true strain of one; the *strain hardening exponent* (n) is the true strain at the onset of necking or plastic instability, which corresponds to the ultimate point in engineering stress–strain (u, Fig. 3.2a).

Some materials may exhibit discontinuities in the stress–strain curve during plastic deformation (Fig. 3.4f). For example, sudden drops in stress may occur in some metals due to localized slip or dislocation motion. In steels, this behavior may be pronounced and is often characterized by an upper yield stress (σ_{yu}) and lower yield stress (σ_{yl}). Brittle materials and highly porous materials (e.g., scaffolds, foams or honeycomb structures) may similarly exhibit sudden drops in stress due to localized fracture events. This can result in "noisy" stress–strain curve data and increased error when measuring mechanical properties using the preceding constitutive models. Continuous fiber-reinforced composites often exhibit a large drop in stress coinciding with fiber failure prior to matrix failure (not shown). Finally, after an initial yield and ultimate point, semicrystalline polymers commonly exhibit extreme ductility (cold drawing) due to the unravelling and alignment of crystalline molecules (cf., Section 3.2.3), and this region can be modeled as perfectly plastic. Orientation strengthening subsequently occurs when strong intramolecular covalent bonds begin to bear the load in place of weak intermolecular van der Waals bonds.

3.2.5. Anisotropy

Materials with directional structural features (e.g., fiber reinforcements, grain shape, crystallographic texture, etc.) exhibit anisotropic mechanical properties – that is, mechanical properties that are directionally dependent. For example, an orthotropic material exhibits three planes of symmetry such that properties differ in each of three orthogonal specimen directions. In contrast, isotropic properties are equivalent in all directions. Interestingly, engineers have historically preferred to design and use isotropic materials in man-made technology, although exceptions such a carbon fiber-reinforced plastics exist. In contrast, biological systems almost always employ anisotropic biomaterials. Bone, tendon, cartilage, tooth enamel, blood vessels, skin, wood, seashells, etc. are all composites with hierarchical and directional structural features [18,19]. The hierarchical and directional structure of natural biomaterials is critically important in enabling the biological organism to achieve the required mechanical properties while minimizing the "cost" of material mass and metabolic energy [18]. Therefore, *mechanical characterization of natural biomaterials, as well as synthetic biomaterials engineered to optimize properties and/or mimic the anisotropy of natural biomaterials, may require characterization of property anisotropy using methods described in this chapter.*

3.2.6. Fracture Mechanics

The field of fracture mechanics originated in the shipbuilding trade in the early 1900s due to a realization that engineering analysis based largely on the preceding concepts of this chapter was unable to predict prevalent catastrophic fractures of ship hulls. Cracks were observed to preferentially form around portholes and door openings due to an inhomogeneous distribution of stress, later known as a stress concentration [20]. *Stress concentration factors* are now tabulated for a number of geometric discontinuities as the ratio of a local maxima in stress at a specific geometric location relative to a nominal or remote applied stress [21]. For example, consider the addition of a circular hole to the flat specimen in Fig. 3.2 such that the size of the hole is small relative the specimen width (Fig. 3.5a). The stress concentration factor at the lateral edge of the hole is three times the remote applied tensile stress. Yet the consideration of stress concentration factors for macroscopic geometric discontinuities, while extremely important in the analysis and prediction of failure to this day, still underestimated failure. However, Inglis recognized that as the radius of curvature of an elliptical hole became increasingly sharp ($a \gg b$, Fig. 3.5a), approaching that of a sharp crack tip, the stress concentration approached infinity.

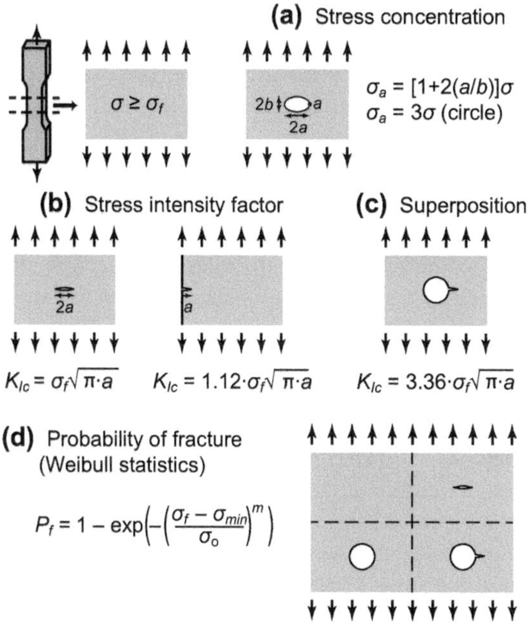

FIGURE 3.5 The development and differences between fundamental concepts in fracture mechanics beginning from a basic strength of materials analysis for a right rectangular prism of a material loaded in quasistatic uniaxial tension, including (a) stress concentrations, (b) the critical mode I stress intensity factor or plane strain fracture toughness (K_{Ic}), (c) superposition, and (d) the probability of fracture (P_f), where σ is the remote applied stress, σ_f is the fracture strength, a and b are the major and minor dimensions of an elliptical hole or crack, σ_a is the stress concentration at point a, σ_{min} is the minimum fracture strength, σ_o is a constant, and m is the Weibull modulus.

Griffith's law introduced an energy criterion for fracture based on equilibrium between the change in applied strain energy per change in crack area (energy input) and the change in surface energy per change in crack area (energy release) [22]. Therefore, catastrophic fracture occurs when crack growth is unstable due to the rate of energy input (G, termed the *strain energy release rate*) exceeding the material's ability to dissipate that energy (R, termed the *crack growth resistance*). For brittle materials, the crack growth resistance typically has a constant value or "flat R-curve." In tough materials, the crack growth resistance can increase with increasing crack length, exhibiting a "rising R-curve." Thus, one measure of a material's resistance to fracture is the *critical strain energy release rate* (G_c) for unstable crack propagation, which occurs when $dG/da > dR/da$. For example, consider the addition of a sharp elliptical through-thickness crack in the center of the flat specimen in Fig. 3.5 such that the size of the crack is small relative to the specimen width. The critical strain energy release rate for tensile (mode I) loading of a material with a flat R-curve can be derived as $G_{Ic} = (\pi \cdot \sigma_f^2 \cdot a)/E$, where σ_f is the remote applied critical stress or fracture stress, a is the crack length, and E is the elastic modulus [23].

Alternatively and more commonly, a *stress intensity factor* (K) can be used to characterize the stress field ahead of a crack tip for either crack opening (mode I), shearing (mode II), or tearing (mode III). Therefore, the fracture criterion is when K exceeds a *critical stress intensity factor* (K_c), also termed the *fracture toughness*, which is reported in units of $\text{MPa}\cdot\text{m}^{1/2}$. Note that mode I is most common and relevant since the mode I critical stress intensity factor (K_{Ic}) is typically much lower than other modes for most materials. Considering again a sharp elliptical through-thickness crack in the center of a flat specimen such that the size of the crack is small relative to the specimen width, the critical stress intensity factor, or plane strain fracture toughness, for tensile (mode I) loading is $K_{Ic} = \sigma_f \cdot (\pi \cdot a)^{1/2} = (E \cdot G_{Ic})^{1/2}$ (Fig. 3.5b). The *plane strain fracture toughness* is generalized as $K_{Ic} = Y \cdot \sigma_f \cdot (\pi \cdot a)^{1/2}$ where Y is a dimensionless geometric factor which accounts for the crack, specimen, and loading geometry such that K_{Ic} is a true material property. For an edge crack in a semi-infinite plate, $Y = 1.12$ (Fig. 3.5b). The principle of superposition can be applied to consider combinations of loading modes, cracks, and specimen geometry. For example, a simple solution for a crack at the edge of a through-thickness hole in the center of a semi-infinite flat plate is shown in Fig. 3.5c as the superposition of the stress concentration for a circular hole (Fig. 3.5a) and the stress intensity factor for an edge crack (Fig. 3.5b). Stress intensity factors for numerous geometries and standard test specimens are available in textbooks, handbooks, and ASTM standards [23–28]. Specimen requirements for a valid K_{Ic} test are described in further detail in Section 3.3.

Measurement of K_{Ic} is accomplished by loading a precracked specimen to failure while monitoring the load and displacement. A provisional fracture toughness is measured as $K_Q = P_Q \cdot f(a/W)/(B \cdot W^{1/2})$, where B is the specimen breadth and $f(a/W)$ is a dimensionless function for the ratio of the crack length (a) to specimen width (W) known for a particular specimen geometry. The crack length is typically measured directly from the specimen surface. The load at failure (P_Q) is measured from the load–displacement curve at either the first maxima encountered prior to unstable crack growth or the intersection of a secant line at 95% of the initial elastic loading slope, whichever occurs first [23,27]. If P_Q is within 10% of the maximum load measured in the test and the specimen dimensions meet the requirements detailed in Section 3.5, $K_Q = K_{Ic}$.

To summarize thus far, K characterizes local conditions near the crack tip and thus primarily governs crack initiation, while G characterizes the energetic driving force for an increment of crack extension and thus primarily governs crack propagation. Therefore, while K_{Ic} is most readily and commonly measured, the initiation fracture toughness may underestimate the true fracture toughness for tough materials that exhibit a rising R-curve. Rising R-curve behavior may be caused by a number of toughening mechanisms – including crack tip plasticity, crack deflection, crack bridging, phase transformations, and microcracking [29] – which expend strain energy for processes other than crack propagation. In such materials, the fracture toughness during crack propagation (K_R) can be measured using displacement-controlled loading, which is amenable to stable crack propagation because unlike load control, the strain-energy release rate decreases with increased crack length. The stable instantaneous crack length is typically inferred using changes in the specimen compliance upon unloading or changes in the specimen electrical resistance, but can also be measured or confirmed directly using a traveling microscope. K_R is thus measured using similar methods to those described above for K_{Ic}, except that K_Q is measured instantaneously at multiple crack lengths, which may require plastic zone size corrections for moderate crack tip plasticity.

The preceding methods were limited to linear elastic materials, and are thus known as *linear elastic fracture mechanics* (LEFM). While LEFM can be extended to account for moderate crack tip plasticity using plastic zone size correction factors [23], tough materials exhibiting extensive crack tip plasticity or *R*-curve behavior may require the use of *elastic–plastic fracture mechanics* (EPFM). The most common EPFM methods include the measurement of a *crack-tip opening displacement* (CTOD) [30] or a nonlinear energy release rate, known as the *J-integral* [31]. A CTOD measured with an appropriate transducer (see Section 3.4.1) provides a direct assessment for the extent of crack tip plasticity in order to more accurately characterize deviation from LEFM solutions using a number of possible models. The *J*-integral assumes that elastic–plastic deformations can be idealized as nonlinear elastic and is thus analogous to *G* in practice. Therefore, J_{Ic} and *J-R* curves can be measured somewhat analogously to K_{Ic} and *K-R*, respectively, as described in detail elsewhere [23,31,32].

Fracture mechanics analyses using stress concentrations and stress intensity factors suggest that failure only requires one flaw (or crack) of a critical size and shape. However, real materials contain a statistical distribution of flaws such that a greater stressed specimen volume has a greater probability of containing a critically sized flaw and thus a greater probability of failure. This effect is most pronounced in brittle materials that are more sensitive to flaws (e.g., acrylic bone cements and ceramics), but, interestingly, has been shown to also characterize failure in "tough" biocomposites like bone [33]. *Weibull statistics* can be used to determine the *probability of failure* (P_f) using the equation shown in Fig. 3.5d, where σ_f is the fracture strength, σ_{min} is the minimum fracture strength, σ_o is a constant in units of stress, and m is the *Weibull modulus*. In the simplest method of analysis, the above parameters can be determined by a mean rank regression or other ranking methods [6,34,35]. Fracture strengths measured for each of N specimens are rank ordered and assigned integer (n) values from 1 to N. The Weibull parameter, $\log(\log(1/(1-P_f)))$, where $P_f = n/(N+1)$, is plotted versus $\log(\sigma_f)$ for each specimen. The data points are fit by a line as $m \cdot \log(\sigma_f - \sigma_{\min}) - m \cdot \log(\sigma_o)$, to find the slope m and the intercept $m \cdot \log(\sigma_o)$. Thus, a relatively high Weibull modulus (m) indicates a narrow distribution of fracture strengths. Weibull analysis can be used with sample sizes as small as several specimens, but, as with any statistical method, accuracy is improved with an increased number of specimens, especially for samples exhibiting a low Weibull modulus.

The effects of differences in specimen size and loading mode can be analyzed using more complex variations of the above analysis, which include terms for specimen volume and a nondimensional parameter called the stress-volume integral that accounts for differences in the stressed volume of a specimen for different loading modes [6,34,36]. Simplified, $<\sigma_{f1}>/<\sigma_{f2}> = (V_2/V_1)^{1/m}$, where $<\sigma_f>$ is the mean fracture strength and V is the effective stressed volume for specimens of two different geometries (1) and (2). For example, an existing data set for uniaxial tensile strength can be used to determine the probability of failure for changing the specimen size or testing the same material in three-point bending.

In summary, stress concentrations, energy criteria for crack initiation and propagation, and statistical methods for the probability of fracture were introduced as three primary methods of analysis for fracture under quasistatic loading (Fig. 3.5). Readers interested in further technical detail, including material behavior and toughening mechanisms, can readily consult numerous textbooks [5–7,23,24] on the subject matter.

3.3. SPECIMENS

3.3.1. Specimen Size and Geometry

As should be clear by now, *a standardized specimen geometry may be critical to obtaining useful test results with accurate measurements of material properties that can be compared to reference data or measurements performed by others.* To review, there are three primary reasons for this. (1) Many modes of loading result in a nonuniform stress or strain distribution within the specimen. Special considerations for common modes of loading are therefore described in Section 3.4.2 and references are provided for numerous standardized specimen geometries. (2) As described above, the stresses ahead of a crack tip in a fracture toughness test are inherently dependent on the specimen size, geometry and crack length. (3) Specimen size can influence the likelihood of encountering critically sized defects and thus the statistical distribution (Weibull modulus) of a measured material property.

Test specimens are typically prepared with a *gauge section* of reduced cross-sectional area in order to control the region of the specimen that is characterized and to provide greater surface area for gripping the specimen outside the gauge section (see Section 3.4.1). For example, extensometers or strain gauges are placed within the gauge section of tensile specimens. However, a gauge section may not be possible for all modes of loading (see Section 3.4.2).

Fracture toughness measurements require perhaps the greatest care in specimen size and geometry, and numerous standardized and nonstandardized specimen geometries are commonly utilized [23,27,28]. A valid test for K_{Ic} or

J_{Ic} requires that a, B, $(W-a) \geq 2.5(K_{Ic}/\sigma_y)^2$ or $B, (W-a) \geq 25 \cdot K_{Ic}^2/(E \cdot \sigma_y)$, respectively [27,31]. The crack length (a) and uncracked ligament length ($W-a$) criteria ensure a sufficiently small plastic zone for the implicit assumptions of each approach. The specimen breadth (B) criteria ensure plane strain conditions ahead of the crack, although the criteria in this standard have been noted to be more stringent than necessary [23]. Three-dimensional stress analysis of a crack front revealed that greater triaxiality in the specimen center results in predominate plane strain conditions and a smaller plastic zone size compared to the specimen edges where biaxial stresses (plane stress) allow shear deformation and a greater plastic zone size. True plane stress conditions exist only for infinitely thin specimens (perhaps a metal foil) and increasing specimen breadth (B) results in an increased plane strain region such that K_I asymptotically approaches a constant value (K_{Ic}), which is independent of specimen breadth.

Despite the above reasons, there can be unique situations where a standardized specimen size and geometry is simply not possible or feasible. One common example is the mechanical characterization of biological specimens (e.g., whole animal bones) possessing an inherently variable and complex size and shape, which may itself be an important consideration in the mechanical characterization. In this case, measurements may be limited to force, displacement, and stiffness, or approximations may be used to estimate stress, strain, and moduli. Despite these limitations, useful conclusions can be reached regarding biological influences (e.g., genetics [37]) on the biomaterial properties, but these conclusions must be kept within proper context. Another common example is when only a small volume of material is available, requiring deviation from standard specimen sizes [32]. In this situation, one must take special care to clearly define the measured properties, identify deviations from existing standards, and place results in an appropriate context, avoiding comparisons that are simply not valid.

3.3.2. Specimen Preparation

Test specimens are most often machined using cutting and/or grinding tools on mills, lathes, and saws. A standardized specimen size and geometry can be replicated from a computer-aided design drawing using computer numerical control machining or from a template using a pattern-following cutting tool. Care must be taken during the machining process to achieve a smooth surface finish free from defects and to prevent the specimen material from significant alterations due to the generation of heat or surface damage by cutting tools. Lubricants and cutting fluids can be used to reduce heat but must be compatible with the material. Tissue specimens should always remain fully hydrated during preparation. The cutting speed, cutting depth, tool material, and tool geometry should be appropriately selected for the material, specimen geometry, and desired surface finish. Consultation and guidelines can be sought from experienced machinists, commercial suppliers of comparable materials, and machining handbooks [38–40].

Specimens may also be prepared by a variety of near-net-shape processes, such as injection molding plastics, that are beyond the scope of this chapter. In general, if the goal is to measure or compare material properties, care should be taken to minimize undesirable influences of the process on the material. Standards may exist for the preparation of test specimens [41,42] and textbooks on manufacturing processes describe fundamental influences [43].

The surface finish of as-machined or as-molded specimens is often inadequate for a particular material or mechanical test. Tool and die marks remaining on the surface can act as stress concentrations, which is especially problematic for brittle (flaw-sensitive) materials, flexural loading, and/or fatigue testing. Therefore, specimen surfaces, especially within the gauge section or at locations of maximum stress, are often polished using abrasives or diamond tooling of a specified grit size, and sharp corners or mold lines may be removed [28,44]. Quality test results are well worth the added labor and cost of careful specimen preparation.

Fracture toughness (Section 3.2.6) and fatigue crack propagation (Section 3.4.4.3) tests typically require the presence of a sharp crack. Various methods may be used depending on the material, specimen, and preference, but all begin with a stress concentration created by machining a notch using a diamond saw, razor blade, or electric discharge machining. A preload is subsequently applied using either a rigid fixture to "pop-in" a stable crack [28] or cyclic loading to initiate a stable fatigue crack [27]. Microhardness indentations (Section 3.4.5.1) may also be used to either directly initiate a sharp half-penny crack below the indentation or to subsequently preload as described above [28]. Grooves are also occasionally added to the sides of fracture toughness and fatigue specimens to guide crack propagation during *R*-curve and fatigue crack propagation measurements, respectively [27]. Side grooves removing no more than 20% of the specimen breadth can also act to remove the influence of plane stress conditions at the specimen surface. However, side grooves may also forcibly remove the effects of crack deflection on *R*-curve behavior and overly deep grooves can lead to decreased fracture toughness.

3.4. APPLICATION AND MEASUREMENT OF LOAD AND DEFORMATION

3.4.1. Equipment

The application of load(s) for mechanical characterization of materials is accomplished using an *actuator* with feedback to a control system (Fig. 3.6). Actuation has historically been accomplished using either a screw-driven (via an electric motor) or servo-hydraulic *load frame*; however, recent instruments have migrated towards electromechanical systems. Each has advantages and

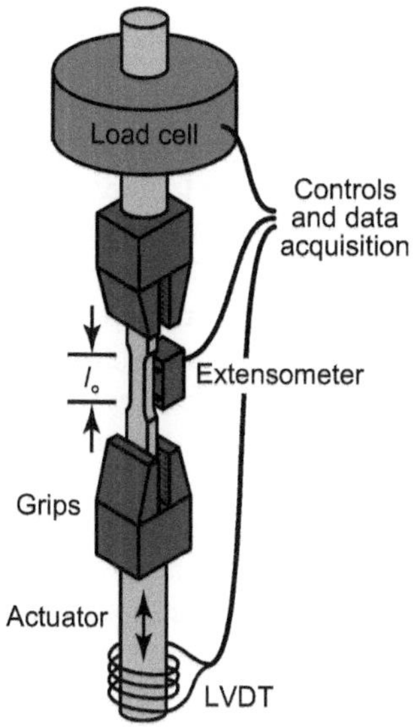

FIGURE 3.6 Schematic diagram showing a common setup for the application of load to a specimen using an actuator with feedback from transducers to a control system. Load is sensed using an in-line load cell. Actuator displacement is sensed using a linear variable displacement transducer (LVDT). Displacement within the specimen gauge length (l_o) is sensed using an extensometer or strain gauge (not shown).

limitations. Screw-driven instruments are economical but not suitable for applying high-frequency cyclic loading. Servo-hydraulic instruments are able to apply the greatest load magnitude and are well suited for cyclic loading, but typically require greater cost and maintenance. In most cases, electromechanical systems are now best suited for testing small biomaterial specimens due to their relatively high precision, adaptability, and low cost.

The applied load magnitude is sensed using an in-line *load cell* (Fig. 3.6) The most appropriate load cell for a particular application will primarily consider the load-cell capacity, sensitivity and loading mode. The load-cell capacity should not be exceeded during a test and control system shut-off switches should be set to prevent an overload. On the other hand, a load cell may be too insensitive or "noisy" at less 2% of the load-cell capacity. Load cells should be calibrated upon installation and as part of a regular maintenance schedule in order to ensure accuracy [45]. Commercial suppliers provide state-of-the-art on-site or mail-in calibration and certification against NIST traceable standards. Calibration can be readily verified by users at low cost using a set of calibration weights. Load cells may be rated for tension-only, compression-only, tension–compression, or multiaxial (axial and torsion) loading, as well as quasistatic versus cyclic loading. Finally, fully submersible load cells are also available for testing in a fluid bath.

Grips are used to transfer load from the actuator to the specimen (Fig. 3.6). The appropriate grips and gripping method are chosen to prevent slippage or failure at the grips. The most common general purpose grips clamp the specimen using a screw, a hydraulic or pneumatic piston, or a self-tightening wedge or scissors. Slippage is prevented by increasing the clamping force and/or increasing the surface area of contact between the grip and the specimen, where the latter can be adjusted by the grip size and shape, and the use of serrated surfaces. However, an excessive clamping force, overly serrated surfaces, misaligned grips, or misaligned specimen may lead to specimen failure at the grip rather than within the specimen gauge section. Clevis pins are commonly used to grip fracture toughness specimens in order to prevent crack propagation during gripping and to allow out-of-plane rotational freedom during loading. Other specialized grips may be advantageous or necessary for testing fibers or soft materials. For example, cryogrips are useful for gripping soft, hydrated tissues such as tendon. Finally, other specialized *load fixtures* may be required for different modes of loading described in the following section.

Actuator or crosshead displacement is measured using a *linear variable displacement transducer* (LVDT) (Fig 3.6). However, the crosshead displacement includes load-line or system compliance due to the grips, specimen shoulders adjacent to the grips, couplings, load cell, etc. Load-line compliance is usually negligible in magnitude relative to the specimen compliance. However, rigid specimens and precise measurements of specimen deformation should utilize an *extensometer* to directly measure deformation within the specimen gauge length (l_0), cross-sectional area (A_0), or a crack opening. The most appropriate extensometer for a particular application will primarily consider mounting, specimen dimensions, range (or travel), and sensitivity. Contact extensometers are most common and economical, but must be carefully mounted on the specimen gauge section using clips, rubber bands, or springs. Noncontact extensometers eliminate potential influence on the specimen, and damage to the extensometer upon failure or use in extreme environmental conditions. The extensometer should be appropriately sized for the specimen dimensions. The range of a contact extensometer should not be exceeded during a test, or else the test must be paused to remove the extensometer. Moreover, the extensometer should have sufficient sensitivity within the range of displacement to be measured. Extensometers should be regularly calibrated using a micrometer in order to ensure accuracy. A limitation of an extensometer is that deformation within the specimen gauge section must be reasonably assumed to be uniform.

A *strain gauge* enables direct and highly sensitive measurement of local material strains on a specimen surface through differences in electrical resistance of a small, serpentine metal foil. The signal from a strain gauge should be measured and calibrated using a Wheatstone bridge [46]. Moreover, the gauge must be properly bonded to the specimen surface using an adhesive, such as cyanoacrylate. A strain gauge rosette can be used to determine strain directionality or principal strains on a surface. While multiple strain gauges may be used to characterize a strain field at selected points, strain gauges are not a practical means to measure nonuniform strains over a relatively large surface area. Moreover, strain gauges still require specimen contact. Popular noncontact methods for measuring strain fields include *photoelasticity*, *digital image correlation*, and *laser interferometry*, which are described in detail elsewhere [47].

Additional information and specifications for various load frames, load cells, grips and other load fixtures, displacement transducers, extensometers, and strain gauges are readily available on the websites of companies such as Applied Test Systems, Bose Test Systems, Cooper Instruments and Systems, Epsilon, Instron, Interface, Mechanical Test Systems, Omega, Test Resources, and Vishay Precision Group, among others. A basic introduction to transducers, calibration, signal processing, and related fundamentals is also available in textbooks and handbooks [48,49].

3.4.2. Modes of Loading

A number of common loading modes may be used for different purposes described below (Fig. 3.7). Note that brittle materials (such as most ceramics, some refractory metals and hardened alloys, and some thermosetting polymers) tend to fracture along planes normal to the maximum (tensile) principal stress, even when applied loads are compressive. On the other hand, ductile materials (such as metals and thermoplastic polymers) tend deform under shear and

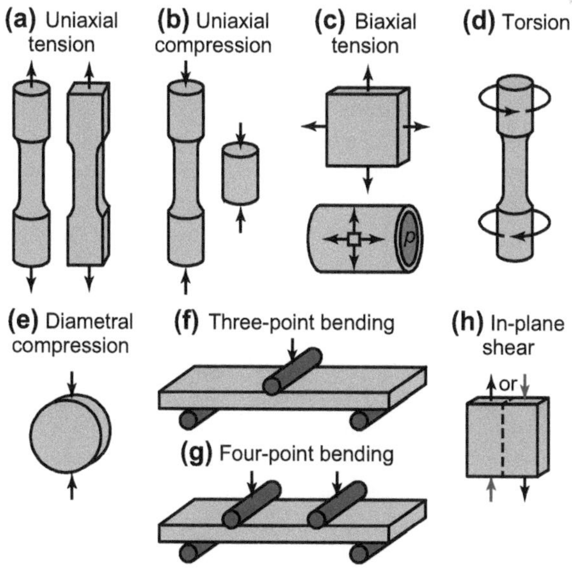

FIGURE 3.7 Common modes of loading materials for mechanical characterization include (a) uniaxial tension, (b) uniaxial compression, (c) biaxial tension, either by in-plane loading of a sheet or pressurizing (*p*) a vessel, (d) torsion, (e) diametral compression, (f) three-point bending, (g) four-point bending, and (h) in-plane shear.

fracture along planes of maximum shear stress. These general rules hold for all materials in all modes of loading, as described below.

Uniaxial tension (Fig. 3.7a), a.k.a., tensile testing, is the most fundamental of all tests for measuring material properties [12,50]. All preceding concepts and methods in this chapter were introduced for uniaxial tension (mode I at a crack tip) or uniaxial compression as the mode of loading. Uniaxial tension, and to a lesser extent uniaxial compression, are indeed most common and most suited for measuring material properties that are free of influences from the test method. A major reason is that the specimen cross-section experiences uniform stresses and strains. Uniaxial tension may also be useful for characterizing biomaterial-tissue interfacial strength or the bond strength between an adhesive and tissue. Microtensile bond strength is a common method for evaluating dental adhesives [51,52].

Uniaxial compression (Fig. 3.7b) is often utilized for brittle or porous materials that can be difficult to grip [42,53–55]. Moreover, brittle materials are typically designed for compressive loading in service due to exhibiting a much greater strength in compression compared to tension. Brittle materials loaded in uniaxial compression fracture at internal cone cracks. Ductile materials loaded in uniaxial compression deform by bulging at the center (termed "barreling") due to frictional stresses at the loading platens, increasing the cross-sectional area. The effects of friction and contact stresses can be minimized using highly polished loading platens, possibly with a lubricant, or by preparing specimens with a gauge section of reduced cross-sectional area, similar to tensile specimens. Porous materials (e.g., scaffolds) loaded in uniaxial compression, whether exhibiting brittle or ductile behavior, begin to densify after reaching a stress maxima or an inflection point corresponding to elastic buckling, yielding, or crushing [55,56]. Therefore, stress–strain data after this maxima and during densification are dependent on a changing level of porosity and generally not useful.

Biaxial tension (Fig. 3.7c) is particularly relevant to compliant or ductile sheet materials (e.g., skin graft, hermetic casings, and manufacturing processes for metal or polymer sheets) and thin-walled pressure vessels (e.g., the bladder, blood vessels, intravascular balloons, breast implants, etc.). Elastic and plastic deformation of materials under multiaxial loading behaves differently than under uniaxial deformation due to the effects of Poisson's ratio, yield surfaces, and possibly anisotropy. Soft tissue sheets are often gripped for planar biaxial tension through suturing [57,58], while vessels and tubes are readily loaded via internal fluid pressure [59]. Another application of biaxial loading is a *bulge test*, where a punch or applied fluid pressure is used to form a hemispherical dome in a flat sheet of a ductile material [60,61]. The utility of the bulge test lies in characterizing plastic deformation (ductility) and fracture when only a small volume of material is available due to cost or availability (e.g., implant retrieval).

Torsion (Fig. 3.7d) produces a stress state of pure shear and is thus useful for measuring material properties in shear, including nearly any that were described above for axial loading (e.g., *shear modulus*, *G*). However, shear stress (τ) and strain (γ) must be treated with some caution in torsion because shear stresses and strains vary from zero at the axis of rotation to maximum values at the specimen perimeter. Torsion is particularly relevant to biomaterials used in catheters, bone screws [62], and various surgical instruments, as well as tissues [63,64]. Moreover, anisotropic properties may be characterized by combination of uniaxial tension and torsion tests [63,64]. Ductile materials loaded in torsion exhibit a flat, transverse fracture surface, while brittle materials exhibit a spiral fracture surface that is readily demonstrated with a piece of chalk.

Diametral compression (Fig. 3.7e), a.k.a., the Brazilian disk test, is perhaps a misnomer as it is used to measure the *diametral tensile strength* (*DTS*) of brittle materials as, $\sigma_{DTS} = 2P/(\pi \cdot D \cdot t)$, where *D* is the specimen diameter and *t* is the specimen thickness [65,66]. Note that the *DTS* is a surrogate for the *UTS* but is not equivalent. A proper *DTS* test specimen fails by a crack plane initiating at the center of the disk and parallel to the direction of applied load. The maximum tensile stress acts normal to the applied compressive load due to the Poisson effect. The material must exhibit sufficiently high strength and hardness such that the specimen does not permanently deform at a near-line of contact with the loading platens [66], though correction can be made to the above equation for some flattening at the contact points [67]. Diametral compression is also useful when only a small volume of material is available due to cost or availability. Biomaterial applications have included characterization of calcium phosphate cements [67] and dental cements [68].

Three-point bending (Fig. 3.7f) and *four-point bending* (Fig. 3.7g), or flexural loading, are primarily used for testing materials that are either expected to be similarly loaded in service, difficult to grip for uniaxial tension, or only available as small specimens. However, three- and four-point bendings produce multiple stress states in the specimen that are not homogeneous [5–7,69]. Portions of specimens located between the outer- and inner-loading supports are under constant shear stress/strain lengthwise, but shear stress/strain varies parabolically within the cross-section height from zero at the upper and lower surface to a maximum value at the neutral axis. For an initially flat, isotropic beam with symmetric cross-section, the neutral axis is located at the geometric centroid. Normal stress/strain increases linearly from zero at the outer load support to a maximum at the inner load support, and also from zero at the neutral axis to a maximum in tension and compression on the bottom and top specimen surfaces, respectively (Fig. 3.7f and g). Thus, for a uniform specimen cross-section, the maximum stress is $\sigma_{max} = P \cdot l \cdot h/(2 \cdot I)$, where *P* is the applied load, *l* is the distance between the outer and inner load support, *h* is the specimen height, and *I* is the cross-sectional moment of inertia (e.g., $I = b \cdot h^3/12$ for a rectangular cross-section and $I = \pi \cdot d^4/64$ for circular cross-section, where *b* is the specimen breadth and *d* is the specimen diameter). The cross-sectional distribution of maximum normal stresses remains constant within the portion of the specimen location between the inner load supports in four-point bending.

The preceding discussion of stress analysis was necessary to highlight important considerations for flexural loading. The strength measured for a material in bending is not equivalent to that measured in uniaxial tension, and should thus be reported as the *flexural strength* or *modulus of rupture*. Moreover, even three- and four-point bending tests may exhibit differences due to the Weibull effect as the stressed specimen volume is lower in three-point bending compared to four-point bending. Thus, the strength of a material measured in three-point bending will be greater than for the same material measured in four-point bending, which is still greater than for the same material measured in uniaxial tension. The above equations, and related solutions for the elastic modulus, assume linear elastic deformation. Permanent deformation of the specimen either in bulk or due to local contact stresses at load supports can compromise the validity of measured properties and prevent the use of any stress–strain data beyond the yield point. Moreover, recall that regions of shear loading are present in the specimen. Therefore, ductile materials may undergo plastic deformation during bending even before a yield point is observed in stress–strain data. In summary, material properties measured using bending are quite sensitive to the test methods, perhaps more than any other common loading mode, highlighting the importance of the following standardized methods [44,70,71]. While there are valid reasons to use flexural tests, they should be used with caution and results should be interpreted carefully.

In-plane shear (Fig. 3.7h) is used to characterize material properties and interfaces. In-plane shear tests can be advantageous over torsion tests due to producing a reasonably uniform shear stress/strain within the specimen, but require relatively complex test fixtures and may load a relatively small volume of material. The shear properties of materials and interfaces between bulk materials can be measured using the *Iosipescu test*, *Arcan test*, *V-notch rail test* and related variations in loading fixtures and specimen geometry [72–74]. These methods were developed and are most commonly used for testing composite materials, but have also been applied to testing tissues [75].

The *biomaterial-tissue interfacial shear strength* after *in vivo* implantation is a key measurement for characterizing biocompatibility and tissue integration, particularly in load-bearing musculoskeletal tissues [76]. Several common test methods all employ in-plane shear loading to characterize the biomaterial-tissue interfacial strength. In a *push-out test*, a flat punch is used to push a biomaterial disk out of the surrounding tissue (e.g., from a cylindrical cranial, femoral, or tibial defect) using an annular fixture for the specimen. The specimen thickness and clearance (difference between the punch and annulus diameters) must be properly controlled such that failure occurs at the biomaterial-tissue interface [77]. The interfacial shear strength can be calculated by dividing the maximum load by the nominal interfacial area (cylindrical surface). Measurements are influenced by the biomaterial stiffness and should therefore be compared only between biomaterials of similar elastic modulus [77]. In a *pull-out test*, a screw, rod, or similar biomaterial device/specimen that can be gripped is pulled out of the surrounding tissue, which must be rigidly gripped using the standard specimen grips or an annular fixture. The mechanical analysis and considerations are otherwise similar to a push-out test. However, a pull-out test can also be employed nondestructively on percutaneous devices [78]. Finally, a *torque removal test* can also be applied in similar situations as a pull-out test [62] with the load applied using torsion rather than tension.

All the above methods of loading may be applied dynamically (e.g., cyclically) in addition to quasistatic loading, as discussed in Section 3.4.4. Indentation methods of loading are discussed under nondestructive tests in Section 3.4.5.1. Mechanical characterization of *coatings* typically utilizes variants of the above methods in uniaxial tension (*pull-off test* or *peel test*), bending, or shear, and are discussed in greater detail in Chapter 4 of this book.

In summary, material properties free of the influence of test methods should be preferably measured in uniaxial tension, uniaxial compression, and (for pure shear) torsion, although other modes of loading may offer advantages relative to material properties, intended application, material form, and availability. *Therefore, material properties measured in different modes of loading should neither be treated as equivalent nor directly compared. The loading mode should always be clearly denoted when reporting material properties measured by different modes of loading.*

3.4.3. Strain Rate and Time-Dependent Behavior

The preceding methods and discussion of mechanical characterization assumed quasistatic loading and time-independent material behavior. The strain rate of a typical quasistatic mechanical test is 10^{-2} to 10^{-3} s^{-1}. However, strain rates in service may range from much higher (e.g., ~10^2 to 10^3 s^{-1} due to impact from a fall to ~10^2 to 10^6 s^{-1} due to a ballistic impact) to lower (~10^{-6}, e.g., creep under a constant load) extremes. Moreover, elastic deformation can be coupled with rate or time-dependent viscous deformation in metals and ceramics at elevated temperature, in polymers near or above the glass-transition temperature, and in hydrated tissues due to poroelasticity. Therefore, this section will overview the characterization of time-dependent mechanical behavior, including strain rate sensitivity, creep, and stress relaxation.

Differences in strain rate can profoundly influence the stress–strain behavior (and thus measured mechanical properties) for the same material in an otherwise identical test method, which is known as *strain rate sensitivity*. Strain rate sensitivity can be modeled using a power-law (similar to strain hardening in Fig. 3.4e) as $\sigma = C \cdot \dot{\varepsilon}^m$, where $\dot{\varepsilon}$ is the true strain rate, C is a strength coefficient, and m is the strain rate sensitivity. Note that this constitutive equation can also be coupled with the constitutive law for power-law plasticity. The strain rate sensitivity (m) can range from 0 to 1. Most materials exhibit $m \approx 0.2$, but materials with $m > 0.3$ exhibit high strain-rate sensitivity. For example, silly putty has a strain rate sensitivity approaching the upper bound of one such that a rope of silly putty can be stretched to a nearly infinitesimally small cross-section at a low strain rate but will exhibit brittle failure when stretched at a high strain rate. Characterizing these effects in materials usually requires testing over orders of magnitude differences in strain rate.

3.4.3.1. Impact (High Strain Rates)

At increased strain rates, the effects of stress wave propagation in the specimen and acceleration on load measurement must be considered. A conventional servohydraulic or electromechanical load frame can achieve strain rates up to $10^0\ s^{-1}$, although some specialized instruments can achieve up to $10^1\ s^{-1}$. Note that the actual strain rate is dependent on the instrument crosshead speed, system compliance, specimen compliance, specimen length, and specimen cross-sectional area. Slack grips may be required to ensure that the instrument reaches the desired strain rate before loading the specimen. Moreover, the use of a conventional load cell and extensometer may be limited by inertial effects and vibrations, which can be mitigated using a stiffer piezoelectric load transducer and noncontact methods for strain measurement (Section 3.4.1).

Strain rates from 10^0 to $10^2\ s^{-1}$ can be achieved with specialized machines utilizing a pendulum or drop tower. Classic *impact testing* includes the *Charpy impact test* (or Charpy V-notch test) and *Izod impact test* [79–82]. Both tests measure the *impact energy absorption* (or "impact toughness"), typically reported in J/m or J/m^2, as a swinging mass on a pendulum strikes the specimen. The Charpy test loads a notched specimen in three-point bending, while the Izod test loads a notched specimen as a cantilevered beam. Similar impact tests can be performed using a drop tower instead of a pendulum. For example, the *Gardner test* or *free-falling dart test* is primarily used to measure the impact resistance of thin film and sheet materials [83,84]. The dart typically has a hemispherical tip, making this a dynamic punch test, but other striker tip geometries are also used. Pendulum and drop tower tests are useful for economical screening of materials (e.g., commercial quality control) and characterizing ductile–brittle transitions (e.g., due to temperature) in materials. Limitations include semiquantitative (comparative) results and a lack of strain rate control, though the strain rate can be measured with instrumentation. There has been a recent growth of interest in instrumented impact testing, as well as the development of instrumented drop weight towers, which enable high strain rate tensile testing of compliant or ductile materials at 10^1 to $10^3\ s^{-1}$ [85].

High strain rates from 10^2 to $10^4\ s^{-1}$ can be achieved using a *split Hopkinson pressure bar* (a.k.a., *Kolsky bar*), where a specimen is placed in between the ends of two straight, perfectly aligned bars called the incident and transmitted bars [86,87]. At the end of the incident bar away from the specimen, an incident stress wave is created as the striker bar impacts the incident bar after being accelerated by a compressed gas gun. At the specimen interface, part of the stress wave is reflected back down the incident bar and another part is transmitted through the specimen and into the transmitted bar. All three bars are composed of the same material of known properties, typically a high strength steel, although softer and harder materials may be used for testing softer materials and harder materials, respectively. Strain gauges placed on the incident and transmitted bar allow the measurement of strains, strain rates, and ultimately, using known properties for the bar, a true stress–strain curve [86]. The strain rate can be controlled by varying the impact velocity (via gas pressure in the gun) and specimen size. Specimens are typically loaded in compression, although tension and torsion are also possible. Bending configurations may be used to measure dynamic fracture toughness [88,89]. Direct measurements of specimen deformation and crack growth require the use of high speed videography. While split Hopkinson bar testing is quantitative, unlike most pendulum or drop tower tests, the lack of a feedback control system typically results in an iterative design of experiments and a lack of standards. Nonetheless, split Hopkinson bar testing can be used to characterize the unique effects of high strain-rate deformation on biomaterials and tissues such as bone [89,90], which is not possible by other methods.

Very high strain rates from 10^4 to 10^6 s^{-1} can be achieved using the acceleration of projectiles or the detonation of explosives to generate shock waves [87], but are generally not relevant or necessary for biomaterials characterization.

3.4.3.2. Creep (Low Strain Rates)

At decreased strain rates, creep deformation must be considered. *Creep* is time-dependent permanent deformation under a constant stress and possibly an elevated temperature (see also Section 3.5.1), typically occurring at 10^{-3} to 10^{-8} s^{-1}. In general, all materials can exhibit creep at temperatures greater than one-third of their melting temperature, although ceramics may require a temperature closer to one-half the melting temperature, and polymers may creep at temperatures near or above their glass transition temperature. The deformation behavior in creep can be divided into three regimes after an initial elastic strain (ε_0): (1) primary or transient creep exhibiting a decreasing strain rate, (2) secondary or steady-state creep exhibiting a constant strain rate, and (3) tertiary creep or rupture exhibiting an unstable increasing strain rate preceding failure (Fig. 3.8a and b). Constitutive models for steady-state creep are useful for predicting lifetime and vary by the deformation mechanism [6–8]. Creep deformation is facilitated by a number of mechanisms including dislocation plasticity in metals, diffusion in metals and ceramics, grain

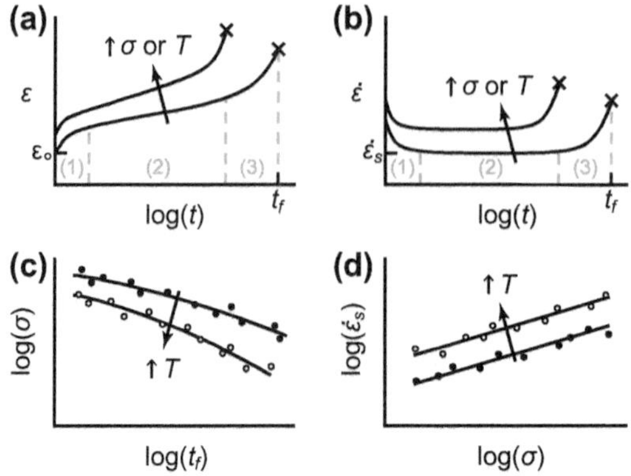

FIGURE 3.8 Schematic diagram for the time-dependent (a) strain and (b) strain rate ($\dot{\varepsilon}$) response of a material loaded under a constant stress, often at elevated temperature (*T*), showing three regimes of creep deformation behavior: (1) primary or transient creep exhibiting a decreasing strain rate, (2) secondary or steady-state creep exhibiting a constant strain rate, and (3) tertiary creep or rupture exhibiting an unstable increasing strain rate, preceding failure. Creep lifetime can be predicted by measuring and extrapolating (c) the creep rupture time (t_f) for accelerated tests using the Larson–Miller relation or other empirical methods, or (d) the steady-state strain rate ($\dot{\varepsilon}_s$).

boundary sliding in ceramics, and intermolecular motions (viscous flow) in viscoelastic polymers.

Superplasticity can be considered as a special case of creep deformation where a material exhibits a high resistance to necking at low strain rates (10^{-1} to 10^{-3} s^{-1}, but typically $< 10^{-2}$ s^{-1}), enabling large deformations ($\varepsilon > 1000\%$) without failure. Therefore, superplasticity is advantageous for forming of complex metallic parts as the material can exhibit Newtonian flow (viscous deformation) rather than strain-hardening plasticity and elastic recovery. In most engineering materials, superplasticity also requires elevated temperature and small, equiaxed grains.

Strain rates as low as 10^{-6} s^{-1} can be achieved using a conventional servohydraulic or electromechanical load frame. Therefore, these instruments are suitable for characterizing superplasticity in uniaxial or biaxial tension (bulge test) [91,92]. However, such an instrument is not required to apply a constant load, not to mention the impractical amount of time required for a creep test. For creep tests, a constant applied load/stress is applied using a dead weight in uniaxial tension, compression, or bending, possibly at an elevated temperature [93–95]. The strain rate is thus dependent on the specimen size, material behavior (e.g., diffusion), and influence of temperature, and must be measured by intermittent dimensional changes or *in situ*. Noncontact methods are often necessary for *in situ* strain measurements due to large deformations and/or elevated temperatures.

Given the impracticality of conducting creep rupture tests over months or years, tests are often accelerated to seconds or minutes using an increased stress magnitude. The applied stress and *creep rupture time* (t_f) for accelerated tests can be plotted as data points on a log scale (Fig. 3.8c) and extrapolated using the Larson–Miller relation, or other empirical methods, to predict the creep rupture lifetime [6–8]. Alternatively, failure in a particular application may not require creep rupture, but only excessive deformation. In this case, the steady-state strain rate ($\dot{\varepsilon}_s$) is measured (Fig. 3.8a and b) and extrapolated either using various empirical and/or mechanistic models (Fig. 3.8d) [6–8].

3.4.3.3. Viscoelasticity

Viscoelasticity is observed as a combination of both recoverable elastic deformation and permanent viscous deformation. In the previous section, the contribution of viscous deformation was shown to result in time-dependent permanent deformation (*creep*) under an instantaneous, constant stress. Similarly, under an instantaneous constant deformation or strain, viscous deformation can cause *stress relaxation* (Fig. 3.9). Nearly all synthetic and natural polymers used as biomaterials, as well as biological tissues, exhibit viscoelasticity to varying degrees.

Noncrystalline or semicrystalline polymers and glasses exhibit free volume and a *glass transition temperature* (T_g) due to the inability of relatively large molecules to organize into a perfectly crystalline network. Below T_g, molecular motions are restricted by time, and the material exhibits glassy or brittle elastic behavior. Near T_g, or more precisely within a temperature range surrounding T_g, the free volume allows some viscous molecular rearrangement and the material exhibits a viscoelastic response, which becomes increasingly elastomeric (nonlinear elastic) with increased temperature above T_g. Near the melting temperature, elastic deformation is lost and the material exhibits only viscous flow. Additional background on structure-property relationships – including the effects of molecular weight, crystallinity, cross-linking, etc. – for noncrystalline and semicrystalline materials (polymers and glasses) is available in a number of textbooks [5–8,96]. Hydrated biological tissues can also exhibit viscoelasticity due to the permeability of the tissue to fluid flow (poroelasticity).

Constitutive models for creep and stress relaxation in viscoelastic materials utilize numerous combinations of two simple elements used as mechanical analogs (Fig. 3.9). A spring element models time-independent elastic behavior, as described in Section 3.2. A dashpot element models time-dependent viscous behavior as $\dot{\varepsilon} = \sigma/\eta$, where $\dot{\varepsilon}$ is the strain rate (normal or shear), σ is the applied stress (normal or shear), and η is the viscosity. In polymers, elastic behavior is due to strong covalent intramolecular bonds, while viscous behavior is due to weak intermolecular van der Waals bonds. Thus, the viscoelastic behavior of thermoplastic polymers lies somewhere in-between these two extremes.

The classic *Maxwell model* combines a spring and dashpot in series to model viscoelastic stress relaxation, but the prediction of Newtonian flow

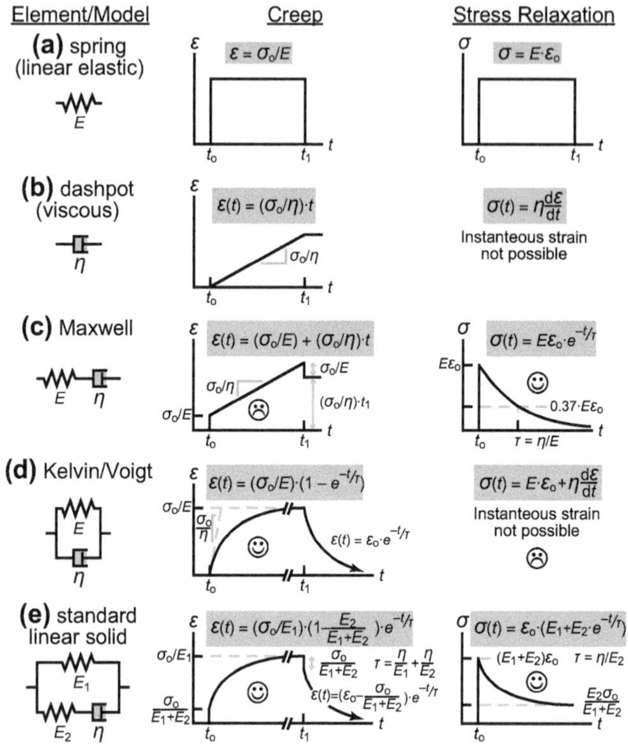

FIGURE 3.9 Common viscoelastic constitutive models for viscoelastic creep and stress relaxation, due to an instantaneous and constant applied stress (σ_o) and strain (ε_o), respectively, including (a) a spring element, (b) a dashpot element, (c) the Maxwell model, (d) the Kelvin or Voigt model, and (e) the standard linear solid model, where t is time, E is the elastic modulus of spring elements, η is the viscosity of dashpot elements, and τ is the relaxation time.

during creep or strain recovery is not realistic, and stress relaxation does not necessarily decay to zero (Fig. 3.9c). Conversely, the classic *Kelvin* or *Voigt model* combines a spring and dashpot in parallel to model viscoelastic creep, but stress relaxation is not possible and creep deformation may be erroneous at short times (Fig. 3.9d). By combining the strengths of the Maxwell and Kelvin models, the *standard linear solid model* (a.k.a., the three-parameter model) provides a good approximation of both creep and stress relaxation for the most linear viscoelastic polymers (Fig. 3.9e), although the stress relaxation may be more rapid than reality as has been shown for bone tissue [97]. In each of the above models, creep and stress relaxation are most accurately modeled by an exponential rise and decay, respectively, characterized by a time constant or *relaxation time* (τ), which is a function of spring stiffness and dashpot viscosity. The *Burgers model* (a.k.a., four-parameter model) also combines the Maxell and Voigt models in series and is popular. The *generalized Maxwell model* combines a spring in parallel with n Maxwell elements, and the *generalized Kelvin model* combines a series of n Kelvin elements such that $n = 4$ to 6 is sufficient to fit the stress relaxation and creep of nearly any data. However, models with more than several parameters become physically less meaningful, beyond experimental capabilities, and increasingly difficult to calculate.

Reliable creep (see also Section 3.4.3.2) and stress-relaxation [98,99] characterization requires particularly well-defined specimens (see Section 3.3), precise instrumentation (see Section 3.4.1) suitable for small deformations and low rates, and precise control or monitoring of temperature and humidity (see Section 3.5). Specimens are typically preconditioned at the highest temperature to be analyzed by applying the maximum load for the maximum time period, followed by unloading to allow recovery over a period 10 times longer than the loading period, and repeating this procedure until achieving reproducible behavior. Preconditioning promotes fully recoverable deformation and erases memory of prior thermomechanical history. Due to the likelihood of relatively large strains in viscoelastic materials, load levels should be chosen or altered such that the specimen cross-section remains under constant stress. Step loads and deformations are typically applied in uniaxial tension, uniaxial compression, bending, or torsion [98]. Compression can be problematic due to buckling or barreling, and bending can be problematic due to nonuniform stress/strain. However, bending can be advantageous for load fixtures, applying small loads, and measuring small strains by amplified deflections. In torsion, shear stress and strain can be substituted in Fig. 3.9 in place of normal stress and strain [97]. A strain rate that is too high may result in damped vibrations in a creep experiment, while a low strain rate may result in a low, noninstantaneous stress for stress relaxation. Acceptable deviation from the assumption of an initial instantaneous stress or strain should be compensated by shifting t_0 (Fig. 3.9) to the point where the maximum strain or stress, respectively, is achieved in a creep or stress relaxation test [98,99]. Finally, creep tests should be performed over a sufficient time scale to examine the possibility of accelerated creep at long times or nonlinear viscoelasticity requiring corresponding constitutive equations [96].

A full understanding of viscoelastic behavior requires testing over a wide range of time (or frequency) and temperature. Fortunately, time–temperature equivalence may be used to measure, for example, creep or relaxation data over a short time period at several temperatures, enabling the effective response to be known over several orders of magnitude of time [5,96]. Experimental characterization may not only utilize transient measurements of creep and stress relaxation, as described above, but also cyclic loading, which will be discussed in the following section. Creep and stress-relaxation tests are conducted at low frequency (<1 Hz), while higher frequency analysis may employ free or forced oscillations (10^{-2} to 10^{2} Hz, e.g., DMA), resonant vibrations (10^{2} to 10^{4} Hz), and wave propagation ($>10^{4}$ Hz).

3.4.4. Frequency and Time-Dependent Behavior

The frequency of loading, or cyclic loading, is another aspect of time-dependent behavior which is also critical for mechanical characterization of biomaterials. Biomaterials and biomedical devices are typically subjected to cyclic mechanical loading in service due to repetitive voluntary activities, such as running or walking, and involuntary muscle contractions, such as in the cardiovascular system. Cyclic loading can lead to fatigue deformation and fracture, as well as creep and stress relaxation in viscoelastic materials.

3.4.4.1. Cyclic Loading and Fatigue

Cyclic mechanical loading can lead to *fatigue* failure at significantly lower stress levels than would be expected to cause failure under quasistatic loading. Fatigue can be divided into a three-stage process including (I) *crack initiation* at local stress concentrations due to defects and microstructural inhomogeneities (e.g., inclusions, grain boundaries, pores, surfaces, etc.), (II) *crack growth* when a stage I crack reaches a critical size, and (III) *fatigue fracture* when the remaining uncracked cross-sectional area is unable to sustain the maximum applied tensile stress in a loading cycle. Predictive methods for fatigue failure primarily consider stage II crack growth rates, as described below. Examination of fatigue failure surfaces reveals *beachmarks* that are visible to the naked eye and show intermittent bursts of crack growth due to discontinuous loading, and *striations*, which are visible with an electron microscope and show increments of crack growth for each loading cycle during stage II (Fig. 3.10).

Fatigue tests specify a waveform, frequency, cyclic stress (stress-controlled) or strain (strain-controlled) levels, loading mode, and test duration. Sinusoidal waveforms (Fig. 3.11a) are most commonly employed due to the relative ease of controlling gradual changes in stress/strain. However, a square or triangular waveform may be employed when impulse loading or a constant strain rate, respectively, is necessary, but can pose greater difficulty with control. In all cases, additional precautions must be taken to properly calibrate load cells for cyclic loading [100]. Moreover, most test instruments utilize proportional-integral-derivative control systems that must be properly tuned for the desired waveform and frequency [101]. The loading frequency may range from 10^{-2} to 10^{2} Hz, but is most commonly of the order of 1 Hz for biomaterials and biomedical devices as this frequency is similar to that of walking or a beating heart. Greater frequencies may be employed for accelerated tests or simulating higher levels of activity, but a greater loading frequency for a fixed stress imposes a greater strain rate. Therefore, the fatigue behavior of the strain rate-sensitive materials (e.g., viscoelastic materials) will be highly sensitive to the loading frequency. Moreover, a loading frequency above 5 Hz can result in the heating and melting of thermoplastic polymers. On the other hand, loading frequency can be leveraged to measure viscoelastic material properties, as discussed further in Section 3.4.4.4.

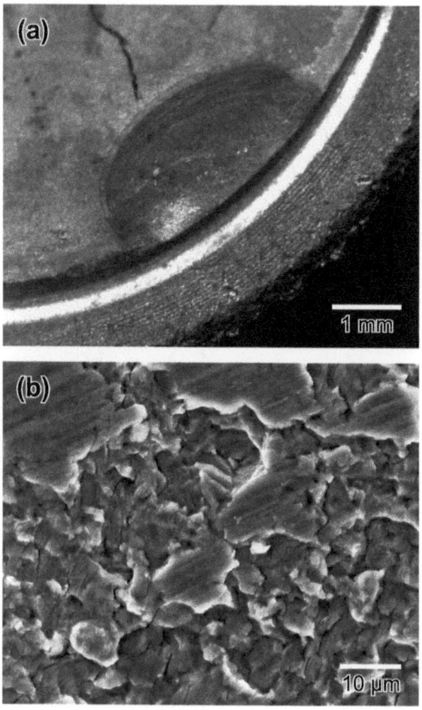

FIGURE 3.10 (a) An optical micrograph showing beachmarks and (b) an SEM micrograph showing striations on the failure surface of a Ti-6Al-4V hip stem retrieved after fracturing 26 months into service.

Cyclic stress (or strain) levels are characterized by the *maximum stress* (σ_{max}), *minimum stress* (σ_{min}), *stress amplitude* (σ_a), *mean stress* ($\overline{\sigma} = (\sigma_{max} + \sigma_{min})/2$), *stress ratio* ($R = \sigma_{min}/\sigma_{max}$), and *stress range* ($\Delta\sigma = \sigma_{max} - \sigma_{min} = 2 \cdot \sigma_a$) (Fig. 3.11a). In *fully reversed fatigue*, the stress amplitude oscillates between tensile and compressive stresses equal in magnitude and the mean stress is zero ($R = -1$), while *tension–tension fatigue* has a mean stress greater than zero ($0 < R < 1$). Fatigue tests commonly utilize $R = 0.1$ such that that specimen is never fully unloaded in order to avoid slack in loading fixtures, but yet $\sigma_a \approx \sigma_{max}$ [102]. Fatigue testing of materials normally utilizes fixed stress (stress-controlled) or strain (strain-controlled) levels, e.g., a constant stress amplitude and waveform [102–107]. The effects of variable loading amplitude and/or irregular loading history can be approximated by Miner's rule [5,7], which may be useful for failure analysis but is not recommended for biomaterials characterization. Thus, constant-amplitude fatigue is most useful for biomaterials characterization, and the following discussion will consider stress-controlled fatigue, although strain-controlled fatigue is also used and discussed in detail elsewhere [7,108].

The simplest and most common fatigue tests utilize specimens with uniform cross-section loaded axially (Fig. 3.7), either in tension–tension or fully

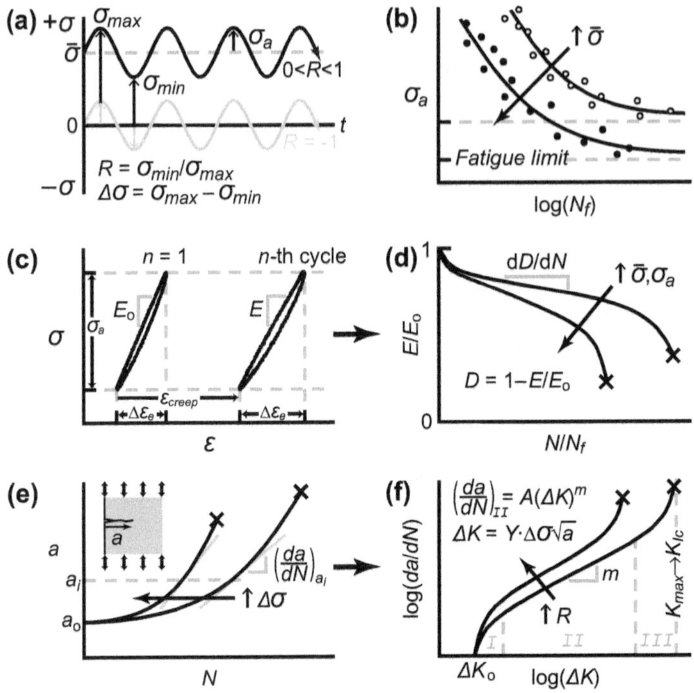

FIGURE 3.11 (a) Schematic diagram showing sinusoidal cyclic stresses applied in stress-controlled fatigue tests where σ_a is the stress amplitude, σ_{max} is the maximum applied stress, σ_{min} is the minimum applied stress, $\overline{\sigma}$ is the mean applied stress, R is the stress ratio, and $\Delta\sigma$ is the stress range. (b) "*S-N* curves" showing the fatigue life (N_f) and fatigue limit for a material after conducting numerous fatigue tests at varying, controlled levels of stress amplitude (σ_a). (c,d) Degradation, or damage (D), and creep deformation (ε_{creep}) during a fatigue test may be measured by changes in the secant or tangent modulus (E) and translation of cyclic stress–strain hysteresis loops during a test, where $\Delta\varepsilon_e$ is the recoverable elastic strain for each loading cycle. (e,f) Fatigue crack propagation tests measure the stage II crack growth rate (da/dN) from a pre-existing notch or crack (a_0) for a stress-intensity factor range (ΔK), where A and m are empirical constants, and Y is geometric factor for the specimen (see Section 3.2.6).

reversed tension–compression [102–107]. Flexural fatigue testing most commonly utilizes a rotating beam [109] and cyclic four- or three-point bending [110]. Rotating beam tests can be useful as a lower cost option for material screening. Flexural fatigue tests may also be used for the same reasons described above for quasistatic loading, but are also subject to the same potential complications and limitations (see Section 3.4.2). Thus, flexural fatigue may [111] or may not [112] provide a suitable option and should be used judiciously. Biaxial loading [113,114] and torsion [115] may also be useful for specific devices but are less common for biomaterials characterization.

3.4.4.2. Fatigue Life and Damage

Fatigue lifetime test data are most commonly reported as an *S-N curve*, which plots the stress amplitude (σ_a) versus the *number of cycles to failure* (N_f) for a number of test specimens on log–log or semi-log axes (Fig. 3.11b). The *fatigue life* is the number of cycles to failure at a given stress level. The *fatigue strength* is the stress corresponding to a given number of cycles to failure. The *fatigue limit* is the stress level at which the fatigue failure will never (at least practically speaking) occur. Thus, an increased stress level results in a decreased fatigue life. Increasing the mean stress results in a shorter fatigue life for a given stress amplitude and a lower fatigue limit. Specimens that survive a designated number of cycles without failure (often 10^7 or 10^8) are designated as a run-out, which often serves as a default fatigue limit, but run-out data points must be excluded from statistical or empirical analyses of *S-N* curves. *S-N* curves are commonly fit empirically using a power-law (Basquin equation) as $\sigma_a = \sigma_f \cdot N_f^b$, where σ_f is the fatigue strength coefficient which is comparable to the quasistatic fracture strength and b is the Basquin exponent, or using other empirical models [116]. Moreover, a number of empirical models (e.g., Goodman, Gerber, Soderberg) can be used to relate the fatigue limit to the stress amplitude and mean stress [5–8]. Statistical analysis of fatigue life and fatigue strength data typically requires the use of a logarithmic distribution or nonparametric tests [103,116]. Since fatigue behavior is inherently variable and sensitive to defects, confidence intervals are often reported for *S-N* data [103,105,116]. Furthermore, a Weibull analysis for the probability of failure is often very useful for characterizing fatigue life and fatigue strength data [107,116,117], including data sets with run-out specimens, similar to methods used for quasistatic fracture strength (Section 3.2.6).

The presence of a stress concentration can significantly decrease the fatigue life of a material, similar to the effects previously described for strength and fracture toughness (Section 3.2.6). Therefore, the above methods may also be applied to testing notched specimens. The behavior of notched versus un-notched specimens is characterized by a *fatigue notch factor*, $k_f = \sigma_a/\sigma_{a,notched}$, where σ_a and $\sigma_{a,notched}$ are the stress amplitudes or fatigue strengths for the un-notched and notched specimens, respectively, at the same number of loading cycles. Metals and polymers that are otherwise ductile can exhibit a shorter fatigue life and more brittle behavior in the presence of a notch or stress concentration [109,118,119].

Mechanical properties are degraded during the course of a fatigue test, as evidenced from changes in the stress–strain hysteresis loops (Fig. 3.11c). The secant or tangent modulus (E) may decrease due to various *damage* processes, including microcracking in quasi-brittle materials (e.g., bone [112,120,121] and acrylic bone cement [117]), or cyclic softening and void formation (including crazing) in ductile materials. Note that while most metals and alloys exhibit cyclic softening, some ductile metals can exhibit cyclic hardening. Moreover, crazing can also lead to an increased modulus early in the fatigue life of thermoplastics due to orientation strengthening [111]. A damage parameter can be defined as $D = 1 - E/E_0$, where is E_0 is the initial secant or tangent modulus, such that the amount and rate (dD/dN) of damage accumulation can be measured during a test as a function of the number of cycles or fatigue life (Fig. 3.11d) [120,121]. In stress-controlled fatigue, the entire hysteresis loop may also translate to higher strains indicative of a permanent *creep strain* (ε_{creep}) (Fig. 3.11c), which may also be measured as a function of the number of cycles [111,112,121–123]. Finally, *strain energy dissipation* can be measured by the area of the hysteresis of each loading cycle (Fig. 3.11c), due to processes such as microcracking, plasticity, adiabatic heating, etc., and may also be measured as a function of the number of cycles [111,120].

3.4.4.3. Fatigue Crack Propagation

Fatigue crack propagation integrates fracture mechanics concepts with fatigue in order to characterize the stage II crack growth rate. Standardized specimens (e.g., compact tension) are precracked (see Section 3.3.2), and crack growth (da) is measured as a function of loading cycles (N) using a traveling microscope, specimen compliance, and/or electrical resistance [23,124]. The crack growth rate increases throughout the fatigue lifetime until failure, and increased stress levels increase the crack growth rate at a given number of cycles (Fig. 3.11e). Fracture mechanics is integrated by calculating a stress intensity factor range ($\Delta K = K_{max} - K_{min}$) from the stress range (see Section 3.2.6). The stress intensity factor range is related to the *crack growth rate* by the *Paris Law*, $da/dN = A \cdot \Delta K^m$, where A and m are empirical constants dependent on the material properties, environment, loading frequency, and stress ratio [23]. The exponent is usually in the range of $2 < m < 10$. A log–log plot of da/dN versus ΔK reveals the three stages of fatigue described above (Fig. 3.11f), initiated by a threshold stress intensity range (ΔK_0) for fatigue crack propagation and ending with fracture as K_{max} approaches K_{Ic}, although typically only stage II is plotted. An increased stress ratio results in increased da/dN (or m) and decreased fatigue life. Once the empirical constants of the Paris Law are known, the fatigue life can be predicted by integrating the Paris Law from an initial to final crack length [5–8,23]. Moreover, the striation spacing (s) measured from electron micrographs of failure specimens or retrievals can be used to estimate or confirm ΔK as $s = A_s \cdot (\Delta K/E)^2$, where A_s is the Bates–Clark constant, which is a material-dependent empirical constant [125,126]. Finally, the stress intensity range in fatigue crack propagation tests can be modified to account for plasticity using plastic-zone size correction factors or the J-integral [23], similar to fracture toughness measurements, and the Paris Law can also be modified to account for combined creep [127].

3.4.4.4. Dynamic Mechanical Analysis

Cyclic loading of a viscoelastic material will result in a time delay or phase shift between the applied stress and strain response. DMA utilizes this behavior to characterize the viscoelastic properties of materials, typically polymers [5,96,128]. A sinusoidal stress takes the form, $\sigma(t) = \sigma_0 \cdot \sin(\omega t + \delta)$, where t is time, ω is the angular frequency ($\omega = 2\pi f$) in rad/s, f is the ordinary frequency in Hz, and δ is the *phase angle* of the time delay for the lagging strain response, $\varepsilon(t) = \varepsilon_0 \cdot \sin(\omega t)$. The stress can be expanded into an elastic in-phase component reflecting strain energy stored and released, $\sigma_0 \cdot \cos(\delta)$, and a viscous out-of-phase component reflecting dissipated strain energy, $\sigma_0 \cdot \sin(\delta)$. Therefore, the *storage modulus* is measured as $E' = (\sigma_0/\varepsilon_0) \cdot \cos(\delta)$; the *loss modulus* is measured as $E'' = (\sigma_0/\varepsilon_0) \cdot \sin(\delta)$; and $E'/E'' = \tan(\delta)$. The storage and loss moduli can also be determined as shear moduli (G', G''). Storage and loss moduli for polymers can vary dramatically with changes in the loading frequency and temperature, enabling determination of the glass transition temperature (as well as other secondary transitions) and relaxation times, respectively. DMA is advantageous over creep and stress relaxation tests in obviating the need for long time periods to measure time constants.

Similar to creep and stress relaxation tests (Section 3.4.3), DMA requires well-defined specimens, precise instrumentation, precise control or monitoring of temperature and humidity, and possibly preconditioning [129]. DMA instruments employ either free oscillation, most commonly using a torsion pendulum, or forced oscillation in a variety of possible loading modes (uniaxial tension, compression, torsion, bending, and in-plane shear). The torsion pendulum is a simple test but limited by decreasing stress/strain amplitude with time and a dependence of frequency on the material properties. Forced oscillation tests are more complex but also more reproducible and able to investigate a wider frequency and temperature range.

3.4.5. Nondestructive Tests

Almost all of the preceding tests discussed in this chapter are considered *destructive tests*, as the specimen is loaded until failure occurs by permanent deformation or fracture. While single cycle elastic deformation of a specimen or DMA could be considered nondestructive, *nondestructive testing* is usually used to describe tests that can be performed on a material or device prior to or during service without any detrimental affect on performance. Nondestructive tests for mechanical characterization of biomaterials primarily include indentation and ultrasound, and these will be introduced below. Other methods of nondestructive evaluation are primarily used to detect flaws or aspects of material structure associated with, and likely causal for, changes in mechanical properties. Broadly categorized, these methods utilize X-rays (e.g., radiography, computed tomography), electromagnetics (e.g., magnetic resonance imaging, magnetic flux leakage, magnetic particle testing, eddy current testing), sound and vibration (e.g., ultrasound, acoustic emission, resonance), visible and infra-red light (e.g., penetrants, interferometry, thermography) to detect flaws or changes in microstructure [130].

3.4.5.1. Indentation

Indentation methods utilize an indenter of specified geometry (Table 3.1) that upon loading and elastic unloading leaves a permanent indentation in the specimen. Thus, indentation tests could be excluded from a strict definition of nondestructive testing due to localized permanent deformation providing a stress concentration that could lead to failure. If used prior to service, indentations must be verified to not compromise the biomaterial or device performance, e.g., by placement in locations experiencing little or no stress in service. Thus, indentation tests have been historically utilized for quality control, but their utility for biomaterials characterization is primarily found in sampling small volumes of material and/or characterizing gradients or heterogeneity in mechanical properties. Moreover, many methods for specimen preparation prior to indentation tests are also necessarily destructive.

Indentation tests measure the *hardness* (*H*) of a material, which is the resistance to permanent deformation by the indenter and can thus be correlated with the yield strength (e.g., $H \approx 3 \cdot \sigma_y$ for many metals), elastic modulus, and other mechanical properties [131]. A number of different indenters, corresponding hardness scales, and common applications are summarized in Table 3.1. Spherical indenters (Hertzian contact) and spheroconical diamond indenters are used with relatively high applied loads for traditional *Brinell and Rockwell hardness* tests for bulk material properties [132–134]. Brinell tests measure the "true hardness" as the ratio of applied load to the projected contact area (A_c) of the indentation, but produce large indentations requiring large loads. Rockwell tests have enjoyed decades of commercial popularity due to providing rapid, reliable results, a small indentation area, and a wide range of scales (A, B, C, D, etc.) suitable for almost all materials. However, Rockwell hardness is measured as a nondimensional number corresponding to the difference in the depth of indenter penetration between an applied load and a preload, after a specified hold time, and must be empirically converted to true hardness [135].

Microhardness tests are particularly useful for sampling a small volume of material (e.g., films, coatings, and microstructural features) and/or characterizing gradients or heterogeneity in mechanical properties at the microscale. *Knoop and Vickers hardness* utilize sharp diamond pyramid indenters (Table 3.1) that require much lower loads and leave indentations requiring an optical microscope for measuring the indentation size [136–138]. Knoop and Vickers can also be used at higher loads to characterize bulk material hardness and to characterize anisotropy via differences in indentation diagonal lengths relative to the specimen orientation. Knoop indenters are well suited for measuring hardness in brittle materials due to producing a small indentation depth. Conversion tables are available for relating Brinell, Rockwell, Knoop, and Vickers hardness, but are material-specific [135]. Brinell, Knoop, and

TABLE 3.1 Common Indenters, Hardness Scales, and Applications for the Mechanical Characterization of Materials by Indentation and Instrumented Indentation

Indenter Type/Hardness Scale			α	P	Indent	H	A_c	Common Applications
Sphere (Hertzian)	WC or steel $D = 1, 2.5, 5, 10$ mm (Brinell)	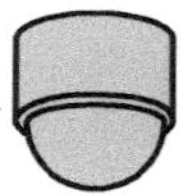	n/a	1–3000 kgf	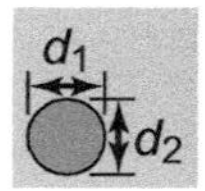	$BHN = \frac{2P}{\pi d \cdot (D - \sqrt{D^2 - d^2})}$	$\pi(D \cdot h_c - h_c^2)$	Ductile metals polymers
	WC or steel $D = 1/16, 1/8, 1/4, 1/2$ in (Rockwell B,E,F,G,K,L,M,R)		n/a	60–150 kgf	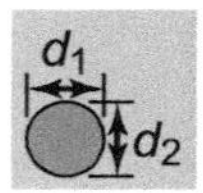	$HRX = 130 - (h_c/0.002)$	$\pi(D \cdot h_c - h_c^2)$	Ductile metals polymers
Spheroconical diamond (Rockwell A,C,D)			60°	60–150 kgf	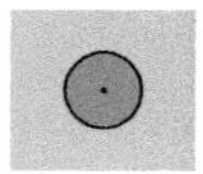	$HRX = 100 - (h_c/0.002)$	$\pi h_c^2 \cdot \tan^2(\alpha)$	Hard metals ceramics
Diamond pyramids	Knoop	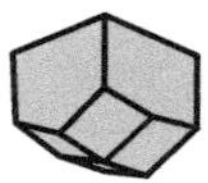	$\alpha_1 = 86.25°$ $\alpha_2 = 65°$	1–120 kgf (macro) 1–1000 gf (micro)	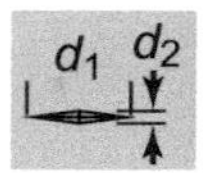	$KHN = \frac{2P}{d_1^2 \cdot \cot(\alpha_1) \cdot \tan(\alpha_2)}$	$2h_c^2 \cdot \tan(\alpha_1) \cdot \tan(\alpha_2)$	Hard metals ceramics microhardness

(Continued)

TABLE 3.1 Common Indenters, Hardness Scales, and Applications for the Mechanical Characterization of Materials by Indentation and Instrumented Indentation—cont'd

Indenter Type/Hardness Scale		α	P	Indent	H	A_c	Common Applications
Vickers	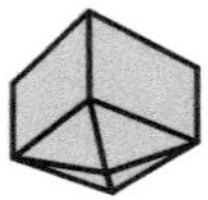	68°	1–120 kgf (macro) 1–1000 gf (micro) ~10 nN – 10 N (nano)	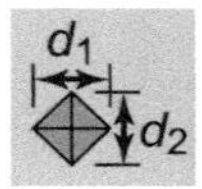	$VHN = (2P/d^2)\cdot\sin(a)$ $H_{nano} = P/A_c$	$4h_c^2\cdot\tan^2(\alpha)$	Hard metals ceramics microhardness toughness nanoindentation
Berkovich	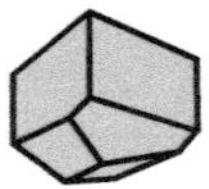	65.3°	~10 nN – 10 N	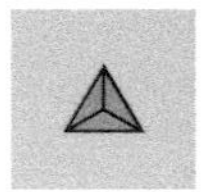	$H_{nano} = P/A_c$	$3\sqrt{3}\cdot h_c^2\cdot\tan^2(\alpha)$	Nanoindentation
Cube corner	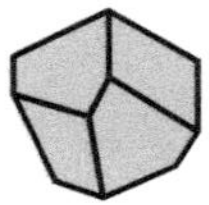	35.3°	~10 nN – 10 N	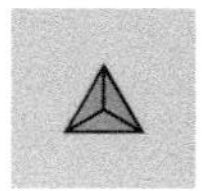	$H_{nano} = P/A_c$	$3\sqrt{3}\cdot h_c^2\cdot\tan^2(\alpha)$	Nanoindentation toughness
Circular flat punch hardened steel $D = 0.25–1.5$ mm		90°	~10 nN – 10 N	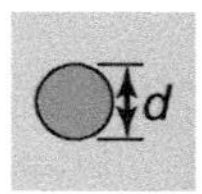	$H_{nano} = P/A_c$	$\pi d^2/4$	Viscoelastic and/or highly compliant materials

D is the indenter diameter, α is the angle between the vertical centerline and contact face(s) of the indenter, P is the applied load, d_1 and d_2 are orthogonal measurements of the indentation diameter or diagonal, d is the mean indentation diameter or diagonal, H is hardness, BHN is the Brinell hardness number, HRX is the Rockwell hardness number for any scale (X = A, B, C, D, etc.), KHN is the Knoop hardness number, VHN is the Vickers hardness number, H_{nano} is the hardness measured by nanoindentation, A_c is the projected contact area of the indentation, and h_c is the indentation depth.

Vickers hardness are typically reported in MPa by multiplying units of kgf/mm^2 by 9.807 m/s^2.

There has been a resurgence of interest in indentation tests over the last 20 years due to the development of *instrumented indentation* and *nanoindentation* [139–141]. Rather than simply applying a fixed load, instrumented indentation enables load and displacement measurement throughout loading and elastic unloading of the indenter. Therefore, a *reduced modulus* (E_r) for the specimen and indenter materials can be measured as $(S/2\beta)\cdot(\pi/A_c)^{1/2}$, where S is the unloading stiffness measured by taking the derivative of a nonlinear least-squares power-law fit to the unloading curve at P_{max}, β is a nondimensional parameter used to account for any deviations from the ideal solution (e.g., $\beta \approx 1.034$ for a Berkovich indenter), and A_c is the projected contact area of the indention [139]. The contact area is calculated from the ideal indenter geometry as a function of the contact depth (h_c), using the equations in Table 3.1 [140]. The contact depth is calculated as $h_c = h_{max} - (\varepsilon\cdot P_{max})/S$, where h_{max} is the indenter depth at P_{max} and ε is a geometric constant ($\varepsilon \approx 0.75$ for most indenters) [139]. Note that the equations for A_c in Table 3.1 are ideal and are calibrated in practice to account for the indenter tip rounding and other deviations [139]. The *indentation modulus* (E_i) of the specimen can be determined from the reduced modulus using the relation, $1/E_r = (1-\nu^2)/E_i + (1-\nu_I^2)/E_I$, where ν is Poisson's ratio of the specimen material, and ν_I and E_I are the Poisson's ratio and elastic modulus of the indenter material. For diamond, $\nu_I = 0.07$ and $E_I = 1141$ GPa. If Poisson's ratio is not known for the specimen material, the plane strain indentation modulus can be reported as $E_i' = (1-\nu^2)/E_i$.

The use of sharp diamond indenters (Berkovich, cube corner, and Vickers) combined with precise instrumentation and control (via electrostatic or electromagnetic actuation and capacitive or inductive transducers) enables nanoscale resolution within the range of both the applied load (~10 nN – 10 N) and indentation depth (~1 nm – 20 μm), hence the term "nanoindentation." *Nanoindentation hardness* (H_{nano}) is thus measured as P_{max}/A_c (Table 3.1). The three-sided Berkovich and cube-corner indenters are more readily machined to a sharp tip compared to Vickers and Knoop indenters, making them best suited for instrumented nanoindentation, as well as imaging and mechanical characterization using atomic force microscopy.

In all indentation tests, the indenter material must have greater hardness than the specimen material. An appropriate load must be selected based on the specimen material, indenter type, and hardness scale (Table 3.1). The size of the indentation (depth) can influence the measured hardness or modulus and should thus be controlled and documented for comparison of results, regardless of whether the size effect is artifactual or real. The applied load is also always held for a specified period of time to remove or control effects of strain-rate sensitivity or creep. Other important requirements include flat specimen surfaces, surface roughness less than 10% of the indentation depth, a specimen thickness at least 10-times the indentation depth, spacing between the center of indentations and the specimen edge of at least 2.5 indentation diameters, and spacing between the center of indentations of at least 3 indentation diameters, although the recommended indentation spacing varies slightly between standards [132,133,136–138].

Indentation can also be used to characterize viscoelastic and compliant biomaterials. *Durometer hardness* is used to characterize the bulk hardness of viscoelastic and elastomeric polymers [142]. *Creep* and *stress relaxation* time constants can be measured by varying holding times (typically 3–120 s), loading and unloading rates, or by fitting data to viscoelastic models (see Section 3.4.3.3) [97,141]. *Dynamic nanoindentation* may also be used to measure the storage and loss moduli [140]. A spherical or *circular flat punch indenter* (Table 3.1) may be useful for characterizing viscoelastic and highly compliant biomaterials [143,144]. Note that fluid cells are also available to maintain specimens in a hydrated environment.

Indentation testing has also been used to measure *indentation fracture toughness* in relatively brittle materials. A sufficiently high load applied to a Vickers indenter in a brittle material was observed to result in cracks propagating away from indentation edges at and below the surface as "half-penny" cracks. Numerous solutions have been proposed and used for Vickers indentation fracture toughness ($K_{c,i}$) but the most common is $K_{c,i} = 0.16 \cdot P \cdot (E/H)^{1/2} \cdot (c_o)^{-3/2}$, where P is the applied load, E/H is the elastic modulus to hardness ratio, and c_o is the radius of the half penny crack measured at the specimen surface from the center of the indent to the crack tip [145,146]. The E/H ratio can be experimentally measured on the same specimen using Knoop indentations as $E/H = 0.45/(0.14 - d_2/d_1)$, where d_2 and d_1 are the short and long diagonals, respectively, of the indentation [147]. The lower centerline-face angle (α) of a cube corner indenter relative to a Vickers indenter results in crack formation at lower loads and thus smaller indentations. Therefore, cube corner indenters have also been used for measuring indentation fracture toughness with similar equations [145]. However, indentation fracture toughness measurements are not equivalent to plane strain fracture toughness (K_{Ic}), are subject to large errors, and are difficult to perform in a hydrated environment [145]. Therefore, indentation fracture toughness should be used cautiously and only compared between tests using identical methods and similar materials. Nonetheless, indentation fracture toughness can occasionally be useful for materials and situations where plane strain fracture toughness measurements are not possible due to the available specimen size, or for characterizing heterogeneity over small distances, provided that results are treated as semiquantitative.

In summary, indentation testing enables the mechanical characterization of biomaterials at spatial scales not possible by other methods. However, there are also a number of inherent limitations for indentation testing. (1) Hardness is an aggregate property reflecting the effects of a number of material properties.

(2) Indentation tests tend to produce semiquantitative results due to the need for approximations and correction factors to describe the true indentation shape. (3) Indentations tests are highly sensitive to the indenter geometry and test parameters. Therefore, instrument calibration is critical using test blocks for hardness and microhardness, and using detailed procedures for instrumented nanoindentation that are described elsewhere [139,140]. Finally, material properties (e.g., E_i) measured by indentation should not be treated as equivalent nor directly compared between different indentation methods or with other test methods.

3.4.5.2. Ultrasound

Ultrasound is primarily thought of as an imaging technique, but ultrasonic wave propagation can also be used to measure the elastic constants, or stiffness coefficients, of a material nondestructively and precisely [148]. Thus, elastic anisotropy can be characterized on a single specimen [149,150], whereas destructive tests require multiple tests (e.g., uniaxial tension and torsion) on multiple specimens to fully characterize anisotropy. The high frequency of ultrasound ensures purely elastic deformation, even in viscoelastic materials. Ultrasound can also be used with small specimens (mm-scale) [151], enabling the characterization of gradients or heterogeneity in the magnitude and anisotropy of elastic constants at the mesoscale [152]. Moreover, scanning acoustic microscopy can be used to generate acoustic images with up to 1 μm spatial resolution on thin, planar specimens [153]. Finally, ultrasound is well-suited for testing biomaterials and tissues in a hydrated environment.

Ultrasonic waves are generated and received by piezoelectric transducers operating in the kHz–MHz range. The transducer frequency should be chosen considering the specimen size and material properties. The minimum specimen distance that can be resolved is $n \cdot \lambda/2$, where n is the number of wavelengths, λ is the wavelength ($\lambda = v/f$), v is the wave velocity, and f is the frequency [154]. Waves may be continuous [150], or pulsed [149,151,152] to generate an intense burst to aid detection. The pulse width is the most critical parameter governing the minimum specimen distance that can be resolved [154]. The "through-transmission" method uses two transducers to separately transmit and receive the wave on opposite sides of the specimen (Fig. 3.12a). The reflection ("pulse-echo") method uses a single transducer to transmit and receive a wave after reflecting on the opposite specimen surface. Transducers can generate either normal (a.k.a., longitudinal) or shear (a.k.a., transverse) waves (Fig. 3.12b). Signal transmission from the transducer to the specimen is aided by using a liquid couplant. Water generally suffices for normal waves, which is advantageous, if not necessary, for biomaterials characterization. However, shear waves typically require a high-viscosity couplant, such as honey, glycerin, or propriety products, which are water soluble and are readily removed from the specimen surface after testing.

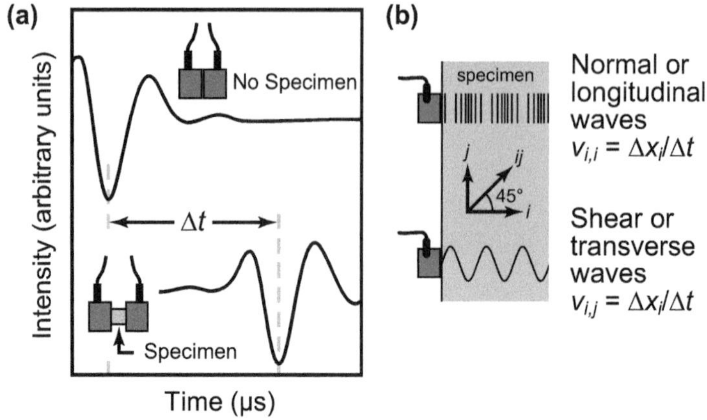

(c)

$$C_{ij} = \begin{bmatrix} C_{11} & C_{12} & C_{13} & 0 & 0 & 0 \\ C_{13} & C_{22} & C_{23} & 0 & 0 & 0 \\ C_{13} & C_{23} & C_{33} & 0 & 0 & 0 \\ 0 & 0 & 0 & C_{44} & 0 & 0 \\ 0 & 0 & 0 & 0 & C_{55} & 0 \\ 0 & 0 & 0 & 0 & 0 & C_{66} \end{bmatrix} \qquad \begin{aligned} C_{11} &= \rho \cdot v_{1,1}^2 \\ C_{22} &= \rho \cdot v_{2,2}^2 \\ C_{33} &= \rho \cdot v_{3,3}^2 \\ C_{44} &= \rho \cdot v_{2,3}^2 = \rho \cdot v_{3,2}^2 \\ C_{55} &= \rho \cdot v_{1,3}^2 = \rho \cdot v_{3,1}^2 \\ C_{66} &= \rho \cdot v_{1,2}^2 = \rho \cdot v_{2,1}^2 \end{aligned}$$

$$C_{12} = (C_{11} + C_{66} - 2\rho \cdot v_{12,12}^2) \cdot (C_{22} + C_{66} - 2\rho \cdot v_{12,12}^2) - C_{66}$$
$$C_{13} = (C_{11} + C_{55} - 2\rho \cdot v_{13,13}^2) \cdot (C_{33} + C_{55} - 2\rho \cdot v_{13,13}^2) - C_{55}$$
$$C_{23} = (C_{22} + C_{44} - 2\rho \cdot v_{23,23}^2) \cdot (C_{33} + C_{44} - 2\rho \cdot v_{23,23}^2) - C_{44}$$

$$\left. \begin{aligned} C_{55} + C_{66} &= 2\rho \cdot v_{23,1}^2 \\ C_{44} + C_{66} &= 2\rho \cdot v_{13,2}^2 \\ C_{44} + C_{55} &= 2\rho \cdot v_{12,3}^2 \end{aligned} \right] \rightarrow \textit{used to test orthotropy}$$

FIGURE 3.12 Characterization of anisotropic elastic constants using the pulse transmission method for ultrasonic wave propagation between two transducers: (a) Schematic diagram showing the time delay (Δt) for transmission of an ultrasonic wave pulse through a specimen as measured on an oscilloscope. (b) Schematic diagram showing the propagation of normal (or longitudinal) and shear (or transverse) waves in a specimen by corresponding ultrasonic transducers and the calculation of the wave velocity (v), where Δx is the specimen distance traversed by the wave. Note that the directions of wave propagation and displacement are coincident for normal waves but orthogonal for shear waves ($v_{i,j}$), where i is the direction of wave propagation and j is the direction of displacement. (c) Equations used to measure all elastic constants (C_{ij}) in an orthotropic material and to test the condition of orthotropy, where ρ is the apparent density of the specimen material, $v_{ij,ij}$ is the velocity for wave propagation in the ij direction and displacement in the ij plane, and $v_{ij,k}$ is the velocity for wave propagation in the ij direction and displacement in the k direction.

The use of normal and shear waves propagating in various specimen directions can be used to directly measure all the *elastic constants* of an orthotropic material (see Section 3.2.5) using the equations in Fig. 3.12c. The wave velocity (v) for a normal or shear wave is measured by the time delay (Δt)

for wave transmission through the specimen (Fig. 3.12a) in a particular direction over a distance, Δx (Fig. 3.12b). The distance traversed by the wave is usually measured by digital calipers, and the time delay is measured using an oscilloscope connected to the transducers and wave signal generator. The general solution for elastic wave propagation in a one-dimensional solid is $C = \rho \cdot v^2$, where C is the elastic constant or stiffness coefficient, ρ is the apparent density of the specimen material, and v is velocity of a normal or shear wave in the appropriate specimen direction and orientation (Fig. 3.12c). The apparent density of the specimen is typically measured by Archimedes principle using standardized methods [155,156]. Engineering constants (elastic moduli, shear moduli, and Poisson's ratios) can be extracted from the compliance tensor (S) after inverting the measured elastic constant tensor [6,149,150].

The advantage of nondestructive characterization is also the cause of the main limitation of ultrasound, which is that only elastic constants (or moduli) can be measured. Moreover, while ultrasound is nondestructive for the specimen, measurements of elastic heterogeneity and anisotropy typically require destructive sectioning of the device or a larger sample. Parallel specimen surfaces are required for both the through transmission and reflection techniques. The most fundamental and common methods for ultrasonic mechanical characterization of material were described above, but many other methods are available for mechanical characterization, including oblique through transmission, ultrasonic wave attenuation, ultrasonic resonance spectroscopy, Rayleigh and Lamb surface waves, etc. [148], as well as the closely related technique of acoustic emission.

3.5. ENVIRONMENT

Tabulated mechanical properties of engineering materials are typically characterized in air at ambient temperature and pressure. However, biomaterials are typically subjected to various sterilization methods prior to service and an aqueous physiological environment in service. Therefore, mechanical characterization of biomaterials often warrants the consideration of temperature, pressure, aqueous media containing various chemicals, and irradiation, which may each lead to degradation of mechanical properties.

3.5.1. Temperature and Pressure

Deviation from ambient temperature and pressure may occur due to any of three common extremes. Implantable biomaterials function at body temperature (37 °C) in service. Steam sterilization occurs in an autoclave at temperatures and pressures up to 134 °C and 207 kPa, respectively. Biomedical materials that may be subjected to steam sterilization includes not only the materials used in implantable devices and surgical instruments but also the materials used in cases, trays, and packaging for the implantable devices and instruments. Finally, cryopreservation occurs at temperatures as low as −196 °C. Typically only cells and tissues are subjected to cryopreservation, but biomaterials may also be used for packaging and as a cryoprotectant or cell carrier.

Increased temperature typically results in a decreased elastic modulus and strength, but increased ductility and toughness, with the opposite effects for decreased temperature. These effects are typically small for short-term exposure and small temperature differences, but can become significant near the melting temperature or glass transition temperature of a material. Long-term- or cyclic exposure to elevated temperature may result in creep of low melting temperature and viscoelastic materials, thermal degradation of polymers, increased crystallinity of polymers, or accelerated chemical degradation of materials. In general, polymers are more sensitive to temperature [157] than metals, and ceramics are least sensitive to temperature. However, select polymers such as polyphenylsulfone are routinely used for non-implantable devices subjected to more than a 1000 steam sterilization cycles without loss of mechanical properties. Hydrostatic pressure can increase the elastic modulus, strength, and more significantly, the ductility of materials, but is most likely of secondary importance relative to the effects of temperature.

Biomaterial specimens may be subjected to elevated temperature in an oven, elevated temperature and pressure in an autoclave, or low temperature in a refrigerator or freezer either continuously or cyclically prior to mechanical characterization. *In situ* testing requires a test instrument outfitted with a temperature-controlled fluid bath, oven, or environmental chamber. Environmental chambers are commercially available to facilitate *in situ* testing and thermal cycling ranging from roughly liquid nitrogen temperatures to several hundred degrees Celsius. Conditions should be carefully documented and standardized whenever possible [157].

Shape-memory alloys (e.g., Nitinol) and polymers are designed to undergo reversible changes in shape and mechanical properties (elastic modulus) upon temperature change and thus warrant special considerations for the mechanical characterization [158]. For example, Nitinol endovascular stents are designed to be easily deformed and placed into a delivery system at or below ambient temperature, but expand and exhibit superelastic properties when heated above the transformation temperature (typically 30°C) during implantation.

3.5.2. Aqueous Media

Exposure to aqueous media is generally detrimental to mechanical properties and thus used to provide a more realistic appraisal of performance in a physiological environment. Specific effects on mechanical properties are numerous and can only be introduced here. Common media utilized for mechanical characterization of biomaterials include water, PBS, calcium-supplemented media, and FBS or synovial fluid [159–161].

Exposure to water or a humidified atmosphere may be important to characterize the effects of water uptake (swelling), interfacial degradation, and hydrolytic degradation [157,159–161]. The uptake of water in a hydrophilic biomaterial may be desirable and essential to mechanical and biological functions (e.g., soft contact lens or tissue graft). Mechanical characterization of biological tissues should always take place while fully hydrated. Hydration generally leads to decreased elastic modulus and strength, but increased ductility and toughness in such materials. For example, characterization of dehydrated collagenous tissue and scaffolds results in increased stiffness (modulus) and decreased toughness compared to characterization while hydrated [162,163]. On the other hand, the uptake of water into a synthetic biomaterial, either in bulk or at interfaces (e.g., between the matrix and fiber reinforcements), can also cause swelling or form voids that act as a precursor to mechanical failure. The combination of water uptake, corrosion, and stresses in metals can lead to stress corrosion cracking, which will be discussed further below. The combination of water uptake and stresses can lead to environmental stress crazing or cracking (ESC) in polymers and interfacial degradation in composites. Polymers and some ceramics may also intentionally (e.g., poly-α-ortho esters such as polylactide, polyglycolide and copolymers thereof, and calcium phosphates) or unintentionally degrade by hydrolysis. In polymers, ESC acts to degrade intermolecular bonds, while hydrolytic degradation degrades intramolecular bonds.

PBS provides ionic species (Na, K, Cl, and phosphates), which maintain constant pH (7.4) and more closely match levels found in the human body [159–161]. The presence of these salts can accelerate corrosion of metals. Corrosion can lead to pits and other defects that act as stress concentrations, leading to premature failure. Moreover, internal variations in stress due to applied bending, stress concentrations, and residual stresses can lead to differences in electrochemical potential, which further accelerates corrosion via galvanic corrosion. Thus, corrosion can have a profound effect on fatigue and wear such that the combined effect is much greater than either effect alone. The combination of corrosion and fatigue can eliminate the presence of a fatigue limit entirely.

Calcium-containing media is used here to describe various media containing similar salts as PBS with the addition of calcium salts. Examples include synthetic body fluid and variations of minimum essential medium (e.g., Dubecco's modified Eagle medium, DMEM), where the latter also includes various vitamins and amino acids. These media may be used to examine desirable (e.g., calcification of polymer scaffolds for bone tissue) or undesirable (e.g., calcification of cardiovascular biomaterials) calcification of biomaterials. For example, calcification of artificial and bioprosthetic heart valves occurs synergistically in highly stressed regions resulting in loss of function due to increased biomaterial rigidity and possibly mechanical failure or thrombosis [164].

FBS provides a rich variety of proteins found in blood plasma, which may interact with biomaterial surfaces. Protein surface adsorption guides the response of cells to biomaterials, and is thus important for mechanical characterization of the interface between a biomaterial and cells/tissue in an *in vitro* or *in vivo* study. Synovial fluid provides a replication of the proteins and concentrations present in joint capsules and is thus used exclusively for simulated *in vivo* wear testing, described in Chapter 7. Protein-containing media may more accurately mimic the physiological environment, but also requires significantly greater expense and care. Without proper care, protein solutions may degrade, coagulate, or become colonized by bacteria during the course of a test, which may compromise the test entirely.

Other less common but potentially important considerations also deserve mention. Exposure to lipids may lead to swelling and degradation in otherwise stable biomaterials. For example, the fatigue properties of polysulfone were found to be degraded in the presence of lipids despite otherwise excellent mechanical properties for cardiovascular and load-bearing orthopaedic devices [165,166]. Cells and bacteria can produce an acidic microenvironment that may enhance the corrosion of metals or the degradation of some polymers and ceramics. For example, initial polyurethane insulation on pacemaker leads was highly resistant to hydrolytic degradation but often failed prematurely due to ESC initiated by direct surface oxidation mediated by a combination of host phagocytic cells and metal ions released from the conductor.

Degradation that is intended or unintended is often investigated by subjecting biomaterial samples to elevated temperature, elevated pressure, and/or aqueous media for a period of time before testing [4,157,159–161]. These tests are necessarily laborious and time intensive. Therefore, tests may be accelerated using greater than physiological levels with proper validation.

Biomaterial specimens may be subjected to any of the above aqueous media for a defined period of time prior to mechanical characterization. *In situ* testing requires a test instrument outfitted with a fluid bath or environmental chamber. Simultaneous exposure to elevated temperature and/or pressure may require a temperature-controlled fluid bath or autoclave. Sterile conditions may need to be maintained for media-containing biological molecules (e.g., DMEM or FBS). When the use of nonsubmersible transducers (e.g., clip-on extensometer) prevents the use of a fluid bath during testing, hydration may be adequately maintained over a short time period using a gravity-fed drip with a catch basin.

3.5.3. Irradiation

Sterilization by gamma or electron beam irradiation may be advantageous over steam sterilization. However, ionizing radiation can cause cross-linking or oxidative chain scission in polymers resulting in significant, potentially deleterious, changes to mechanical properties. For example, gamma irradiation in air was discontinued for sterilization of UHMWPE orthopaedic implants due to evidence of oxidative chain scission and long-term mechanical degradation [167]. However, gamma irradiation has also been used to cross-link UHMWPE for improved wear resistance, but highly cross-linked UHMWPE incurs a loss of tensile strength, ductility, toughness, and fatigue resistance [168]. Due to the long-term nature of irradiation damage, accelerated aging conditions are often utilized prior to mechanical characterization.

3.6. DATA ACQUISITION AND ANALYSIS

When acquiring signals from transducers (e.g., load cell, LVDT, extensometer, etc.) for either data collection or system control [48,49], the appropriate *data acquisition rate* and *bandwidth* should consider the strain rate, loading frequency, and material behavior. The bandwidth reflects the frequency sensitivity. As a general guideline, the data acquisition rate should not exceed 10-times the bandwidth. An excessive data acquisition rate (oversampling) and bandwidth can result in unnecessary data and noise, respectively. A low data acquisition rate (undersampling) and bandwidth can result in the loss or filtering of meaningful data (e.g., peaks) and potentially incorrect results. For example, a split Hopkinson bar test for high strain rate testing typically requires a frequency response of 100 kHz [86].

Nearly all modern digital control and data acquisition systems include user-friendly *software* enabling the creation of custom test methods, automatic calculation of mechanical properties, and generation of reports. However, these advantages can also cause problems if a negligent user is not aware that the wrong data are being used for the calculation of a mechanical property. Unfortunately, I've consulted for a client only to determine that a previous "professional" consulting firm hired to provide accurate data had made this egregious error. The moral of the story is that one must take care and responsibility for all that is happening "behind" the user interface window. Computers and software have not yet replaced the importance of a knowledgeable and skilled test engineer!

Data analysis for the mechanical characterization of biomaterials can be greatly aided by a few basic *statistical methods* [169,170]. Force-displacement or stress–strain data can be fit using *linear or nonlinear least squares regression*. The correlation coefficient (R^2) provides a measure of the goodness of fit. Biological systems and interactions with biological systems generally produce greater variability than most engineering systems. Therefore, *analysis of variance* (ANOVA) or regression models should be used to examine whether an experimental factor (or variable) exhibits a statistically significant influence on a mechanical property. The criteria for significance are (by convention) taken at a level of 95% confidence or a p-value less than 0.05. Interactions between multiple factors and their relative influence can also be considered statistically using these methods. In the case of grouped data, post hoc comparisons should be used after ANOVA has determined that a factor is significant. The *t*-test is most common, although Tukey's HSD test is considered more conservative. These tests assume that the data are normally distributed. Therefore, data exhibiting a non-normal distribution or an unknown distribution due to a small sample size are best examined using nonparametric methods (e.g., Kruskal–Wallis ANOVA and Wilcoxon post hoc tests).

ANOVA and post hoc tests indicate whether an effect or difference is statistically significant, but note that statistical equivalence is an entirely different matter and should not be claimed. Additionally, in tests where the measured effect of a variable or the difference between two groups is not statistically significant, further testing and analysis may be required. Retrospective *power analysis* using the test data can determine whether an effect or difference could have been measured using a greater sample size and the number of additional samples needed. The claim that an effect or difference is "not statistically significant" can only be made when the *p*-value is greater than the level of significance and the statistical power is sufficiently high (sometimes taken as greater than 0.80). A prospective power analysis can be extremely useful in the design of experiments when possible. Given reasonable estimates of the difference to be detected, error standard deviation, and level of significance, the statistical power of various sample sizes can be estimated.

Thorough statistical analysis often requires methods and capabilities beyond what is available in test instrument control and data acquisition packages. A number of software packages are commercially available or open source, ranging from products that are very powerful and comprehensive (e.g., SAS, SPSS, R, Matlab) to more user friendly (e.g., JMP, GraphPad, AcaStat). Specific packages and modules are also available for Weibull analysis, fatigue crack propagation, and other methods.

ACKNOWLEDGMENTS

This chapter was made possible by 40 years of learning facilitated by federal and private funding, which has supported my research, clients who trust my judgment, students who teach me more than I could ever learn on my own, colleagues who constantly sharpen my skills, professional and personal mentors who encouraged me to pursue my career path, college and graduate school instructors who introduced me to the mechanical behavior of materials, manual mechanically-oriented labor that taught the necessity of hard work and motivated study, and parents who encouraged youthful curiosity. Words cannot express my gratitude for the constant love and support of my wife Lisa, and children Zachary and Eliana. Finally, this chapter was also made possible by the ordered complexity of condensed matter (a.k.a., the science of materials) and the laws of mechanics, which are ignored only at great personal and societal risk!

REFERENCES

[1] Sumner DR, Turner TM, Igloria R, Urban RM, Galante JO. Functional adaptation and ingrowth of bone vary as a function of hip implant stiffness. J Biomech 1998;31:909–17.

[2] Neumann A, Jahnke K. Biomaterials for ossicular chain reconstruction. A Review. Mat wiss u Werkstofftech 2003;34:1052–7.

[3] Nicolson PC, Vogt J. Soft contact lens polymers: an evolution. Biomaterials 2001;22:3273–83.

[4] Middleton JC, Tipton AJ. Synthetic biodegradable polymers as orthopedic devices. Biomaterials 2000;21:2335–46.

[5] Pruitt LA, Chakravartula AM. Mechanics of biomaterials. Cambridge, UK: Cambridge University Press; 2011.

[6] Bowman K. Mechanical behavior of materials. John Wiley and Sons, Inc; 2004.

[7] Dowling NE. Mechanical behavior of materials. Upper Saddle River, NJ: Pearson Prentice Hall; 2012.

[8] Meyers MA, Chawla KK. Mechanical behavior of materials. Cambridge, UK: Cambridge University Press; 2008.

[9] McAfee OC. Interbody fusion cages in reconstructive operations on the spine. J Bone Joint Surg 1999;81A:859–80.

[10] Vadapalli S, Sairyo K, Goel VK, Robon M, Biyani A, Khandha A, et al. Biomechanical rationale for using polyetheretherketone (PEEK) spacers for lumbar interbody fusion – a finite element study. Spine 2006;31:E992–8.

[11] Roeder RK, Smith SM, Conrad TL, Yanchak NJ, Merrill CH, Converse GL. Porous and bioactive PEEK implants for interbody spinal fusion. Adv Mater Process 2009;167:46–8.

[12] ASTM D638-01. Standard test method for tensile properties of plastics. West Conshohocken, PA: American Society for Testing and Materials; 2001.

[13] Chawla KK. Composite materials science and engineering. 2nd ed. New York, NY: Springer; 1998.

[14] Halpin JC. Primer on composite materials analysis. 2nd ed. Lancaster, PA: Technomic Publishing Co; 1992.

[15] Hinchliffe A. Molecular modeling for beginners. 2nd ed. Chichester, UK: John Wiley and Sons Ltd; 2011.

[16] Holzapfel GA. Nonlinear solid mechanics: a continuum approach for engineering. Chichester, UK: John Wiley and Sons Ltd; 2000.

[17] Hollingsworth NT, Wagner DR. Modeling shear behavior of the annulus fibrosus. J Mech Behav Biomed Mater 2011;4:1103–14.

[18] Vincent JFV. Survival of the cheapest. Mater Today 2002;5:28–41.

[19] Chen P-Y, Lin AYM, Lin Y-S, Seki Y, Stokes AG, Peyras J, et al. Structural and mechanical properties of selected biological materials. J Mech Behav Biomed Mater 2008;1:208–26.

[20] Gordon JE. The new science of strong materials. Princeton, NJ: Princeton University Press; 1976.

[21] Pilkey WD, Pilkey DF. Peterson's stress concentration factors. Hoboken, NJ: John Wiley and Sons, Inc; 2008.

[22] Griffith AA. The phenomena of rupture and flow in solids. Phil Trans R Soc 1920; 221A:163–98.

[23] Anderson TL. Fracture mechanics: fundamentals and applications. 3rd ed. Boca Raton, FL: Taylor and Francis Group; 2005.

[24] Sun CT, Jin Z-H. Fracture mechanics. Waltham, MA: Academic Press; 2012.

[25] Tada H, Paris PC, Irwin GR. The stress analysis of cracks handbook. 3rd ed. New York, NY: American Society of Mechanical Engineers; 2000.
[26] Murakami Y. Stress intensity factors handbook. New York, NY: Pergamon Press; 1987.
[27] ASTM E399-09. Standard test method for linear-elastic plane-strain fracture toughness K_{Ic} of metallic materials. West Conshohocken, PA: American Society for Testing and Materials; 2009.
[28] ASTM C1421-10. Standard test method for determination of fracture toughness of advanced ceramics at ambient temperature. West Conshohocken, PA: American Society for Testing and Materials; 2010.
[29] Zimmermann EA, Barth HD, Ritchie RO. The multiscale origins of fracture resistance in human bone and its biological degradation. JOM 2012;64:486–93.
[30] ASTM E1290-08. Standard test method for crack-tip opening displacement (CTOD) fracture toughness measurement. West Conshohocken, PA: American Society for Testing and Materials; 2008.
[31] ASTM E1820-11. Standard test method for measurement of fracture toughness. West Conshohocken, PA: American Society for Testing and Materials; 2011.
[32] Ritchie RO, Koester KJ, Ionova S, Yao W, Lane NE, Ager III JW. Measurement of the toughness of bone: a tutorial with special reference to small animal studies. Bone 2008;43:798–812.
[33] Taylor D, Kuiper J-H. The prediction of stress fractures using a stressed volume concept. J Orthop Res 2001;19:919–26.
[34] Creyke WEC, Sainsbury IEJ, Morrell R. Design with non-ductile materials. New York, NY: Applied Science Publishers; 1982.
[35] ASTM C1239-07. Standard test method for reporting uniaxial strength data and estimating Weibull distribution parameters for advanced ceramics. West Conshohocken, PA: American Society for Testing and Materials; 2007.
[36] ASTM C1683-10. Standard test method for size scaling of tensile strengths using Weibull statistics for advanced ceramics. West Conshohocken, PA: American Society for Testing and Materials; 2010.
[37] Turner CH, Roeder RK, Wieczorek A, Foroud T, Liu G, Peacock M. Variability in skeletal mass, structure, and biomechanical properties among inbred strains of rats. J Bone Miner Res 2001;16:1532–9.
[38] Oberg E. Machinery's handbook. 28th ed. New York, NY: Industrial Press; 2008.
[39] Davis JR. Metals handbook. In: Machining, vol. 16. Materials Park, OH: ASM International; 1989.
[40] Marinescu ID. Handbook of advanced ceramics machining. Boca Raton, FL: CRC Press; 2006.
[41] ASTM D3641-12. Standard practice for injection molding test specimens of thermoplastic molding and extrusion materials. West Conshohocken, PA: American Society for Testing and Materials; 2012.
[42] ASTM F451-08. Standard specification for acrylic bone cement. West Conshohocken, PA: American Society for Testing and Materials; 2008.
[43] Kalpakjian S, Schmid SR. Manufacturing processes for engineering materials. 5th ed. Upper Saddle River, NJ: Pearson Education; 2008.
[44] ASTM C1161-02. Standard test method for flexural strength of advanced ceramics at ambient temperature. West Conshohocken, PA: American Society for Testing and Materials; 2002.
[45] ASTM E4-10. Standard test method for force verification of testing machines. West Conshohocken, PA: American Society for Testing and Materials; 2010.

[46] ASTM E251-92. Standard test methods for performance characteristics of metallic bonded resistance strain gages. West Conshohocken, PA: American Society for Testing and Materials; 2009.
[47] Shelton JC, Orr JF, editors. Optical measurement methods in biomechanics. London, UK: Chapman and Hall; 1996.
[48] Dunn PF. Measurement and data analysis for engineering and science. 2nd ed. Boca Raton, FL: CRC Press; 2010.
[49] Morris AS, Langari R. Measurement and instrumentation: theory and application. Waltham, MA: Academic Press; 2011.
[50] Davis JR. Tensile testing. 2nd ed. Materials Park, OH: ASM International; 2004.
[51] ISO 11405. Testing of adhesion to tooth structure. International organization for standardization. Geneva: Switzerland; 2003.
[52] Sano H, Shono T, Sonoda H, Takatsu T, Ciucchi B, Carvalho R, et al. Relationships between surface area for adhesion and tensile bond strength – evaluation of a micro-tensile bond test. Dent Mater 1994;10:236–40.
[53] ASTM C1424-10. Standard test method for monotonic compressive strength of advanced ceramics at ambient temperature. West Conshohocken, PA: American Society for Testing and Materials; 2010.
[54] ASTM D695-10. Standard test method for compressive properties of rigid plastics. West Conshohocken, PA: American Society for Testing and Materials; 2010.
[55] ASTM D1621-10. Standard test method for compressive properties of rigid cellular plastics. West Conshohocken, PA: American Society for Testing and Materials; 2010.
[56] Gibson LJ, Ashby MF. Cellular solids: structure and properties. Cambridge, UK: Cambridge University Press; 2001.
[57] Sacks MS, Sun W. Multiaxial mechanical behavior of biological materials. Annu Rev Biomed Eng 2003;5:251–84.
[58] Hollingsworth NT, Wagner DR. The stress and strain states of the posterior annulus fibrosus under flexion. Spine 2012;37:E1134–9.
[59] Roeder R, Wolfe J, Lianakis N, Hinson T, Geddes LA, Obermiller J. Compliance, elastic modulus, and burst pressure of small-intestine submucosa (SIS), small-diameter vascular grafts. J Biomed Mater Res 1999;47:65–70.
[60] Jaekel DJ, MacDonald DW, Kurtz SM. Characterization of PEEK biomaterials using the small punch test. J Mech Behav Biomed Mater 2011;4:1275–82.
[61] ASTM F2183-02. Standard test method for small punch testing of ultra-high molecular weight polyethylene used in surgical implants. West Conshohocken, PA: American Society for Testing and Materials; 2002.
[62] ASTM F543-07. Standard specification and test methods for metallic medical bone screws. West Conshohocken, PA: American Society for Testing and Materials; 2007.
[63] Reilly DT, Burstein AH. The elastic and ultimate properties of compact bone tissue. J Biomech 1975;8:393–405.
[64] Dong XN, Guo XE. The dependence of transversely isotropic elasticity of human femoral cortical bone on porosity. J Biomech 2004;37:1281–7.
[65] ASTM C496-11. Standard test method for splitting tensile strength of cylindrical concrete specimens. West Conshohocken, PA: American Society for Testing and Materials; 2011.
[66] Procopio AT, Zavaliangos A, Cunningham JC. Analysis of the diametral compression test and the applicability to plastically deforming materials. J Mater Sci 2003;38:3629–39.
[67] Pittet C, Lemaître J. Mechanical characterization of brushite cements: a Mohr circles' approach. J Biomed Mater Res Appl Biomater 2000;53:769–80.

[68] Ferracane JL, Greener EH. The effect of resin formulation on the degree of conversion and mechanical properties of dental restorative resins. J Biomed Mater Res 1986;20:121–31.

[69] Gere JM, Goodno BJ. Mechanics of materials. 8th ed. Stamford, CT: Cengage Learning; 2013.

[70] ASTM D790-10. Standard test methods for flexural properties of unreinforced and reinforced plastics and electrical insulating materials. West Conshohocken, PA: American Society for Testing and Materials; 2010.

[71] ASTM D6272-10. Standard test method for flexural properties of unreinforced and reinforced plastics and electrical insulating materials by four-point bending. West Conshohocken, PA: American Society for Testing and Materials; 2010.

[72] ASTM D5379-05. Standard test method for shear properties of composite materials by V-notched beam method. West Conshohocken, PA: American Society for Testing and Materials; 2005.

[73] Lee S, Munro M. Evaluation of in-plane shear test methods for advanced composite materials by the decision analysis technique. Composites 1986;17:13–22.

[74] Adams DF. A comparison of shear test methods. High Perform Compos, www.compositesworld.com September 2005.

[75] Turner CH, Wang T, Burr DB. Shear strength and fatigue properties of human cortical bone determined from pure shear tests. Calcif Tissue Int 2001;69:373–8.

[76] An YH, Draughn RA. Mechanical testing of bone and the bone-implant. Boca Raton, FL: Interface CRC Press; 2000.

[77] Dhert WJA, Verheyen CCPM, Braak LH, de Wijn JR, Klein CPAT, de Groot K, et al. A finite element analysis of the push-out test: influence of test conditions. J Biomed Mater Res 1992;26A:119–30.

[78] Berzins A, Shah B, Weinans H, Sumner DR. Nondestructive measurements of implant-bone interface shear modulus and effects of implant geometry in pull-out tests. J Biomed Mater Res 1997;34A:337–40.

[79] Siewert TA, Manahan Sr. MP, editors. Pendulum impact testing: a century of progress. STP 1380. West Conshohocken, PA: American Society for Testing and Materials; 2000.

[80] ASTM E23-07. Standard test methods for determining the notched bar impact testing of metallic materials. West Conshohocken, PA: American Society for Testing and Materials; 2007.

[81] ASTM D256-10. Standard test methods for determining the Izod pendulum impact resistance of plastics. West Conshohocken, PA: American Society for Testing and Materials; 2010.

[82] ASTM D6110-10. Standard test methods for determining the Charpy impact resistance of notched specimens of plastics. West Conshohocken, PA: American Society for Testing and Materials; 2010.

[83] ASTM D5420-10. Standard test method for impact resistance of flat, rigid plastic specimen by means of a striker impacted by a falling weight (Gardner impact). West Conshohocken, PA: American Society for Testing and Materials; 2010.

[84] ASTM D1709-09. Standard test methods for impact resistance of plastic film by the free falling dart method. West Conshohocken, PA: American Society for Testing and Materials; 2009.

[85] Mott PH, Twigg JN, Roland DF, Schrader HS, Pathak JA, Roland CM. High-speed tensile test instrument. Rev Sci Instrum 2007;78:045105.

[86] Chen W, Song B. Split Hopkinson (Kolsky) bar design, testing and applications. New York, NY: Springer; 2010.

[87] Meyers MA. Dynamic behavior of materials. New York, NY: John Wiley and Sons; 1994.

[88] Ravi-Chandar K. Dynamic fracture. Amsterdam, The Netherlands: Elsevier; 2004.

[89] Kulin RM, Jiang F, Vecchio KS. Effects of age and loading rate on equine cortical bone failure. J Mech Behav Biomed Mater 2011;4:57–75.

[90] Pilcher A, Wang X, Kaltz Z, Garrison JG, Niebur GL, Mason J, et al. High strain rate testing of bovine trabecular bone. J Biomech Eng 2011;132:081012. p. 7.

[91] ASTM E2448-11. Standard test method for determining the superplastic properties of metallic sheet materials. West Conshohocken, PA: American Society for Testing and Materials; 2011.

[92] ASTM E2712-09. Standard test methods for forming superplastic metallic sheet. West Conshohocken, PA: American Society for Testing and Materials; 2009.

[93] ASTM E139-11. Standard test methods for conducting creep, creep-rupture, and stress-rupture tests of metallic materials. West Conshohocken, PA: American Society for Testing and Materials; 2011.

[94] ASTM C1291-00. Standard test method for elevated temperature tensile creep strain, creep strain rate, and creep time-to-failure for advanced monolithic ceramics. West Conshohocken, PA: American Society for Testing and Materials; 2010.

[95] ASTM D2990-09. Standard test methods for tensile, compressive, and flexural creep and creep-rupture of plastics. West Conshohocken, PA: American Society for Testing and Materials; 2009.

[96] Ward IM, Sweeney J. The mechanical properties of solid polymers. 2nd ed. Chichester, UK: John Wiley and Sons; 2004.

[97] Shepherd TN, Zhang J, Ovaert TO, Roeder RK, Niebur GL. Direct comparison of nano-indentation and macroscopic measurements of bone viscoelasticity. J Mech Behav Biomed Mater 2011;4:2055–62.

[98] ASTM E328-02. Standard test methods for stress relaxation for materials and structures. West Conshohocken, PA: American Society for Testing and Materials; 2008.

[99] ASTM D6048-07. Standard practice for stress relaxation testing of raw rubber, unvulcanized rubber compounds, and thermoplastic elastomers. West Conshohocken, PA: American Society for Testing and Materials; 2007.

[100] ASTM E467-08. Standard practice for verification of constant amplitude dynamic forces in an axial fatigue testing system. West Conshohocken, PA: American Society for Testing and Materials; 2008.

[101] ASTM E1942-98. Standard guide for evaluating data acquisition systems used in cyclic fatigue and fracture mechanics testing. West Conshohocken, PA: American Society for Testing and Materials; 2010.

[102] ASTM C1361-10. Standard practice for constant amplitude, axial, tension-tension cyclic fatigue of advanced ceramics at ambient temperature. West Conshohocken, PA: American Society for Testing and Materials; 2010.

[103] ASTM F2118-10. Standard test method for constant amplitude of force controlled fatigue testing of acrylic bone cement materials. West Conshohocken, PA: American Society for Testing and Materials; 2010.

[104] ASTM E466-07. Standard practice for conducting force controlled constant amplitude axial fatigue tests of metallic materials. West Conshohocken, PA: American Society for Testing and Materials; 2007.

[105] ASTM E468-11. Standard practice for presentation of constant amplitude fatigue test results for metallic materials. West Conshohocken, PA: American Society for Testing and Materials; 2011.

[106] ASTM D7791-12. Standard test method for uniaxial fatigue properties of plastics. West Conshohocken, PA: American Society for Testing and Materials; 2012.

[107] ASTM D3479-96. Standard test method for tension-tension fatigue of polymer matrix composite materials. West Conshohocken, PA: American Society for Testing and Materials; 2007.

[108] ASTM E606-04. Standard practice for strain-controlled fatigue testing. West Conshohocken, PA: American Society for Testing and Materials; 2004.

[109] Cook SD, Georgette FS, Skinner HB, Haddad Jr RJ. Fatigue properties of carbon- and porous-coated Ti-6Al-4V alloy. J Biomed Mater Res 1984;18:497–512.

[110] ASTM D7774-12. Standard test method for flexural fatigue properties of plastics. West Conshohocken, PA: American Society for Testing and Materials; 2012.

[111] Kane RJ, Converse GL, Roeder RK. Effects of the reinforcement morphology on the fatigue properties of hydroxyapatite reinforced polymers. J Mech Behav Biomed Mater 2008;1:261–8.

[112] Landrigan MD, Roeder RK. Systematic error in mechanical measures of damage during four-point bending fatigue of cortical bone. J Biomech 2009;4:1212–7.

[113] ASTM F2477-07. Standard test methods for in vitro pulsatile durability testing of vascular stents. West Conshohocken, PA: American Society for Testing and Materials; 2007.

[114] Pelton AR, Schroeder V, Mitchell MR, Gong X-Y, Barney M, Robertson SW. Fatigue and durability of nitinol stents. J Mech Behav Biomed Mater 2008;1:153–64.

[115] Runciman A, Xu D, Pelton AR, Ritchie RO. An equivalent strain/Coffin-Manson approach to multiaxial fatigue and life prediction in superelastic Nitinol medical devices. Biomaterials 2011;32:4987–93.

[116] ASTM E739-10. Standard practice for statistical analysis of linear or linearized stress-life (*S-N*) and strain-life (ε-*N*) fatigue data. West Conshohocken, PA: American Society for Testing and Materials; 2010.

[117] Kane RJ, Yue W, Mason JJ, Roeder RK. Improved fatigue life of acrylic bone cements reinforced with zirconia fibers. J Mech Behav Biomed Mater 2010;3:504–11.

[118] Niinomi M. Mechanical biocompatibilities of titanium alloys for biomedical applications. J Mech Behav Biomed Mater 2008;1:30–42.

[119] Sobieraj MC, Murphy JE, Brinkman JG, Kurtz SM, Rimnac CM. Notched fatigue behavior of PEEK. Biomaterials 2010;31:9156–62.

[120] Pattin CA, Caler WE, Carter DR. Cyclic mechanical property degradation during fatigue loading of cortical bone. J Biomech 1996;29:69–79.

[121] Cotton JR, Winwood K, Zioupos P, Taylor M. Damage rate is a predictor of fatigue life and creep strain rate in tensile human cortical bone samples. J Biomech Eng 2005;127:213–9.

[122] Cotton JR, Zioupos P, Winwood K, Taylor M. Analysis of creep strain during tensile fatigue of cortical bone. J Biomech 2003;36:943–9.

[123] ASTM E2714-09. Standard test method for creep-fatigue testing. West Conshohocken, PA: American Society for Testing and Materials; 2009.

[124] ASTM E647-11. Standard test method for measurement of fatigue crack growth rates. West Conshohocken, PA: American Society for Testing and Materials; 2011.

[125] Connors WC. Fatigue striation spacing analysis. Mater Char 1994;33:245–53.

[126] Bajaj D, Sundaram N, Arola D. An examination of fatigue striations in human dentin: in vitro and in vivo. J Biomed Mater Res 2008;85B:149–59.

[127] ASTM E2760-10. Standard test method for creep-fatigue crack growth testing. West Conshohocken, PA: American Society for Testing and Materials; 2010.

[128] Menard KP. Dynamic mechanical analysis. 2nd ed. Boca Raton, FL: CRC Press; 2008.

[129] ASTM D4065-06. Standard practice for plastics: dynamic mechanical properties: determination and report of procedures. West Conshohocken, PA: American Society for Testing and Materials; 2006.

[130] Hellier CJ. Handbook of nondestructive evaluation. New York, NY: McGraw-Hill; 2001.

[131] Herrmann K. Hardness testing: principles and applications. Materials Park, OH: ASM International; 2011.

[132] ASTM E10-12. Standard test method for Brinell hardness of metallic materials. West Conshohocken, PA: American Society for Testing and Materials; 2012.

[133] ASTM E18-11. Standard test methods for Rockwell hardness of metallic materials. West Conshohocken, PA: American Society for Testing and Materials; 2011.

[134] ASTM D785-08. Standard test method for Rockwell hardness of plastics and electrical insulating materials. West Conshohocken, PA: American Society for Testing and Materials; 2008.

[135] ASTM E140-07. Standard hardness conversion tables for metals: relationship among Brinell hardness, Vickers hardness, Rockwell hardness, Superficial hardness, Knoop hardness, and Scleroscope hardness. West Conshohocken, PA: American Society for Testing and Materials; 2007.

[136] ASTM E384-11. Standard test method for Knoop and Vickers hardness of materials. West Conshohocken, PA: American Society for Testing and Materials; 2011.

[137] ASTM C1326-08. Standard test method for Knoop indentation hardness of advanced ceramics. West Conshohocken, PA: American Society for Testing and Materials; 2008.

[138] ASTM C1327-08. Standard test method for Vickers indentation hardness of advanced ceramics. West Conshohocken, PA: American Society for Testing and Materials; 2008.

[139] Oliver WC, Pharr GM. Measurement of hardness and elastic modulus by instrumented indentation: advances in understanding and refinements to methodology. J Mater Res 2004;19:3–20.

[140] Fischer-Cripps AC. Nanoindentation. 3rd ed. New York, NY: Springer; 2011.

[141] Ebenstein DM, Pruitt LA. Nanoindentation of biological materials. Nano Today 2006;1:26–33.

[142] ASTM D2240-05. Standard test method for rubber property–durometer hardness. West Conshohocken, PA: American Society for Testing and Materials; 2010.

[143] Slaboch CL, Alber MS, Rosen ED, Ovaert TO. Mechano-rheological properties of the murine thrombus determined via nanoindentation and finite element modeling. J Mech Behav Biomed Mater 2012;10:75–86.

[144] Blum MM, Ovaert TC. A novel polyvinyl alcohol hydrogel functionalized with organic boundary lubricant for use a low-friction cartilage substitute: synthesis, physical/chemical, mechanical, and friction characterization. J Biomed Mater Res 2012;100B:1755–63.

[145] Kruzic JJ, Kim DK, Koester KJ, Ritchie RO. Indentation techniques for evaluating the fracture toughness of biomaterials and hard tissues. J Mech Behav Biomed Mater 2009;2:384–95.

[146] Anstis GR, Chantikul P, Lawn BR, Marshall DB. A critical evaluation of indentation techniques for measuring fracture toughness: I, direct crack measurements. J Am Ceram Soc 1981;64:533–8.

[147] Marshall DB, Noma T, Evans AG. A simple method for determining elastic-modulus-to-hardness ratios using Knoop indentation measurements. J Am Ceram Soc 1982;65:C175–6.

[148] Ensminger D, Bond LJ. Ultrasonics: fundamentals, technology, and applications. 3rd ed. Boca Raton, FL: CRC Press; 2011.

[149] Van Buskirk WC, Cowin SC, Ward RN. Ultrasonic measurement of orthotropic elastic constants of bovine femoral bone. J Biomech Eng 1981;103:67–72.

[150] Ashman RB, Cowin SC, Van Buskirk WC, Rice JC. A continuous wave technique for the measurement of the elastic properties of cortical bone. J Biomech 1984;17:349–61.

[151] Kohles SS, Bowers JR, Vailas AC, Vanderby Jr R. Ultrasonic wave velocity measurement in small polymeric and cortical bone specimens. J Biomech Eng 1997;119:232–6.
[152] Espinoza Orías AA, Deuerling JM, Landrigan MD, Renaud JE, Roeder RK. Anatomic variation in the elastic anisotropy of cortical bone tissue in the human femur. J Mech Behav Biomed Mater 2009;2:255–63.
[153] Hofmann T, Heyroth F, Meinhard H, Franzel W, Raum K. Assessment of composition and anisotropic elastic properties of secondary osteon lamellae. J Biomech 2006;39:2282–94.
[154] Bhardwaj MC. High-resolution ultrasonic characterization. Ceram Bull 1990;69:1490–6.
[155] ASTM C373-88. Standard test method for water absorption, bulk density, apparent porosity, and apparent specific gravity of fired Whiteware products. West Conshohocken, PA: American Society for Testing and Materials; 2006.
[156] ASTM D792-08. Standard test methods for density and specific gravity (relative density) of plastics by displacement. West Conshohocken, PA: American Society for Testing and Materials; 2008.
[157] ASTM D618-08. Standard practice of conditioning plastics for testing. West Conshohocken, PA: American Society for Testing and Materials; 2008.
[158] ASTM F2516-07. Standard test method for tension testing of nickel-titanium superelastic materials. West Conshohocken, PA: American Society for Testing and Materials; 2007.
[159] ASTM F1634-95. Standard practice for in-vitro environmental conditioning of polymer matrix composite materials and implants. West Conshohocken, PA: American Society for Testing and Materials; 2008.
[160] ASTM F1635-11. Standard test method for in vitro degradation testing of hydrolytically degradable polymer resins and fabricated forms for surgical implants. West Conshohocken, PA: American Society for Testing and Materials; 2011.
[161] ASTM F2150-07. Standard guide for characterization and testing of biomaterial scaffolds used in tissue-engineered medical products. West Conshohocken, PA: American Society for Testing and Materials; 2007.
[162] Nyman JS, Roy A, Shen X, Acuna RL, Tyler JH, Wang X. The influence of water removal on the strength and toughness of cortical bone. J Biomech 2006;39:931–8.
[163] Harley BA, Leung JH, Silva ECCM, Gibson LJ. Mechanical characterization of collagen-glycosaminoglycan scaffolds. Acta Biomaterialia. 2007;3:463–74.
[164] Schoen FJ, Levy RJ. Calcification of tissue heart valve substitutes: progress toward understanding and prevention. Ann Thorac Surg 2005;79:1072–80.
[165] Asgian CM, Gilbertson LN, Blessing NN, Crowninshield RD. Environmentally induced fracture of polysulfone in lipids. Trans. of the 15th annual meeting of the society for biomaterials. Birmingham, AL, 17; 1989.
[166] Trentacosta JD, Cheban JC. Lipid sensitivity of polyaryl-ether-ketones and polysulfone. Trans Orthop Res Soc 1995;41:783.
[167] Kurtz SM, Muratoglu OK, Evans M, Edidin AA. Advances in the processing, sterilization, and crosslinking of ultra-high molecular weight polyethylene for total joint arthroplasty. Biomaterials 1999;20:1659–88.
[168] Pruitt LA. Deformation, yielding, fracture and fatigue behavior of conventional and highly cross-linked ultra high molecular weight polyethylene. Biomaterials 2005;26:905–15.
[169] NIST/SEMATECH e-Handboook of Statistical Methods. National Institute of Standards and Technology. Gaithersberg, MD, http://www.itl.nist.gov/div898/handbook/; 2012.
[170] McKillup S. Statistics explained: an introductory guide for life scientists. Cambridge, UK: Cambridge University Press; 2005.

Chapter 4

Surface Characterization of Biomaterials

Huaiyu Wang and Paul K. Chu
Department of Physics and Materials Science, City University of Hong Kong, Kowloon, Hong Kong, China

Chapter Outline

Characterization of Biomaterials. http://dx.doi.org/10.1016/B978-0-12-415800-9.00004-8

4.1. X-RAY PHOTOELECTRON SPECTROSCOPY

X-ray photoelectron spectroscopy (XPS), which is also called electron spectroscopy for chemical analysis, is the most widely used analytical technique to monitor the surface chemistry of solid materials due to its simplicity, flexibility, and sound theoretical basis. The typical XPS instrument includes an ultra-high vacuum system, X-ray source, electron energy analyzer, and data acquisition system. In XPS, the sample surface is irradiated by monochromatic X-ray and the emitted photoelectrons are detected. Figure 4.1 describes the general mechanism of photoelectron creation [1,2] and the energy of the photoelectron is given by:

$$E_B = h\nu - \mathrm{KE},$$

where E_B is the binding energy of the electron in the atom, $h\nu$ is the energy of the X-ray source (a known value in the experiment), and KE is the kinetic energy of the emitted electron that is measured. E_B is usually expressed in electron volts (eV) and $1\ \mathrm{eV} = 1.6 \times 10^{-19}$ J. Since there are different electrons and binding energies in an atom, each element produces a set of unique peaks in the photoelectron spectrum and the peak intensity is a direct measure of the elemental concentration. As XPS is extremely surface sensitive, surface contamination can lead to stray results. XPS can provide qualitative and quantitative information on all elements except hydrogen and helium and furthermore, the shape of each peak and the exact binding energy can be slightly altered by the chemical state of the emitting atom. Hence, XPS can provide chemical bonding information as well.

XPS is typically performed by first taking a survey scan covering a range of 1000 eV and then, smaller energy ranges indicative of specific features are subsequently scanned in higher resolution. Survey scans are often in low resolution and are used to identify as well as quantify in terms of atomic percentage of major elements present on the materials surface. The x-axis is generally the 'binding energy' and the y-axis is typically 'intensity' or 'number of counts'. Readers are referred to excellent papers in the literature on how to identify and quantify elements from survey XPS data [3–10]. As an example, the typical survey spectra acquired from biomedical polytetrafluoroethylene films before and after plasma surface modification are illustrated in Fig. 4.2a

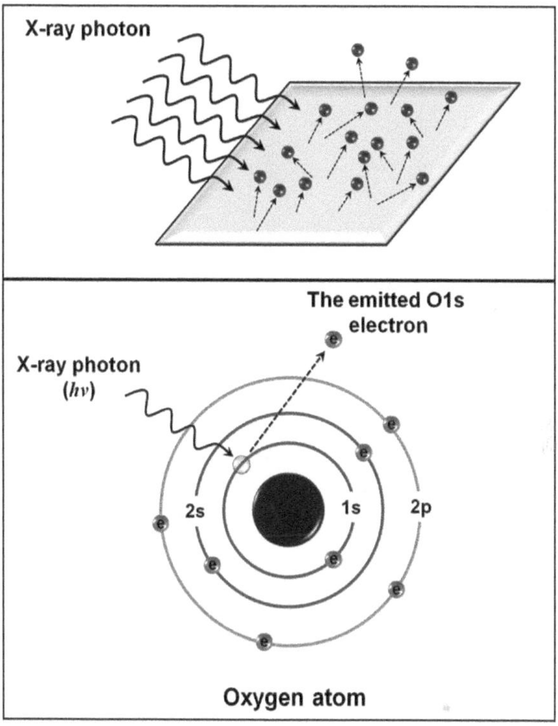

FIGURE 4.1 (Top) A surface irradiated by X-ray photons resulting in emission of photoelectrons and (bottom) the X-ray photon transfers its energy to a core-level electron imparting enough energy for the electron to leave the atom.

and b. As shown in the survey spectra, there are photoemission peaks associated with core-level photoionization events and X-ray-induced Auger electron peaks. Based on XPS handbooks [11,12], the photoelectron peak locations can be identified readily. The Auger lines are also usually listed in photoelectron peak tabulations and they can be readily distinguished from those of photoelectrons by changing the incident X-ray energy, for instance, using a Mg Kα source instead of an Al Kα source. The kinetic energies of Auger lines are invariable with incoming X-ray energy, whereas those of photoelectrons shift by the energy difference of the two X-ray sources. To obtain chemical bonding information, high-resolution scans are performed to cover a range of typically 20 eV or less and there are many associated publications in the literature [7–10,13]. Figure 4.2c and d displays the high-resolution C1s spectra acquired from the samples used to generate Fig. 4.2a and b, respectively. The spectrum in Fig. 4.2d is composed of a number of sub-peaks that convey chemical-state information about the carbon atom and details about the deconvolution can be

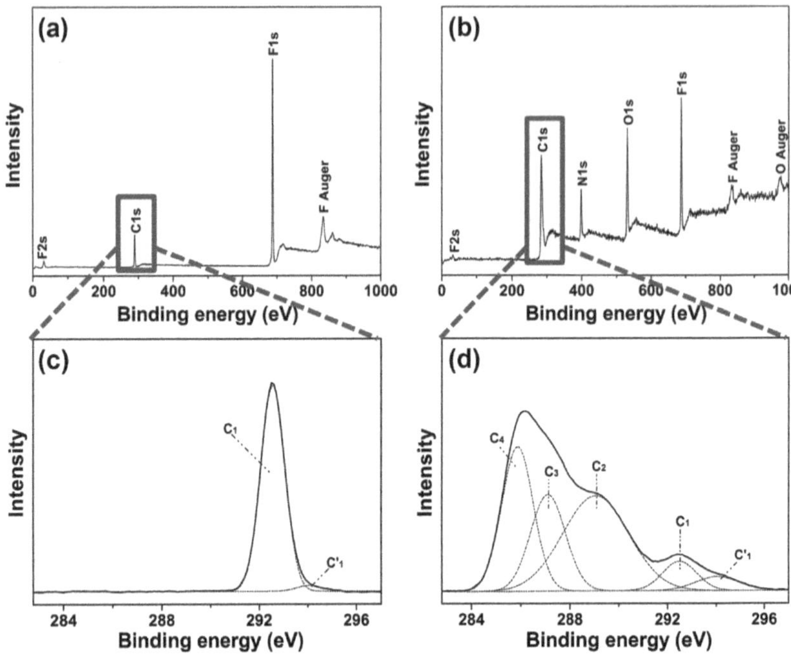

FIGURE 4.2 (a, b) XPS survey spectra and (c, d) high-resolution C1s spectra (c, d) acquired from biomedical polytetrafluoroethylene (PTFE) films before (a, c) and after (b, d) nitrogen/ammonia plasma immersion ion implantation: C_1 (F–C*–F), C'_1 (C_1 at the end of polymer chains), C_2 (–C*=O, –C*=NH, –C*–F), C_3 (–C*–NH_2, –C*–OH), and C_4 (–C*–H, –C*–C–).

found in the literature [14–18]. In addition to organic biomaterials, surface chemical information from inorganic biomaterials or inorganic coatings on organic biomaterials can be obtained [19–22]. However, the spectral features of other elements may be more complex. Heavier elements have other electronic orbitals such as p, d, and f and magnetic interaction between the electron spin (up or down) and orbital angular momentum may split the degenerate state into two components producing spin-orbit doublets. Figure 4.3 depicts the high-resolution Ti2p spectra acquired from TiO_2 before and after ultraviolet ozone treatments [22]. The Ti2p signals consist of both the $2p_{1/2}$ and $2p_{3/2}$ components with the separation resulting from spin-orbit coupling. In other cases involving transition elements, multiplet splitting can be observed.

Since only electrons emitted from the outermost surface of a few nanometres can escape without energy loss and be quantified, XPS is a much more surface sensitive technique than attenuated total reflection Fourier transform infrared (ATR-FTIR) spectroscopy, in which the IR beam can penetrate from several hundred nanometres to more than 1 μm. In order to collect elemental depth profiles, argon ion sputtering is performed in concert [23,24].

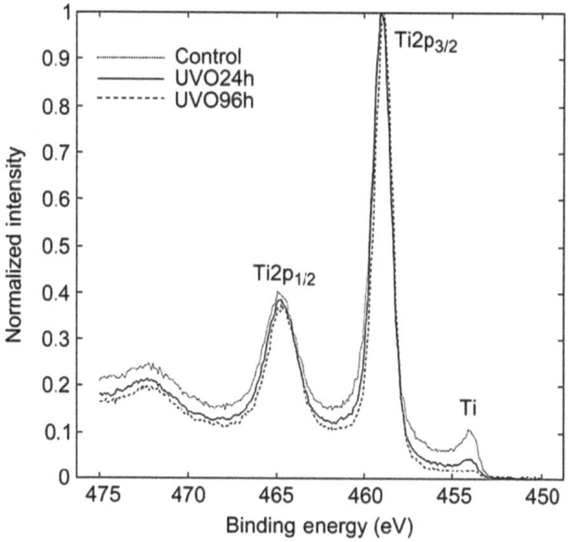

FIGURE 4.3 High-resolution Ti2p spectra obtained from TiO_2 before and after ultraviolet ozone (UVO) treatment. *Reprinted with permission from Ref. [22]. Copyright (2010) Elsevier.*

A representative depth profiling example is shown in Fig. 4.4 for a biomedical NiTi alloy after anodization and hydrothermal treatment. However, since argon ion sputtering can create damage, interpretation of chemical information requires care. Furthermore, ion beam mixing and associated effects such as atomic knock-on and sample roughening degrade the depth resolution, which is worse for a longer sputtering time (that is, deeper crater). Recently, it has been reported that accurate XPS depth profiles can be obtained from organic materials by using C_{60}^+ [25–27] or coronene ($C_{24}H_{12}^+$) sputtering [28]. Figure 4.5 shows the XPS depth profile acquired from a drug-loaded PLA film using coronene as the sputtering ions [28,29].

In addition to depth profiling, recent improvements in the spatial resolution offered by commercial XPS systems enable more accurate and better XPS imaging capabilities [30–33]. Imaging XPS has been applied to biomaterials surface characterization and Fig. 4.6 illustrates an example [34]. As shown in the XPS images in Fig. 4.6, the P, N, and Na components are significantly increased after hybridization.

In conclusion, new instrumentation, technique development, and enhanced data analysis continue to expand the utility of XPS in biomaterials surface characterization. XPS provides survey spectra, high-resolution spectra, elemental depth profiles, and elemental maps using imaging. Although XPS is quite mature, new applications to biomaterials continue to be uncovered and the technique is expected to be a workhorse in biomaterials research, especially organic biomaterials such as polymers.

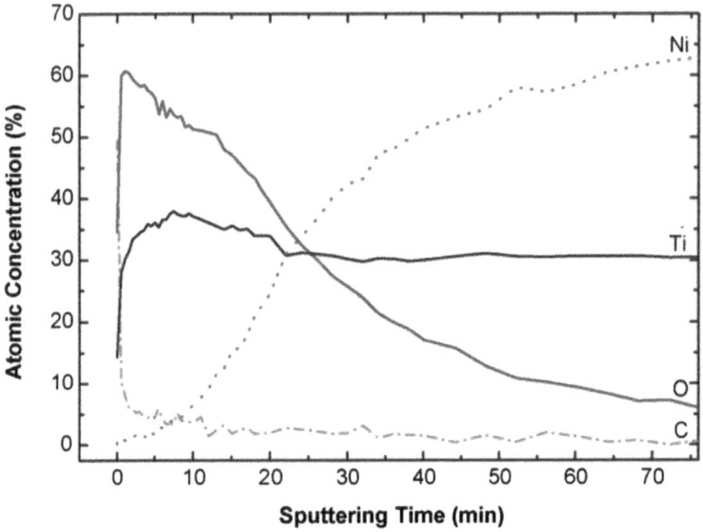

FIGURE 4.4 XPS elemental depth profiles obtained from nickel titanium alloy after anodization and subsequent hydrothermal treatment in water. *Reprinted with permission from Ref. [24]. Copyright (2007) Elsevier.*

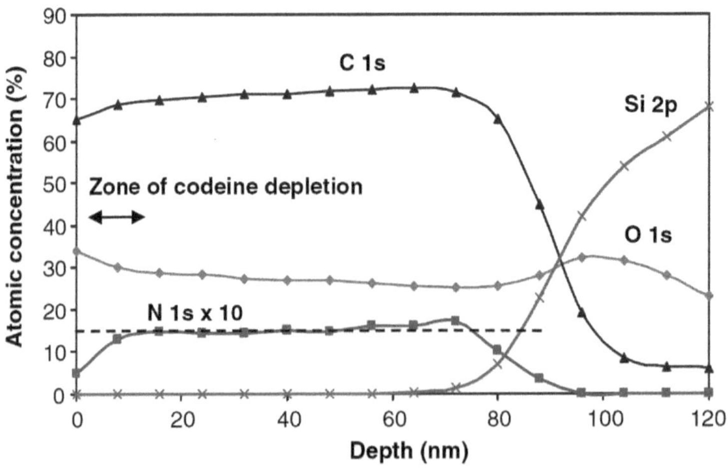

FIGURE 4.5 XPS depth profile acquired with a coronene ion source from a codeine-loaded PLA film showing atomic concentrations of C, O, Si, and N × 10. The N signal is unique to codeine and shows its depth distribution. The dashed line shows the theoretical N (×10) signal assuming uniform drug distribution through the film thickness. *Reprinted with permission from Refs [28,29]. Copyright (2009, 2011) Elsevier.*

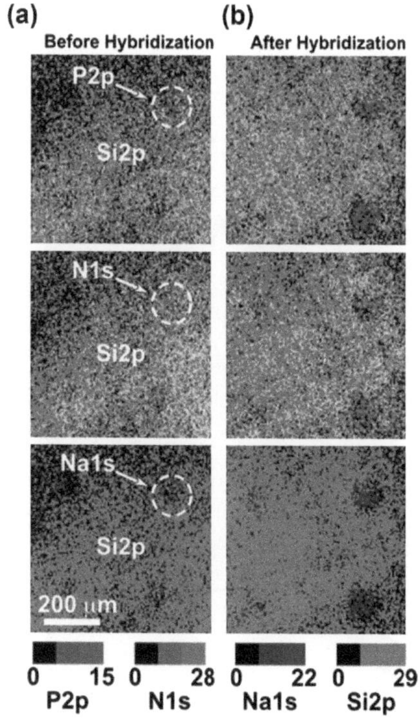

FIGURE 4.6 Superimposed XPS images of phosphorus (P2p), nitrogen (N1s), sodium (Na1s) with the substrate silicon (Si2p) signal intensity images (800 × 800 μm) from printed DNA probes on CodeLink microarray slides (a) before and (b) after target hybridization. *Reprinted with permission from Ref. [34]. Copyright (2007) American Chemical Society.*

4.2. AUGER ELECTRON SPECTROSCOPY

When electrons with energy between 3 and 30 keV bombard a solid surface, the Auger transition described in Fig. 4.7 takes place and Auger electrons carrying information about the host atom are emitted from the surface. The Auger process is related to three electrons at two different levels and the kinetic energy of an auger electron is determined by the energy difference of the singly ionized state and the double ionized final state. For an arbitrary ABC transition in an atom of atomic number z, the Auger electron kinetic energy is given by the difference in the binding energies of energy levels A, B and C [35] as

$$E_{\mathrm{ABC}}(z) = E_{\mathrm{A}}(z) - E_{\mathrm{B}}^{*}(z) - E_{\mathrm{C}}^{*}(z) - W_{\mathrm{s}},$$

where W_{s} is the spectrometer work function, A, B and C are the three energy levels of the Auger process involved (i.e. KLL, LMM, MNN – omitting the sub-levels), and E^{*} is the binding energy of a level in the presence of a core hole and is greater than the binding energy of the same level in a neutral atom. The kinetic energy of Auger electron depends on the atom and elemental information and can be obtained by Auger electron spectroscopy (AES).

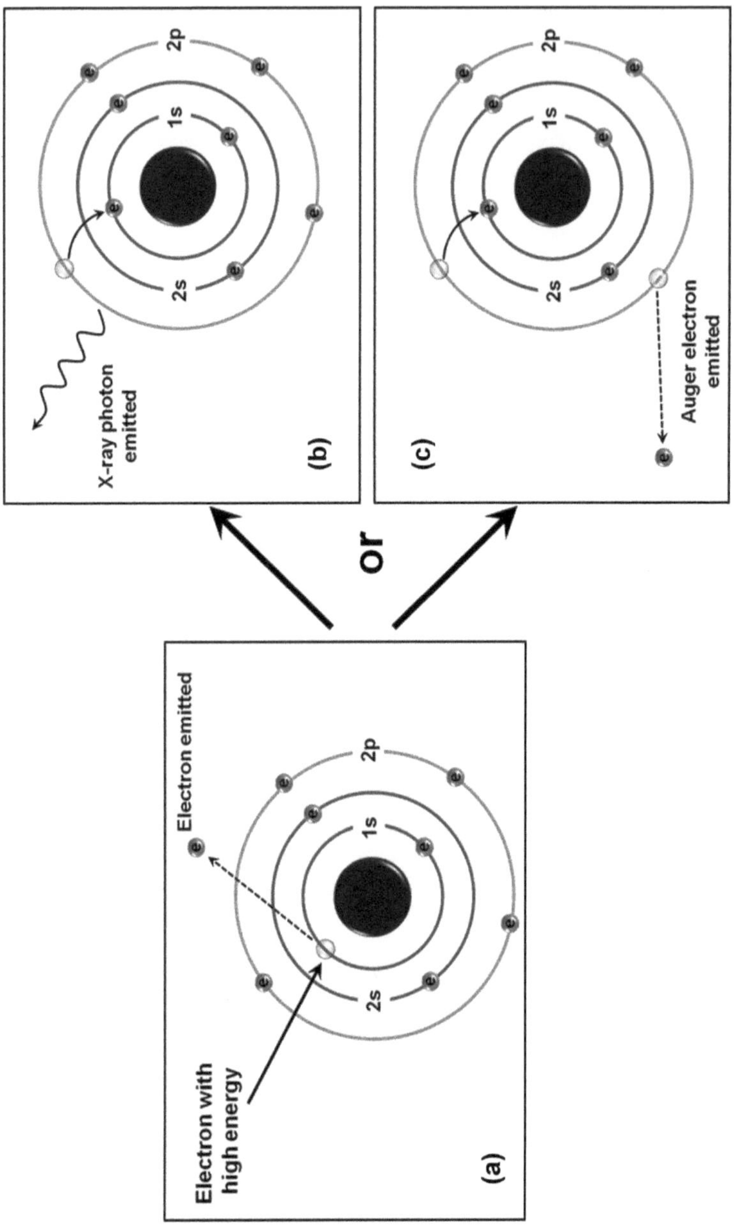

FIGURE 4.7 Schematic diagram illustrating the Auger electron emission process. (a) High-energy electron induces target atom with a core-level electron emission. (b, c) An electron drops from a higher energy level to the vacant core by (b) emitting X-ray or (c) ejecting an Auger electron.

AES is an important analysis tool for surface elemental and chemical identification. Similar to XPS, AES can detect all elements except hydrogen and helium in the outermost surface of a sample and depth profiles can be obtained when combined with ion beam sputtering. However, chemical shifts are less evident in AES compared to XPS although useful bonding information can be obtained in some cases. AES differs from XPS in several aspects. For instance, the electron spot size in AES is much smaller than the X-ray spot size in XPS and so, high-resolution imaging can be accomplished. Coupled with scanning, AES can be used to map the elemental distributions in the near-surface region and in-depth (with ion sputtering). However, because of charging, AES is not versatile as XPS in the analysis of insulating samples such as polymers and electron damage of organic samples can be severe and affects the results. Hence, contrary to XPS, AES is primarily used to study inorganic biomaterials [36–39]. For example, Hanzi et al. [36] treated the biomedical Mg–Y–RE alloy with thermal oxidation in air for different periods of time and as shown in Fig. 4.8, the elemental depth distributions are quite different. As aforementioned, owing to the excellent focusability of electron beams, AES can be used in elemental mapping of inorganic materials at very high spatial resolution [40–42].

4.3. SECONDARY ION MASS SPECTROMETRY (SIMS)

Secondary ion mass spectrometry (SIMS) is a sensitive technique to determine the elemental and molecular contents. Besides popular applications in semiconductors, new SIMS advances such as the time-of-flight technique enable the analysis of organic biomaterials. The surface sensitivity of low-bombardment-energy SIMS is quite high as the sampling depth is only about 1–2 nm compared to that of about 5–10 nm for XPS. Furthermore, it can detect all elements including hydrogen and helium as well as atomic ions and molecular ions at low concentrations. A basic SIMS instrument consists of three main components: the primary ion source, ion filtering system which selects ions with a defined energy, and mass spectrometer in vacuum.

No special sample preparation is required for SIMS, but similar to XPS and AES, the sample surface should be relatively free of adventitious contaminants from storage or transport. In SIMS, the sample surface is bombarded by ions or neutrals (normally between 1 and 40 keV). The momentum transfer initiates a collisional cascade, in which some atomic and molecular species are emitted as secondary ions. Ion mixing, knock-on, and sample roughening limit the depth resolution which depends on several instrumental factors such as primary ion energy, primary ion mass, and incident angle. Typical secondary ions originate approximately from the top two or three layers of the solid and have kinetic energy on the order from several to hundreds of electron volt. After leaving the solid surface, the positive or negative secondary ions are filtered by energy and mass-to-charge (m/z) ratio and detected by an electron multiplier or

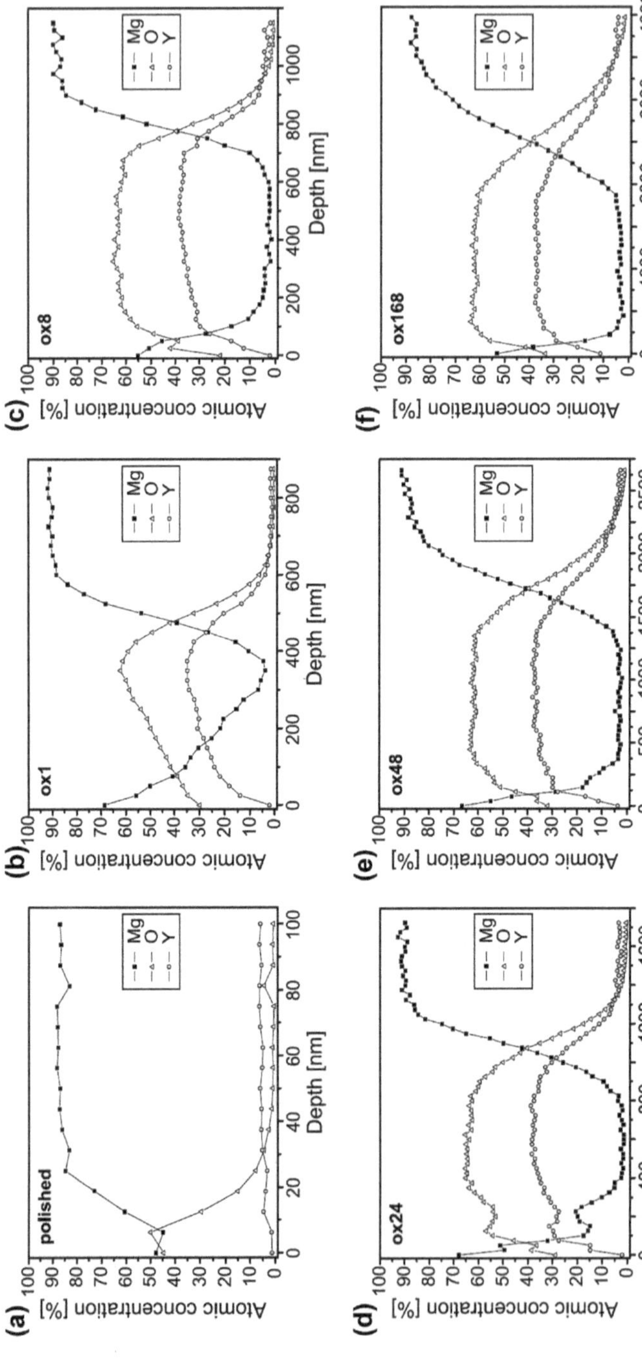

FIGURE 4.8 (a–f) AES depth profiles of Mg–Y–RE alloy: (a) Polished and after thermal oxidation in air at for (b) 1 h, (c) 8 h, (d) 24 h, (e) 48 h, and (f) 168 h. *Reprinted with permission from Ref. [36]. Copyright (2009) Elsevier.*

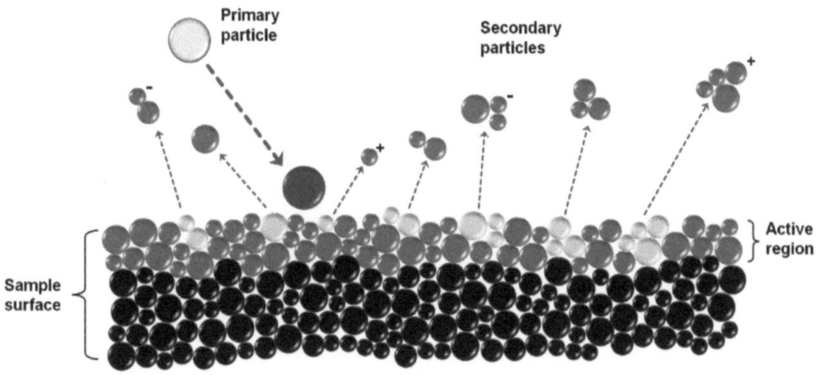

FIGURE 4.9 Schematic illustration of the secondary ion emission process initiated by the impact of a primary particle. Extensive fragmentation occurs near the collision site producing mainly atomic particles. Away from the point of impact collision, the process becomes less energetic resulting in the emission of larger molecular fragments.

image detector composed of micro-channel plates. The secondary ion fragments often impart characteristic information about molecular information and sometimes chemical information. A pictorial presentation of the sputtering process in SIMS is illustrated in Fig. 4.9 and more details can be found in the literature [43–49].

4.3.1. Static SIMS

There are two types of SIMS, dynamic and static. Dynamic SIMS, which is used in depth profiling analysis, is especially suitable for semiconductors. On the other hand, static SIMS uses a low-current primary ion beam (1 nA/cm^2 or less) in order to produce the least amount of surface damage in order to acquire molecular information to disclose the surface chemical structure. To avoid sample damage, there is a dose limit for static SIMS as 10^{13} ions cm^{-2}, but a value $\leq 10^{12}$ ions cm^{-2} is recommended. Static SIMS is a useful technique for biomaterials due to the excellent surface sensitivity and chemical (mass) selectivity. For instance, in protein adsorption experiments, static SIMS not only identifies the type of proteins present but also reveals the conformation, orientation, degree of denaturation, and other characteristics of the adsorbed protein not possible by XPS. Most proteins of interest have dimensions of 4–10 nm and the static SIMS spectrum of an adsorbed protein represents an amino acid assay of the outer 10–15 Å of the protein. Hence, if the distribution of amino acids in a protein is heterogeneous, the relative intensities of the amino acid fragments detected in the static SIMS spectrum are sensitive to the orientation of the protein on the surface and degree of conformational

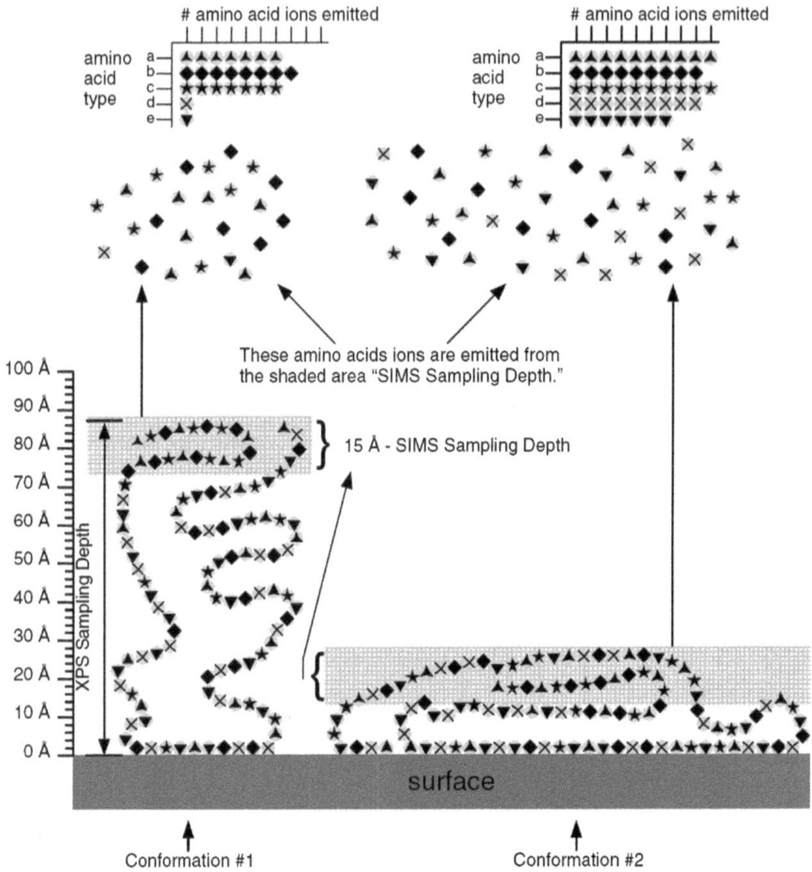

FIGURE 4.10 Schematic diagram of static SIMS investigation of the same protein with different conformation absorbed on the surface. A nonuniform distribution of those amino acids can produce different distributions of the detected amino acid fragments. *Reprinted with permission from Ref. [50]. Copyright (2001) John Wiley and Sons.*

alteration. When a protein adjusts to the surface and changes its conformation or orientation, new regions of the protein with different amino acid compositions are exposed for static SIMS. The schematic diagram of static SIMS study on the conformation of absorbed protein is depicted in Fig. 4.10 [50].

There are many published papers on the use of static SIMS on biomaterials characterization [51–54]. As shown in Fig. 4.11, Salerno et al. [53] compared the positive SIMS spectra acquired from a polymeric biomaterial (PEEK-WC-PU) before and after H_2/NH_3 plasma treatment. The new and more intense mass peaks at m/z of 18 (NH_4^+), 28 (CH_2N^+), and 30 (CH_4N^+) from the modified surface indicate that N-containing functional groups are successfully grafted onto the substrate. However, when studying a more complicated system such as

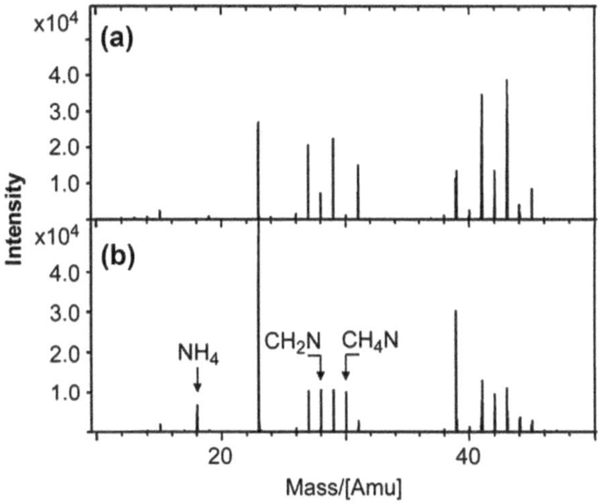

FIGURE 4.11 Positive-ion SIMS spectra of (a) pristine and (b) H_2/NH_3 plasma-modified PEEK-WC-PU membrane in the 10–50 m/z region. *Reprinted with permission from Ref. [53]. Copyright (2009) Elsevier.*

different conformation of the same adsorbed protein or different proteins absorbed on the same substrate, interpretation of all the secondary ions is complex and requires professional expertise. To overcome this difficulty, multivariate data analysis methods such as principal component analysis (PCA) are used to identify the major differences between SIMS spectra. With PCA, the dimensions of the SIMS spectra are reduced to a small number of variables called principal components, which are used to identify the differences among samples. A more rigorous description of the PCA can be found elsewhere [55] and applications of PCA analysis of static SIMS data to biomaterials research are increasing [56–58]. An example is illustrated in Fig. 4.12, in which the SIMS data are acquired from the ECM proteins (laminin, Matrigel, collagen I and collagen IV, respectively) on mica substrates [56]. Figure 4.12a shows that principal component 1 (PC1), which describes 60% of the total variance in the data set, successfully distinguishes collagen IV from the other three proteins. Principal component 2 (PC2) (describing 29% of total variances) separates the two collagens from laminin and Matrigel. Thus, by employing PC1 and PC2, these four kinds of proteins can be well differentiated. Figure 4.12b and c denotes the PCA loadings, which represent the relative abundance of one fragment versus another. As an illustration, a high positive loading for one fragment indicates that this fragment is present in a higher concentration in the sample with a high positive score. Likewise, fragments with high negative loadings indicate that the fragments are present in higher concentrations in samples with high negative scores.

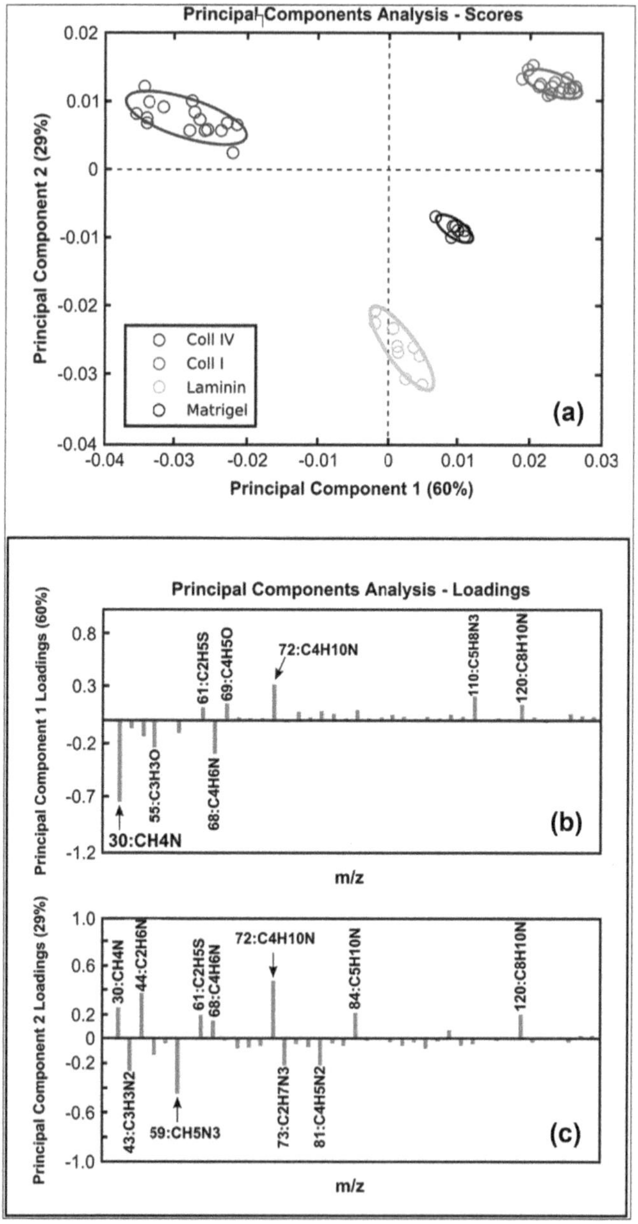

FIGURE 4.12 Principal component analysis (PCA) showing scores and loadings for SIMS data from protein films on mica. The protein samples are: collagen I (blue circles), collagen IV (red circles), laminin (green circles), and Matrigel (black circles). (a) is the PC1 versus PC2 scores plot, (b) shows the PC1 loadings, and (c) shows the PC2 loadings. *Reprinted with permission from Ref. [56]. Copyright (2011) Elsevier.*

On the heels of recent developments of SIMS instrumentations and applications of primary ions with higher secondary ion yields such as Bi^+, Bi_3^+, C_{60}^+, etc., static SIMS can provide chemical maps of biological or other complex materials in high spatial resolution [59–64]. The image resolution can be even higher in association with multivariate analysis methods (PCA and so on) to highlight the differences [65,66]. SIMS imaging is a powerful method to probe materials surface with complex chemistry, even chemical contrast can be verified in the sub-micrometre range [63,64]. For example, Ogaki et al. [64] fabricated chemically complex and patterned substrates (Au-Polymer-Au-SiO_2) by combining different processes including particle self-assembling, plasma eching (optional), plasma polymerization, sputtering, evaporating, and finally ultrasonication. The Au and SiO_2 regions on the substrates are selectively functionalized with fluorinated thiol and chlorinated silane, respectively, to yield local specific signals (F^- for Au and $^{35}Cl^-$/$^{37}Cl^-$ for SiO_2). As shown in the SIMS images in Fig. 4.13, the different regions are identified well by the unique fragments ions generated (HC^-, F^- and $^{35}Cl^-$/$^{37}Cl^-$) with high spatial resolution.

4.3.2. Dynamic SIMS

In dynamic SIMS, a high-current primary ion beam (1 $\mu A\ cm^{-2}$ or greater) bombards the sample to erode the materials surface continuously while the real-time signals are recorded. The signal is plotted against sputtering time (or depth with proper calibration) to disclose the change in the elemental concentration with depth. Dynamic SIMS boasts several advantages such as high-depth resolution, high lateral resolution, excellent sensitivity, and high dynamic range. Unlike semiconductors, conventional dynamic SIMS is not popular in biomaterials characterization, especially for organic or protein absorbed biomaterials as sample damage from the primary ions is too severe resulting in information loss. There have been more applications to inorganic biomaterials [67] and dynamic SIMS can also be used to image cells labelled by different isotopes [68,69].

Polyatomic cluster ions as C_{60}^+ has been applied to SIMS characterization as the resulting damage cross-sections are much less than that of metal and metal cluster ions. This new technique expands the applications of dynamic SIMS to the characterization of organic materials not possible before. By using C_{60}^+ [70,71] or other ions [72] as the primary ions, the molecular depth profiles of the organic substrate can be obtained. As shown in Fig. 4.14 [70], the pure and different peptides deposited on silicon with trehalose films were characterized by dynamic SIMS with C_{60}^+ sputtering. It is clear that the signals of trehalose and the contained peptides remain constant until the entire film is removed and the Si substrate is reached, thus indicating the success of molecular depth profiling. With this new dynamic SIMS technique, 3D images of cells or organic substrates can be obtained without isotope labelling and more details can be found in the literature [73,74].

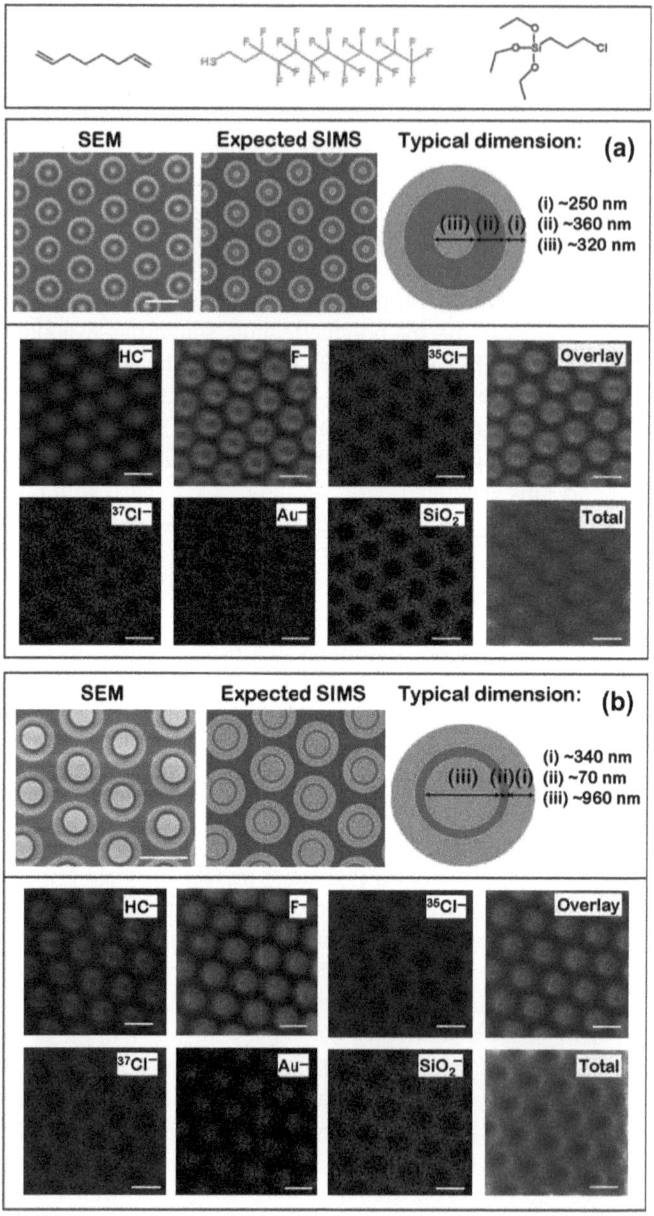

FIGURE 4.13 High-spatial-resolution negative-ion SIMS images of the Au-polymer-Au-SiO_2 surfaces with (a) 'small' Au and (b) 'large' Au central spots. The Au and SiO_2 regions are selectively functionalized with fluorinated thiol (top-center structure in green) and chlorinated silane (top-right structure in blue), respectively, and the monomer for polymer region is octadiene (top-left structure in red). The scale bar is 2 μm. *Reprinted with permission from Ref. [64]. Copyright (2011) John Wiley and Sons.*

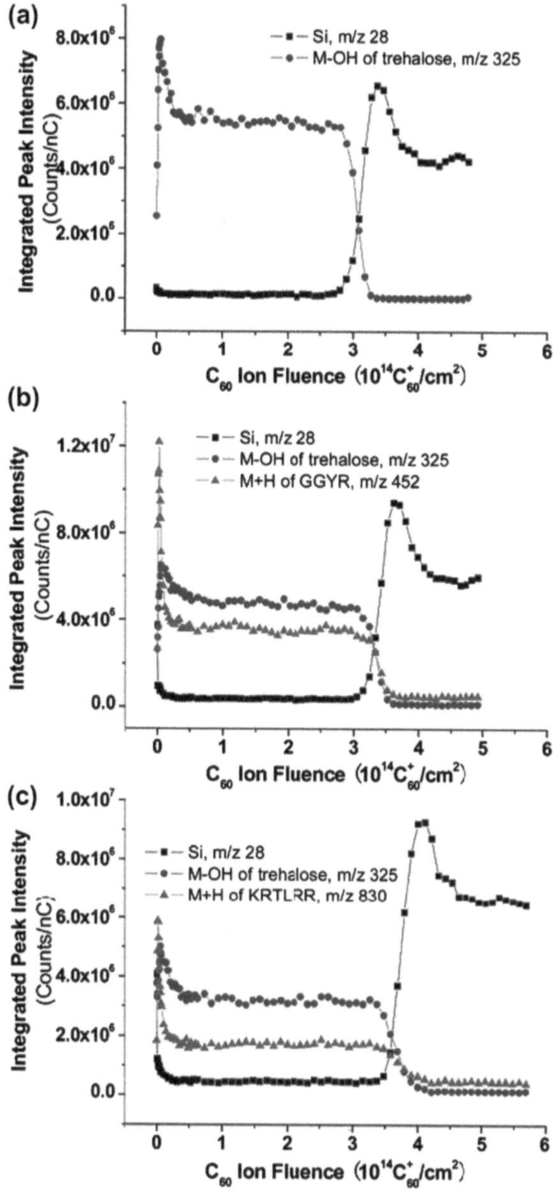

FIGURE 4.14 Secondary-ion signal intensities versus accumulated C_{60}^{+} ion fluence during depth profiling of (a) a pure trehalose film (270 nm), (b) a trehalose film doped with 1% GGYR (Gly–Gly–Tyr–Arg), and (c) a trehalose film doped with 1% KRTLRR (Lys–Arg–Thr–Leu–Arg–Arg). *Reprinted with permission from Ref. [70]. Copyright (2006) American chemical Society.*

In recapitulation, SIMS is a useful technique to obtain elemental and molecular weight information. By using static SIMS, the conformation, orientation, degree of denaturation, and other characteristics of adsorbed proteins can be studied. With the aid of multivariate data analysis, the major differences between complex SIMS spectra can be readily identified. Static SIMS is also effectively used to map biomaterials surface with complex chemical states with a spatial resolution better than that of XPS imaging. Molecular depth profiles and even 3D images can be obtained from organic biomaterials by dynamic SIMS using C_{60}^{+} as the sputtering ions. Although applications of SIMS to biomaterials are not as prevalent as those of XPS, the impact of SIMS is increasing with new technological development.

4.4. SURFACE MATRIX-ASSISTED LASER DESORPTION IONIZATION MASS SPECTROMETRY

Surface matrix-assisted laser desorption ionization mass spectrometry (Surface-MALDI-MS) is an extension of conventional MALDI-MS. In conventional MALDI-MS, a solution consisting of the analyte and a suitable matrix is laid on an MALDI plate, which is usually a specially designed metal plate. After the solvent evaporates, the analyte and matrix are co-crystallized. As the matrix usually has strong optical absorption in either the UV or the IR range, UV or IR lasers are employed and the matrix absorbs heavily the laser light, and then facilitates protonation (deprotonation) of the analyte molecules via a series of processes such as matrix ablation and ionization. MALDI-MS is a mass spectrometric method utilizing soft ionization and suitable for the analysis of biomolecules such as DNA and proteins as well as organic molecules like polymers and dendrimers, which tend to be fragile and easily fragmented if conventional ionization techniques are employed. A common problem for conventional MALDI-MS in that most biomaterials are solid and may not be easily dissolved by a solvent. On the other hand, surface-MALDI-MS can analyze adsorbed multicomponent bimolecular layers on the biomaterials by applying a direct coating with the matrix molecules to the sample surface. The matrix molecules that form on the surface are subsequently irradiated by a UV or an IR laser resulting in protonation (deprotonation) of the surface adsorbed entities. A pictorial representation of surface-MALDI-MS is illustrated in Fig. 4.15.

Surface-MALDI-MS is a relatively new surface analytical method for biomaterials. By desorbing the adsorbed macromolecules from the biomaterial surface and determining their molecular ions with high mass resolution and at levels below monolayer coverage, the detection limitation plaguing conventional MALDI-MS can be avoided. However, surface-MALDI-MS is not quantitative and requires control experiments and parallel XPS analyses for absolute quantification. In recent years, more work concerning surface-MALDI-MS has been performed. For example, Clarke et al. [75] employed surface-MALDI-MS to characterize albumin adsorption on different prepared

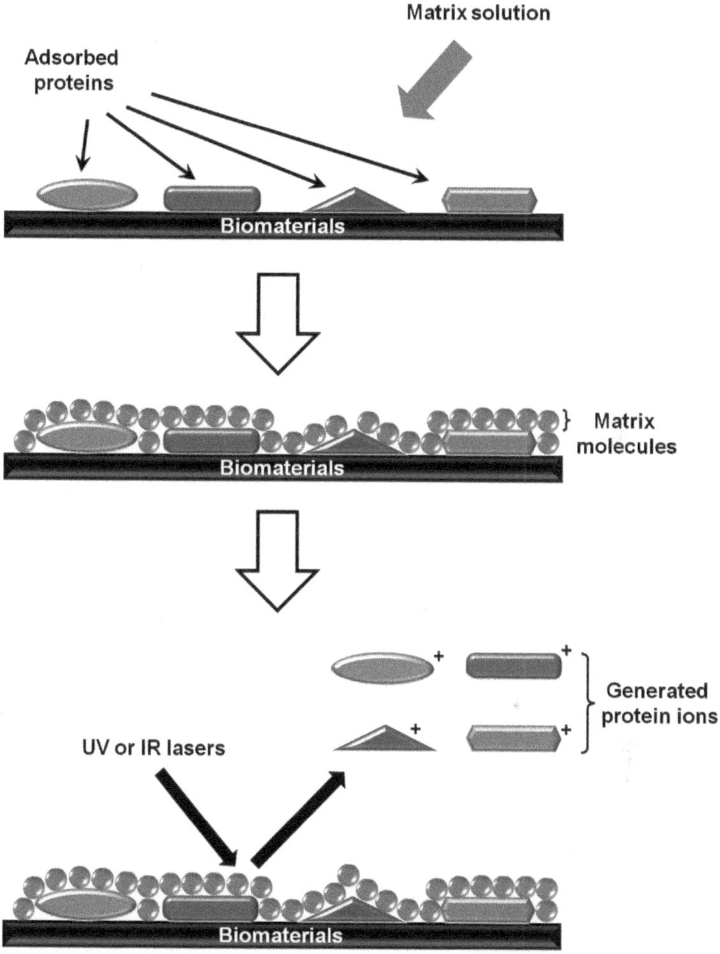

FIGURE 4.15 Schematic diagram of surface-MALDI-MS.

nitinol wires and observed that a nitinol wire with high nickel concentration and low oxygen concentration in the outer oxide layer was desirable for albumin adsorption. Heuts et al. [76] prepared various functionalized polyvinylidene fluoride (PVDF) substrates by ammonia plasma treating, star-poly(ethylene glycol) (PEG) coating, and star-PEG coating/Gly-Arg-Gly-Asp-Ser (GRGDS) peptide coupling and then assessed the surface protein repelling efficacy by surface-MALDI-MS. As shown in Fig. 4.16, insulin can only be detected on pure PVDF and ammonia-plasma-treated PVDF surfaces and albumin can only be detected on the pure PVDF surface. On the star PEG and biofunctionalized star PEG coatings, no insulin and albumin can be detected. There are other

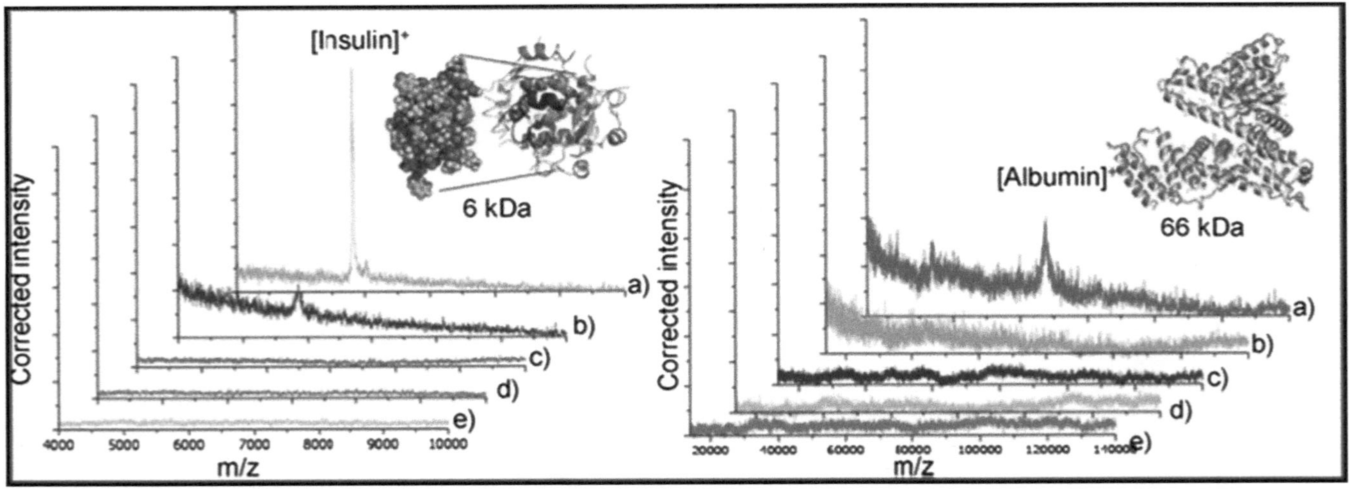

FIGURE 4.16 Surface-MALDI mass spectra of insulin (left hand side) and albumin (right hand side) on (a) pure PVDF, (b) ammonia plasma-treated PVDF, (c) star PEG-coated PVDF, (d) star PEG-coated PVDF biofunctionalized by consecutive coupling of 0.1 μmol mL^{-1} GRGDS, and (e) star PEG-coated PVDF biofunctionalized by consecutive coupling of 1.0 μmol mL^{-1} GRGDS. *Reprinted with permission from Ref. [76]. Copyright (2009) John Wiley and Sons.*

publications about Surface-MALDI-MS in biomaterials characterization [77–79] and interested readers are referred to a related review [80] for more information.

4.5. INFRARED (IR) SPECTROSCOPY

When a sample is irradiated by IR light, IR absorption occurs if the IR frequency coincides with the vibrational frequency of a bond. Hence, by monitoring the transmitted or absorbed IR light, information about the molecular structure can be obtained and this method is termed as IR spectroscopy. IR spectroscopy is the most widely used surface vibrational spectroscopy. As a surface chemistry studying tool, IR spectroscopy is attractive due to its versatility (applicable to almost any surface), compatibility with both high- and low-pressure conditions, and low cost compared to techniques requiring vacuum conditions. Most modern surface IR instrumentations, regardless of sampling mode, utilize a Fourier transform IR spectrometer rather than a grating monochromator in order to increase the signal level, enhance the photometric accuracy, and so on.

There are several modes of IR spectroscopy. The first one is transmission IR spectroscopy, which involves passing the IR beam through the sample. This mode requires at least partial transmission of the IR through the sample and so the acquired spectrum measures the bulk properties in lieu of the surface. Consequently, the transmission mode is rarely used in surface analysis of biomaterials, which also tend to be opaque to IR radiation. As an alternative, reflecting methods such as ATR IR spectroscopy and reflection absorption infrared spectroscopy (RAIRS) are more widely applied to surface characterization of biomaterials.

4.5.1. ATR IR Spectroscopy

ATR IR spectroscopy couples IR spectroscopy with total internal reflection to restrict the analysis volume. The IR spectrum is obtained from the substrate in contact with the internal reflection elements (IRE) such as zinc selenide or germanium. The IR radiation is first focused onto the end of the IRE at angles greater than the critical angle, enters the IRE, and is reflected down the length of the IRE. As shown in Fig. 4.17, at each internal reflection, the IR radiation actually penetrates a short distance from the surface of the IRE into the specimen. ATR IR spectroscopy is a powerful tool for the surface chemical characterization of opaque samples. However, it should be noted that as the penetration depth ranges from several hundred nanometres to more than 1 μm, ATR IR spectroscopy is not as surface specific as other analytical tools such as AES, XPS, and SIMS.

There are many applications of ATR IR spectroscopy to biomaterials [81–83] and generally, ATR IR spectroscopy is used qualitatively. For instance,

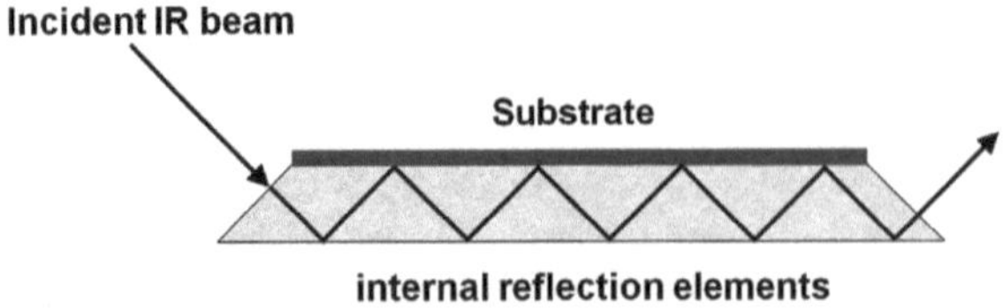

FIGURE 4.17 Schematic diagram showing the sampling IR beam passing through the internal reflection elements in contact with the opaque substrate.

Blaker et al. [82] produced biodegradable poly(D,L-lactide) with different proportional Bioglass® particles to form various composite samples with bioactivity and the ATR IR spectra reveal different chemistry. Rebollar et al. [83] altered the surface chemistry and topography of biomedical polystyrene foils by UV irradiation and as shown in Fig. 4.18, after irradiation, an intense and broad band centered at around 1720 cm^{-1} appears and can be attributed to the formation of carbonyl (C=O) containing species. Interpretation of the IR spectra is normally according to the IR handbooks [84,85].

4.5.2. Reflection Absorption Infrared Spectroscopy (RAIRS)

Greenler [86] demonstrated that for adsorbates on a metallic or conducting film, absorption of IR radiation by the adsorbate overlayer was enhanced at high angles of incidence (near grazing) and involved only one polarization of

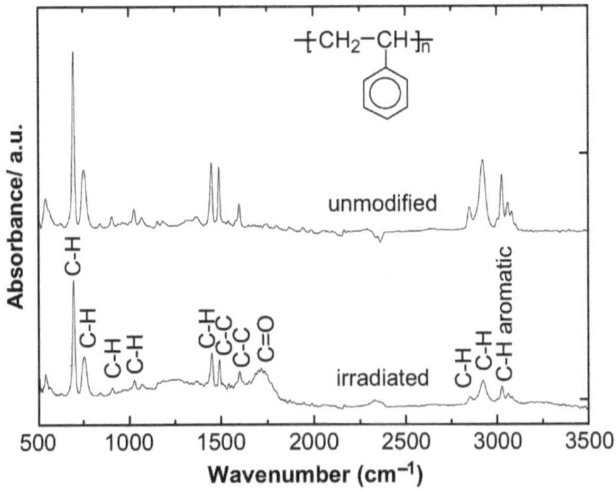

FIGURE 4.18 ATR IR spectra of unmodified polystyrene foil and UV irradiated polystyrene foil. The inset shows the chemical structure of polystyrene. *Reprinted with permission from Ref. [83]. Copyright (2008) Elsevier.*

the incident IR beam. RAIRS is based on this principle. RAIRS is very surface sensitive and particularly useful to the study of thin layers on a metal surface. Many types of biomaterials are surface-modified metals and so, RAIRS can be very useful [87–89]. Humblot et al. [88] functionalized gold surfaces with chemical groups and then covalently bound the anti-bacterial peptide (Magainin I). The RAIR spectra indicate successful binding of the functional agents. Briand et al. [89] fabricated a sample to serve as an immunosensor for goat anti-rabbit immunoglobulin (anti-rIgG) by immobilizing rIgG on a gold-coated substrate. After the sample was treated with solutions of labelled antigen with increasing concentrations, RAIRS was performed. As shown in Fig. 4.19, the characteristic amide I and amide II vibrational peaks (1660 and 1550 cm^{-1}) of the adsorbed antigen increase with antigen concentration, indicating that the sensitivity and specificity of this immunosensor are high. There are other IR methods available such as far-IR spectroscopy [90,91] and IR microscopy [92,93], but so far they are seldom used to characterize biomaterials.

4.6. RAMAN SPECTROSCOPY

When a light quantum $h\nu_0$ impacts a surface, an elastic scattering process termed *Rayleigh scattering* ensues. This process has the highest probability but there are also inelastic processes in which the vibrational energy is altered by $h\nu_s$. The inelastic process is called Raman scattering and an energy of $h\nu_0 \pm h\nu_s$

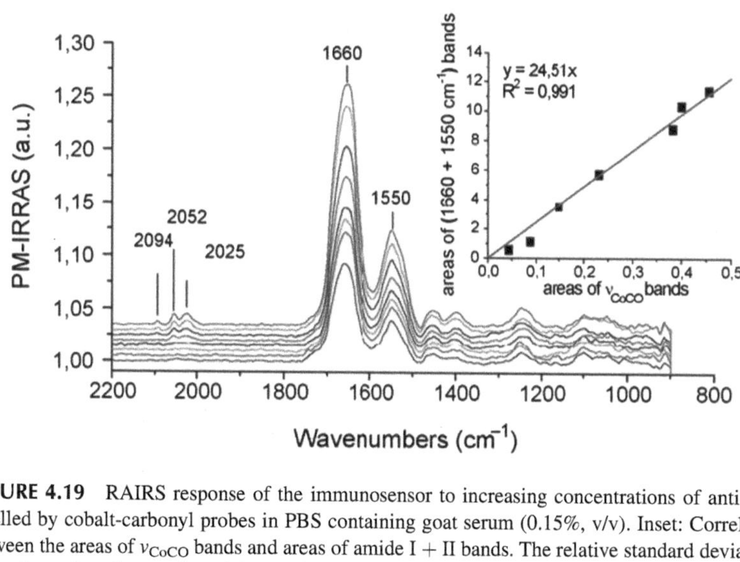

FIGURE 4.19 RAIRS response of the immunosensor to increasing concentrations of anti-rIgG labelled by cobalt-carbonyl probes in PBS containing goat serum (0.15%, v/v). Inset: Correlation between the areas of ν_{CoCo} bands and areas of amide I + II bands. The relative standard deviations are estimated to be equal to 0.1 a.u. for each IR area value. *Reprinted with permission from Ref. [89]. Copyright (2007) Elsevier.*

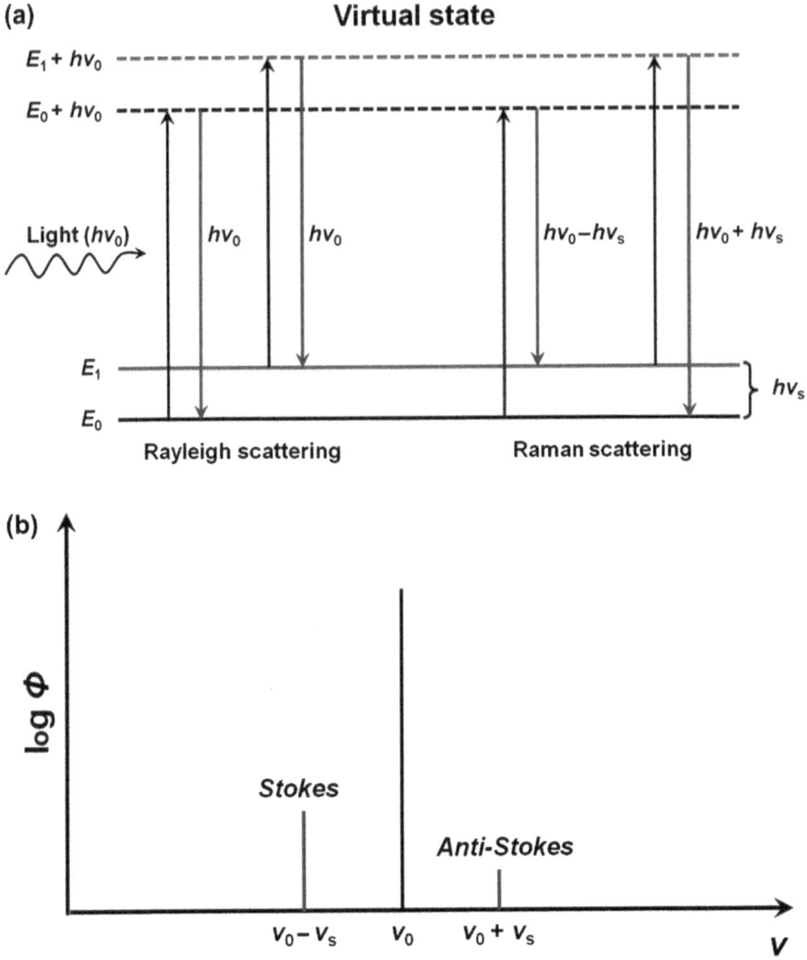

FIGURE 4.20 Schematic diagram showing the principle of Raman scattering: (a) term diagram and (b) related Raman spectra. Because the population of atoms in the excited state is much less than that of the ground state, the Stokes line is more intense than the anti-Stokes line.

is emitted. As shown in Fig. 4.20, when the incident radiation is absorbed by a virtual electronic state of the molecule, followed by emission back to the excited vibrational state, the emitted energy is $h\nu_0 - h\nu_s$ and the corresponding Raman lines are referred to as the *Stokes lines*. Alternatively, when the molecular vibration is already in the excited state and upon re-emission back to the ground state, the emitted energy is $h\nu_0 + h\nu_s$ and the Raman lines are called the anti-Stokes lines. As the population of molecules in the excited vibrational state is much less than that in the ground state at ambient temperature, the emitted quanta having energy of $h\nu_0 - h\nu_s$ are more prevalent than those with energy of $h\nu_0 + h\nu_s$. As a result, the intensity of the Stokes lines is much higher than that of the anti-Stokes ones and normally, only the Stokes lines are monitored in Raman spectrum [94].

Raman spectroscopy is a powerful tool to investigate the surface chemistry of materials. Similar to IR spectroscopy, Raman spectroscopy is one of the molecular vibrational spectroscopies. IR and Raman are highly complementary as Raman spectroscopy is based on polarizability and IR spectroscopy is concerned with dipole moments. Sometimes, Raman spectroscopy is sensitive to those vibrational modes, which are either not observed by IR or give rise to only weak IR absorption bands. In Raman spectroscopy, the wavelength of the excitation source can be ultraviolet, visible, near-IR, and even IR. Its surface sensitivity is not only related to the excitation wavelength (exciting light with shorter wavelength penetrates a lower depth) but also concerned with the properties of the materials. Some modern Raman instruments are equipped with a microscope (particularly confocal microscope) to achieve higher spatial resolution and to eliminate fluorescence. Raman instruments can also be equipped with Fourier transform accessories to eliminate the fluorescence interferences but in this mode, the wavelength of the excitation source is fixed in the IR region and so, the advantages gained by working with visible or ultraviolet radiation are lost. Interpretation of Raman spectra is typically based on handbooks [85,94,95].

There are many examples of Raman spectroscopy applications to biomaterials and they are mainly divided into two parts. First and mainly, Raman

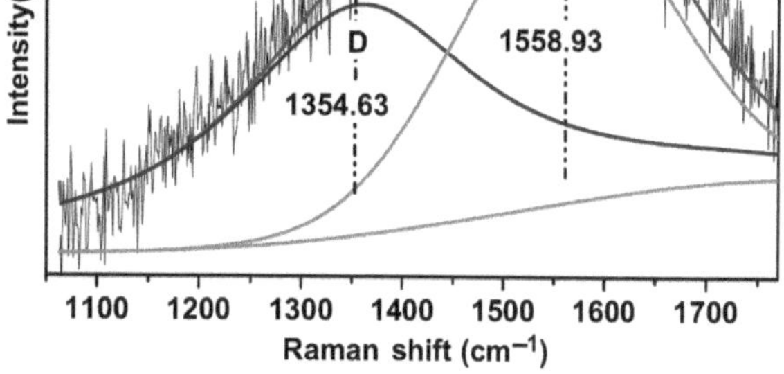

FIGURE 4.21 Gaussian–Lorentzion fitted Raman spectrum obtained from a biocompatible DLC film deposited on PEEK. *Reprinted with permission from Ref. [96]. Copyright (2010) Elsevier.*

spectroscopy is used to identify the chemical structure of biomaterials [96–99]. For instance, Wang et al. [96] deposited a diamond-like carbon (DLC) coating on poly aryl-ether-ether-ketone (PEEK) substrate to obtain better properties as a bone replacement. As shown in Fig. 4.21, the typical G band at 1558.93 cm^{-1} and D band at approximately 1354.63 cm^{-1} with the $I(D)/I(G)$ ratio of 1.2 are observed indicating successful DLC deposition on PEEK. Secondly, surface enhanced Raman scattering (SERS) by metals such as silver and gold has become a popular area and there are many publications about the use of metallic micro- or nano-structures in biosensors [100–102]. For example, Fang et al. [100] fabricated a series of Ag_2O mesocages with different shape such as mesosphere, mesododecahedron, mesocube and mesooctahedron. These Ag_2O mesocages were subsequently reduced to Ag with the same mesostructures and then deposited with crystal violet to serve as biosensors with SERS effects. As shown in Fig. 4.22, the Raman intensity

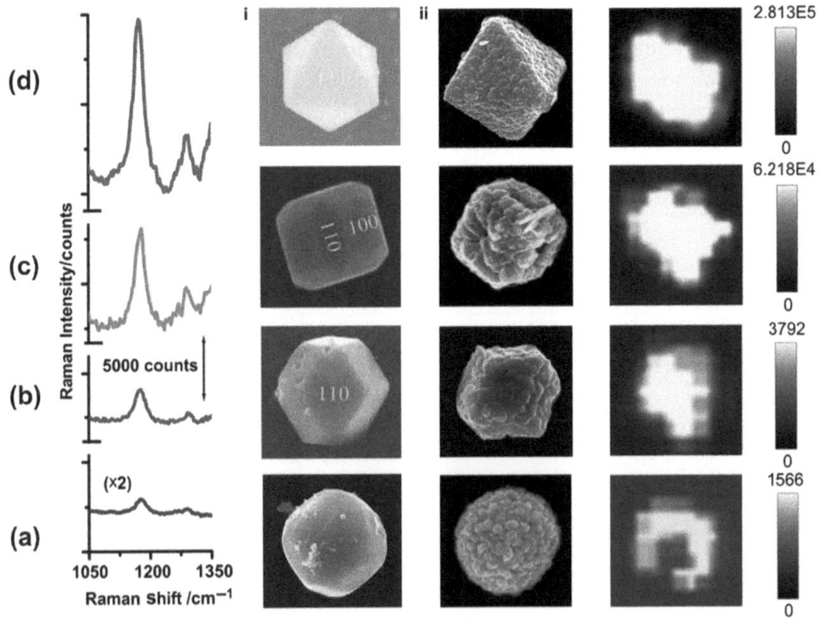

FIGURE 4.22 Single-particle SERS spectra and images of crystal violet (CV) on a polyhedral silver mesocages. a, mesosphere, b, mesododecahedron, c, mesocube, d, mesooctahedron. The color-coded signals of the right column correspond to the intensity of the Raman band of crystal violet (CV) at 1172 cm^{-1} integrated over 1120–1250 cm^{-1} after background subtraction. The spectra were acquired from individual silver mesostructures on Si substrates under the same conditions. CV was deposited onto the mesocages on Si substrates by drop coating a 10^{-9} M CV solution. Columns i and ii show the typical SEM images of individual polyhedral Ag_2O (i) and Ag (ii) mesostructures. *Reprinted with permission from Ref. [100]. Copyright (2011) Elsevier.*

of a single Ag mesooctahedron is much higher than that of the other three mesostructures, indicating the highest SERS effect. There are other techniques for biomaterials characterization based on Raman scattering such as Raman mapping which is generally used *in vitro* [102] or *in vivo* [103] and will be discussed in more detail in Chapters 5 and 6.

4.7. ELECTRON ENERGY LOSS SPECTROSCOPY

When a specimen is exposed to a beam of electrons with a narrow range of kinetic energy, some electrons undergo inelastic scattering accompanied by energy loss besides the more probable elastic scattering. As inelastic scattering is related to the sample properties, the amount of energy loss can impart useful information about the materials and this technique is termed electron energy loss spectroscopy (EELS). EELS as an analytical tool makes use of a series of inelastic interactions including phonon excitation, inter- and intra-band transitions, plasmon excitation, inner shell ionization, and so on. According to the geometry and kinetic energy of the incident electrons, EELS can be classified commonly into high-resolution EELS and transmission EELS.

In high-resolution EELS, electronic excitation of vibrational mode of the surface or molecules adsorbed on a surface is monitored by bombarding the sample with low-energy (1–10 eV) electrons and measuring the energy loss in the range of 10^{-3} eV to 1 eV. High-resolution EELS is a surface-sensitive vibrational spectroscopy, which plays an important role in the investigation of surface structure, catalysts, dispersion of surface phonons as well as monitoring of epitaxial growth. However, it is not often applied to biomaterials due to the rigorous experimental requirements and poor spectral resolution compared to IR spectroscopy and laser Raman scattering.

In comparison with high-resolution EELS, transmission EELS detects electrons with higher kinetic energy (typically 100–300 keV) and is typically coupled to a transmission electron microscope (TEM) to attain high spatial resolution. Because of the higher energy, the incident electrons can pass through specimens several hundreds of nanometres thick. Strictly speaking, transmission EELS is not a surface analytical technique, but nonetheless, transmission EELS is a unique tool for structural and chemical characterization of materials on the nanometre scale and can be combined with TEM, electron diffraction, and X-ray spectroscopy in the same instrument. According to the energy loss, the transmission EEL spectrum is usually divided into two regions, low-loss spectrum and core-loss spectrum. Low-loss spectroscopy is mainly used to investigate the changes in the plasmon peak. Not only the plasmon energy but also the plasmon peak widths are different for different materials. Core-loss spectroscopy is usually used to provide elemental information. The variations in the energy-loss near-edge structure of the element of interest can also be investigated. Transmission EELS is even capable of measuring the sample thickness based on $t/\lambda = \ln(I_t/I_0)$, where t is the absolute thickness, λ is the mean free path, I_0 is the area under zero-loss peak, and I_t is the total area under the spectrum [104–106]. This equation can be implemented on each pixel yielding a semi-quantitative thickness map.

Despite high spatial resolution and useful information, transmission EELS is not commonly used in biomaterials characterization so far. This is probably due to the difficult sample preparation (the sample must be thin enough) and lack of hardware/operation expertise. Moreover, interpretation of EELS spectra may be complicated. Nevertheless, there are publications about characterization of biomaterials by transmission EELS [107–110]. For example, Ma et al. [107] prepared DLC coatings for blood interfacing application, then characterized the coatings by EELS with respect to sp^3 content investigation. Liou et al. [110] synthesized calcium-deficient hydroxyapatite nano-crystals with Ca/P ratios from 1.5 to 1.67 and EELS was subsequently used to characterize the structural disorder in samples.

4.8. ULTRAVIOLET–VISIBLE SPECTROSCOPY

Ultraviolet–visible (UV–vis) spectroscopy encompasses absorption spectroscopy and reflectance spectroscopy in the UV–vis spectral region. Molecules containing π-electrons or non-bonding electrons (n-electrons) can absorb ultraviolet or visible light energy and be excited to higher anti-bonding molecular orbitals. UV–vis spectroscopy is different from IR spectroscopy in the excitation wavelengths and that molecules undergo electronic transitions in the ultraviolet or visible region, whereas they undergo vibrational transitions in the IR region. Generally, UV–vis spectroscopy is used to determine elemental concentrations quantitatively in a solution according to Beer–Lambert law:

$$A = \mathrm{Log}_{10}(I_0/I) = \varepsilon c L,$$

where A is the measured absorbance, I_0 is the intensity of the incident light at a given wavelength, I is the transmitted intensity, L is the path length through the sample, c is the concentration of the absorbing species, and ε is a constant known as the molar absorptivity or extinction coefficient for each species and wavelength. Normally, ε at the wavelength of maximum absorption (λ_{max}) is employed in quantitative analysis as errors resulting from instrumental wavelength uncertainty are minimized at the peak of the absorbance curve. According to this relationship, the concentration of the analyte can be calculated when ε is known, L is fixed, and I_0 and I are measured. When a calibration curve of the analyte is available, the concentration of the analyte can be determined more accurately. Nevertheless, this application of UV–vis spectroscopy to biomaterials is not widespread because most biomaterials are not in a solution, the molar extinction coefficients of some biomaterials in a solution are unknown, and calibration curves are hard to get. Instead, a solution containing biomaterials is characterized by UV–vis spectroscopy to obtain the absorbance spectra rather than concentrations [111–113]. For example, Zhou et al. [111] explored and evaluated the utilization of water-soluble nano-crystal CdTe quantum dots capped with negatively charged 3-mercapitalpropionic acid (MPA-CdTe QDs) to enhance the drug (daunorubicin) uptake into the targeted cancer cells. The UV–vis absorption spectra of the daunorubicin residue after cancer cells incubation reveal that MPA-CdTe QDs can effectively enhance the cellular uptake of this drug.

UV–vis spectroscopy is also available for the characterization of solid biomaterials on the nanometre scale when these nano-biomaterials are dispersed homogeneously in a solvent [114–117]. Lin et al. [115] fabricated magnetic nanoparticle (MNP) clusters by chemically linking Pluronic® F127 (PF127) to water-soluble polyacrylic acid-bound iron oxide (PAAIO) and further modified PF127-PAAIO with folic acid (FA) to obtain MNPs with tumor targeting properties. The content of FA decorated on MNPs is clearly indicated by the UV–vis spectrum of FA-PF127-PAAIO, in which a UV absorbance peak around 270 nm is attributed to the aromatic ring of FA. Mao et al. [117] employed UV–vis spectroscopy to measure the cloud points of the thermoresponsive copolymer [trimethyl chitosan chloride-g-poly(*N*-isopropylacrylamide)] which is used for gene transfection.

Besides the aforementioned examples concerning solvent-dispersed biomaterials, UV–vis spectroscopy can be utilized to measure the solid biomaterials in their natural states [118–120]. In this mode, only the surface regions of solid biomaterials are measured. For instance, van den Beucken et al. [118] fabricated two types of multilayered coatings, poly-D-lysine/DNA (PDL/DNA) and poly(allylamine hydrochloride)/DNA (PAH/DNA), on quartz substrates and then monitored the build-up of DNA coatings by UV–vis spectroscopy by measuring the absorbance maximum of the nucleic base chromophores in the DNA at 260 nm. Lee et al. [120] developed a multifunctional 4-bit biomemory chip that consisted of recombinant azurin variants

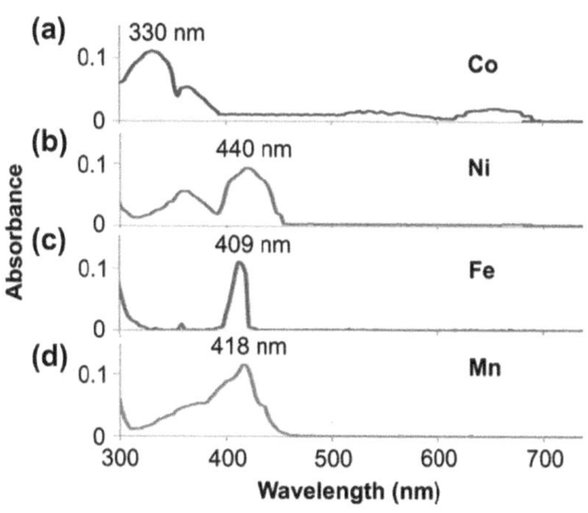

FIGURE 4.23 UV–vis absorption spectra of (a) Co-type azurin (b) Ni-type azurin (c) Fe-type azurin and (d) Mn-type azurin. *Reprinted with permission from Ref. [120]. Copyright (2011) Elsevier.*

when the Cu ion in this recombinant azurin protein was substituted with other metal ions such as Co, Mn, Fe, and Ni ion to allow the protein to perform various memory functions. UV–vis spectroscopy was subsequently performed to confirm the metal substitution, and the absorption spectra of different azurin proteins are shown in Fig. 4.23.

4.9. LIGHT MICROSCOPY AND CONFOCAL MICROSCOPY

Light microscopy, often referred to as the 'optical microscopy', utilizes visible light and a system of lenses to obtain magnified optical images. As one of the oldest analytical techniques, it is very popular in the scientific and industrial communities and applications include biomaterials characterization [121,122]. For instance, Ji et al. [121] fabricated 3,3′-dithiobis(propanoic dihydrazide)-modified HA (HA-DTPH) nanofibrous scaffolds by electrospinning and then employed light microscopy to map the surface after equilibrium swelling in air. The images in Fig. 4.24 reveal the morphological change on the sample surface after water evaporation for different time. However, because of basic physics, the spatial resolution of conventional light microscopy is limited and the intrinsic properties of focus (in-focus and out-of-focus information points contribute equally to the images) of this technique can lead to blurred images.

Confocal microscopy (CM), an improved light microscopy technique, can provide blur-free optical sectioning of a specimen by eliminating out-of-focus information through spatial filtering using a point source of light for excitation. Sample preparation is easier and high-resolution images can be more easily obtained by CM. The actual useful depth of CM is limited to approximately 100–200 μm and the image resolution is 0.18 μm in the *xy*-plane and about 0.35 μm in the *z*-plane. Three-dimensional structures can be reconstructed by carefully compiling two-dimensional images obtained.

CM is a powerful tool in various fields such as life science, semiconductors, and materials characterization. In the biomaterials field, it not only is a popular tool for the *in vitro* characterization (e.g. imaging cells seeded on the samples) [123–125] but also can be used on samples before cell culturing by mapping the surface and structure [122,126–129]. For example, Cross et al. [126] prepared hydrated collagen gels with different concentrations by mixing an acidic collagen solution with alkali solution, and then polymerized them into poly (dimethylsiloxane) (PDMS) moulds. CM was subsequently used to image the fiber structure of the collagen gels. Da Silva et al. [128] fabricated periodic hydrogel scaffolds with a 3D honeycomb-like structure from colloidal crystal templates and then examined the surface and structure by CM. As shown in Fig. 4.25, a series of confocal slices (a) can be obtained at different depths (the surface, middle of the first layer, and even deeper) of the fabricated hydrogel scaffolds. CM can also map the z-projection (b) and construct the three-dimensional structure (c) of a segment from the same stack. The use of

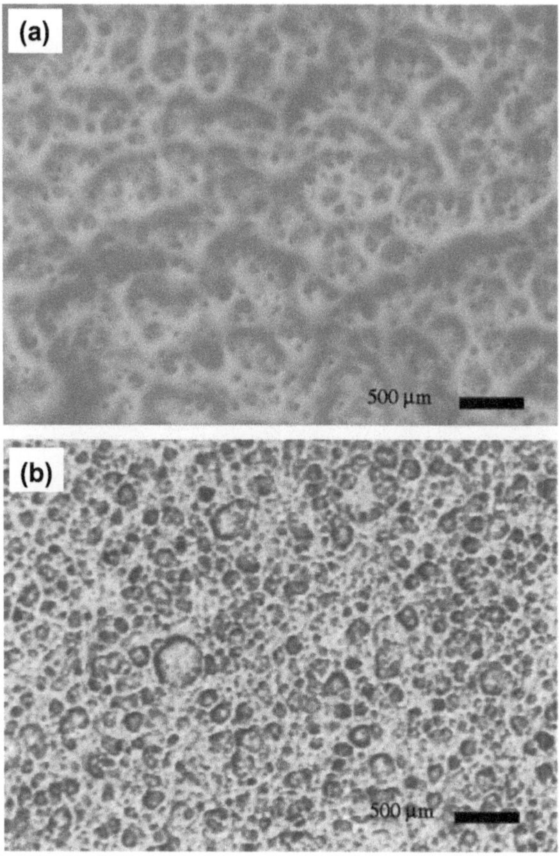

FIGURE 4.24 Optical images of HA-DTPH nanofibrous scaffold after equilibrium swelling in air for (a) 30 min and (b) 120 min. The scale bar is 500 μm. *Reprinted with permission from Ref. [121]. Copyright (2006) Elsevier.*

conventional light microscopy and CM in the *in vitro* and *in vivo* characterization of biomaterials will be discussed in more detail in Chapters 5 and 6.

4.10. SCANNING ELECTRON MICROSCOPY

Scanning electron microscopy (SEM) is an electron microscopic technique that images a sample surface by scanning a focused energetic electron beam. SEM is one of the most common tools to study surface morphology and identify small areas that cannot be resolved by optical microscopy. The primary electrons interact with atoms in the sample surface and various signals are emitted. The primary electrons are emitted from an electron gun. The common types are

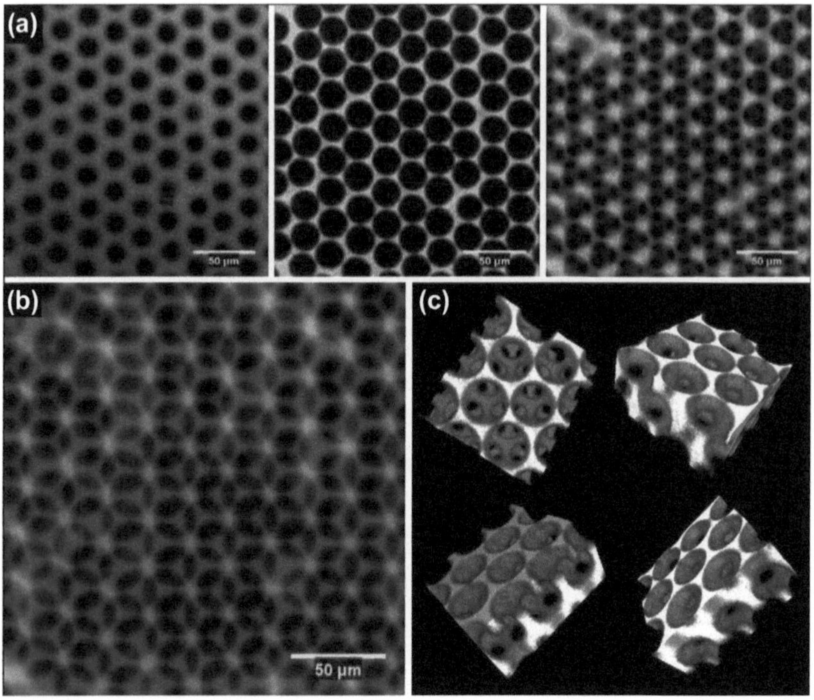

FIGURE 4.25 Images obtained from a confocal image stack of a periodic hydrogel scaffold before annealing: (a) Images showing the surface, the middle of the first layer and the pores between the first and second layer of the scaffold, (b) Z-projection of the three first layers of the scaffold, and (c) three-dimensional reconstruction of a scaffold segment. *Reprinted with permission from Ref. [128]. Copyright (2010) Elsevier.*

thermionic electron emitters as tungsten filament cathodes and lanthanum hexaboride cathodes and field emission sources such as a cold-cathode type using a tungsten single crystal emitter or thermally assisted Schottky type emitter of zirconium oxide. The electrons are typically accelerated to energy of 0.2 keV–40 keV and spatially focused by one or two condenser lenses to a spot about 0.4 nm–5 nm in diameter. The electron beam then passes through scanning coils or deflector plates in the electron column to achieve deflection in the *x* and *y* directions for scanning. The scanned area is usually square of rectangle. When the primary electrons strike the sample, the electrons lose energy by repeated random scattering and absorption within a volume of the specimen known as the interaction volume, which resembles a tear drop or balloon with a depth of 1–5 μm into the surface depending on the electron energy. Various signals originate from different depths of the interaction volume and secondary electrons are emitted from the top 5–10 nm of the sample surface and contain topographical information. In addition, reflection of high-energy electrons by elastic scattering and emission of electromagnetic radiation are some of the phenomena. The emitted signals are detected by specialized detectors and the signals are amplified and displayed on a cathode ray tube (CRT). The CRT display is synchronized with the rastering of the electron beam and the resulting image represents an intensity map of signals emitted within the scanned area.

4.10.1. Secondary Electron Imaging

There are several SEM imaging modes according to the signals detected. Among these modes, secondary electron imaging is the most common. Secondary electrons are low-energy electrons (<50 eV) ejected from atoms via inelastic scattering. As aforementioned, they originate within a few nanometres of the surface due to their low energy. Secondary electrons are detected by an Everhart–Thornley detector which is a type of the scintillator–photomultiplier system. After being attracted towards an electrically biased grid at about +400 V, the secondary electrons emitted are further accelerated towards a scintillator positively biased to about +2000 V and causing the scintillator to emit flashes of light. These electrical signals are amplified by a connected photomultiplier and displayed as a two-dimensional intensity distribution that can be viewed on an analogue video display and saved as a digital image. The brightness of the signal depends on the number of secondary electrons reaching the detector and is sensitive to the surface topography of the sample. For instance, as the electron beam enters the sample perpendicular to the surface, the activated region is uniform about the axis of the beam and a certain number of electrons 'escape' from within the sample. When the angle of incidence increases, the 'escape' distance of one side of the beam will decrease, and more secondary electrons will be emitted. Consequently, steep surfaces and edges of the samples tend to be brighter than flat surfaces in the second electrons images, which results in images with a well-defined, three-dimensional appearance.

Secondary electron imaging is very useful in examining surface topography with high spatial resolution. Topographical images with resolution of <1 nm can even be obtained [130]. However, such high spatial resolution is uncommon for biomaterials as images with this resolution are hard to get and most biomaterials have topographical features larger than 1 nm. There have been many reports on the characterization of biomaterials by secondary electron imaging [131–137]. For instance, Park et al. [131] generated titania (TiO_2) nanotubes with defined diameters between 15 and 100 nm on titanium to direct different behavior of mesenchymal stem cells. The diameters and topographies of various TiO_2 nanotubes were determined by the secondary electron imaging. Rujitanaroj et al. [133] encapsulated small-interfering RNA (siRNA) and transfection reagent (TKO) complexes within nanofibers comprising a copolymer of caprolactone and ethyl ethylene phosphate (PCLEEP) to control the release of siRNA complexes in long-term gene-silencing applications. The secondary electron images reveal that the SiRNA-encapsulated PCLEEP fiber are successfully fabricated and both the siRNA- and GAPDH siRNA/TKO-encapsulated nanofibers are uniform and bead-free. He et al. [134] fabricated mineralized nanofibrous poly (L-lactic acid) (PLLA) scaffolds for bone regeneration by depositing calcium phosphate on the PLLA scaffolds using an electrodeposition process. As shown in the SEM images in Fig. 4.26, by varying the voltage in electrodeposition, the surface topography of the calcium phosphate coatings on the nanofibrous PLLA scaffolds is quite different. Flower-like topography, flake-like structure, and fiber-like coating are obtained by electrodeposition at 2, 3, and 5 V, respectively. Secondary electrons imaging is also widely used in the *in vitro* characterization of biomaterials after cell seeding [138–140] and more details on secondary electron imaging can be found in Chapter 5.

4.10.2. Backscattered Electron Imaging

Besides secondary electrons, backscattered electrons produced by the electron beam can be used to image the sample surface for a different purpose. Backscattered electrons consist of high-energy electrons, which are reflected or back-scattered out of the specimen interaction volume by elastic scattering with the substrate atoms. Since heavy elements (high atomic number) backscatter electrons more efficiently than light elements (low atomic number), the signal is higher and the spot looks brighter in the image. Hence, backscattered electrons are usually used to monitor the contrast among areas with different chemical compositions. The detector for backscattered electrons is different from that of secondary electrons as the backscattered electrons have high energy (in keV) and the positively biased detection grid has little ability to attract them. The detector used for secondary electron detection is normally positioned on one side of the specimen and is inefficient in the detection of backscattered electrons because few such electrons are emitted in the solid angle subtended by the detector. Therefore, dedicated backscattered electron detectors are positioned above the sample in a 'doughnut' arrangement concentric with the electron beam in order to maximize the solid angle of collection. When all parts of the detector are used to collect electrons, an atomic number contrast is produced. Furthermore, strong topographic contrast can be produced by collecting backscattered electrons from one side above the specimen using an asymmetrical, directional BSE detector. The resulting contrast appears as illumination of the topography from that side. Semiconductor detectors can be made in radial segments and can be switched in or out to control the type of contrast produced (atomic number contrast or topographic contrast) and directionality.

In most cases, backscattered electrons are used to image biomaterials after implantation. There are reports about the *in vivo* characterization of bone replacement biomaterials by backscattered electron imaging [141–143], and

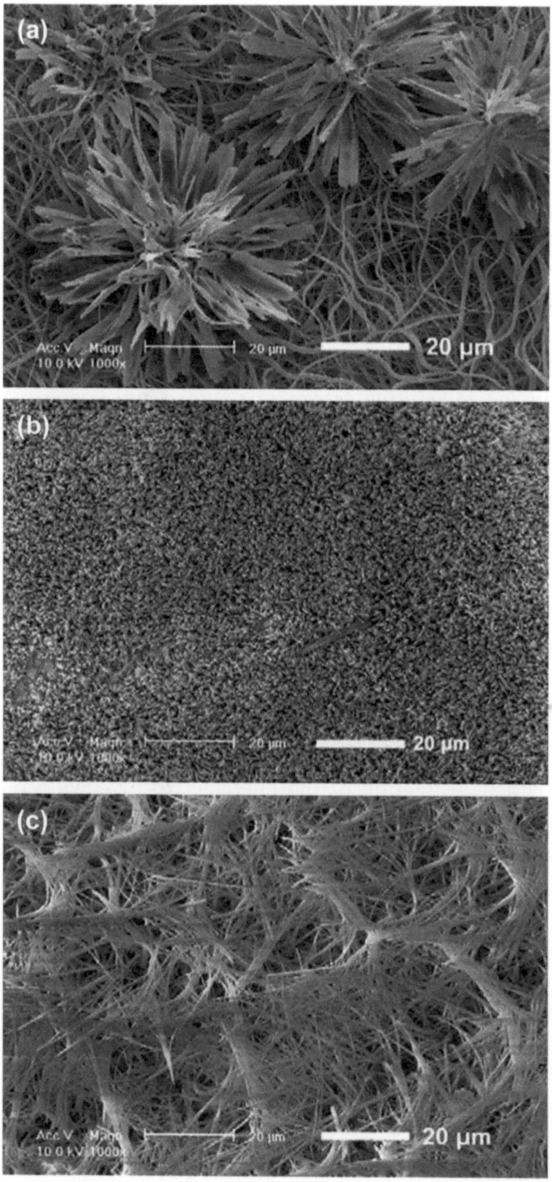

FIGURE 4.26 Secondary electron images of calcium phosphate deposition on electrospun PLLA nanofibrous scaffolds. Electrodeposition is performed for 60 min at 60 °C and (a) 2, (b) 3, and (c) 5 V, respectively. *Reprinted with permission from Ref. [134]. Copyright (2010) John Wiley and Sons.*

the implant materials and newly formed bone can be well differentiated in the images (see Chapter 6 for more details). Before *in vitro* and *in vivo* characterization, backscattered electrons imaging can also be used to characterize the cross-section [144,145] and surface [146,147] of biomaterials by revealing the elemental contrast. Zhang et al. [146] prepared silver-loaded coral hydroxyapatites as bone tissue engineering scaffolds with anti-bacterial properties and observed from the backscattered electron images that silver particles existed on the sample surface, whereas this feature was not observed from the secondary electron images. Variola et al. [147] modified the surface of Ti6Al4V alloy by etching with a H_2SO_4/H_2O_2 solution to improve the surface osteo-compatibility of the sample. The untreated and modified Ti6Al4V alloys were characterized by SEM in the backscattered mode. As shown in Fig. 4.27, the β-phase is initially present as interstitial grains surrounded by α-phase grains in the untreated sample (Fig. 4.27a) and the β-phase grains are preferentially removed in the course of etching (Fig. 4.27b). The resolution of backscattered electron images is not as high as secondary electron ones because the interaction volume for backscattered electrons is much larger than that of secondary electrons.

4.10.3. Energy-Dispersive X-ray Spectroscopy/Wavelength-Dispersive X-ray Spectroscopy

Generally, SEM is equipped with an energy-dispersive X-ray spectroscopy (EDS) and/or wavelength-dispersive X-ray spectroscopy (WDS) detector for elemental analysis or chemical characterization. Both EDS and WDS rely on the detection and investigation of the characteristic X-rays produced by the primary electrons. During electron illumination, an electron in the inner shell of the target atom can be ejected from the shell and an electron hole is created. Subsequently, an electron from an outer shell fills the hole, and the difference in

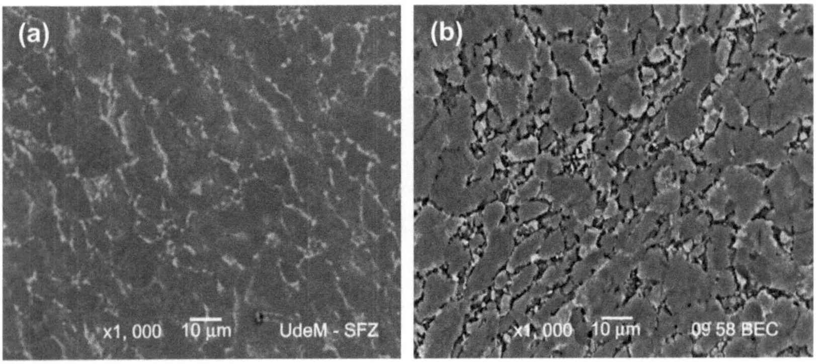

FIGURE 4.27 Backscattered SEM image of (a) polished untreated Ti-alloy disk and (b) the sample after 4 h of exposure to the etching solution. *Reprinted with permission from Ref. [147]. Copyright (2008) Elsevier.*

the energy between the outer- and inner-shell electrons can be released in the form of X-ray [see illustration in Fig. 4.7b]. As the energy of this X-ray is characteristic of the atomic structure of the element from which it is emitted, by measuring the quantity and energy of the X-rays emitted from the sample, the elemental composition of the sample can be determined. WDS differs from EDS in the way the X-ray is detected and that it uses the diffraction patterns created by light–matter interaction as its raw data and the wavelength of the X-ray is calculated by the Bragg' law: $n\lambda = 2d \sin \theta$, where n is an integer, d is the spacing between the planes in the atomic lattice which is known, θ is the measured angle between the incident ray and the scattering planes, and λ is the wavelength of the incident X-ray can be calculated. The spectral resolution of WDS (about 5 eV) is much higher than that of EDS (above 100 eV). However, data collection in WDS is slower than that of EDS as only one element can be monitored at a time in WDS while EDS can gather a spectrum of all the elements in the sample simultaneously.

Characteristic X-rays are not as surface sensitive as secondary electrons because the interaction extent of characteristic X-rays is even larger than that of backscattered electrons. Consequently, as the means for chemical characterization of sample surface, the surface sensitivity of EDS and WDS is not good enough. Nevertheless, EDS is widely used for the elemental analysis of biomaterials 'surface' [148–150] due to the small-area capability. For example, Han et al. [148] synthesized different kinds of calcium hydroxide-loaded biodegradable microcapsules by phase separation and utilized EDS to determine the elemental percentages of the samples. Khaled et al. [150] synthesized strontium-modified titania nanotubes (n-SrO-TiO_2 tubes) by using the alkaline hydrothermal technique and characterized them with EDS mapping. As shown in Fig. 4.28, the elemental maps of the atomic concentrations of Ti, Sr, and O reveal that Ti and Sr are homogeneously dispersed throughout the nanotube samples. It should be noted that there are some limitations in characterizing the sample chemistry with EDS. The first three elements of the periodic table, namely H, He, and Li, do not have enough electrons to produce characteristic X-ray and so they are not detectable by EDS. Furthermore, owing to the low spectral resolution, many elements have overlapping peaks in the EDS spectrum. In spite of the better resolution, WDS is seldom used for biomaterials probably because WDS is not as common as EDS in SEM instruments.

4.10.4. Sample Preparation

Since the typical SEM instrument is under vacuum, the sample is normally required to be dry. To avoid charging, the sample should also be electrically conductive, at least on the surface, and should be properly grounded to prevent accumulation of electrostatic charges on the surface. Metallic samples usually require little special preparation except cleaning and mounting. Non-conductive specimens can charge when bombarded by the electron beam and

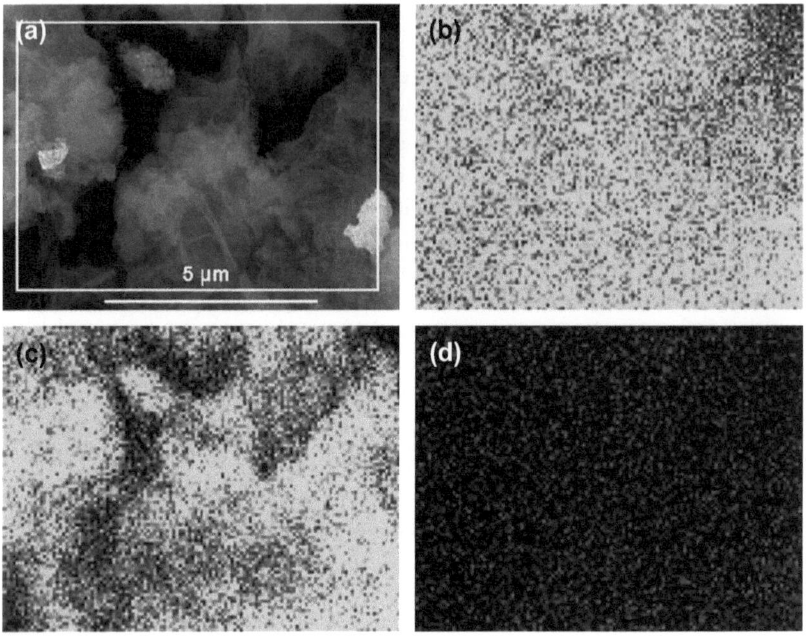

FIGURE 4.28 EDS elemental mapping: (a) Selected area on the sample (n-SrO-TiO_2 tubes); (b) Ti mapping; (c) O mapping; (d) Sr mapping. *Reprinted with permission from Ref. [150]. Copyright (2010) Elsevier.*

therefore, they are usually coated with a thin coating of electrically conductive materials such as gold or carbon. Because some biomaterials are hydrated and non-conductive, e.g. hydrogel, before they are inserted into the SEM instrument, a series of pre-treatments such as structure fixing, sample drying, and deposition of a conductive coating are required.

4.10.5. Environmental Scanning Electron Microscopy

Conventional SEM requires samples that are compatible with vacuum conditions. For hydrated and non-conductive biomaterials, the pre-treatments can be complicated and the natural state of the samples may even be altered after the pre-treatments. Environmental scanning electron microscopy (ESEM) [151] has the same working principle as conventional SEM but is specialized for imaging hydrated and non-conductive biomaterials in their natural states. There are differential pumping systems in ESEM instruments, which allow the transfer of the electron beam from the high vacuum part in the gun area to the high-pressure region in the specimen chamber. By placing the sample in the high-pressure specimen chamber where the working distance is short, the electron beam can produce various signals such as secondary electrons, backscattered electrons, and characteristic X-rays from the sample. Although the signals used in ESEM are the same as those in conventional SEM, the detectors are quite different. A gas detection device is specially designed for the detection of signals in the gaseous environment of ESEM.

Compared to conventional SEM, ESEM boasts several advantages in biomaterials characterization. First of all, hydrated biomaterials can be examined naturally in ESEM, whereas in conventional SEM, the specimens are normally desiccated under the vacuum conditions. Secondly, accumulation of electric charges on the surface of non-conductive specimens can be avoided in ESEM because positively charged ions generated by beam interactions with the gas help to neutralize the negative charges on the specimen surface. In conventional SEM, the insulating specimens without conductive coating will charge up during electron bombardment making imaging problematic or even impossible. Accordingly, hydrated and non-conductive biomaterials can be examined faster and more easily by ESEM as the complex and time-consuming pre-treatments can be avoided. More importantly, the natural surface of the specimens can be imaged by ESEM. Hence, ESEM is widely used for the surface topographical characterization of the hydrated or/and non-conductive biomaterials such as hydrogels [152,153], micelles [154,155], nanofibers [156], and so on. For instance, Fundueanu et al. [152] functionalized the pullulan microspheres (hydrogel) with pendant thermosensitive units to serve as the carriers for drug control and release. ESEM was subsequently used to characterize the extension and collapse of the pendant thermosensitive units below and above the lower critical solution temperature. Shi et al. [156] synthesized the copolymer (PLA-co-MPC) of L-lactide with a monomer containing propargyl group and then electrospun it into nanofibers for proteins immobilization. The natural morphology of the PLA-co-MPC nanofibers was evaluated by ESEM, which indicates that the nanofibers are smooth and continuous with a diameter of 2–4 μm (Fig. 4.29). Backscattered electrons [157] and

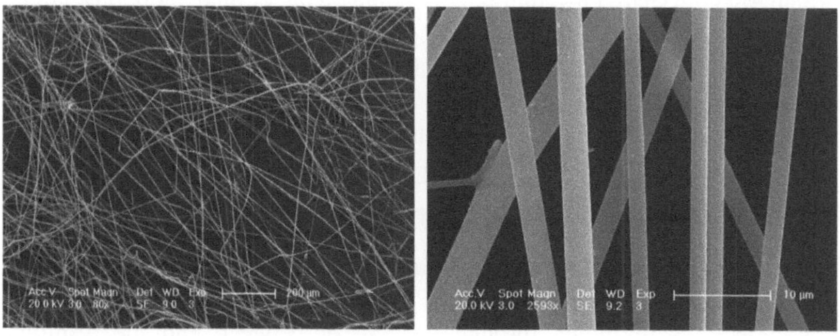

FIGURE 4.29 ESEM images of the electron-spun PLA-co-MPC nanofibers. *Reprinted with permission from Ref. [156]. Copyright (2008) Elsevier.*

characteristic X-rays [158] in ESEM can also be used for the characterization of biomaterials.

However, there are some disadvantages of ESEM. The main one arises from the limitation of the distance in the specimen chamber (1–10 mm) over which the electron beam remains usable in the gaseous environment. The presence of gas may cause unwanted effects in certain applications and the mechanism of these side-effects is not always clear.

In summary, SEM is a widely used tool in the characterization of biomaterials. Various signals such as secondary electrons, backscattered electrons, and characteristic X-rays emitted from the sample surface can be used to determine different properties of the samples including surface topography and elemental distributions. However, because of the vacuum conditions in conventional SEM, hydrated and/or non-conductive biomaterials must be pre-treated and the process can be complicated and sometimes destructive. Hence, for these biomaterials, ESEM is preferable as the complex and time-consuming pre-treatments are avoided and the samples can be imaged in their natural state. Owing to space limitation, we only describe selected examples before *in vitro* and *in vivo* experiments in this chapter. The *in vitro* and *in vivo* characterization of biomaterials by SEM will be discussed in more detail in the following chapters (see Chapters 5 and 6).

4.11. SCANNING TUNNELLING MICROSCOPY AND ATOMIC FORCE MICROSCOPY

Scanning probe microscopy (SPM), which was adopted with the invention of the scanning tunnelling microscopy (STM) in 1981, encompasses a family of microscopic techniques that image surfaces with a physical probe that scans the specimen. An image of the surface is obtained by mechanically scanning the probe line by line and recording the probe–surface interaction as a function of position. Among the various surface imaging techniques, SPM has the advantages of easy sample preparation, operation in a liquid environment, and particularly high spatial resolution. Depending on the sample, individual atoms can even be imaged by SPM. According to the different probe–surface interactions, SPM is classified into various sub-types such as STM, AFM, frictional force microscopy, and so on. Of these techniques, STM and AFM are most commonly used for the evaluation of sample surface morphology.

4.11.1. Scanning Tunnelling Microscopy

STM is based on the concept of quantum tunnelling. When a conducting tip is brought very close to the surface to be examined, a bias (voltage difference) allows electrons to tunnel through the gap in between and the resulting tunnelling current is a function of the tip position, applied voltage, and local density of states on the sample surface. Lateral information is acquired by monitoring the current as a function of the tip position during scanning across the surface. Since the tunnelling current varies exponentially upon the separation between the STM tip and sample surface, a vertical resolution of 0.01 Å with a lateral resolution of 0.1 Å is attainable by STM for electronically inhomogeneous surfaces. However, the STM image obtained, even at atomic resolution, is far from being a simple visual representation of the spatial locations of atoms. Instead, it is a complex image determined by a convolution of the electronic structures of the tip and the sample surface. Only for surfaces which have relatively uniform electronic properties (and with dimensions >10 Å), the STM image can effectively represents the surface topography.

Despite the outstanding spatial resolution of STM, publications about the use of STM in biomaterials characterization are few and only related to selective substrates such as gold [159,160] and carbon [161] because of practical obstacles. First and mainly, the samples for STM measurement should be conductive, but many biomaterials are organic or inorganic non-metallic materials which are insulating. Even for metallic biomaterials such as titanium, the surface usually has a native oxide which is too thick for electron tunnelling. Second, STM is a tool to determine the surface electronic structure. The image contrast does not necessarily bear a simple relationship with surface topography, and interpretation of STM images may not be straightforward. Although the application of STM to biomaterials is limited at present, STM is the preferred method to investigate the atomic structures of the novel conductive materials such as carbon nanotubes [162,163], silicon nanowires [164], and graphene [165,166] and more applications are being discovered.

4.11.2. Atomic Force Microscopy

At high tunnelling currents, the STM tip may interact physically with the surface in such a way that disruption of the surface structure occurs. In fact, there is usually a finite interaction force between the tip and sample surface and even at relatively low tunnelling currents, the interaction force may be substantial when measured against the strengths of molecular interactions. Consequently, Binnig et al. [167] invented the first atomic force microscope in 1986 and 3 years later, the first commercial atomic force microscope became available.

A commercial AFM instrument consists of a cantilever with a sharp tip (probe) at its end that is used to scan the specimen surface. The cantilever is typically made of silicon or silicon nitride (100–200 μm long and about 1 μm thick) with a tip radius of curvature on the order of nanometres. When the tip is brought close to the sample surface, the force between the tip and sample leads to a deflection of the cantilever according to Hooke's law. The deflection is measured using a laser reflected from the top surface of the cantilever into an array of photodiodes. A schematic diagram of a typical AFM instrument is depicted in Fig. 4.30. There are other AFM methods available involving optical interferometry, capacitive sensing, and piezoresistive AFM cantilevers.

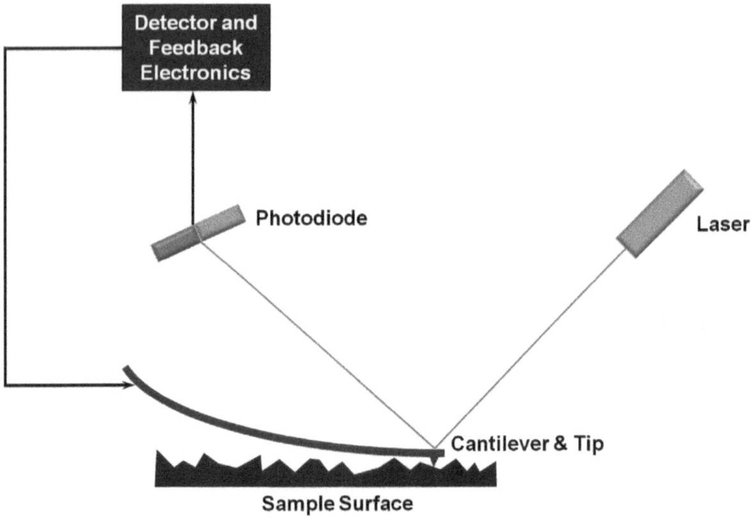

FIGURE 4.30 Schematic diagram of a typical commercial AFM instrument.

Depending on the information sought, the force measured in AFM can be different and it includes mechanical contact force, van der Waals force, capillary force, chemical bonding, electrostatic force, magnetic force, and so on.

Despite the short history, AFM has become one of the essential tools for surface imaging, particularly on the nanoscale. With regard to biomaterials, AFM boasts some advantages in comparison with other imaging techniques. Sample preparation for AFM is substantially simpler than that required by SEM or TEM and AFM can produce detailed three-dimensional maps rather than two-dimensional images without expensive and time-consuming sample preparation. For non-conducting samples, AFM can provide a direct view of surface features in high resolution, whereas in SEM, a conductive coating is needed and it may hide some surface details and even damage or alter the surface features. Furthermore, a vacuum is not required and samples can be imaged in ambient air or liquid. These advantages have rendered AFM suitable for biomaterials research in the native or quasi-native environment as well as in real time.

There are three common modes of AFM: contact mode, non-contact mode, and tapping mode. In the contact mode, the AFM tip scans across a surface at a very low force and is deflected by the repulsive force between the tip and surface atoms. A feedback loop maintains constant deflection of the cantilever by vertically moving the scanner as it scans laterally across the surface and then the topographic image of the sample surface is constructed. As the tip always exerts a mechanical load on the sample which may lead to damage, contact AFM is not suitable for soft biomaterials.

In the non-contact mode, the tip of the cantilever does not make direct contact with the sample surface. Instead, the cantilever is oscillated at a frequency slightly above its resonant frequency where the amplitude of oscillation is typically a few nanometres (<10 nm). The van der Waals force, which is strongest from 1 nm to 10 nm above the surface, or any other long-range force which extends above the surface decreases the resonance frequency of the cantilever. This decrease in resonant frequency combined with the feedback loop system maintains a constant oscillation amplitude or frequency by adjusting the average tip-to-sample distance. In measuring soft samples, non-contact AFM is preferable over contact AFM as it does not suffer from tip or sample degradation effects that are sometimes observed with contact AFM. However, non-contact AFM is also not commonly used for biomaterials because most biomaterials are hydrophilic and develop a liquid meniscus layer under ambient conditions. Keeping the probe tip close enough to these samples may possibly lead to the tip sticking resulting in low-resolution imaging.

In the tapping mode, also called intermittent contact mode, the cantilever is driven to oscillate up and down at near its resonant frequency similar to that in the non-contact mode. However, the amplitude of this oscillation is much greater than 10 nm, typically 100–200 nm. Owing to the force (van der Waals force, electrostatic force, etc.) acting on the cantilever when the tip comes close to the surface, the amplitude of this oscillation is decreased as the tip gets closer to the sample. As the cantilever is scanned over the sample, an electronic servo adjusts the height of the cantilever to maintain constant oscillation amplitude or frequency. Among the three AFM modes described here, tapping-mode AFM is the most widely used in biomaterials characterization. In tapping mode AFM, although there is still mechanical contact, damage to the sample surface is lessened as the contact is intermittent and no lateral frictional force is applied to the sample. Tapping-mode AFM is even gentle enough for the visualization of supported lipid monolayers under liquid medium [168]. Moreover, the problem of non-contact AFM in hydrophilic biomaterials characterization can be effectively avoided.

Tapping-mode AFM can provide topographical information with superior resolution on soft biomaterials such as biopolymers. In a typical example reported by Zhou et al. [169], tapping mode AFM was performed to acquire the height image of self-assembled peptide-based hydrogels in high resolution. Besides the height images, the phase images were also obtained in the tapping mode simultaneously. The z-scale of the phase images was in degree (°), which is quite different from that of the height images. In phase imaging, the phase lag between the driving oscillation and cantilever response is measured. The magnitude of this lag provides an indication of the amount of energy dissipated in the tip–sample interaction, which is related to the surface mechanical properties of the sample. Therefore, phase imaging can be used to determine the size, shape, and spacing of different materials domains that cannot otherwise be discerned from heights alone. Many types of biomaterials are modified to produce heterogeneous surface properties and so, phase imaging is a useful tool [170–173]. For example, Ye et al. [170] fabricated self-assembled RADA16-I peptide on mica at different pH values (pH = 1 or 4) in peptide solutions and investigated the samples by tapping-mode AFM. The height and phase images are depicted together in Fig. 4.31. The nanofibers produced at pH = 1 (Fig. 4.31D) are branched, whereas the nanofibers formed at pH = 4

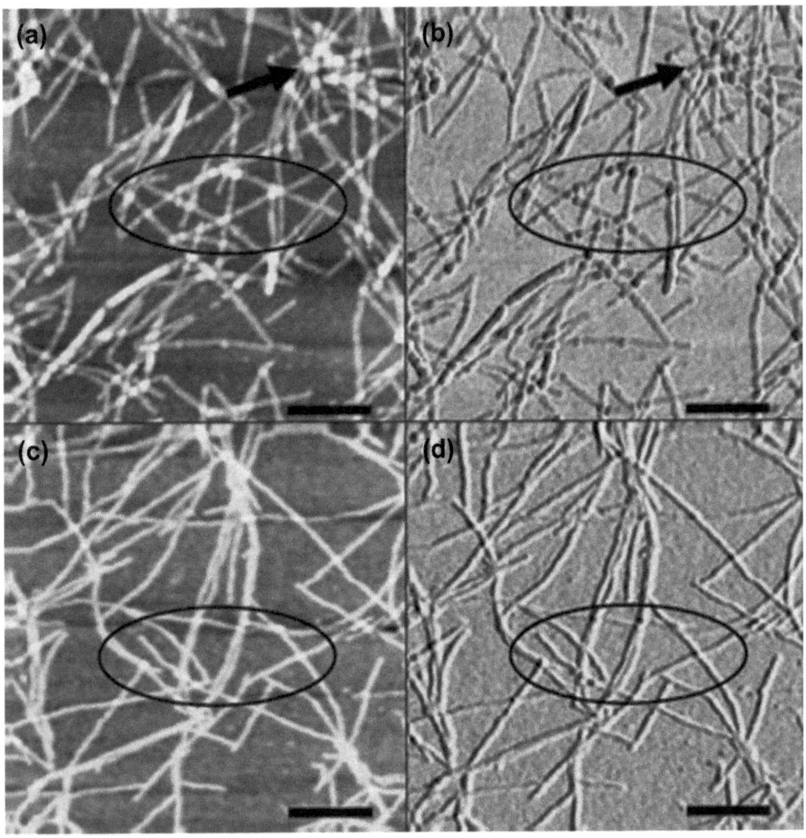

FIGURE 4.31 Typical AFM morphological images and corresponding phase images of self-assembling peptide RADA16-I denatured at pH = 4 and 1 on a freshly cleaved mica surface. After denaturing at pH = 4, the height image (a) with the maximum 3.0 nm scale and the phase image (b), with the maximum 7° scale. At pH = 1, the height image (c) with the maximum 2.6 nm scale and the phase image (d) with the maximum 0.9° scale. The scale bar is 500 nm. *Reprinted with permission from Ref. [170]. Copyright (2010) John Wiley and Sons.*

have no branch (Fig. 4.31B), although these features are not so clear in the corresponding height images (Fig. 4.31A and C).

In some situations, contact-mode AFM is used to study biomaterials, particularly when the lateral force images are desired. Lateral force images, also called friction images, map the lateral bending of the cantilever in the contact mode. The signals measured are closely related to the frictional interaction between the tip and the sample and so are useful in the identification of chemical differences on the samples. Similar to phase images in the tapping mode, lateral force images can be obtained together with height images simultaneously while the unit of *z*-scale is volt (V). There have been studies on the use of lateral force imaging to probe the chemical functionalities of potential biomaterials [174–176]. For example, Denis et al. [175] fabricated samples with a monolayer of CH_3-terminated alkylsilanes in nanoscale tracks surrounded by a monolayer of PEG-terminated alkylsilanes (named as CH_3-*X*/PEG-*Y*, *X* and *Y* are the widths of the CH_3 and PEG tracks in nanometres, respectively). These chemical nanopatterns were quite different in collagen adsorption and could lead to collagen nanopatterns on them by solution immersion. Figure 4.32 presents the height (a) and lateral force (b and c) AFM images obtained on sample CH_3-50/PEG-550 before collagen adsorption. The lateral force images show strong contrast between the tracks and matrix when the topographical image obtained is smooth along the *z*-scale (10 nm). The reversal of forward and backward lateral force images evidently indicates that sample CH_3-50/PEG-550 is chemically anisotropic.

Besides sample imaging, commercial AFM apparatus can be used to characterize the mechanical properties where the AFM tip serves as an indenter. This technique is called AFM indentation. When the indenter is pressed against the sample surface, load versus displacement curves are acquired. Since the load is applied by the cantilever, only mechanical characterization of compliant samples is allowed. In contrary, for stiff materials, the cantilever is more easily deflected than indenting the sample surface. Thus, this AFM-based technique can be applied to compliant biomaterials such as biopolymers. The force–displacement curves can even be used to evaluate the elasticity (resilience) of the substrates by converting them into force–penetration/retraction curves and calculating the area ratio of penetration curve to retraction curve [177,178].

There are other AFM-related techniques for biomaterials characterization. With magnetic or acoustic accessories, more details about the materials properties can be obtained. A typical example was presented by Block et al. [179] who suggested a new approach (called nanoparticle identification on magnetic imaging atomic force microscopy, NIOMI AFM), which was capable of distinguishing superparamagnetic maghemite and diamagnetic gold nanoparticles from a mixture by means of the magnetic interactions. As shown in Fig. 4.33, it is impossible to differentiate both types of nanoparticles by only relying on the information provided by the height map (Fig. 4.33 (top)).

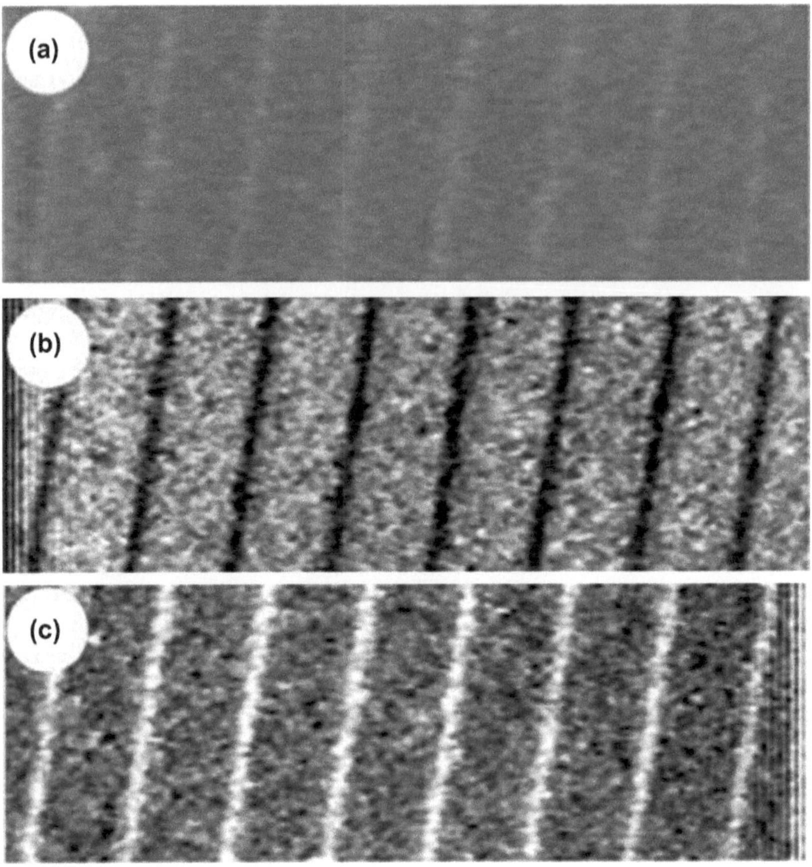

FIGURE 4.32 AFM height image (a; z range: 10 nm) and lateral force images (b, c; z range: 0.3 V) simultaneously recorded by forward (a, b) or backward (c) scanning of a CH_3-50/PEG-550 area (5 × 1.7 μm^2) in air. *Reprinted with permission from Ref. [175]. Copyright (2005) John Wiley and Sons.*

However, in the amplitude map created by NIOMI AFM, the maghemite and gold nanoparticles exhibit different contrast (Fig. 4.33 (middle)) and they are easily distinguished (Fig. 4.33 (bottom)). NIOMI AFM can even be used to characterize biofunctionalized nanoparticles (protein-tagged maghemite nanoparticles) [179]. The AFM tip can also be functionalized by protein adsorption or chemical covalent bonding and this AFM approach is called chemical force microscopy. Interested readers are referred to a review [180] for more details. Furthermore, AFM is a powerful tool for the *in vitro* characterization of biomaterials (mapping cells, viruses even their proteins) and this topic will be discussed in detail in Chapter 5.

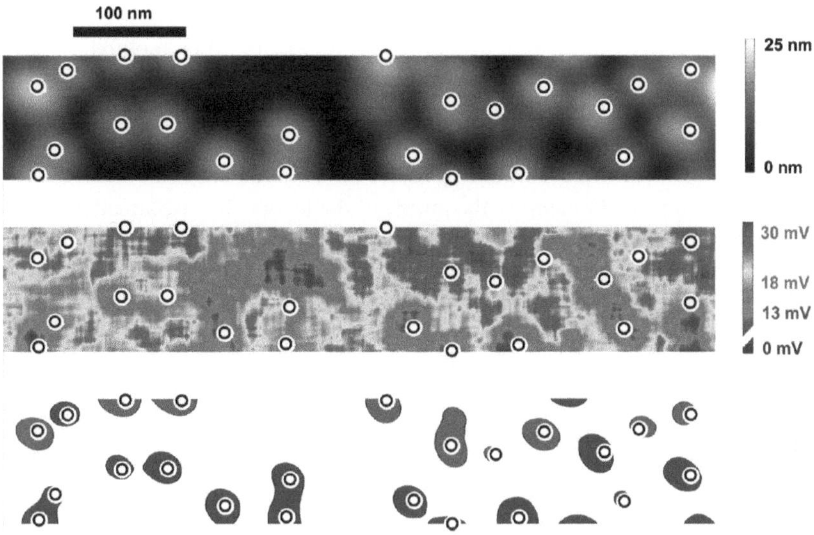

FIGURE 4.33 NIOMI AFM imaging of a sample containing superparamagnetic maghemite and diamagnetic gold nanoparticles. Image size, 110 × 640 nm^2; top, height profile; middle, amplitude map of the AFM cantilever oscillation (after line wise flattening and low pass filtering for noise reduction); bottom, correlation of height and amplitude map for nanoparticle identification (diamagnetic nanoparticles are marked in blue, while red color is used for the superparamagnetic ones). The centers of the particles are recognized by an algorithm in the height map (top, monochrome circles) and then transferred to the amplitude map for comparison. Correlation of the nanoparticle position with the color in the amplitude map identifies the nanoparticle upon their magnetization (cf. bottom). *Reprinted with permission from Ref. [179]. Copyright (2011) American Chemical Society.*

4.12. PROFILOMETRY

Profilometer, which is historically similar to the classical phonograph, is used to determine the surface topography of a solid sample. By tracking the surface changes using a profilometer, the surface roughness and morphological images of the samples can be determined. Modern profilometers can be roughly classified as contact profilometers and non-contact profilometers. Generally, contact profilometry is similar to contact-mode AFM, except that the measurement range is wider and the spatial resolution is lower. In contact profilometry, a stylus typically made of diamond vertically in contact with a sample is moved laterally across the sample surface for a specified distance at a specified contact force. The surface variations as monitored by the vertical stylus displacements are measured as a function of position to obtain the profile of the sample surface. Contact profilometry is a direct technique as no modelling is required. The stylus tip can be as small as 20 nm and so the lateral resolution is much better than that of white-light optical profilometry (the resolution of optical profilometry is determined by the optical wavelength). More importantly, compared to optical profilometry, this method is less affected by contaminants on the sample surface and it is more preferable than optical profilometry when the measurement environment is complex. Contact profilometry is widely used to study biomaterials [97,181–183]. For instance, Sima et al. [181] formed fibronectin layers on a silicon substrate and utilized contact profilometry to measure the profile of the layers. Saber-Samandari et al. [97] deposited calcium phosphate on titanium by thermal printing and conducted contact profilometry to determine the profiles of the deposits. Barranco et al. [183] fabricated oxide barriers on biomedical Ti6Al4V by blasting with different particles and thermal treatment. Contact profilometry was subsequently performed to determine the roughness of the samples.

As an alternative to stylus-based profilometry, optical profilometry is a non-contact method that can provide similar information as contact profilometry. There are also advantages associated with optical profilometry such as high vertical resolution and convenient operation. A big advantage of this method is that it is a non-contact technique and hence, the instrument and samples cannot be damaged by surface wear or careless operators. Optical profilometry is also commonly used to measure the surface topography of biomaterials [184–186]. Park et al. [184] modified Ti–6Al–4V alloys by a hydrothermal treatment in a strontium-containing solution and used optical profilometry to determine the surface roughness of the modified samples. Truong et al. [186] achieved ultrafine crystallinity in the bulk of titanium by severe plastic deformation by means of equal channel angular pressing (ECAP). By means of optical profilometry, the surface roughness and the topographical image of polished ECAP-Ti were obtained. Besides measuring the surface profiles, roughness, and topography, profilometry can even be used to determine the thickness [187] and wear depth [188] of biomaterials, and interested readers are referred to the cited articles for more information.

4.13. CONTACT ANGLE MEASUREMENT

Contact angle (CA) measurement is a commonly used thermodynamic method for the characterization of surface hydrophilicity/hydrophobicity of biomaterials. As perhaps one of the oldest methods used to investigate biomaterials surfaces, it yields very useful information as the surface hydrophilicity of biomaterials is crucial to the functions of biomaterials such as cell adhesion and spreading. CA measurement is a relatively simple method for solid samples and typically no special sample preparation is required. Anyway, the samples characterized should be clean enough and do not swell or dissolve in the test liquid. For an immobile surface, CA measurement is very surface method as the CA is only related to the outermost 3–10 Å of the surface. There are several types of CA measurements and the more common Wilhelmy method, sessile drop method, and captive bubble method are described here.

4.13.1. Wilhelmy Method

Wilhelmy method is a dynamic method by measuring the equilibrium surface or interfacial tension at the air–liquid or liquid–liquid interface. In this method, the solid sample has same properties on both sides and is oriented perpendicularly to the interface. The wetting force on the solid is measured as the solid is immersed in or withdrawn from a liquid of a known surface tension. The average advancing and receding CAs (θ) is calculated according to the equation:

$$\cos\theta = (F - F_b)/I\sigma,$$

where F is the total force measured by the force metre, F_b is the force of buoyancy due to the solid sample displacing the liquid, I is the wetted length, and σ is the known surface tension (force per unit length) of the liquid.

As a lot of modern biomaterials are surface modified to obtain different properties on both sides, the Wilhelmy method is not a popular tool. However, there is a key advantage in that the Wilhelmy method is not limited by the sample size and shape and so single fiber can be used to determine the CAs [189,190]. The Wilhelmy method can even be adopted on the nanoscale by making use of a nanotube [191] or AFM probe [192,193] as the cantilever to measure the wetting forces of nanotubes and nanofibers. Hence, the Wilhelmy method will attract more attention in the future as more biomaterials have a small size.

4.13.2. Sessile Drop Method

There are two kinds of sessile drop method, namely the static sessile drop method and dynamic sessile drop method. In the static sessile drop method, a droplet of pure liquid is deposited vertically onto the sample surface and CAs are measured by a CA goniometer using an optical sub-system. The angle formed between the solid/liquid interface and liquid/gas interface is determined as the liquid CA (Fig. 4.34). The dynamic sessile drop method is similar to the static sessile drop method but requires the drop to be modified. Both advancing angle (the largest CA possible without increasing its solid/liquid interfacial area by adding volume dynamically) and receding angle (the smallest possible angle by removing the volume without decreasing its solid/liquid interfacial area) are measured in the dynamic sessile drop method. The difference between the advancing and receding angle is the CA hysteresis. The advancing and receding angle determined by the dynamic sessile drop method can also be used to calculate the equilibrium CA [194,195].

The sessile drop method is the most commonly used CA measurement for biomaterials [196–199] because it is a straightforward approach revealing surface energetics of the samples. Typically, pure water serves as the test liquid. For instance, Yamanlar et al. [196] described a simple approach to functionalize

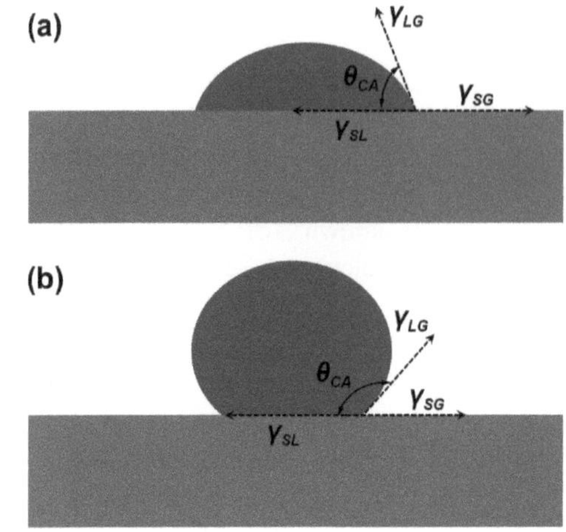

FIGURE 4.34 Illustration of the sessile drop method by a solid substrate with good wettability (a) or poor wettability (b). θ_{CA} is the liquid contact angle, and γ_{SL}, γ_{SG}, and γ_{LG} represent the solid/liquid, solid/gas, and liquid/gas interfaces, respectively.

photo-cross-linked HA hydrogels by deposition of poly(L-lysine) (PLL) and HA multilayer films layer by layer (LBL). After each layer deposition, they measured the water CA on the substrate using the static sessile drop method and the results directly indicate that HA reduce the CA, whereas PLL layer increases the CA on the HA hydrogel surface in an alternating manner. Alauzun et al. [197] covalently linked hyaluronic acid (HA) to biomedical PDMS elastomer surfaces using a series of steps to achieve better biocompatibility and used the dynamic sessile drop method to characterize the hydrophilicity. The results reveal that both the advancing water CA and receding water CA are reduced after each step of chemical modification.

Besides the surface chemistry, the surface roughness is also crucial to the CA in the sessile drop method [200]. When the liquid and rough substrate are in a wetted contact (liquid completely fills the grooves of the rough surface where they contact), the Wenzel's theory provides correlation between the surface roughness and the apparent CA of liquid droplet with the following relationship:

$$\cos \theta_w = r \cos \theta,$$

where θ_w is the apparent CA in the Wenzel mode, θ is the true CA, and r is the surface roughness factor. As the roughness factor (r) is always above 1, if the CA of a liquid on a smooth surface is less than 90°, the apparent angle on a rough surface will be smaller, whereas for a true CA > 90°, the CA on a rough surface will be larger. In other words, in the Wenzel mode, there is an amplified effect of surface roughness to the CA (for $\theta < 90°$, $\theta_w < \theta$; for $\theta > 90°$, $\theta_w > \theta$). Another theory, Cassie's theory, is applicable when the liquid and rough substrate are in non-wetted contact (vapour pockets are assumed to be trapped underneath the liquid). In the Cassie mode, the apparent CA is always larger than the true CA, no matter whether the true CA is above 90° or not.

For solid biomaterials with surface micro-/nano-structures, the sessile drop method can be used to characterize the hydrophilicity and hydrophobicity when the surface roughness is taken into consideration [98,201–203]. Ma et al. [201] prepared polyethylene terephthalate (PET) nanofibers by electrospinning and modified the surface with polymerization of methacrylic acid (PMMA) grafting followed by gelatin grafting to achieve better surface biocompatibility. Both the static and dynamic sessile drop methods were employed to characterize the wettability of PET nanofibers before and after surface modification. The PET films before and after PMMA and gelatin grafting were also analyzed and compared to PET nanofibers. As shown in Table 4.1, the PET nanofibers are totally different from the PET films based on the water CAs. The original PET nanofibers have the much higher advancing and static water CAs than the original PET film, whereas the receding angle is much smaller. Both PMMA and gelatin modification enhance the hydrophilicity of the PET surface. The decrease in the water CA is much more significant on the PET nanofibers due to its surface structure. According to the theories described above, the water CA on the original PET nanofibers is in the Cassie's mode (air is trapped in the pores and the CA appearance is larger than the true CA). With regard to the PET nanofibers after PMMA and gelatin grafting, the apparent water CA is changed to the Wenzel's mode (true CA is $<90°$ and apparent CA is even smaller as water is sucked into the nanofibers).

By using this method, the surface energy of the solid biomaterials can be calculated when the surface energy of the test liquid is known and the liquid-sample CAs are measured. Generally, the use of different test liquids is required to calculate the surface energy. The surface energy has the unit of Joules per area, which is equivalent to that in the case of liquids that surface tension is

TABLE 4.1 Water Contact Angles of the Original and Surface-modified PET Films and PET nanofibers

Sample	Advancing (deg)	Receding (deg)	Static (deg)
Original PET films	97 ± 3	53 ± 4	80 ± 2
PMAA-grafted PET films	74 ± 2	19 ± 1	53 ± 2
Gelatin-grafted PET films	71 ± 3	17 ± 2	50 ± 3
Original PET nanofibers	144 ± 3	15 ± 2	128 ± 3
PMAA-grafted PET nanofibers	–	–	0
Gelatin-grafted PET nanofibers	–	–	0

Reprinted with permission from [201]. Copyright (2005) Elsevier.

measured in Newton per metre. It is known that surface energy can be subdivided according to the various interactions such as interactions due to dispersive (van der Waals) force, hydrogen bonding, polar interactions, acid/base interactions, and so on. Generally, dispersive interactions and polar interactions are the major contributions to the overall surface energy. In this case, the overall surface energy can be described as:

$$\sigma_S = \sigma_S^D + \sigma_S^P \text{ and } \sigma_L = \sigma_L^D + \sigma_L^P,$$

where σ_S is the total surface energy of the solid, σ_S^D and σ_S^P are, respectively, the dispersive and polar components of the solid surface energy, σ_L is the total surface tension/energy of the liquid, and σ_L^D and σ_L^P are, respectively, the dispersive and polar components of the liquid surface energy. Many theories have been employed to determine the surface energy from CAs and they differ in derivation, convention, and number of parameters. The common theories used such as the Owens/Wendt theory [204–207] and the Fowkes theory [208] divide the surface energy into two components: surface energy due to dispersive interactions and surface energy due to polar interactions. The Fowkes theory's principle equation is mathematically equivalent to that of Owens/Wendt:

$$\sigma_L(1 + \cos\theta) = 2(\sigma_S^D)^{1/2}(\sigma_L^D)^{1/2} + 2(\sigma_S^P)^{1/2}(\sigma_L^P)^{1/2}.$$

Thus, by utilizing two test liquids with known surface energies (σ_L^D and σ_L^P) to measure the CAs on a solid sample, the surface energy of the sample (σ_S^D and σ_S^P) can be calculated.

The sessile drop method has advantages and disadvantages. Aside from the straightforward nature, another advantage of the sessile drop method is that with a large-enough solid surface, multiple droplets can be deposited in various locations on the sample to determine the surface heterogeneity. However, when the sample is only large enough for one droplet, the determination of sample's heterogeneity will be difficult. The CA measurement by the sessile drop method is impractical if the sample is even smaller (not large enough for only one droplet). In addition, this measurement is hampered by the inherent subjectivity as the placement of the lines is determined either by the user looking at the picture or by the image analysis software's definition of the lines.

4.13.2.1. Captive Bubble Method

The captive bubble method measures the wettability of samples in a liquid (probably in water). In this method, an air bubble is placed in contact with the sample immersed in a solution. After the contact, the drop profile of the air bubble is imaged using a charge-coupled device (CCD) camera and the CA is calculated from the image. As shown in Fig. 4.35, the interface between the solid sample and droplet is the solid/gas interface. As the CA (liquid, θ_{CA}) is defined as the angle formed between the solid/liquid interface and the liquid/gas interface,

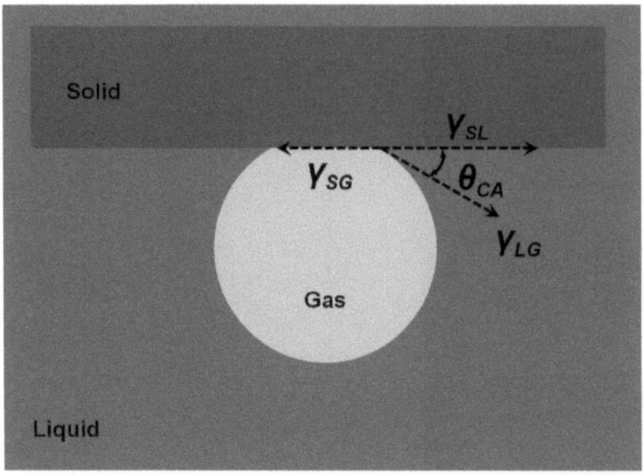

FIGURE 4.35 Illustration of the captive bubble method by a solid substrate with good wettability. θ_{CA} is the liquid contact angle, and γ_{SL}, γ_{SG}, and γ_{LG} represent the solid/liquid, solid/gas, and liquid/gas interfaces, respectively.

the location of θ_{CA} is different from that in the sessile drop method (comparing Fig. 4.35 with Fig. 4.34). The captive bubble method can also be performed in the dynamic mode by using a syringe plunger to increase or decrease the air bubble volume and both the advancing and receding angles can be measured as a result.

Although it is not as popular as sessile drop method, the captive bubble method is useful to the biomaterials with hydrated surfaces [209–212]. For instance, it is good for the investigation of biomedical hydrogels since the materials are immersed in a liquid throughout the process and, therefore, does not undergo dehydration. It has also been pointed out [209,210] that for hydrated surfaces, the water CA measured by the captive bubble method does not agree with that measured by the sessile drop method, probably due to sample dehydration when the sample is exposed in air.

In conclusion, the CA measurement is a useful analytical tool for biomaterials not only revealing the surface energetics but also relating to the surface structure. Comparison of CA measurements on the control and surface-modified biomaterials is usually adopted to confirm successful alteration of the sample surface. Aside from the three types mentioned above, there are other types of CA measurements such as the capillary penetration method but owing to space limitation, they are not described here.

4.14. ELLIPSOMETRY

Ellipsometry is a specular optical technique, which provides unequalled capabilities for thin film metrology. In this technique, electromagnetic radiation is emitted by a light source and linearly polarized by a polarizer, passes through an

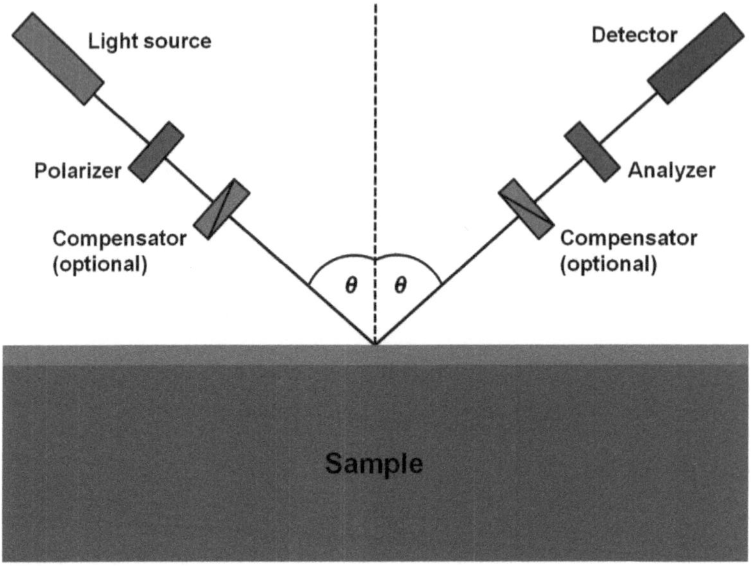

FIGURE 4.36 Schematic diagram of ellipsometry.

optional compensator (retarder, quarter wave plate), and irradiates the sample. After reflection, the radiation passes a compensator (optional) and a second polarizer (analyzer) and is detected. As schematically illustrated in Fig. 4.36, the incident and reflected beams span the plane of incidence. The polarization state of the light incident upon the sample may be decomposed into the s and p components (the s component oscillates perpendicular to the plane of incidence and parallel to the sample surface and the p component oscillates parallel to the plane of incidence). The amplitudes of the s and p components, after reflection and normalized to their initial value, are denoted by r_s and r_p, and are used to calculate the complex reflectance ratio (ρ) by the following equation:

$$\rho = r_p/r_s = \tan(\Psi)e^{i\Delta}$$

where tan(Ψ) is the amplitude ratio upon reflection, and Δ is the phase shift (difference). By performing a suitable model analysis on the data, Ψ and Δ, which match the experimental data best, are calculated and provide the optical constants and thickness of the sample.

Based on the analysis of the change of polarization of light reflected off a sample, ellipsometry can yield information about layers that are thinner than the wavelength of the probing light itself, even down to a single atomic layer. As an optical technique, the non-destructive and contactless characteristics of ellipsometry are desirable for soft layers on biomaterials. Furthermore, since ellipsometry measures the intensity ratio instead of pure intensity, it is less affected by intensity instability of the light source or atmospheric absorption, and no reference measurement is necessary thus yielding very good accuracy.

Ellipsometry has many applications in many different fields such as semiconductor physics, microelectronics, biomaterials, and so on. In the biomaterials field, ellipsometry is commonly used to characterize the thickness of biofunctional layers on the substrates ranging from a few Angstroms to several micrometres for layers which are optically homogeneous and isotropic and when a significant refractive index discontinuity exists at the interface. Both monochromatic ellipsometry [213,214] and spectroscopic ellipsometry [215–217] are utilized to determine the thickness of the layers. For instance, Toworfe et al. [213] used monochromatic ellipsometry with a HeNe laser source (632.8 nm) to characterize the thickness and homogeneity of calcium phosphate coated on various surface functionalized oxidized silicon substrates. Pei et al. [217] fabricated surface density gradients of PEG on titanium dioxide surfaces by a controlled dipping process and variable-angle spectroscopic ellipsometry was employed to determine the thickness of the gradient polymer coverage.

Besides conventional characterization of film thickness, ellipsometry can be employed *in situ* (referred to as *in situ* ellipsometry) in dynamic measurements [218–220]. Santonicola et al. [219] grafted polymeric brushes of poly (methacrylic acid) on silicon substrates and then made use of *in situ* ellipsometry to quantify the pH-induced swelling and collapse of the grafted polymer layers in the liquid environment. Blacklock et al. [220] produced the assembly of LBL films on a flexible stainless steel substrate by using plasmid DNA and reducible hyperbranched poly(amido amine) (RHB) polycation. *In situ* ellipsometry was performed subsequently to confirm the disassembly of the films under reducing conditions. As shown in Fig. 4.37, the thickness of the

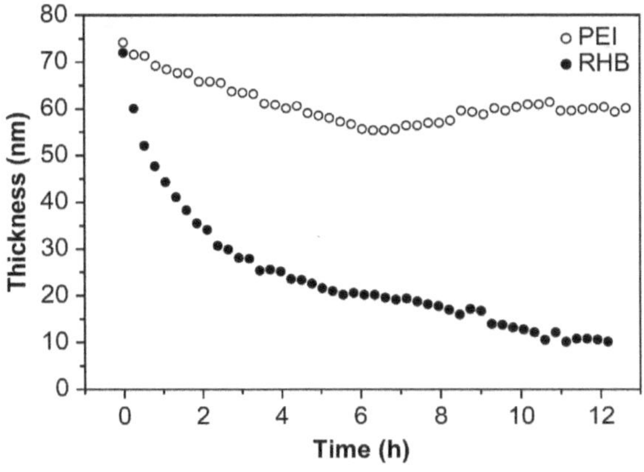

FIGURE 4.37 Film disassembly in reducing conditions. Thickness of $(DNA/RHB)_{15}$ (solid circles) and $(DNA/PEI)_{15}$ (hollow circles) films deposited on the surface of silicon wafer was determined by *in situ* ellipsometry in 20 mM DTT solution. *Reprinted with permission from Ref. [220]. Copyright (2009) Elsevier.*

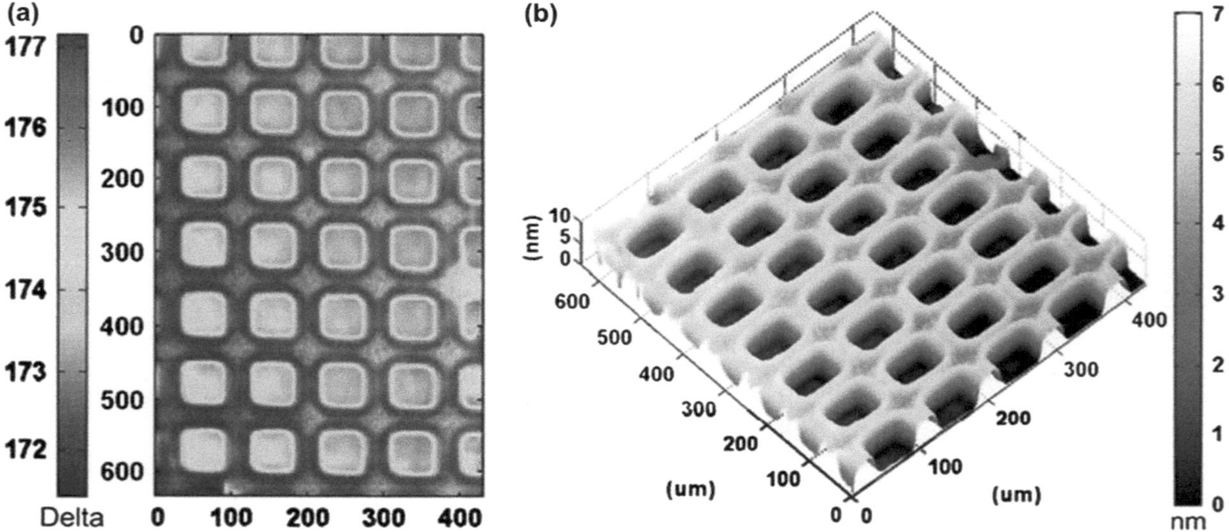

FIGURE 4.38 Ellipsometric characterization of thickness, uniformity, and patterns of supported DMPC membranes. (a) Spatially resolved map of ellipsometric phase shift, Δ, for a micropatterned DMPC bilayer on SiO_2/Si substrate and (b) corresponding thickness map derived using a refractive index of 1.44. *Reprinted with permission from Ref. [224]. Copyright (2007) Elsevier.*

DNA/RHB films decreases exponentially over a period of 12 h when the control DNA/poly(ethylenimine) (DNA/PEI) films show only about 10 nm decrease in the thickness throughout the same period of time.

Ellipsometry can even be extended to imaging ellipsometry by using a monochromatic laser as light source and a CCD camera as the detector. Good contrast among images can be achieved [221–224]. Sapuri-Butti et al. [222] used imaging ellipsometry to characterize the spatially patterned phospholipid bilayers on PDMS substrates with their thickness images. Howland et al. [224] fabricated a micropatterned 1,2-dimyristoyl-*sn*-glycero-3-phosphocholine (DMPC) bilayer on a silicon substrate and then characterized the surface coverage by imaging ellipsometry. As shown in Fig. 4.38, both the phase shift (Δ) and thickness maps indicate the checked pattern of DMPC bilayer.

4.15. CONCLUSIONS

The surface properties can impact critically the success or failure of biomaterials. Consequently, the appropriate surface analytical techniques are crucial to the revelation of important surface properties such as chemistry, topography, hydrophilicity/hydrophobicity, and so on. Generally, a combination of surface characterization methods is recommended to address a problem from different perspectives and to provide more comprehensive information about the biomaterials surface. By taking advantage of these techniques, the *in vitro* and *in vivo* performances of biomaterials can be assessed reliably. In this chapter, we only describe the common surface analytical techniques due to the length limitation and readers are urged to check other references. For instance, colorimetric methods can be utilized to determine the functional groups and immobilized proteins on biomaterials by using specific reagents (coomassie brilliant blue [225], ninhydrin [226], etc.) or kits (bicinchoninic acid kit [227], etc.). Conventional bulk analytical techniques such as X-ray diffraction can also be adopted to determine the crystal structure of biomaterials surface by resorting to a grazing configuration.

REFERENCES

[1] Berkowitz J. Photoabsorption, photoionization, and photoelectron spectroscopy. Academic Press; 1979.

[2] Feldman LC, Mayer JW. Fundamentals of surface and thin film analysis. North Holland-Elsevier; 1986.

[3] He W, Ma ZW, Yong T, Teo WE, Ramakrishna S. Fabrication of collagen-coated biodegradable polymer nanofiber mesh and its potential for endothelial cells growth. Biomaterials 2005;26:7606–15.

[4] Liu XH, Won YJ, Ma PX. Porogen-induced surface modification of nano-fibrous poly (L-lactic acid) scaffolds for tissue engineering. Biomaterials 2006;27:3980–7.

[5] Shen H, Hu XX, Yang F, Bei JZ, Wang SG. Combining oxygen plasma treatment with anchorage of cationized gelatin for enhancing cell affinity of poly(lactide-co-glycolide). Biomaterials 2007;28:4219–30.

[6] Citeau A, Guicheux J, Vinatier C, Layrolle P, Nguyen TP, Pilet P, et al. In vitro biological effects of titanium rough surface obtained by calcium phosphate grid blasting. Biomaterials 2005;26:157–65.

[7] Swan EEL, Popat KC, Desai TA. Peptide-immobilized nanoporous alumina membranes for enhanced osteoblast adhesion. Biomaterials 2005;26:1969–76.

[8] Zhang F, Zhang ZB, Zhu XL, Kang ET, Neoh KG. Silk-functionalized titanium surfaces for enhancing osteoblast functions and reducing bacterial adhesion. Biomaterials 2008;29:4751–9.

[9] Steinmueller-Nethl D, Kloss FR, Najam-U-Haq M, Rainer M, Larsson K, Linsmeier C, et al. Strong binding of bioactive BMP-2 to nanocrystalline diamond by physisorption. Biomaterials 2006;27:4547–56.

[10] Wang HY, Ji JH, Zhang W, Wang W, Zhang YH, Wu ZW, et al. Rat calvaria osteoblast behavior and antibacterial properties of O_2 and N_2 plasma-implanted biodegradable poly(butylene succinate). Acta Biomater 2010;6:154–9.

[11] Wagner CD, Riggs WM, Davis LE, Moulder JF, Muilenberg GE. Handbook of x-ray photoelectron spectroscopy: a reference book of standard data for use in x-ray photoelectron spectroscopy. Perkin-Elmer Corporation; 1979.

[12] Moulder JF, Stickle WF, Sobol PE, Bomben KD. Handbook of x-ray photoelectron spectroscopy: a reference Book of standard spectra for identification and interpretation of XPS data. Perkin-Elmer Corporation; 1992.

[13] Lee KS, Won MS, Noh HB, Shim YB. Triggering the redox reaction of cytochrome c on a biomimetic layer and elimination of interferences for NADH detection. Biomaterials 2010;31:7827–35.

[14] Briggs D, Seah MP. Practical surface analysis by auger and x-ray photoelectron spectroscopy. John Wiley & Sons, Ltd; 1983.

[15] Briggs D, Seah MP. Practical surface analysis (second editon): volume 1- auger and x-ray photoelectron spectroscopy. John Wiley & Sons, Ltd; 1990.

[16] Briggs D, Grant JT. Surface analysis by Auger and x-ray photoelectron spectroscopy. SurfaceSpectra and IM Publications; 2003.

[17] Andrade JD. Surface and interfacial aspects of biomedical polymers: volume 1- surface chemistry and physics. Plenum Press; 1985.

[18] Beamson G, Briggs D. High resolution XPS of organic polymers: the Scienta ESCA300 Database. John Wiley & Sons, Ltd; 1992.

[19] Cao HL, Liu XY, Meng FH, Chu PK. Biological actions of silver nanoparticles embedded in titanium controlled by micro-galvanic effects. Biomaterials 2011;32:693–705.

[20] Arnould C, Volcke C, Lamarque C, Thiry PA, Delhalle J, Mekhalif Z. Titanium modified with layer-by-layer sol-gel tantalum oxide and an organodiphosphonic acid: a coating for hydroxyapatite growth. J Colloid Interface Sci 2009;336:497–503.

[21] Pisarek M, Lewandowska M, Roguska A, Kurzydlowski KJ, Janik-Czachor M. SEM, scanning Auger and XPS characterization of chemically pretreated Ti surfaces intended for biomedical applications. Mater Chem Phys 2007;104:93–7.

[22] Linderback P, Harmankaya N, Askendal A, Areva S, Lausmaa J, Tengvall P. The effect of heat- or ultra violet ozone-treatment of titanium on complement deposition from human blood plasma. Biomaterials 2010;31:4795–801.

[23] Barrere F, van Blitterswijk CA, de Groot K, Layrolle P. Nucleation of biomimetic Ca-P coatings on Ti6Al4V from a SBF X 5 solution: influence of magnesium. Biomaterials 2002;23:2211–20.
[24] Wong MH, Cheng FT, Man HC. Characteristics, apatite-forming ability and corrosion resistance of NiTi surface modified by AC anodization. App Surf Sci 2007;253:7527–34.
[25] Sanada N, Yamamoto A, Oiwa R, Ohashi Y. Extremely low sputtering degradation of polytetrafluoroethylene by C_{60} ion beam applied in XPS analysis. Surf Interface Anal 2004;36:280–2.
[26] Chen YY, Yu BY, Wang WB, Hsu MF, Lin WC, Lin YC, et al. X-ray photoelectron spectrometry depth profiling of organic thin films using C_{60} sputtering. Anal Chem 2008;80:501–5.
[27] Yu BY, Chen YY, Wang WB, Hsu MF, Tsai SP, Lin WC, et al. Depth profiling of organic films with X-ray photoelectron spectroscopy using C_{60}^{+} and Ar^{+} co-sputtering. Anal Chem 2008;80:3412–5.
[28] Rafati A, Davies MC, Shard AG, Hutton S, Mishra G, Alexander MR. Quantitative XPS depth profiling of codeine loaded poly(L-lactic acid) films using a coronene ion sputter source. J Control Release 2009;138:40–4.
[29] Kingshott P, Andersson G, McArthur SL, Griesser HJ. Surface modification and chemical surface analysis of biomaterials. Curr Opin Chem Biol 2011;15:667–76.
[30] Artyushkova K, Farrar JO, Fulghum JE. Data fusion of XPS and AFM images for chemical phase identification in polymer blends. Surf Interface Anal 2009;41:119–26.
[31] Hajati S, Tougaard S, Walton J, Fairley N. Noise reduction procedures applied to XPS imaging of depth distribution of atoms on the nanoscale. Surf Sci 2008;602:3064–70.
[32] Hajati S, Coultas S, Blornfield C, Tougaard S. Nondestructive quantitative XPS imaging of depth distribution of atoms on the nanoscale. Surf Interface Anal 2008;40:688–91.
[33] Vohrer U, Blomfield C, Page S, Roberts A. Quantitative XPS imaging-new possibilities with the delay-line detector. Appl Surf Sci 2005;252:61–5.
[34] Lee CY, Harbers GM, Grainger DW, Gamble LJ, Castner DG. Fluorescence, XPS, and TOF-SIMS surface chemical state image analysis of DNA microarrays. J Am Chem Soc 2007;129:9429–38.
[35] Hedberg CL. Handbook of Auger electron spectroscopy: a book of reference data for identification and interpretation in Auger electron spectroscopy. Physical Electronics, Inc; 1995.
[36] Hanzi AC, Gunde P, Schinhammer M, Uggowitzer PJ. On the biodegradation performance of an Mg-Y-RE alloy with various surface conditions in simulated body fluid. Acta Biomater 2009;5:162–71.
[37] Shabalovskaya SA, Tian H, Anderegg JW, Schryvers DU, Carroll WU, Van Humbeeck J. The influence of surface oxides on the distribution and release of nickel from Nitinol wires. Biomaterials 2009;30:468–77.
[38] Xue WC, Liu XY, Zheng XB, Ding CX. In vivo evaluation of plasma-sprayed titanium coating after alkali modification. Biomaterials 2005;26:3029–37.
[39] Nakai M, Niinomi M, Akahori T, Ohtsu N, Nishimura H, Toda H, et al. Surface hardening of biomedical Ti-29Nb-13Ta-4.6Zr and Ti-6Al-4V ELI by gas nitriding. Mater Sci Eng A 2008;486:193–201.
[40] Jiang XR, Chen R, Bent SF. Spatial control over atomic layer deposition using micro-contact-printed resists. Surf Coat Technol 2007;201:8799–807.

[41] Wang ZX, Liu JY, Zhang K, Cai HB, Zhang GH, Wu YK, et al. Fabrication of well-aligned and highly dense cadmium sulfide nanowires on DNA scaffolds using the poly (dimethylsiloxane) transfer method. J Phys Chem C 2009;113:5428–33.
[42] Watanabe S, Shibata H, Sakamoto F, Azumi R, Sakai H, Abe M, et al. Directed self-assembly of gold nanoparticles and gold thin films on micro- and nanopatterned templates fabricated from mixed phase-separated Langmuir-Blodgett films. J Mater Chem 2009;19:6796–803.
[43] Vickerman JC, Brown A, Reed NM. Secondary ion mass spectrometry, principles and applications. Oxford University Press; 1989.
[44] Behrisch R. Sputtering by particle bombardment I: physical sputtering of single-element solids. Springer-Verlag; 1981.
[45] Behrisch R. Sputtering by particle bombardment II: sputtering of alloys and compounds, electron and neuron sputtering, surface topography. Springer-Verlag; 1983.
[46] Behrisch R, Wittmaack K. Sputtering by particle bombardment III: characteristics of sputtered particles, technical applications. Springer-Verlag; 1991.
[47] Behrisch R, Eckstein W. Sputtering by particle bombardment: experiments and computer calculations from threshold to MeV energies. Springer-Verlag; 2007.
[48] Benninghoven A, Rüdenauer FG, Werner HW. Secondary ion mass spectrometry: basic concepts, instrumental aspects, applications, and trends. John Wiley & Sons, Ltd; 1987.
[49] Vickerman JC, Briggs D. ToF-SIMS: surface analysis by mass spectrometry. SurfaceSpectra and IM Publications; 2001.
[50] Tidwell CD, Castner DG, Golledge SL, Ratner BD, Meyer K, Hagenhoff B, et al. Static time-of-light secondary ion mass spectrometry and X-ray photoelectron spectroscopy characterization of adsorbed albumin and fibronectin films. Surf Interface Anal 2001;31:724–33.
[51] von Vacano B, Xu R, Hirth S, Herzenstiel I, Ruckel M, Subkowski T, et al. Hydrophobin can prevent secondary protein adsorption on hydrophobic substrates without exchange. Anal Bioanal Chem 2011;400:2031–40.
[52] Killian MS, Krebs HM, Schmuki P. Protein denaturation detected by time-of-flight secondary ion mass spectrometry. Langmuir 2011;27:7510–5.
[53] Salerno S, Piscioneri A, Laera S, Morelli S, Favia P, Bader A, et al. Improved functions of human hepatocytes on NH_3 plasma-grafted PEEK-WC–PU membranes. Biomaterials 2009;30:4348–56.
[54] Pidhatika B, Moller J, Benetti EM, Konradi R, Rakhmatullina E, Muhlebach A, et al. The role of the interplay between polymer architecture and bacterial surface properties on the microbial adhesion to polyoxazoline-based ultrathin films. Biomaterials 2010;31:9462–72.
[55] Vickerman JC, Gilmore IS. Surface analysis-the principal techniques. John Wiley & Sons, Ltd; 2009.
[56] Barnes CA, Brison J, Michel R, Brown BN, Castner DG, Badylak SF, et al. The surface molecular functionality of decellularized extracellular matrices. Biomaterials 2011;32:137–43.
[57] Brown BN, Barnes CA, Kasick RT, Michel R, Gilbert TW, Beer-Stolz D, et al. Surface characterization of extracellular matrix scaffolds. Biomaterials 2010;31:428–37.
[58] Yang J, Mei Y, Hook AL, Taylor M, Urquhart AJ, Bogatyrev SR, et al. Polymer surface functionalities that control human embryoid body cell adhesion revealed by high throughput surface characterization of combinatorial material microarrays. Biomaterials 2010;31:8827–38.

[59] Tyler BJ, Bruening C, Rangaranjan S, Arlinghaus HF. TOF-SIMS imaging of adsorbed proteins on topographically complex surfaces with Bi^{3+} primary ions. Biointerphases 2011;6:135–41.

[60] Vaezian B, Anderton CR, Kraft ML. Discriminating and imaging different phosphatidylcholine species within phase-separated model membranes by principal component analysis of TOF-secondary ion mass spectrometry images. Anal Chem 2010;82:10002–14.

[61] Liu F, Dubey M, Takahashi H, Castner DG, Grainger DW. Immobilized antibody orientation analysis using secondary ion mass spectrometry and fluorescence imaging of affinity-generated patterns. Anal Chem 2010;82:2947–58.

[62] Dubey M, Emoto K, Takahashi H, Castner DG, Grainger DW. Affinity-based protein surface pattern formation by ligand self-selection from mixed protein solutions. Adv Funct Mater 2009;19:3046–55.

[63] Ogaki R, Lyckegaard F, Kingshott P. High resolution surface chemical analysis of a tri-functional pattern made by sequential colloidal shadowing. Chem Phys Chem 2010;11:3609–16.

[64] Ogaki R, Cole MA, Sutherland DS, Kingshott P. Microcup arrays featuring multiple chemical regions patterned with nanoscale precision. Adv Mater 2011;23:1876–81.

[65] Tyler BJ, Rayal G, Castner DG. Multivariate analysis strategies for processing ToF-SIMS images of biomaterials. Biomaterials 2007;28:2412–23.

[66] Park JW, Min H, Kim YP, Shon HK, Kim J, Moon DW, et al. Multivariate analysis of ToF-SIMS data for biological applications. Surf Interface Anal 2009;41:694–703.

[67] Lamolle SF, Monjo M, Rubert M, Haugen HJ, Lyngstadaas SP, Ellingsen JE. The effect of hydrofluoric acid treatment of titanium surface on nanostructural and chemical changes and the growth of MC3T3-E1 cells. Biomaterials 2009;30:736–42.

[68] Lechene C, Hillion F, McMahon G, Benson D, Kleinfeld AM, Kampf JP, et al. High-resolution quantitative imaging of mammalian and bacterial cells using stable isotope mass spectrometry. J Biol 2006;5:20.

[69] Tyler BJ, Takeno MM, Hauch KD. Identification and Imaging of ^{15}N labeled cells with ToF-SIMS. Surf Interface Anal 2011;43:336–9.

[70] Cheng J, Wucher A, Winograd N. Molecular depth profiling with cluster ion beams. J Phys Chem B 2006;110:8329–36.

[71] Cheng J, Winograd N. Depth profiling of peptide films with TOF-SIMS and a C_{60} Probe. Anal Chem 2005;77:3651–9.

[72] Shard AG, Green FM, Brewer PJ, Seah MP, Gilmore IS. Quantitative molecular depth profiling of organic delta-layers by C_{60} ion sputtering and SIMS. J Phys Chem B 2008;112:2596–605.

[73] Fletcher JS, Lockyer NP, Vaidyanathan S, Vickerman JC. TOF-SIMS 3D biomolecular imaging of xenopus laevis oocytes using buckminsterfullerene (C_{60}) primary ions. Anal Chem 2007;79:2199–206.

[74] Wucher A, Cheng J, Winograd N. Protocols for three-dimensional molecular imaging using mass spectrometry. Anal Chem 2007;79:5529–39.

[75] Clarke B, Kingshott P, Hou X, Rochev Y, Gorelov A, Carroll W. Effect of nitinol wire surface properties on albumin adsorption. Acta Biomater 2007;3:103–11.

[76] Heuts J, Salber J, Goldyn AM, Janser R, Moller M, Klee D. Bio-functionalized star PEG-coated PVDF surfaces for cytocompatibility-improved implant components. J Biomed Mater Res A 2009;92A:1538–51.

[77] Boyd AR, Burke GA, Duffy H, Holmberg M, Kane CO, Meenan BJ, et al. Sputter deposited bioceramic coatings: surface characterization and initial protein adsorption studies using surface-MALDI-MS. J Mater Sci mater Med 2011;22:71–84.
[78] Groll J, Ademovic Z, Ameringer T, Klee D, Moeller M. Comparison of coatings from reactive star shaped PEG-stat-PPG prepolymers and grafted linear PEG for biological and medical applications. Biomacromolecules 2005;6:956–62.
[79] Ademovic Z, Gonera A, Mischnick P, Klee D. Biocompatible surface preparation using amino-functionalized amylase. Biomacromolecules 2006;7:1429–32.
[80] Griesser HJ, Kingshott P, McArthur SL, McLean KM, Kinsel GR, Timmons RB. Surface-MALDI mass spectrometry in biomaterials research. Biomaterials 2004;25:4861–75.
[81] Ma DHK, Lai JY, Cheng HY, Tsai CC, Yeh LK. Carbodiimide cross-linked amniotic membranes for cultivation of limbal epithelial cells. Biomaterials 2010;31:6647–58.
[82] Blaker JJ, Bismarck A, Boccaccini AR, Young AM, Nazhat SN. Premature degradation of poly(α-hydroxyesters) during thermal processing of Bioglass®-containing composites. Acta Biomater 2010;6:756–62.
[83] Rebollar E, Frischauf I, Olbrich M, Peterbauer T, Hering S, Preiner J, et al. Proliferation of aligned mammalian cells on laser-nanostructured polystyrene. Biomaterials 2008;29:1796–806.
[84] Simons WW. The Sadtler handbook of infrared spectra. Sadtler Research Laboratories; 1978.
[85] Chalmers JM, Griffiths PR. Handbook of vibrational spectroscopy. John Wiley & Sons, Ltd; 2002.
[86] Greenler RG. Infrared study of adsorbed molecules on metal surfaces by reflection techniques. J Chem Phys 1966;44:310–5.
[87] Freitas SC, Barbosa MA, Martins MCL. The effect of immobilization of thrombin inhibitors onto self-assembled monolayers on the adsorption and activity of thrombin. Biomaterials 2010;31:3772–80.
[88] Humblot V, Yala JF, Thebault P, Boukerma K, Hequet A, Berjeaud JM, et al. The antibacterial activity of Magainin I immobilized onto mixed thiols self-assembled monolayers. Biomaterials 2009;30:3503–12.
[89] Briand E, Salmain M, Compere C, Pradier CM. Anti-rabbit immunoglobulin G detection in complex medium by PM-RAIRS and QCM influence of the antibody immobilisation method. Biosens Bioelectron 2007;22:2884–90.
[90] Harding DJ, Gruene P, Haertelt M, Meijer G, Fielicke A, Hamilton SM, et al. Probing the structures of gas-phase rhodium cluster cations by far-infrared spectroscopy. J Chem Phys 2010;133:214304.
[91] Prati S, Sciutto G, Mazzeo R, Torri C, Fabbri D. Application of ATR-far-infrared spectroscopy to the analysis of natural resins. Anal Bioanal Chem 2011;399:3081–91.
[92] Meyer T, Bergner N, Bielecki C, Krafft C, Akimov D, Romeike BFM, et al. Nonlinear microscopy, infrared, and Raman microspectroscopy for brain tumor analysis. J Biomed Opt 2011;16:021113.
[93] Lee ES, Lee JY. High resolution cellular imaging with nonlinear optical infrared microscopy. Opt Express 2011;19:1378–84.
[94] Schrader B. Infrared and Raman spectroscopy: methods and applications. Wiley-VCH; 1995.
[95] Lewis IR, Edwards HGM. Handbook of Raman spectroscopy: from the research laboratory to the process line. Marcel Dekker, Inc; 2001.

[96] Wang HY, Xu M, Zhang W, Kwok DTK, Jiang J, Wu ZW, et al. Mechanical and biological characteristics of diamond-Like carbon coated poly aryl-ether-ether-ketone. Biomaterials 2010;31:8181–7.
[97] Saber-Samandari S, Gross KA. The use of thermal printing to control the properties of calcium phosphate deposits. Biomaterials 2010;31:6386–93.
[98] Ku SH, Park CB. Human endothelial cell growth on mussel-inspired nanofiber scaffold for vascular tissue engineering. Biomaterials 2010;31:9431–7.
[99] Lopez-Heredia MA, Sohier J, Gaillard C, Quillard S, Dorget M, Layrolle P. Rapid prototyped porous titanium coated with calcium phosphate as a scaffold for bone tissue engineering. Biomaterials 2008;29:2608–15.
[100] Fang JX, Liu SY, Li ZY. Polyhedral silver mesocages for single particle surface-enhanced Raman scattering-based biosensor. Biomaterials 2011;32:4877–84.
[101] Qian J, Jiang L, Cai FH, Wang D, He SL. Fluorescence-surface enhanced Raman scattering co-functionalized gold nanorods as near-infrared probes for purely optical in vivo imaging. Biomaterials 2011;32:1601–10.
[102] Noh MS, Jun BH, Kim S, Kang H, Woo MA, Minai-Tehrani A, et al. Magnetic surface-enhanced Raman spectroscopic (M-SERS) dots for the identification of bronchioalveolar stem cells in normal and lung cancer mice. Biomaterials 2009;30:3915–25.
[103] Liu XW, Tao HQ, Yang K, Zhang SA, Lee ST, Liu ZA. Optimization of surface chemistry on single-walled carbon nanotubes for in vivo photothermal ablation of tumors. Biomaterials 2011;32:144–51.
[104] Egerton RF, Cheng SC. Measurement of local thickness by electron energy-loss spectroscopy. Ultramicroscopy 1987;21:231–44.
[105] Malis T, Cheng SC, Egerton RF. EELS log-ratio technique for specimen-thickness measurement in the TEM. J Electron Microsc Tech 1988;8:193–200.
[106] Egerton RF. Electron energy-loss spectroscopy in the electron microscope. Plenum; 1996.
[107] Ma WJ, Ruys AJ, Mason RS, Martin PJ, Bendavid A, Liu ZW, et al. DLC coatings: effects of physical and chemical properties on biological response. Biomaterials 2007;28:1620–8.
[108] Gries K, Kroger R, Kubel C, Fritz M, Rosenauer A. Investigations of voids in the aragonite platelets of nacre. Acta Biomater 2009;5:3038–44.
[109] Heimann RB, Wirth R. Formation and transformation of amorphous calcium phosphates on titanium alloy surfaces during atmospheric plasma spraying and their subsequent in vitro performance. Biomaterials 2006;27:823–31.
[110] Liou SC, Chen SY, Lee HY, Bow JS. Structural characterization of nano-sized calcium deficient apatite powders. Biomaterials 2004;25:189–96.
[111] Zhou YY, Shi LX, Li QN, Jiang H, Lv G, Zhao J, et al. Imaging and inhibition of multi-drug resistance in cancer cells via specific association with negatively charged CdTe quantum dots. Biomaterials 2010;31:4958–63.
[112] Matsusaki M, Yoshida H, Akashi M. The construction of 3D-engineered tissues composed of cells and extracellular matrices by hydrogel template approach. Biomaterials 2007;28:2729–37.
[113] Murphy AR, John PS, Kaplan DL. Modification of silk fibroin using diazonium coupling chemistry and the effects on hMSC proliferation and differentiation. Biomaterials 2008;29:2829–38.
[114] Ding YJ, Liu J, Wang H, Shen GL, Yu RQ. A piezoelectric immunosensor for the detection of α-fetoprotein using an interface of gold/hydroxyapatite hybrid nanomaterial. Biomaterials 2007;28:2147–54.

[115] Lin JJ, Chen JS, Huang SJ, Ko JH, Wang YM, Chen TL, et al. Folic acid-Pluronic F127 magnetic nanoparticle clusters for combined targeting, diagnosis, and therapy applications. Biomaterials 2009;30:5114–24.
[116] Tong J, Zimmerman MC, Li SM, Yi X, Luxenhofer R, Jordan R, et al. Neuronal uptake and intracellular superoxide scavenging of a fullerene (C60)-poly(2-oxazoline)s nanoformulation. Biomaterials 2011;32:3654–65.
[117] Mao ZW, Ma L, Yan J, Yan M, Gao CY, Shen JC. The gene transfection efficiency of thermoresponsive N, N, N-trimethyl chitosan chloride-g-poly(N-isopropylacrylamide) copolymer. Biomaterials 2007;28:4488–500.
[118] Van den Beucken JJJP, Vos MRJ, Thune PC, Hayakawa T, Fukushima T, Okahata Y, et al. Fabrication, characterization, and biological assessment of multilayered DNA-coatings for biomaterial purposes. Biomaterials 2006;27:691–701.
[119] Yang MH, Yang Y, Yang HF, Shen GL, Yu RQ. Layer-by-layer self-assembled multilayer films of carbon nanotubes and platinum nanoparticles with polyelectrolyte for the fabrication of biosensors. Biomaterials 2006;27:246–55.
[120] Lee T, Min J, Kim SU, Choi JW. Multifunctional 4-bit biomemory chip consisting of recombinant azurin variants. Biomaterials 2011;32:3815–21.
[121] Ji Y, Ghosh K, Shu XZ, Li BQ, Sokolov JC, Prestwich GD, et al. Electrospun three-dimensional hyaluronic acid nanofibrous scaffolds. Biomaterials 2006;27:3782–92.
[122] Tai BCU, Wan ACA, Ying JY. Modified polyelectrolyte complex fibrous scaffold as a matrix for 3D cell culture. Biomaterials 2010;31:5927–35.
[123] Pang YG, Ucuzian AA, Matsumura A, Brey EM, Gassman AA, Husak VA. The temporal and spatial dynamics of microscale collagen scaffold remodeling by smooth muscle cells. Biomaterials 2009;30:2023–31.
[124] Thibault M, Astolfi M, Tran-Khanh N, Lavertu M, Darras V, Merzouki A, et al. Excess polycation mediates efficient chitosan-based gene transfer by promoting lysosomal release of the polyplexes. Biomaterials 2011;32:4639–46.
[125] Kim SJ, Ise H, Goto M, Komura K, Cho CS, Akaike T. Gene delivery system based on highly specific recognition of surface-vimentin with N-acetylglucosamine immobilized polyethylenimine. Biomaterials 2011;32:3471–80.
[126] Cross VL, Zheng Y, Choi NW, Verbridge SS, Sutermaster BA, Bonassar LJ, et al. Dense type I collagen matrices that support cellular remodeling and microfabrication for studies of tumor angiogenesis and vasculogenesis in vitro. Biomaterials 2010;31:8596–607.
[127] Blakeney BA, Tambralli A, Anderson JM, Andukuri A, Lim DJ, Dean DR, et al. Cell infiltration and growth in a low density, uncompressed three-dimensional electrospun nanofibrous scaffold. Biomaterials 2011;32:1583–90.
[128] Da Silva J, Lautenschlager F, Sivaniah E, Guck JR. The cavity-to-cavity migration of leukaemic cells through 3D honey-combed hydrogels with adjustable internal dimension and stiffness. Biomaterials 2010;31:2201–8.
[129] Pang YG, Wang XL, Ucuzian AA, Brey EM, Burgess WH, Jones KJ, et al. Local delivery of a collagen-binding FGF-1 chimera to smooth muscle cells in collagen scaffolds for vascular tissue engineering. Biomaterials 2010;31:878–85.
[130] Futaba DN, Hata K, Namai T, Yamada T, Mizuno K, Hayamizu Y, et al. 84% catalyst activity of water-assisted growth of single walled carbon nanotube forest characterization by a statistical and macroscopic approach. J Phys Chem B 2006;110:8035–8.
[131] Park J, Bauer S, von der Mark K, Schmuki P. Nanosize and vitality: TiO_2 nanotube diameter directs cell fate. Nano Letters 2007;7:1686–91.

[132] Oh S, Brammer KS, Li YSJ, Teng D, Engler AJ, Chien S, et al. Stem cell fate dictated solely by altered nanotube dimension. Proc Natl Acad Sci U S A 2009;106:2130–5.

[133] Rujitanaroj PO, Wang YC, Wang J, Chew SY. Nanofiber-mediated controlled release of siRNA complexes for long term gene-silencing applications. Biomaterials 2011;32:5915–23.

[134] He CL, Xiao GY, Jin XB, Sun CH, Ma PX. Electrodeposition on nanofibrous polymer scaffolds: rapid mineralization, tunable calcium phosphate composition and topography. Adv Funct Mater 2010;20:3568–76.

[135] Yuk SH, Oh KS, Koo H, Jeon H, Kim K, Kwon IC. Multi-core vesicle nanoparticles based on vesicle fusion for delivery of chemotherapic drugs. Biomaterials 2011;32:7924–31.

[136] Kai D, Prabhakaran MP, Jin GR, Ramakrishna S. Guided orientation of cardiomyocytes on electrospun aligned nanofibers for cardiac tissue engineering. J Biomed Mater Res B 2011;98B:379–86.

[137] Lamers E, Walboomers XF, Domanski M, te Riet J, van Delft FCMJM, Luttge R, et al. The influence of nanoscale grooved substrates on osteoblast behavior and extracellular matrix deposition. Biomaterials 2010;31:3307–16.

[138] Wang HY, Ji JH, Zhang W, Zhang YH, Jiang J, Wu ZW, et al. Biocompatibility and bioactivity of plasma-treated biodegradable poly(butylene succinate). Acta Biomater 2009;5:279–87.

[139] Antunes JC, Oliveira JM, Reis RL, Soria JM, Gomez-Ribelles JL, Mano JF. Novel poly(L-lactic acid)/hyaluronic acid macroporous hybrid scaffolds: characterization and assessment of cytotoxicity. J Biomed Mater Res A 2010;94A:856–69.

[140] Zhang KH, Qian YF, Wang HS, Fan LP, Huang C, Yin AL, et al. Genipin-crosslinked silk fibroin/hydroxybutyl chitosan nanofibrous scaffolds for tissue-engineering application. J Biomed Mater Res A 2010;95A:870–81.

[141] Reis ECC, Borges APB, Araujo MVF, Mendes VC, Guan LM, Davies JE. Periodontal regeneration using a bilayered PLGA/calcium phosphate construct. Biomaterials 2011;32:9244–53.

[142] Daculsi G, Goyenvalle E, Cognet R, Aguado E, Suokas EO. Osteoconductive properties of poly(96L/4D-lactide)/beta-tricalcium phosphate in long term animal model. Biomaterials 2011;32:3166–77.

[143] Bandyopadhyay-Ghosh S, Faria PEP, Johnson A, Felipucci DNB, Reaney IM, Salata LA, et al. Osteoconductivity of modified fluorcanasite glass–ceramics for bone tissue augmentation and repair. J Biomed Mater Res A 2010;94A:760–8.

[144] Kusakabe H, Sakamaki T, Nihei K, Oyama Y, Yanagimoto S, Ichimiya M, et al. Osseointegration of a hydroxyapatite-coated multilayered mesh stem. Biomaterials 2004;25:2957–69.

[145] Zinelis S, Barmpagadaki X, Vergos V, Chakmakchi M, Eliades G. Bond strength and interfacial characterization of eight low fusing porcelains to cp Ti. Dent Mater 2010;26:264–73.

[146] Zhang Y, Yin QS, Zhang Y, Xia H, Ai FZ, Jiao YP, et al. Determination of antibacterial properties and cytocompatibility of silver-loaded coral hydroxyapatite. J Mater Sci Mater Med 2010;21:2453–62.

[147] Variola F, Yi JH, Richert L, Wuest JD, Rosei F, Nanci A. Tailoring the surface properties of Ti6Al4V by controlled chemical oxidation. Biomaterials 2008;29:1285–98.

[148] Han B, Wang XY, Gao XJ, Liu JG, Liang FX, Qu XZ, et al. Synthesis and characterization of biodegradable microcapsules for the controlled delivery of calcium hydroxide. J Biomed Mater Res B 2011;99B:120–6.

[149] Coelho PG, Lemons JE. Physico/chemical characterization and in vivo evaluation of nanothickness bioceramic depositions on alumina-blasted/acid-etched Ti-6Al-4V implant surfaces. J Biomed Mater Res A 2009;90A:351–61.

[150] Khaled SMZ, Charpentier PA, Rizkalla AS. Synthesis and characterization of poly(methyl methacrylate)-based experimental bone cements reinforced with TiO_2-SrO nanotubes. Acta Biomater 2010;6:3178–86.

[151] Danilatos GD. Environmental scanning electron microscopy. In: In-situ microscopy in materials research: leading international research in electron and scanning probe microscopies. Kluwer Academic Publishers; 1997.

[152] Fundueanu G, Constantin M, Oanea I, Harabagiu V, Ascenzi P, Simionescu BC. Entrapment and release of drugs by a strict "on-off" mechanism in pullulan microspheres with pendant thermosensitive groups. Biomaterials 2010;31:9544–53.

[153] Fang Y, Yu H, Chen L, Chen S. Facile glycerol-assisted synthesis of N-vinyl pyrrolidinone-based thermosensitive hydrogels via frontal polymerization. Chem Mater 2009;21:4711–8.

[154] Tian HY, Deng C, Lin H, Sun JR, Deng MX, Chen XS, et al. Biodegradable cationic PEG–PEI–PBLG hyperbranched block copolymer: synthesis and micelle characterization. Biomaterials 2005;26:4209–17.

[155] Ma PA, Liu S, Huang YB, Chen XS, Zhang LP, Jing XB. Lactose mediated liver-targeting effect observed by ex vivo imaging technology. Biomaterials 2010;31:2646–54.

[156] Shi Q, Chen XS, Lu TC, Jing XB. The immobilization of proteins on biodegradable polymer fibers via click chemistry. Biomaterials 2008;29:1118–26.

[157] Taylor JE, Laity PR, Hicks J, Wong SS, Norris K, Khunkamchoo P, et al. Extent of iron pick-up in deforoxamine-coupled polyurethane materials for therapy of chronic wounds. Biomaterials 2005;26:6024–33.

[158] Lusvardi G, Zaffe D, Menabue L, Bertoldi C, Malavasi G, Consolo U. In vitro and in vivo behaviour of zinc-doped phosphosilicate glasses. Acta Biomater 2009;5:419–28.

[159] Zareie HM, Boyer C, Bulmus V, Nateghi E, Davis TP. Temperature-responsive self-assembled monolayers of oligo(ethylene glycol): control of biomolecular recognition. ACS Nano 2008;2:757–65.

[160] Kim SU, Yagati AK, Min JH, Choi JW. Nanoscale protein-based memory device composed of recombinant azurin. Biomaterials 2010;31:1293–8.

[161] Yang NJ, Uetsuka H, Williams OA, Osawa E, Tokuda N, Nebel CE. Vertically aligned diamond nanowires: fabrication, characterization, and application for DNA sensing. Physica Status Solidi A 2009;206:2048–56.

[162] Odom TW, Huang JL, Kim P, Lieber CM. Atomic structure and electronic properties of single-walled carbon nanotubes. Nature 1998;391:62–4.

[163] Ouyang M, Huang JL, Cheung CL, Lieber CM. Energy gaps in "metallic" single-walled carbon nanotubes. Science 2001;292:702–5.

[164] Ma DDD, Lee CS, Au FCK, Tong SY, Lee ST. Small-diameter silicon nanowire surfaces. Science 2003;299:1874–7.

[165] Coraux J, N'Diaye AT, Busse C, Michely T. Structural coherency of graphene on Ir(111). Nano Lett 2008;8:565–70.

[166] Ishigami M, Chen JH, Cullen WG, Fuhrer MS, Williams ED. Atomic structure of graphene on SiO_2. Nano Lett 2007;7:1643–8.

[167] Binnig G, Quate CF, Gerber C. Atomic force microscope. Phys Rev Lett 1986;56:930–3.

[168] Ebeling D, Holscher H, Fuchs H, Anczykowski B, Schwarz UD. Imaging of biomaterials in liquids: a comparison between conventional and Q-controlled amplitude modulation ('tapping mode') atomic force microscopy. Nanotechnology 2006;17:S221–6.

[169] Zhou M, Smith AM, Das AK, Hodson NW, Collins RF, Ulijn RV, et al. Self-assembled peptide-based hydrogels as scaffolds for anchorage-dependent cells. Biomaterials 2009;30:2523–30.

[170] Ye Z, Zhao X. Phase imaging atomic force microscopy in the characterization of biomaterials. J Microsc 2010;238:27–35.

[171] He LM, Liao SS, Quan DP, Ngiam M, Chan CK, Ramakrishna S, et al. The influence of laminin-derived peptides conjugated to Lys-capped PLLA on neonatal mouse cerebellum C17.2 stem cells. Biomaterials 2009;30:1578–86.

[172] Xu LC, Runt J, Siedlecki CA. Dynamics of hydrated polyurethane biomaterials: surface microphase restructuring, protein activity and platelet adhesion. Acta Biomater 2010;6:1938–47.

[173] Hentschel J, Boerner HG. Peptide-directed microstructure formation of polymers in organic media. J Am Chem Soc 2006;128:14142–9.

[174] Pallandre A, Glinel K, Jonas AM, Nysten B. Binary nanopatterned surfaces prepared from silane monolayers. Nano Lett 2004;4:365–71.

[175] Denis FA, Pallandre A, Nysten B, Jonas AM, Dupont-Gillain CC. Alignment and assembly of adsorbed collagen molecules induced by anisotropic chemical nanopatterns. Small 2005;1:984–91.

[176] Baralia GG, Pallandre A, Nysten B, Jonas AM. Nanopatterned self-assembled monolayers. Nanotechnology 2006;17:1160–5.

[177] Qin GK, Lapidot S, Numata K, Hu X, Meirovitch S, Dekel M, et al. Expression, cross-linking, and characterization of recombinant chitin binding resilin. Biomacromolecules 2009;10:3227–34.

[178] Hu XA, Wang XL, Rnjak J, Weiss AS, Kaplan DL. Biomaterials derived from silk-tropoelastin protein systems. Biomaterials 2010;31:8121–31.

[179] Block S, Glockl G, Weitschies W, Helm CA. Direct visualization and identification of biofunctionalized nanoparticles using a magnetic atomic force microscope. Nano Lett 2011;11:3587–92.

[180] Kienberger F, Ebner A, Gruber HJ, Hinterdorfer P. Molecular recognition imaging and force spectroscopy of single biomolecules. Acc Chem Res 2006;39:29–36.

[181] Sima F, Davidson P, Pauthe E, Sima LE, Gallet O, Mihailescu IN, et al. Fibronectin layers by matrix-assisted pulsed laser evaporation from saline buffer-based cryogenic targets. Acta Biomater 2011;7:3780–8.

[182] Park JW, Jang JH, Lee CS, Hanawa T. Osteoconductivity of hydrophilic microstructured titanium implants with phosphate ion chemistry. Acta Biomater 2009;5:2311–21.

[183] Barranco V, Escudero ML, Garcia-Alonso MC. Influence of the microstructure and topography on the barrier properties of oxide scales generated on blasted Ti6Al4V surfaces. Acta Biomater 2011;7:2716–25.

[184] Park JW, Kim HK, Kim YJ, Jang JH, Song H, Hanawa T. Osteoblast response and osseointegration of a Ti-6Al-4V alloy implant incorporating strontium. Acta Biomater 2010;6:2843–51.

[185] Park JW, Kim YJ, Jang JH, Kwon TG, Bae YC, Suh JY. Effects of phosphoric acid treatment of titanium surfaces on surface properties, osteoblast response and removal of torque forces. Acta Biomater 2010;6:1661–70.

[186] Truong VK, Lapovok R, Estrin YS, Rundell S, Wang JY, Fluke CJ, et al. The influence of nano-scale surface roughness on bacterial adhesion to ultrafine-grained titanium. Biomaterials 2010;31:3674–83.

[187] Park JW, Park KB, Suh JY. Effects of calcium ion incorporation on bone healing of Ti6Al4V alloy implants in rabbit tibiae. Biomaterials 2007;28:3306–13.

[188] Turssi CP, Ferracane JL, Vogel K. Filler features and their effects on wear and degree of conversion of particulate dental resin composites. Biomaterials 2005;26:4932–7.

[189] Huang FL, Wei QF, Wang XQ, Xu WZ. Dynamic contact angles and morphology of PP fibres treated with plasma. Polym Test 2006;25:22–7.

[190] Wortmann FJ, Wortmann G, zur Wiesche ES. Spatial probing of the properties of the human hair surface using Wilhelmy force profiles. Langmuir 2010;26:7365–9.

[191] Yum K, Yu MF. Measurement of wetting properties of individual boron nitride nanotubes with the Wilhelmy method using a nanotube-based force sensor. Nano Lett 2006;6:329–33.

[192] Yazdanpanah MM, Hosseini M, Pabba S, Berry SM, Dobrokhotov VV, Safir A, et al. Micro-Wilhelmy and related liquid property measurements using constant-diameter nanoneedle-tipped atomic force microscope probes. Langmuir 2008;24:13753–64.

[193] Stachewicz U, Barber AH. Enhanced wetting behavior at electrospun polyamide nanofiber surfaces. Langmuir 2011;27:3024–9.

[194] Tadmor R. Line energy and the relation between advancing, receding, and young contact angles. Langmuir 2004;20:7659–64.

[195] Chibowski E, Terpilowski K. Surface free energy of sulfur—revisited I. Yellow and orange samples solidified against glass surface. J Colloid Interface Sci 2008;319:505–13.

[196] Yamanlar S, Sant S, Boudou T, Picart C, Khademhosseini A. Surface functionalization of hyaluronic acid hydrogels by polyelectrolyte multilayer films. Biomaterials 2011;32: 5590–9.

[197] Alauzun JG, Young S, D'Souza R, Liu L, Brook MA, Sheardown HD. Biocompatible, hyaluronic acid modified silicone elastomers. Biomaterials 2010;31:3471–8.

[198] Kyomoto M, Moro T, Takatori Y, Kawaguchi H, Nakamura K, Ishihara K. Self-initiated surface grafting with poly(2-methacryloyloxyethyl phosphorylcholine) on poly(ether-ether-ketone). Biomaterials 2010;31:1017–24.

[199] Alibeik S, Zhu SP, Yau JW, Weitz JI, Brash JL. Surface modification with polyethylene glycol-corn trypsin inhibitor conjugate to inhibit the contact factor pathway on blood-contacting surfaces. Acta Biomater 2011;7:4177–86.

[200] Feng XJ, Jiang L. Design and creation of superwetting/antiwetting surfaces. Adv Mater 2006;18:3063–78.

[201] Ma ZW, Kotaki M, Yong T, He W, Ramakrishna S. Surface engineering of electrospun polyethylene terephthalate (PET) nanofibers towards development of a new material for blood vessel engineering. Biomaterials 2005;26:2527–36.

[202] Ma ZW, He W, Yong T, Ramakrishna S. Grafting of gelatin on electrospun poly(-caprolactone) nanofibers to improve endothelial cell spreading and proliferation and to control cell orientation. Tissue Eng 2005;11:1149–58.

[203] Wang HY, Kwok DTK, Wang W, Wu ZW, Tong LP, Zhang YM, et al. Osteoblast behavior on polytetrafluoroethylene modified by long pulse, high frequency oxygen plasma immersion ion implantation. Biomaterials 2010;31:413–9.

[204] Feng B, Weng J, Yang BC, Qu SX, Zhang XD. Characterization of surface oxide films on titanium and adhesion of osteoblast. Biomaterials 2003;24:4663–70.

[205] Michiardi A, Aparicio C, Ratner BD, Planell JA, Gil J. The influence of surface energy on competitive protein adsorption on oxidized NiTi surfaces. Biomaterials 2007;28:586–94.

[206] Ochsenbein A, Chai F, Winter S, Traisnel M, Breme J, Hildebrand HF. Osteoblast responses to different oxide coatings produced by the sol-gel process on titanium substrates. Acta Biomater 2008;4:1506–17.

[207] Ruttermann S, Trellenkamp T, Bergmann N, Raab WHM, Ritter H, Janda R. A new approach to influence contact angle and surface free energy of resin-based dental restorative materials. Acta Biomater 2011;7:1160–5.

[208] Bernard SA, Balla VK, Davies NM, Bose S, Bandyopadhyay A. Bone cell-materials interactions and Ni ion release of anodized equiatomic NiTi alloy. Acta Biomater 2011;7:1902–12.

[209] Bozukova D, Pagnoulle C, De Pauw-Gillet MC, Desbief S, Lazzaroni R, Ruth N. Improved performances of intraocular lenses by poly(ethylene glycol) chemical coatings. Biomacromolecules 2007;8:2379–87.

[210] Wan LS, Xu ZK, Huang XJ, Wang ZG, Ye P. Hemocompatibility of poly(acrylonitrile-co-N-vinyl-2-pyrrolidone)s: swelling behavior and water states. Macromol Biosci 2005;5:229–36.

[211] Read ML, Morgan PB, Kelly JM, Maldonado-Codina C. Dynamic contact angle analysis of silicone hydrogel contact lenses. J Biomater Appl 2011;26:85–99.

[212] Magin CM, Finlay JA, Clay G, Callow ME, Callow JA, Brennan AB. Antifouling performance of cross-linked hydrogels: refinement of an attachment model. Biomacromolecules 2011;12:915–22.

[213] Toworfe GK, Composto RJ, Shapiro IM, Ducheyne P. Nucleation and growth of calcium phosphate on amine-, carboxyl- and hydroxyl-silane self-assembled monolayers. Biomaterials 2006;27:631–42.

[214] Goda T, Konno T, Takai M, Moro T, Ishihara K. Biomimetic phosphorylcholine polymer grafting from polydimethylsiloxane surface using photo-induced polymerization. Biomaterials 2006;27:5151–60.

[215] Lin QK, Ding X, Qiu FY, Song XX, Fu GS, Ji J. In situ endothelialization of intravascular stents coated with an anti-CD34 antibody functionalized heparin-collagen multilayer. Biomaterials 2010;31:4017–25.

[216] Charbonneau C, Liberelle B, Hebert MJ, De Crescenzo G, Lerouge S. Stimulation of cell growth and resistance to apoptosis in vascular smooth muscle cells on a chondroitin sulfate/epidermal growth factor coating. Biomaterials 2011;32:1591–600.

[217] Pei J, Hall H, Spencer ND. The role of plasma proteins in cell adhesion to PEG surface-density-gradient-modified titanium oxide. Biomaterials 2011;32:8968–78.

[218] Lee HS, Eckmann DM, Lee D, Hickok NJ, Composto RJ. Symmetric pH-dependent swelling and antibacterial properties of chitosan brushes. Langmuir 2011;27:12458–65.

[219] Santonicola MG, de Groot GW, Memesa M, Meszynska A, Vancso GJ. Reversible pH-controlled switching of poly(methacrylic acid) grafts for functional biointerfaces. Langmuir 2010;26:17513–9.

[220] Blacklock J, You YZ, Zhou QH, Mao GZ, Oupicky D. Gene delivery in vitro and in vivo from bioreducible multilayered polyelectrolyte films of plasmid DNA. Biomaterials 2009;30:939–50.

[221] Cho WK, Kong B, Park HJ, Kim J, Chegal W, Choi JS, et al. Long-term stability of cell micropatterns on poly((3-(methacryloylamino)propyl)-dimethyl(3-sulfopropyl)ammonium hydroxide)-patterned silicon oxide surfaces. Biomaterials 2010;31:9565–74.

[222] Sapuri-Butti AR, Butti RC, Parikh AN. Characterization of supported membranes on topographically patterned polymeric elastomers and their applications to microcontact printing. Langmuir 2007;23:12645–54.

[223] Szmodis AW, Blanchette CD, Levchenko AA, Navrotsky A, Longo ML, Orme CA, et al. Direct visualization of phase transition dynamics in binary supported phospholipid bilayers using imaging ellipsometry. Soft Matter 2008;4:1161–4.

[224] Howland MC, Szmodis AW, Sanii B, Parikh AN. Characterization of physical properties of supported phospholipid membranes using imaging ellipsometry at optical wavelengths. Biophys J 2007;92:1306–17.

[225] Noel S, Liberelle B, Robitaille L, De Crescenzo G. Quantification of primary amine groups available for subsequent biofunctionalization of polymer surfaces. Bioconjug Chem 2011;22:1690–9.

[226] Ward J, Kelly J, Wang WX, Zeugolis DI, Pandit A. Amine functionalization of collagen matrices with multifunctional polyethylene glycol systems. Biomacromolecules 2010;11:3093–101.

[227] Coutinho DF, Pashkuleva IH, Alves CM, Marques AP, Neves NM, Reis RL. The effect of chitosan on the in vitro biological performance of chitosan-poly(butylene succinate) blends. Biomacromolecules 2008;9:1139–45.

Chapter 5.1

In Vitro Characterization of Cell–Biomaterials Interactions

Y.M. Thasneem and Chandra P. Sharma

Division of Biosurface Technology, Biomedical Technology Wing, Sree Chitra Tirunal Institute for Medical Sciences and Technology, Trivandrum, Kerala, India

Chapter Outline

Characterization of Biomaterials. http://dx.doi.org/10.1016/B978-0-12-415800-9.00005-X

5.1.1. INTRODUCTION

Being at the cutting edge of chemistry, materials science, bioengineering, biology, and medicine, the field of biomaterials has rapidly expanded by building upon advances in the areas it envelopes. An understanding of cell–substratum interactions is the key to the future development of applications including biomolecular and cell analysis, microfluidics, drug delivery, and medical prostheses. How cells adhere and colonize to the biomaterial surface to initiate processes like growth, differentiation, migration, or apoptosis is critical in promoting new tissue formation or regeneration. The process is quite complex in the nature that involves many physicochemical events taking place at different scales ranging from molecular to cell body (organelle) to tissue levels.

The realization of controlled cell–biomaterial interaction requires the logical extrapolation of the understanding of cell biology and cell–extracellular matrix (ECM) interaction to the biomaterial surface. Also important is the understanding of the mechanism in which the cell orchestrates physiological and morphological changes by translating mechanical and structural informations into biochemical signals. Mechanical environment at the cell–biomaterial location and material dimensions on the macro and micro scales, depending on surface chemical composition and surface topography, cannot be overruled in the context of subsequent cell function. The ambitious task can be met after proper designing of intelligent biomaterials and an informative microenvironment mimicking a physiological niche. The spatiotemporal assembly of differentiating cells organized to create functional structures require fine tuning of cell-matrix and cell–cell interactions [1]. The orchestra of tissue growth, development, maintenance, and repair follows the metabolic support, strength, and endurance provided by the biomechanical forces generated by the contact among the differentiating cells within the tissue or with the extracellular matrix [2].

5.1.2. BASICS OF CELL BIOLOGY

The biological cell constitutes the basic unit of life that manoeuvre with the following: the synthesis, sorting, and storage and transport of molecules; the expression of genetic information; the recognition, transmission and transduction of signals; and the powering of molecular motors and machines along with the unique power to replicate. Most of the biological cells are 1–100 μm in size and are covered by phospholipid bilayer membranes reinforced with protein molecules, and the interior of the cell includes a liquid phase (cytosol), a nucleus, the cytoskeleton consisting of networks of microtubules, actin and intermediate filaments, organelles of different sizes and shapes, and other proteins [3].

Deciding which protein to express, when to divide, when to specialize, and when to commit suicide are all ongoing processes that occur within cells depending on the activation of genes in the correct sequence and synchrony, in order to express the numerous proteins needed for proliferation and differentiation to hierarchical organization within organs [4]. Cells are sensitive to a number of input signals from the environment of different nature, which can be classified into mechanical, topographical, and chemical cues and can induce a wide number of cellular behaviors such as growth, differentiation, apoptosis, migration, gene expression, and signal transduction [5]. The knowledge about the regulation of growth and differentiation of cells in multicellular organism by the complex interplay of biochemical and mechanical signals has directed our thoughts towards the roles of mechanical and structural cues in modulating cellular behavior. Increasing evidence regarding the dexterity with cell fates by engineering the physical properties of the microenvironment to which the cells are exposed has inspired the development of intelligent biomaterials that directly and specifically interact with tissue components, and support or even direct the appropriate cellular activities as shown in Fig. 5.1.1 [6].

Any material can be considered informative in the sense that its intrinsic nature, i.e. chemical composition and structure (i.e. macro-micro-nano-architecture), can anyway transmit a signal that can be read and decoded by the adhering cells. The cellular behavior through distinct developmental paths is guided by the combination of intrinsic and extrinsic signals that arises mainly from the nanotopography of the extra cellular matrix [7]. For certainly, cells must respond to nanostructures in

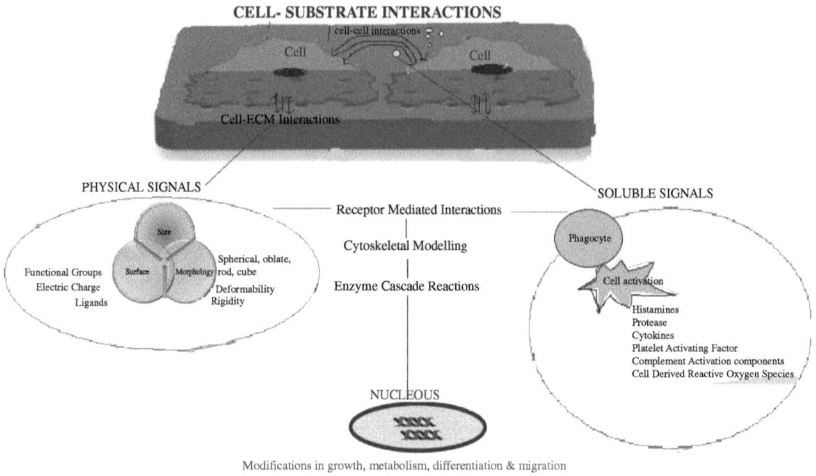

FIGURE 5.1.1 Cell-biomaterial Interactions; Interplay of various factors [84].

a predetermined way, since *in vivo* they live inside an ECM containing nanoscale collagen fibrils and as their own surface structure replicates nanoscale topography with receptors and filopodia [8]. One of the most intriguing concepts in modern biomaterials is capacitating materials to mimic a specific pre-existing microenvironment, inducing cells to differentiate in a predetermined manner and to regenerate by themselves to the desired tissue according to physiological pathways. So the million-dollar question is how to design a surface destined to induce a desired cellular behavior, leave alone the mechanisms underlying cell responses that demand far much-thought process. However, the ever-increasing quest of human knowledge and implant market essentiates the study of the controlled fabrication of model surfaces.

Owing to the dynamic nature of cell structure that adapts to its local mechanochemical environment, it becomes difficult to characterize the mechanical behavior of a living cell in terms of fixed properties. The intricate coupling between the biochemical and mechanical processes of the cell becomes clear with the application of external mechanical stimuli that induce biochemical reactions, including the synthesis of new biomolecules and the enhanced interaction among biomolecules that can generate mechanical forces [9]. Likewise, changes in chemical stimuli like pH, temperature, and biomolecular activity can alter the structure and mechanical integrity of the cell, even in the absence of mechanical stimuli. To achieve controlled responses at the biological interface, biomaterials can be surface modified to precisely interact with biological entities. By altering the surface chemistry of a material, cell adhesion and protein adsorption as well as the direction of cellular responses can be controlled. A material can be made to better-mimic *in vivo* structures by changing the surface topography and architecture, thereby influencing host response [10]. Increasingly, scientists are seeking physics-derived solutions for controlling biological responses as physical attributes, such as size, shape, and mechanical properties, form essential building blocks of biology, just as chemistry and molecular recognition do [11]. Bringing together the interdisciplinary concepts establishes the basis of new design concepts for biomaterials.

5.1.3. MATERIALS FOR BIOMEDICAL APPLICATIONS (BIOMATERIALS)

5.1.3.1. Concept of Biocompatibility

Designing of functional biomaterials capable of modulating specific cellular behavior constitutes an inevitable challenge for the fields of tissue engineering and material science, demanding collaboration with molecular biologists. Professor Bill Bonfield's pioneering research on the interaction of biomaterials with living tissues launched the first successful biocomposite material specially developed for medical devices, trademarked as Hapex [12]. The goal of first generation biomaterials was to 'achieve a suitable combination of physical properties to match those of the replaced tissue with a minimal toxic response in the host' [12]. A shift in the concept from achieving exclusively a bioinert-tissue response to bioactive materials eliciting a controlled action and reaction in the physiological environment marked the beginning of second-generation biomaterials centred around early 1980s. The objective of resorbable biomaterials, another category of second generation, was summarized in 1980, as a second method of manipulating the biomaterial–tissue interface by controlled chemical breakdown of the material facilitating their absorbtion into the host tissues. Resorption of biomaterials appears a perfect solution to the interfacial problems, as the foreign material is ultimately replaced by regenerating tissues. Improvements of the first- and second-generation biomaterials are limited in part as all man-made biomaterials compromise with long-term immune surveillance.

Realizing the bitter truth that no artificial surface can respond to changing physiological loads or biochemical stimuli as the living tissues, the twenty-first century challenge of biomaterial science emphasizes a more biologically based method of repair, replacement, or regeneration of tissues [12]. This concept ultimately led to the upgradation of third-generation biomaterials designed to stimulate specific cellular responses to controlled release of biochemical stimuli at the level of molecular biology.

The attempt to define a 'perfect biomaterial' is still not convincing, and biomaterial scientists tried to provide the definition based on the concept of biocompatibility. Biocompatibility pertains to the fundamental property shared by any material in addition to nontoxicity and noninflammatory. David Williams proposed the definition for biocompatibility, which is one among the most accepted ones. He proposed that 'The biocompatibility of a scaffold or matrix for a tissue-engineering product refers to the ability to perform as a substrate that will support the appropriate cellular activity, including the facilitation of molecular and mechanical signalling systems, in order to optimize tissue regeneration, without eliciting any undesirable local or systemic responses in the eventual host' [13]. The consequence of inflammatory response to any implant material of foreign nature is the formation of fibrotic scar that isolates the polymer from the body, limiting both integration with the host tissue and local supply of blood-borne nutrients [14]. So the goal remains to thoroughly investigate methods to attain low protein adsorption and cell adhesion bestowing some stealth properties to implanted materials along with minimal immune surveillance and clearance.

The biological language to which a cell responds upon approaching a biomaterial/implant is sentenced by the composition and conformation of the adsorbed protein layer [15]. The dictator behind the constructive cellular response favouring wound healing and tissue integration as well as detrimental foreign body response leading to implant failure is none other than the adsorbtion profile and bioactivity set by this primary event. The fathom of its potential consequences can be assessed by the knowledge that cellular behavior, whether to differentiate or proliferate, depends on the nature of interaction with the protein layer. Cell attachment to culture surfaces *in vitro* is usually mediated by adhesion proteins contained in serum-supplemented culture medium, mainly vitronectin and fibronectin, similar to cells in their natural environment adhere to proteins in extra-cellular matrix. The primary interaction between cells and adhesion proteins occurs via integrins (heterodimeric receptors in the cell membrane), involved in intracellular signalling capable of eliciting a diverse range of cell functions. Another process related to adhesion is cell spreading that involves similar extracellular proteins like fibronectin and vitronectin to form focal contacts and essential intracellular structures [16]. Massia and Hubbell estimated that a ligand spacing of 140 nm or less was needed for these structures to form, although response to ligand density may be altered by surface topography [17]. The contractile forces of cell spreading apply tension to the ECM that may remove the adsorbed proteins from the biomaterial surface setting an upper limit to protein adsorption strength for the occurrence of normal cell activity. The motility and migration of cells also follow rules set by the density of adhesion ligands and the strength and flexibility with which it binds the approaching cells to favour both adhesion and release. Adsorbed proteins may provide binding sites for the cooperation of growth factors and biomolecules that promote appropriate signal transduction for cell proliferation, differentiation, or even de-differentiation [18].

The composition of the adsorbed layer is susceptible to change with time in the absence of cellular interactions as faster diffusing molecules like albumin are displaced by high-affinity proteins like kallikrein, often referred to as the Vroman effect [19]. Scaffold architecture for specified cellular responses to a great extend is mediated by the adsorbed proteins that influence the wettability, surface charge, hydrophobic–hydrophilic nature and overall topography of the underlying biomaterial. This plasma protein corona is a major determinant of another aspect of biocompatibility termed as the blood compatibility comprising a complex series of reactions occurring when blood makes contact with the material like the activation of the intrinsic system of coagulation, aggregation of platelets and other blood cells and complement activation, all finalizing to the embolization of thrombin. The major players behind this blood–material interaction are the material surface, its physicochemical nature and structure, blood composition, and blood flow conditions. The parameters routinely checked for hemocompatibility include whole blood assays with RBC, WBC, platelet adhesion, aggregation and activation, coagulation and complement activation along with the inflammatory potential gives an overview of the blood compatibility status (Table 5.1.1).

TABLE 5.1.1 Pros and Cons of the Major Classes of Biomaterials from a Tissue Engineering Perspective

Physicochemical Parameter	Cell Line	Cellular Response	Example
Pore size	Chondrocyte Osteoblasts Fibroblasts	Affect cellular affinity and viability by influencing cellular movement, binding and spreading, intracellular signalling, and transport of nutrients and metabolites	Optimal pore size for chondrocytes and osteoblasts are in 380–405 μm where as for fibroblasts it is 290–310 μm for better bone in growth
Porosity	Osteoblasts	Governs the maximum possible accommodation of cell mass in the scaffold	*In vitro,* low porosity enhances osteogenesis by suppressing cell proliferation and forcing cell aggregation Increased porosity compromise structural integrity and mechanical properties

(*Continued*)

TABLE 5.1.1 Pros and Cons of the Major Classes of Biomaterials from a Tissue Engineering Perspective—cont'd

Physicochemical Parameter	Cell Line	Cellular Response	Example
Stiffness	Fibroblasts Epithelial cells Smooth muscle cells	Affect cellular adhesion, motility, survival and differentiation	Cell contact with the substrate diminishes on increasingly softer substrates Cell migrate from softer region to stiffer region when exposed to substrate stiffness gradient
Wettability/contact angle/surface-free energy	Fibroblasts Epithelial cells Smooth muscle cells	Sigmoid relationship between spread area of cells and surface free energy	Hydrophilic, high surface-free energy biomaterial promote good cell spreading Optimal fibroblast growth was reported on hydrophilic patterns with a water contact angle of around 17°
Charge	Osteoblasts Fibroblasts	Affect cell uptake and spreading	Positive charge resin beads promote osteoblast flattening; negatively charged beads adapted 'strand off' configuration Preferential adhesion and proliferation of 3T3 and 3T12 fibroblasts on more positively charged hydroxyethyl methacrylate polymers than on negative ones
Topography & roughness	Fibroblasts Epithelial Cells	Affect cellular orientation and contact guidance	Special surface topography promote in growth of bone Prevent epithelial down growth along transcutaneous implants.
Surface chemistry	Mesenchymal stem cells	Long-term functional differentiation of cells	Induces changes in conformation of serum proteins like fibronectin leading to alteration in binding of specific integrin receptors

5.1.3.2. Ceramics as Biomaterials

Ceramic tissue engineering revolves around the promises of biocompatibility, tissue conductivity, and biodegradability. Bone–tissue engineering felicit the ceramic field for the imparted mechanical strength required for load-bearing purposes. Being a natural mimic of the mineral component of bone that replicates the calcium to phosphorous ratio along with its excellent osteophilic property, hydroxyl apatite has been the material of choice for the fabrication of inorganic scaffolds as alternatives or additives to autogenous bone for orthopaedic and dental applications. The exciting features of calcium phosphate, whether as nanoparticles, coatings, scaffold, or cement, depend on Ca/P ratio, crystallinity, and phase purity. Nanophase ceramics with grain size less than 100 nm offer enhanced long-term cellular functions like cell proliferation, synthesis of alkaline phosphatase that may result in better cell-population motility, and synthesis of extracellular matrix proteins 16. Studies involving the influence of porosity and pore size on the cell migration, tissue in growth, and tissue regeneration focus on the osteaoblast cells cultured in ceramic scaffolds [20]. Ideal cellular distribution sets the pore size limit to an extent that may hinder proper conformation to bone surface and poor malleability. Re-engineering prospects with bone-inductive proteins such as recombinant human bone morphogenic protein 2 ensure uniform fixation [21]. Tsurga and co-workers have suggested that the optimal pore size of ceramics that supports ectopic bone formation was 300–400 μm. Holmes similarly suggested that the optimal pore size range from 200 to 400 μm with the average human osteon size of approximately 223 μm [22]. Due to the necessity for proper capillary bed formation in successful implants, pore size greater than 300 μm was recommended.

A major concern with hydroxyl apatite prosthesis is the possibility of release of particles capable of modifying cellular responses upon bioactive ceramic dissolution during bone remodelling. A composition-dependent decrease in rat osteoblast growth has been observed *in vitro* using various kinds of calcium phosphate powders (b-tissue culture polystyrene (TCP), hydroxy-apatite (HA) or commercial, and sintered b-dicalcium pyrophosphate) that may compromise the success of bone regeneration. A study by Saldana et al. analyzed the influence of BCP particles and their precursors, calcium-deficient apatite (CDA) particles, on *in vitro* hMSC behavior [23]. This confirmed the efficient internalization of these particles by hMSCs without affecting cell viability, morphology, and actin cytoskeleton reorganization, but the osteogenic maturation was compromised owing to the uptake of Ca from the culture. Further possibility of phagocytosis of these particles by the monocytic-macrophage lineage, triggering an inflammatory response is almost certified. However, the proliferative capacity of osteoblast precursors was not affected by the CDA or BCP particles flashing green light towards the improved bio-reactivity of bioceramics in comparison with metallic particles.

5.1.3.3. Metallic Biomaterials

Manipulation of matter at the nanolevel has led to the development of inorganic biomaterials in the field of implant biology necessitating the elucidation of metal–cell interaction parameters. Major metallic properties that modulate the interaction with biomolecules and cells include (1) the hydrophobic or hydrophilic behavior; (2) the surface charge; (3) the submicrometric topography of the surface (nanotopography); (4) the surface curvature; (5) the occurrence and nature of reactive sites; (6) the chirality; and (7) the dissolution/reprecipitation equilibrium.

Metals, namely, stainless steels, titanium, tantalum and cobalt–chromium-based are more suitable for load-bearing applications compared with ceramics or polymeric materials due to their combination of high mechanical strength and fracture toughness. By virtue of its physical, chemical, and mechanical properties, titanium (Ti) exhibits excellent resistance to corrosion and good biocompatibility, and is currently the best and most widely used material in the manufacture of orthopaedic and dental implants. Their applications are limited by the possible release of toxic metallic ions and/or particles through corrosion or wear processes that can trigger the inflammatory cascades questioning its biocompatibility. Moreover, the elastic moduli of current metallic biomaterials are not well matched with that of natural bone tissue, resulting in stress-shielding effects leading to reduced stimulation of new bone growth and remodelling that compromises the end purpose of implant stability [24]. The principle immunologic concern associated with metallic biomaterials is a delayed-type hypersensitivity (type IV or DTH) that may initiate metabolic alterations, with formation of lymphocyte toxins, and initiation, and/or promotion of chemical carcinogenesis [25].

Tavares et al. relied on comet assay, micronucleus, and chromosomal aberration tests to evaluate and compare the genotoxicity of titanium surfaces with and without plasma treatment using CHO-K1 (Chinese hamster ovary) cells [26]. The studies revealed significant increase in % tail movement, chromosomal aberrations, tetraploidy, micronucleus on the cells exposed to titanium surface when compared to the negative polymeric, and plasma-treated surfaces. The surface roughness and oxide layer thickness are the sole players in conferring hydrophilicity for the conformational stability of adsorbed protein layer and subsequent resistance to the corrosion process. The spontaneous formation of oxide layer, i.e. the guardian angel of titanium, doesn't assist other metallic implants like nickel, which when brought into contact with the human endothelium can induce pathological levels of oxidative stress sufficient to cause breakdown of the barrier function of the endothelium. Surface modification strategies like oxidizing heat treatment can prove beneficial in the application of these metals in contact with cells.

5.1.3.4. Polymeric Biomaterials

The polymeric material satisfies the general biomedical application with its tunable physical and chemical properties that module the way it interacts with biological components, especially proteins and cells. Surface engineering and modification of the base polymer composition with specific proteins or peptides formulate the topography, chemistry, composition, and hydrophilic/hydrophobic balance towards the desired cellular interactions [27]. The proposed end-use dictates the design criteria with the factor polymer longevity deciding on the choice of biostable vs. biodegradable. Application of polymers in biomedical science has seen a steady progress from petroleum-derived synthetic polymers designed to be inert *in vivo* towards the design of bioactive materials that integrate with biological molecules or cells and regenerate tissue. Tissue-engineering expectations on polymer substrate is to perform appropriate cell adhesion, differentiation to extracellular matrix formation, and tissue regeneration.

Currently, polyethylene glycol (PEG) enjoys the status as one of the most widely explored polymers, used either alone or as a surface modifier, for its properties of low cell adhesion and protein adsorption [28]. Probing into atomic realms, we find each PEG chain surrounded by ordered water molecules mimicking hydrated surface that limits protein adsorption and thus recognition by immune cells such as macrophages [29]. Interestingly, this property of limited protein and cell adhesion to PEG has found appliance even in surgery to limit tissue adhesion. In this application, PEG-diacrylates are often used to allow photoinitiated cross-linking. Similarly, cross-linked PEG scaffolds have been synthesized where the inclusion of enzyme-degradable peptide cross-links creates a responsive scaffold, the degradation of which is controlled by the biological milieu. Thus, a PEG-modified polymer guarantees low protein adsorption and cell adhesion with its ability to tune adhesive properties through attachment of specific proteins or peptides.

5.1.3.5. Biodegradable Polymers

The limited life span of implants inside the body invites the use of degradable polymers, as the need for surgical removal is obviated. We must consider the rate and mechanism of degradation process to allow proper cell integration and tissue growth along with the toxic potential of its products to minimise the negative impact on regenerative process. State-of-the-art synthetic degradable polymers are polylactide-co-glycolide and their respective homopolymers [30]. The historical usage as suture materials with the benign nature of its acidic degradation products together with the ease of fabrication of three-dimensional scaffolds triggered its extensive use in tissue engineering. Polyanhydrides and polysebacic acid-co-1,3-bis(p-carboxyphenoxy) propane continue in the line of application.

Coming to natural degradable polymers, we have collagen and chitosan with its biocompatibility and ease of processing and modification. Being a naturally adhesive protein in the extracellular matrix, collagen is cell adhesive and provides a conductive environment to cell viability. Cross-linked porous scaffolds of collagen permits easy penetration of cell populations finding applications in tissue engineering. Insolubility of chitosan at physiological pH has been overcome with its wide modification possibilities and has even been used as nerve guidance channels [31].

5.1.3.6. Biostable Polymers

Biostable or nondegradable polymers are the choice of interest for cell encapsulated therapy where cells are seeded in a polymeric membrane that provides an immune-privileged environment for extended production of therapeutically relevant molecules from transplanted cells [32]. The polymers that have been most thoroughly studied for encapsulated cell therapy include polyacrylonitrile-co-vinylchloride, which has been studied as hollow fibre membranes; poly(2-hydroxyethyl methacrylate-comethyl methacrylate), which has been studied as microcapsules – and alginate/poly(lysine), studied as microcapsules and as scaffolds based on their compatibility and processibility [33].

5.1.3.7. Synthetic vs. Naturally-Derived Polymers

The processing dexterity of synthetic polymers in terms of composition, rate of degradation, mechanical, and chemical properties keep us confused in front of the compositional uniqueness, such as stimulating a specific cellular response provided by the naturally derived polymers. Of the numerous synthetic polymers, a few have been examined in more detail: poly (N-isopropylacrylamide), PEG, and PLGA driven by its interesting thermal and non-cell adhesive properties [34]. Agraose, hyaluronan, collagen, and fibrin possess the most compelling low protein-adsorbtive and cell-adhesive properties among the naturally derived polymers. Being the biochemical players inside the body, these polymers possess recognizable receptors like CD44 for hyaluronan easing its manipulation for multiple platform technology like wound healers [35].

5.1.3.8. Criteria for Designing the Scaffold

Like in the case of polymer choice, the physical aspects of scaffold design are also oriented towards the final application. In simple words, the scaffold must provide the appropriate physical, chemical, and mechanical properties required for cell viability and tissue formation in order to design the definite cellular microenvironment or cell-niche [36]. Three-dimensionality and interconnected porous nature of the scaffold promotes cell proliferation, migration, and/or differentiation through the lucid communication of the dispersed cells. The design criteria of polymeric scaffold must consider the physical structure of the tissue destined for implantation. The celebrated examples include the tissue-engineered ear scaffolds and those possessing grey–white matter tracts for implantation into spinal cord to guide cell growth accordingly [37].

Coming to mechanical properties, it is a common knowledge that hard tissue-implant scaffolds have to be stiff contrasting the scaffold flexibility and malleability of a soft or vascular tissue implant. This even transforms down to cellular level like neural stem cells differentiating on low modulus materials where as the precursors of stiffer tissue like bone require similar scaffolds [38]. Further analysis into the mechanical tension exerted by the cells on polymeric scaffolds is providing valuable insights into the control of differentiation profiles of various cell types. The efficacy of several silk- and collagen-based substrates to support chrondrogenesis of cultured human mesenchymal stem cells (hMSCs) was found to be influenced primarily by scaffold degradation rates rather than by chemical composition [39]. Samples of cross-linked collagen and slowly degrading silk promoted cell differentiation and matrix deposition, whereas uncross-linked collagen samples collapsed prematurely and were unable to support significant cartilaginous tissue formation. PLGA family of polymers has been pampered the most for pore fabrication process meant to enhance cell penetration and three-dimensional tissue formation. For example, PLGA 75/25 (vs. PLGA 85/15 or PLGA 50/50) was found to be the most suitable of the PLGA family for bone–tissue engineering applications.

5.1.3.9. How to Control Cellular Interaction on a Polymeric Scaffold

Exact definition of the biological–material interaction can be realized by modifying the scaffold with specific biomolecules after taking cues from the cascade pathways inside the body. Integrin family of cell-surface receptors provides a range of adhesion ligands not forgetting the ubiquitous tripeptide sequence arginine-glycine-aspartic acid (RGD) [40]. Understanding the cell migration where a series of attractive and repulsive forces guide the cell towards the respective tissue location helped to mimic scaffold design with growth factor concentration gradient and immobilized trophic factors. Scaffolds for nervous system have benefited much with immobilized ligands for spatially controlling and guiding the axons [41]. Agarose and cross-linked PEG gels are the representative images of multiple-cell interaction for 3-D patterning [42]. Research in blood vessel formation progressed much with controlled release strategy of growth factors like cytokines from suitable polymers [43].

Critical event pioneering the sequential cellular interaction to a surface finalizing with differentiated cell function in tissue engineering is the cellular adhesion process. Complications arise upon controlling the multiple parameters for comparative analysis. Inevitable steps for the quantification of extent of adhesion to a surface includes (1) suspension of cells over surface (2) incubation of sedimented cells in culture medium for stipulated time (3) detachment of loosely adherent cells under controlled conditions. Cell migration is the gearing force behind tissue engineering whether it is in association with the material surface or it is through cell populations. Broadly, the migration characterizations can be divided into visual assays confined to the movement of small number of cells to population assays covering larger fractions. Here, the intrinsic cell motility patterns are probed with agarose or filter chamber assays, which evaluate the random motility coefficient and the persistence time.

Correlation of the myriad cellular interaction parameters including migration, differentiation, and subsequent tissue formation can be achieved with characterizing cell aggregates. It provides a miniature replica of tissue microenvironment. Kinetic evaluation and extent of aggregation depend on direct visualization and monitoring aggregate size distribution over time, facilitated by the use of computer image analysis techniques or electronic particle counters. Small-angle light scatterings through rotating sample cuvettes are specialized aggregometers to measure the aggregate growth (Table 5.1.2).

5.1.4. CELL–NANOTOPOGRAPHY RESPONSES ON SYNTHETIC SUBSTRATES

Nanotopography is an inherent feature of cellular niche taking shape as extracellular matrix and cell basement membranes. The distinguished phenomenon of cellular interaction to nanoscale features is secured by contact guidance. Apart from regulating the cell migration, contact guidance can also be important in efficient organelle formation, such as axonal guidance and growth cone motility.

Nanotopography affects basic cell function in almost all types of mammalian cells depending upon many factors including cell type, feature size and geometry, and substrate stiffness. The coupled effect of nanogratings, nanopost arrays, and nanopit arrays with physicochemical properties of the substrate can induce different effects within a single-cell type.

5.1.4.1. Morphology

No other cellular function is easily modified as when compared to cell geometry upon interaction with nanotopography. Almost all the different cell types studied including fibroblasts, endothelial cells, stem cells, smooth muscle cells, epithelial cells, and Schwann cells respond with clear-cut impact on morphology [44]. Cells can sense substrates with features as small as 100 nm and depths as small as 75 nm by exhibiting specific morphological response [45]. An example worth mentioning is the enhanced extension and alignment of neurites from neuroblastoma cells cultured on nanogratings in the presence of nerve growth factor. Variables like decreased pitch and increased depth concur more aggressive

TABLE 5.1.2 Major Physicochemical Parameters and its Influence in Cellular Behavior

Biomaterial	Application	Advantage	Disadvantage	Example
Metal	Load-bearing biomedical prosthesis	Mechanical properties (strength, modulus, fatigue limit)	Susceptible to degradation by corrosion, release by-products causing adverse biological response	316 L stainless steel Cobalt–Cromium–Molybdenum Grade 4 Titanium
Ceramic	Coating on orthopaedic and dental applications to promote tissue integration As implants in load-free anatomical sites	Can be bio-inert, bioactive or biodegradable	Brittleness Poor tensile properties	Alumina Bio-glass Calcium phosphate Pyrolytic Carbon
Polymer	Cardiovascular & orthopaedic implants Drug release Tissue engineering	Tunable mechanical, thermal and degradation behavior	Leaching of additives inducing adverse host reactions	Ultrahigh molecular weight polyethylene, Poly methyl methacrylate Poly(lactide-co-glycolide)

responses. Nanopost and nanopit favour a reduction of cell spreading, adhesion complex, and constant filopodia formation. Human-derived leukocytes, keratinocytes, and monocytes are exceptions in responding to nanogratings [46]. Considering the various cell–biomaterial–geometry combinations, we conclude that the nanogratings are in contrast to nanopost and nanopit responses with initial reduction in the cell attachment and adhesion [45]. Coming to the cell proliferation profile, we cannot make any hypothesis regarding the mechanism for the effect of cell–nanotopography interactions as it is a combinatorial interplay of geometry, length scale, substrate material, and cell types [47]. However, we finalize upon the reduced proliferation rates on nanogratings in comparison to planar surfaces [48]. Equally surprising is the enhanced proliferation at some specific combination of nanopost and pits. Nanotopography has profound effect on the cellular migration, which gets biased towards grating axis [49]. With endothelial cells, epithelial cells, osteoblasts, and C6 glioma cells, an overall increase in migration velocities were observed. Polarization of microtubule organization centres in the work of Yim et al. can be quoted as a direct consequence of nanogratings. Study by Tzvetkova-Chevolleau et al. on the migration of normal (3T3) and malignant (SaI/N) mouse fibroblasts on poly(-dimethylsiloxane) (PDMS) substrates with nanograting and nanopost arrays confirmed the bias in the direction of migration of 3T3 cells alone [50].

5.1.4.2. Genotypic Alteration

Individual and comprehensive genetic analysis of various cell types helped to understand the effect of nanotopogrpahy in altering the gene expression profile. A study by Chou et al. specified the higher expression levels and stability of fibroblasts cultured on titanium nanogratings that even favour better fibronectin incorporation into cell-matrix proteins [51]. Genome-wide expression in fibroblasts and mesenchymal stem cells utilizing gene array techniques scrutinized the downregulation of many genes including those associated with apoptotic initiation, DNA repair, and transcription regulation in parallel to the upregulation of TGF-β2 and those involved in regulating G-protein signalling.

5.1.4.3. Differentiation

Considering the influence of topography on both basic cell function and gene expression, utilization of nanotopography as a signalling modality for directing differentiation cannot be overruled. Work by Yim et al. suggests that hMSCs cultured on nanogratings can be preferentially differentiated into neuronal lineages as determined by the presence of synaptophysin, tuj1, and nestin markers as well as the upregulation of MAP2 [52]. Dalby et al. cultured osteoprogenitor cells and hMSCs for extended periods on poly(methylmethacrylate) nanopit arrays of varying order and established its enhanced differentiation [53]. Another striking variation according to the symmetry and order of the nanopits was observed in expression of two bone-specific ECM proteins, namely osteopontin and osteocalcin. The potential of nanotopography to direct cell fate became obvious with the differential upregulation of genes in hMSCs cultured on nanopits compared to hMSCs cultured with dexamethasone alone on planar substrates [54].

5.1.4.4. Cell Superstructure

Another aftermath of cell–nanotopography interactions rejoiced by tissue engineers is with the potential to generate complex multicellular structures. The hypothesis got further reinforced with the culture of human endothelial progenitor cells (EPCs) on PDMS nanogratings. Even though the protein expression of endothelial markers in EPCs cultured on both nanograting and planar substrates was similar, multicellular band structures were observed in nanogratings after 6 days. The preferential transformation of band structures found in EPCs cultured on nanogratings to organized capillary tubes in an *in vitro* matrigel assay can be the outgrowth of several factors. First, alignment, elongation, and increased migration velocities biased morphology and cell–cell interactions leading to capillary tube precursors. Reduced proliferation of EPCs cultured on nanogratings prevented confluent cell layers resulting in lower cell density enhancing cell mobility thereby controlling the formation of multicellullar structures.

5.1.4.5. Current Theories behind the Mechanisms for Nanotopographic Sensing and Response

The large potential set of experiments and cell-specific outputs cross-examine cell–topography interactions focussing on the mechanism behind signal transduction pathways and the role of organelles including the cytoskeleton. Morphological response eliciting second-order effects in elongation and alignment of the nucleus can elicit genetic alteration of cells experiencing substrate nanotopography. Morphological response can be dated back to anisotropic stress generation. Contact guidance kinetics of fibroblasts lead to time-dependent alignment and rearrangement of microtubules and filaments culminating at focal adhesion contacts. The silent initiators and sequential transmitters of the topographical effects down to cellular functions like stress fibre formation are the organelles, specifically filopodia. Another hypothesis was on the substrate discontinuities featuring patterned protein deposition with apocalyptic changes in cell behavior including morphology, proliferation, differentiation, and apoptosis depending on the type of proteins adsorbed. Focal adhesions play a role in the alignment and elongation of cells contacting nanotopographical substrates through signalling connections between focal adhesions and cytoskeleton proteins. Actin polymerization dynamics involved in cytoskeleton rearrangement are essential for cell attachment and serve as a driving force for directional migration and morphological alterations.

Filopodia serve as topographical sensors in many cellular processes including migration, cytoskeleton rearrangement, and polarization of the cell body, all leading to a gross morphological effect of alignment and elongation. The role of molecule switches composed of Rho family of GTPases including Rac and Cd42 GTPases in the mutual response of cells to substrate nanotopography has been explored recently [55].

5.1.5. TECHNIQUES TO EVALUATE CELL–MATERIAL INTERACTIONS

Optimizing the cell–material interactions are challenging for a number of cell-based applications including fundamental biological study, drug discovery, and development and tissue engineering. This has given way to bioinspired approaches capable of directing specific cell functions with conjugation of biologically active molecules and controlled degradation in a 3-D matrix. As most cell types respond simultaneously to multiple topographical and biochemical aspects of micro- and nanostructural features in three dimensions, the sequential elucidation of structure–function relationship becomes complicated. An integrative understanding and application of cellular and molecular responses to the chemical, mechanical, and topological properties of substrates are indispensible requirements for selective cell attachment, survival, migration, growth, and differentiation (Table 5.1.3).

TABLE 5.1.3 Pros and Cons of the Major Biomaterial Characterization Techniques

Technique	Application	Limitation
Large-scale cell-based screening	Versatile approach used to test cell responses to a wide range of scaffold properties such as surface chemistry, surface energy, surface topology, peptide ligands, pore size, or mechanical properties	Mostly limited to 2-D films or surfaces despite the fact that biomaterials are frequently used to fabricate 3-D scaffolds
Microarray technology	Investigating the changing cellular environment, in particular, the gene- and protein expression pattern, toxicity profile in response to different types of biomaterials	Limited to annotated transcripts along with cost, technical variation and complexity of statistical analysis

TABLE 5.1.3 Pros and Cons of the Major Biomaterial Characterization Techniques—cont'd

Technique	Application	Limitation
Proteomic technology	Enable the profiling of a complex sample for global protein expression profile, including posttranslational modifications and subcellular localization representing the current physiological status	Poor detection of low abundance and hydrophobic membrane proteins with 2-D electrophoresis along with contamination by keratin and the autolysis of trypsin in case of mass spectrometry
Atomic force microscopy	Generates the three-dimensional profiling and imaging of the cells on materials along with providing surface-adhesion force measurements	Provides only surface information and not suitable for three-dimensional scaffolds. Cantilever mechanism cannot be used for surfaces with large changes in surface topography, like macroporous scaffolds with 600 μm deep pores
Four-dimensional elastic light-scattering fingerprinting	Provides pinpoint precision into the variation of nanoscale architecture of living cells and tissue organization upon biomaterial interaction through real time, noninvasive monitoring of these fingerprints	Confined to the assessment of a single process at a time in a very controlled environment.
Bio-Raman spectroscopy	Monitoring the viability, cell cycle, metabolism, mitosis, differentiation, de-differentiation, mineralization and onset of death of single cells and cell assemblages (organoids) in real time without damage to the cells upon biomaterial interaction is realized with bio-Raman spectroscopy	Absorption of radiation from laser beam can cause sample heating and damage

Many questions with *in vitro* tests remains to be addressed before venturing into *in vivo* applications like (1) Cell culture techniques massively use immortal cell lines for the majority of *in vitro* analysis that fail to express the complex arrays of proteins characteristic of mature phenotypes of human cells *in vivo*. (2) *In vitro* response of materials must be correlated to the timeline of cell-cycle events of cells in culture. (3) Sustaining the maintenance of mature cell phenotype in culture during testing. (4) It must shed light on the molecular biological changes taking place in the cells during exposure to the material along with statistical analyses in order to discriminate between small changes in the cell population. Low cost and ease of use are equally important factors. The current advances in cell/molecular biology and material engineering have led to the development of many screening techniques that isolate specific cell response to individual material characteristics that fulfils the above criteria.

5.1.5.1. High-Throughput Technique

Time-consuming interactive processing of traditional assays calls for the requirement of qualitative and quantitative tools capable of conducting biological assays and assessing the dynamic response in high throughput. High-throughput technology is "a semi-automated process whereby, for any given input and the limitations set by the technology in question, the coverage and quantity of data output is comprehensive, normally thousands to millions of data points with a rapid processing speed" [56]. What we understand from high throughput is the rapid rate of production or high speed at which data are received enabling greater reproducibility, reliability, and comparability of studies with some amount of automation. High-throughput screening experiments can play important role in fundamental studies of cell-surface interaction by targeting a particular cell event and its response to substrate surfaces to answer questions like what pathways or networks are altered within the system under study and to what extent high-throughput technologies launched the microarray techniques and combinatorial libraries of polymeric biomaterials along with computational modelling in mind the need to characterize global cellular and molecular changes of biomaterial implantation.

5.1.5.2. Large-Scale Cell-Based Screening

Traditional use of high throughput screening is for hit identification where the effects of performance of large collection of compounds on cell viability, proliferation, differentiation, morphology, and biocompatibility are analyzed through robotic performance in microtitre plates. This provides the most reliable physiological data that can be translated from an *in vitro* to an *in vivo* environment. Blackmore et al. used high-content analysis methods to screen 743 plasmids encoding developmentally regulated genes while functionally assessing neurite outgrowth [57]. Decelerated regenerative capacity of neurons as they mature can be traced back to genetic alterations studied through high-throughput overexpression screen in postnatal CNS neurons. Real-time phenotypic cellular changes replaced static features with the live-cell imaging of primary neurons studies of Daub et al., combined with high content screen. Anderson et al. utilized a high-throughput cell-based screening assay to evaluate a library of 2350 structurally unique, degradable, cationic polymers to identify 46 that can transfect cells with a higher efficiency than conventional delivery systems [58]. Similarly, Alder et al. used a systematic approach to rapidly screen a panel of cell lines (SP2/0 myeloma, CHO, 1929 fibroblasts, and J774 macrophages) treated with biomaterials that have previously been studied as carriers for drugs, proteins, and vaccines [59]. However, it is better to rethink on the sufficiency of these outputs to comprehend the full nature of cell–biomaterial interactions.

5.1.5.3. Expression Studies

5.1.5.3.1. Application of Microarray Technology to Biomaterials

Microarrays have become standard tools to investigate global gene/protein expression patterns in response to different types of biomaterials including metals, titanium, polymers, and textile/fibrin gels. The latest characterization through microarrays is the elucidation of transcriptomic responses of cellular activities with a focus on the ECM proteins, cellular adhesion molecules, and proteins involved in inflammation and apoptosis [60]. This could even extend to search for toxicological and cytotoxicological effects of distinct materials in relation to cells/organisms. However, microarrays can detect only annotated transcripts along with technical variation, cost, and the complexity of statistical analysis involved.

Bioinformatics tools with oligonucleotide and cDNA microarrays contextualize many molecular mechanisms underlying cell–biomaterial interactions and gene ontologies. Pioneering study by Balasubramanian et al. with gold nanoparticles demonstrated the alterations in genetic expression profiles that changed views on the biological inertness of these nanoparticles [61]. These gold nanoparticles after accumulating in liver and spleen of rat models caused the differential expression of around 79 and 62 genes that regulate several pathways including detoxification, lipid metabolism, cell cycle, defence response, and circadian rhythm. DNA microarrays also helped to understand the interaction of vascular cells and ferrous ions, (Fe(II)), from iron stents and to analyze the toxic effects of nickel ions on mouse fibroblast cells enabling genomic-level information beyond routine cytotoxicity tests [62]. Molecular level precision of the pathways that changed upon exposure to metal ions became possible with these sensitive microarrays probing into focal adhesion, cell cycle, insulin signalling, electron transport chain, mRNA processing, and ribosome protein. Microarrays can distinguish potential scaffolds for tissue engineering as with the study of Gundy et al. analysing the influence of PLA in the gene expression profiles for human coronary artery smooth muscle cells seeded within a fibrin gel. As the transcriptomic data provided by micro arrays often don't correlate with the proteomics, the need for protein expression profiles including posttranslational modifications and subcellular localization is on the rise.

5.1.5.4. Protein Expression Profiling: Application in Biomaterials Research

Proteomics highlight a high throughput analysis of proteins with its differential expression. Proteomics offer a more accurate molecular profile of cell–biomaterial interaction than the mRNA expression subjected to regulatory changes and modulations. Traditional analyses of protein expressions like western blotting, ELISA, and radioimmunoassay were limited in providing the complete information about the adsorbed protein along with its respective conformation. This made us to embrace more contemporary proteomic strategies utilizing two-dimensional (2-D) gel electrophoresis and matrix assisted laser desorption ionization time-of-flight mass spectrometry (MALDI-TOF MS). Xu et al. used iTRAQ-coupled 2-D LC-MS/MS to profile human osteoblasts cultured on HA [63]. Dinnes et al. combined two-dimensional electrophoresis with MALDI-TOF to search for protein expression changes in monocyte-derived macrophages (MDM) cultured on different material surfaces [64]. MDMs are key inflammatory cells regulating the host response to a biomaterial. Liquid chromatography MALDI has also been applied to gain insight into the mechanism of monocyte–material interactions. Li et al. used surface-enhanced laser desorption ionization-time of flight mass spectrometry, a protein chip technology in which a protein mixture is spotted on a chip surface modified with different biological or chemical affinities to candidate the marker proteins of regeneration in ear-punched tissue strain, MRL/MpJ-Faslpr (MRL) against C57BL/6J(B6) mice [65]. Derhami et al. attempted to identify proteins adsorbed to the surface of commercial titanium and TCP with the differential expression of intracellular proteins utilizing 2-D gel electrophoresis combined with MALDI-TOF MS. The major limitations of proteomic technologies pointed out by GÖrg et al. include the poor detection of low abundance and hydrophobic membrane proteins with 2-d electrophoresis along with contamination by keratin and the autolysis of trypsin in the case of mass spectrometry, as detailed by Garbis et al.

Data mining and interpretation rely heavily on bioinformatics and systems biology to integrate the data sets from multiple sources with different error rates, biases, sizes, and origins into an optimization algorithm with minimal negatives. The Biomolecular Adsorption Database (BAD) was published in 2009, maximizing the protein adsorption data already produced by archiving in a central location (http://dbweb.liv.ac.uk/bad/). BAD currently comprises 767 records of protein adsorption experiments reported in the literature and, in combination with neural networks and regression analysis, this information can predict the amount of adsorbed protein, the thickness of the adsorbed protein layer, and the surface tension of the protein-covered surfaces.

5.1.5.5. Green Fluorescent Protein as Live Reporters of Cell–Biomaterial Interactions

Immortal cell lines engineered to express green fluorescent protein (GFP) fusion proteins were used as probes to identify differential cellular and molecular morphologies induced by variations in composition and processing of biodegradable, polymer substrates [66]. The major advantage of using GFP-based live reporters to profile biomaterials is the ability to easily obtain dynamic imaging over time compared with the need for multiple, fixed preparations inherent to immunocytochemical methods, and the introduction of artifacts related to the use of vital stains [67,68] Morphometric descriptors quantified using high-resolution, single-cell, confocal images obtained from cells expressing GFP-fusion proteins include (1) cellular morphologic parameters such as area, perimeter, length of major and minor axes of the ellipse with the same geometric area of the cell, mean radius, mean diameter, roundness, protrusions, and feret length; (2) cytoskeletal parameters such as fractal dimension, margination, heterogeneity, clumpiness, and average fibre length/width; (3) focal adhesion parameters such as integrated optical density, heterogeneity, margination, clumpiness, focal adhesions per cell, strength of focal adhesion region, and total area of actinin/paxillin rich structure per cell; (4) texture metrics; and (5) higher order moments such as skewness and kurtosis related to the functional state of a cell.

5.1.5.6. Probing Molecular Recognition Sites on Biosurfaces Using AFM

Ligand–receptor interactions play a central role in biology in the form of interactions between complementary strands of DNA, enzyme–substrate, antigen–antibody, lectin–carbohydrate, and cell-adhesion proteins [69]. Coming to biomaterial science, probing ligand–receptor interactions help us to gain control over cell responses including adhesion process [70]. Techniques like osmotic stress method [71], the surface forces apparatus [72], magnetic beads [73], optical tweezers [74], and the biomembrane force probe [75] have taken the characterization of biomolecular forces into new realms. Yet they lack lateral resolution in mapping recognition sites on living cells. Advent of atomic force microscopy (AFM) has seen the three-dimensional analysis of biomaterial in terms of structure, properties, and biomolecular interactions with pico-Newton sensitivity. In principle, AFM senses the force between a sample and the tip of cantilever and transforms it into a three-dimensional imaging. Instead of using an incident beam as in classical microscopies, AFM relies on laser beam focused on the free end of the cantilever and reflected onto a photodiode.

The AFM tips can be modified with chemical groups like OH, CH3, and COOH for quantifying the cell surface properties. Live yeast cells were coated onto AFM tips after modification with ionizable carboxyl groups to probe the surface charge at nanometre level [76]. While functionalizing the tips for molecular recognition studies, we have to consider (1) the forces which immobilize the molecules should be stronger than the intermolecular force being studied; (2) mobility of the attached biomolecules for free interaction with complementary molecules; (3) minimal contribution of nonspecific adhesion to the measured forces; (4) lower density of biomolecules at the surface is preferred in order to ensure single molecule detection; and (5) site-directed coupling to ensure orientation of all the interacting molecules [77]. Animal and microbial cellular interactions and adhesion forces can be measured after attaching onto AFM cantilevers with various immobilization techniques such as glutaraldehyde treatments, epoxy resin, or via lectins such as wheat germ agglutinin or concanavalin A (Con A) [78]. Adhesion force mapping or dynamic recognition force mapping with AFM give insight into the distribution of individual binding sites with nanoscale resolution.

Single-molecule AFM force spectroscopy helps us to address questions related to molecular forces driving the interaction of cell adhesion molecules like lectins, integrins, cadherins, bacterial adhesions, and their distribution across cell surface. Adhesion force mapping through AFM can complement information provided by fluorescence and electron microscopy methods. AFM-based nanoindendation encourages the study of mechanical properties like elastic modulii, stiffness, and hardness of various materials including tissues such as bone and cartilage, biomolecules such as DNA, proteins, and cell membranes, composite materials and polymers such as hydrogel-based materials. The real-time mechanical characterization of materials mimicking *in vivo* conditions and forced mapping to visualize the topographical distribution of mechanical properties are its critical features. Live cell imaging to probe into the living cell dynamics, intercellular communication, and responses to stimulus, drugs, or toxic substances without the use of potentially interfering fluorescent labels or probes can be obtained in real time with AFM.

5.1.5.7. Four-Dimensional Elastic Light-Scattering Fingerprinting (4D-ELF)

A recently developed light scattering spectroscopy imaging can provide a quantitative assessment of subcellular events, such as organelle enlargement or increased chromatin content, in a real-time and nondestructive manner yielding unprecedented insights into the microarchitectural organization of cells interacting with substrates [79]. Josephine Allen et al. hypothesize that interfacial cell/biomaterial interactions will affect subcellular structures in defined ways, giving rise to specific light-scattering 'fingerprints' that can be used to describe an ongoing cellular process. 4D-ELF provides pinpoint precision into the nanoscale architecture of living cells and tissue organization through real-time, noninvasive monitoring of these fingerprints during proliferation and differentiation of cells on various substrates. The major limitation is in confining to the assessment of a single process at a time in a very controlled environment.

Optical spectroscopy and imaging approaches such as fluorescence and Raman spectroscopy, confocal microscopy, multiphoton microscopy, and optical coherence tomography are other useful tools for tissue engineers depending on the type of information, the resolution/penetration depth requirements, and the cost and complexity of the system required.

5.1.5.8. Bio-Raman Spectroscopy

Raman spectroscopy is a noninvasive technique that fulfils most requirements of an idealized biomaterial–cellular interaction probe [80]. The bio-Raman method uses specific wavelengths of lasers, new spectrometers and detectors to obtain Raman signals from living cells in a special environmental chamber fitted with a microscope and a gas support system. Raman spectral signatures can be obtained noninvasively with a high signal-to-noise ratio on living cells with 2 min exposure to 115 mW of 785 nm laser light. Monitoring the viability, cell cycle, metabolism, mitosis, differentiation, de-differentiation, and mineralization and onset of death of single cells and cell assemblages (organoids) in real time without damage to the cells is realized with bio-Raman spectroscopy [81]. Single cells or three-dimensional assemblages of cells living in a specially designed microscopic chamber can survive for long periods of time while being interrogated as to their state of health without alteration or damage to the cells [82]. The time required for quantifying the spectroscopic signatures upon exposure of cells to test agents introduced into the cell chamber rests within the resolution of the equipment. The biomolecular changes within the cells upon exposure in the form of alterations to DNA, RNA proteins, lipids & carbohydrates transform into spectroscopic signatures on a preselected number of cells inside the chamber. The bio-Raman system offers the potential of monitoring and maintaining cell phenotype by *in situ* spectroscopic analysis of the spectroscopic signatures. The many advantages of bio-Raman microspectroscopy for testing of living cell–biomaterial or cell–nanoparticle interactions include (1) noninvasive, no labels required (2) measurements of cells maintained in physiological conditions (3) rapid data collection (1–2 min per spectrum, less than 10 min per cell) (4) high biomolecular chemical specificity (5) simultaneous monitoring of changes in cell morphology, differentiation, and phenotype of cells being tested and (6) individual cells, cell assemblages, co-cultures, or organoids can be tested. The fact that one doesn't require training in advanced level of molecular biology to conduct the analysis popularizes it as an ideal tool for interdisciplinary sciences like biomaterials.

The mile-stone study done with bio-Raman methodology was the detection and discrimination between two important chemical agents used in terrorist attacks, ricin, and sulphur mustard owing to their differing modes of attack of A549 human lung-like cells.

5.1.5.9. Modular Immune *In Vitro* Construct

A plethora of research in biomaterial science culminates at the point of nano-particles smartening all areas within its reach. Equally challenging is establishing its safety limits and cellular interaction consequences within the human body. A new *in vitro* test system, termed as modular immune *in vitro* construct (MIMIC), has been established that provides a predictive assay model focused on the human inflammatory response to nanoparticles. It interrogates short-term inflammatory response as well as the long-term memory response in either separate or longitudinal studies. The synergistic involvement of blood vein endothelial cells and monocyte-derived dendritic cells enables evaluation of the early immune responses associated with the exposure to a foreign body in terms of elevated levels of pro-inflammatory cytokines and increased maturation and expression of co-stimulatory molecules on dendritic cells. The MIMIC system has been used to assess the immunogenicity of titanium dioxide nanoparticles: anatase (7–10 nm), rutile (15–20 nm), and TiO_2 nanotubes (10–15 nm diameter and 70–150 nm length), characteristic of an *in vivo* inflammatory response [83].

5.1.5.10. Concluding Remarks

Analysing the intricate relation between biomolecules, cells, and biomaterials will continue to haunt us with its element of unpredictability. The advent of nanotechnology has brought forth many state-of-the-art techniques aiming to resolve the conceptual conflicts between cells and biomaterials from the macro-down to the picoscale, circumventing the need for animal studies. This area of research will certainly hold great potential for further breakthroughs in the understanding and exploitation of the cascades of reactions around an implant. Future looks forward to the merger of the sensational stem cells with intelligent materials in harmony, with human biology to welcome the much awaited whole tissue transplants.

ACKNOWLEDGMENTS

Authors wish to thank Dr K. Radhakrishnan, Director, and Dr G. S. Bhuvaneswar, Head, BMT Wing, Sree Chitra Tirunal Institute for Medical Sciences and Technology, Poojapura, Thiruvananthapuram, for providing facilities. This work was supported by the Department of Science & Technology, Government of India through the project "Facility for nano/microparticle based biomaterials – advanced drug delivery systems 8013", under the Drugs & Pharmaceuticals Research Programme, New Delhi. The authors would also like to thank UGC for Junior Research Fellowship.

REFERENCES

[1] Chung BG, Kang L, Khademhosseini A. Micro- and nanoscale technologies for drug discovery and tissue engineering applications. Expert Opin Drug Discov 2007;2(12):1–16.

[2] Ingber DE. Mechanical control of tissue morphogenesis during embryological development. Int J Dev Biol 2006;50(2–3):255–66.

[3] Bao G, Suresh S. Cell and molecular mechanics of biological materials. Nat Mater 2003;2(11):715–25.

[4] Steven MM, George JH. Exploring and engineering the cell surface interface. Science 2005;310(5751):1135–8.

[5] Sanz-Herrera JA, Moreo P, Garcia-Aznar JM, Doblare. M. On the effect of substrate curvature on cell mechanics. Biomaterials 2009;30(34):6674–86.

[6] Tong WY, Liang YM, Tam V, Yip HK, Kao YT, Cheung KM, et al. Biochemical characterization of the cell biomaterial interface by quantitative proteomics. Mol Cell Proteomics 2010;9(10):2089–98.

[7] Tay CY, Irvine SA, Boey FY, Tan LP, Venkatraman S. Micro-/nano-engineered cellular responses for soft tissue engineering and biomedical applications. Small 23 May 2011;7(10):1361–78.

[8] Baker BM, Handorff AM, Lonescu LC, Li WJ, Mauck RL. New directions in nanofibrous scaffolds for soft tissue enginnering and regeneration. Expert Rev Med Devices 2009 Sep;6(5):515–32.

[9] Van Vliet K. The biomechanics tool box: experimental approaches for living cells and biomolecules. Acta Materialia 2003;51(19):5881–905.

[10] Joy A, Cohen DM, Luk A, Anim-Danso E, Chen C, Kohn J. Control of surface chemistry, surface stiffness and cell function in a novel terpolymer methacrylate library. Langmuir 2011;27(5):1891–9.

[11] Mitragotri S, Lahann J. Physical approaches to biomaterial design. Nat Mater 2009;8(1): 15–23.

[12] Hench LL, Thomson I. Twenty first century challenge for biomaterials. J R Soc Interface 2010;7(Suppl. 4):S379–91.

[13] Shoichet MS. Polymer scaffolds for biomaterials applications. Macromol Perspect 2010;43:581–91.

[14] Anderson JM, Schoen FJ. Interactions of blood with artificial surfaces. Current issues in heart valve. In: Butcharl Eric G, Boanar Endre, editors. PP. London: ICR Publishers; 1992. p. 160–71.

[15] Kingshott P, Anderson G, McArthur SL, Griesser SJ. Surface modification and chemical surface analysis of biomaterials. Curr Opin Chem Biol 2011 Oct;15(5):667–76.

[16] Cukierman E, Pankov R, Stevens DR, Yamada KM. Taking cell matrix adhesions to the third dimension. Science 2001;294(5547):1708–12.

[17] Cavalcanti-Adam EA, Daniel A, Hirschfeld-Warneken VC, Spatz JP. Cell adhesion and response to synthetic nanopatterned environments by steering receptor clustering and spatial location. Front Interdiscip Sci Life Sci 2008;2(5):276–85.

[18] Boyan BD, Hummert TW, Dean DD, Schwartz Z. Role of material science in regulating bone and cartilage cell response. Biomaterials 1996;17(2):137–46.

[19] Wilson CJ, Cleqq RE, Leavesley DI, Pearcy MJ. Mediation of biomaterial–cell interaction by adsorbed proteins. Tissue Eng 2005;11(1–2):1–18.

[20] Le Heuc JC, Sckerverbeke T, Clement D, Faber J, Le Rebeller A. Influence of porosity on the mechanical resistance of hydroxyapatite ceramics under compressive stress. Biomaterials 1995;16(2):113–8.

[21] Bose S, Tarafder S. Calcium phosphate ceramic systems in growth factor and drug delivery for bone tissue engineering. A review. Acta Biomater 2011;8(4):1401–21.

[22] Porter JR, Ruckh TT, Popat KC. Bone tissue engineering: a review in bone bio mimetics and drug delivery strategies. Biotechnol Prog 2009;25(6):1539–60.

[23] Saldana L, Sanchez-Salsado S, Izquierdo Barba I, Bensiamar F, Munuera L, Vallet-Regi M, et al. Calcium phosphate based particles influence osteogenic maturation of human mesenchymal stem cells. Acta Biomater 2009;5(4):1294–305.

[24] Liu Y, Li JP, Hunziker EB, de Groot K. Incorporation of growth factors into medical devices via bio mimetic coatings. Philos Transact A Math Phys Eng Sci 2006;364(1838):233–48.

[25] Hallab N, Jacobs JJ, Black J. Hypersensitivity to metallic biomaterials: a review of leukocyte migration inhibition assays. Biomaterials 2000;21(13):1301–14.

[26] Taveres JC, Cornelio DA, da Silva NB, de Moura CE, de Queiroz JD, Sa JC, et al. Effect of titanium surface modified by plasma energy source on genotoxic response in vivo. Toxicology 2009;262(2):138–45.

[27] Vladkova TJ. Surface engineered polymeric biomaterials with improved biocontact properties. Int J Pol Sci 2010.

[28] Merill EW, Salzman EW, Wan S, Mahmud N, Kushner L, Lindon JN, et al. Platelet compatible hydrophilic segmented polyurethanes from polyethylene glycols and cyclohexane diisocyanate. Trans Am Soc Artif Intern Organs 1982;28:482–7.

[29] Sheth SR, Leckband D. Measurements of attractive forces between proteins and end grafted poly ethylene glycol chains. Proc Natl Acad Sci U S A 1997;94(16):8399–404.

[30] Gunatillake PA, Adhikari R. Biodegradable synthetic polymers for tissue engineering. Eur Cells Mater 2003;5:1–16.

[31] Zeilinski BA, Aebischer P. Chitosan as a matrix for mammalian cell encapsulation. Biomaterials 1994;15(13):1049–56.

[32] Lahooti S, Sefton MV. Effect of an immobilization matrix and capsule membrane permeability on the viability of encapsulated HEK cells. Biomaterials 2000;21(10):987–95.

[33] De Vos P, De Haan B, Van Schilfgaarde R. Effect of the alginate composition on the biocompatibility of alginate-poly lysine microcapsules. Biomaterials 1997;18(3):273–8.

[34] Shakesheff K, Cannizaro S, Langer R. Creating biomimetic micro-environments with synthetic polymer peptide hybrid molecules. J Biomater Sci Polym Ed 1998;9(5):507–18.

[35] Jones KS, Sefton MV, Gorenzynski RM. In vivo recognition by the host adaptive immune system of micro encapsulated xenogenic cells. Transplantation 2004;78(10):1454–62.

[36] Madlambayan GJ, Rogers I, Purpura KA, Ito C, Yu M, Kirouac D, et al. Clinically relevant expansion of hematopoietic stem cells with conserved function in a single use, closed system bioprocess. Biol Blood Marrow Transplant 2006;12(10):1020–30.

[37] Cao Y, Vacanti JP, Paige KT, Upton J, Vacanti CA. Transplantation chondrocytes utilizing a polymer-cell construct to produce tissue engineered cartilage in the shape of a human ear. Plast Reconstr Surg 1997;100(2):297–302.

[38] Saha K, Keung AJ, Irwin EF, Li Y, Little L, Schaffer DV, et al. Substrate modulus directs neural stem cell behavior. Biophys J 2008;95(9):4426–38.

[39] Mascarinec SA, Tirrel DA. Protein engineering approaches to biomaterial design. Curr Opin Biotechnol 2005;16:422–6.

[40] Rouslahti E, Pierschbacher MD. New perspectives in cell-adhesion-RGD and integrins. Science 1987;238(4826):491–7.

[41] Moore K, Macsween M, Schoichet M. Immobilized concentration gradients of neurotrophic factors guide neurite outgrowth of primary neurons in macroporous scaffolds. Tissue Eng 2006;12(2):267–78.

[42] Lyo Y, Schoichet MS. Light activated immobilization of biomolecules to agarose hydrogels for controlled cellular response. Biomacromolecules 2004;5(6):2315–23.

[43] Richardson TP, Peters MC, Ennet AB, Mooney DJ. Polymeric system for dual growth factor delivery. Nat Biotechnol 2001;19(11):1029–34.

[44] Kourovkoglou S, Dee KC, Bizios R, McIntire LV, Zygourakis K. Endothelial cell migration on surfaces modified with immobilized adhesive peptides. Biomaterials 2000;21(17):1725–33.

[45] Bettinger CJ, Langer R, Borenstein JT. Engineering substrate topography at the micro- and nanoscale to control cell function. Angew Chem Int Ed Eng 2009;48(30):5406–15.

[46] Saltzman WM. Cell interactions with polymers. In: Lanza R, Langer R, Chick W, editors. Principles of Tissue Engineering. Austin: Academic Press, R.G. Landes Company; 1997. p. 225–46.

[47] Loesberg WA, Riet te J, Delft van FCMJM, Schon P, Figdor CJ, Speller S, et al. The threshold at which substrate nanogroove dimensions may influence fibroblast alignment and adhesion. Biomaterials 2007;28(27):3944–51.

[48] Dalby JM, Gadegaard N, Wilkinson CDW. The response of fibroblasts to hexagonal nano-topography fabricated by electron beam lithography. J Biomed Mater Res A 2008;84:973–9.

[49] Bettinger CJ, Zhang Z, Gerecht S, Borenstein JT, Langer R. Enhancement of in vitro capillary tube formation by substrate nanotopography. Adv Mater 2008;20(1):99–103.

[50] Tzvetkova-Chevolleau T, Stephanou A, Fuard D, Ohayon J, Schiavone P, Tracqui P. The motility of normal and cancer cells in response to the combined influence of the substrate rigidity and anisotropic microstructure. Biomaterials 2008;29(10):1541–51.

[51] Chou LS, Firth JD, Uitto VJ, Brunette DM. Substratum surface topography alters cell shape and regulates fibronectin messenger-RNA level, messenger-RNA stability, secretion and assembly in human fibroblasts. J Cell Sci 1995;108:1563–73.

[52] Yim EK, Pang SW, Leong KW. Synthetic nanostructures inducing differentiation of human mesenchymal stem cells into neuronal lineage. Exp Cell Res 2007;313(9):1820–9.

[53] Dalby MJ, Riehle MO, Yaarwood SJ, Wilkinson CD, Curtis AS. Nucleus alignment and cell signaling in fibroblasts: response to a microgrooved topography. Exp Cell Res 2003;284(2):274–82.

[54] Cheng JY, Mayes AM, Ross CA. Nanostructure engineering by templated self assembly of block copolymers. Nat Mater 2004;3(11):823–8.

[55] Jaffe AB, Hall A. Rho GTPases: biochemistry and biology. Annu Rev Cell Dev Biol 2005;21:247–69.

[56] Power KA, Fitzgerald KT, Gallagher WM. Examination of cell-host biomaterial interactions via high throughput technologies. A reappraisal. Biomaterials 2010;31(26):6667–74.

[57] Blackmore MG, Moore DL, Smith RP, Goldberg JL, Bixby JL, Lemmon VP. High content screening of cortical neurons identifies novel regulators of axon growth. Mol Cell Neurosci 2010;44(1):43–54.

[58] Anderson DG, Lynn DM, Langer R. Semi automated synthesis and screening of a large library of degradable cationic polymers for gene delivery 13. Angew Chem Int Ed 2003;42(27):3153–8.

[59] Alder AF, Peterson LK, Wilson JH, Torres MP, Thorstenson JB, Gardner SW. High throughput cell based screening of biodegradable polyanhydride libraries. Comb Chem High Throughput Screen 2009;12(7):634–45.

[60] Gallagher WM, Lynch I, Allen LT, Miller I, Penney SC, O'Connor DP. Molecular basis of cell biomaterial interaction: insights gained from transcriptomic proteomic studies. Biomaterials 2006;27(35):5871–82.

[61] Balasubramanian SK, Jittiwat J, Manikandan J, Ong CN, Yu LE, Ong WY. Biodistribution of gold nanoparticles and gene expression changes in the liver and spleen after intravenous administration in rats. Biomaterials 2010;31(8):2034–42.

[62] Mueller PP, May T, Perz A, Hauser H, Peuster M. Control of smooth muscle cell proliferation by ferrous irons. Biomaterials 2006;27(10):2193–200.

[63] Xu JL, Khor KA, Sui J, Zhang J, Tan TL, Chen WN. Comparative proteomics profile of osteoblasts cultured on dissimilar hydroxyapatite biomaterials: an iTRAQ-coupled 2-D LC-MS/MS analysis. Proteomics 2008;8(20):4249–58.

[64] Dinnes DLM, Marcal H, Mahler SM, Santerre JP, Labow RS. Material surfaces affect the protein expression patterns of human macrophages: a proteomics approach. J Biomed Mater Res A 2007;80(4):895–908.

[65] Li X, Mohan S, Gu W, Miyakoshi N, Baylink DJ. Differential protein profile in the ear--punched tissue of regeneration and non-regeneration strains of mice: a novel approach to explore the candidate genes for soft tissue regeneration. Biochim Biophys Acta 2000;1524 (2–3):102–9.

[66] Treiser MD, Liu E, Dubin RA, Sung HJ, Kohn J, Moghe PV. Profiling cell biomaterial interactions via cell based fluororeporter imaging. Biotechniques 2007;43(3):361–6. 368.

[67] Miyawaki A. Visualization of the spatial and temporal dynamics of intracellular signaling. Dev Cell 2003;4(3):295–305.

[68] Ting AY, Kain KH, Klemke RL, Tsien RY. Genetically encoded fluorescent reporters of protein tyrosine kinase activities in living cells. Proc Natl Acad Sci U S A 2001;98(26): 15003–8.

[69] Mrksich M. What can surface chemistry do for cell biology. Curr Opin Chem Biol 2002;6:794–7.

[70] Gumbiner BM. Regulation of cadherin mediated adhesion in morphogenesis. Nat Rev Mol Cell Biol 2005;6(8):622–34.

[71] LeNeveu DM, Rand RP, Parsegian VA. Measurement of forces between lecithin bilayers. Nature 1976;259(5544):601–3.

[72] Leckband DE, Israelachvili JN, Schmitt FJ, Knoll W. Long range attraction and molecular rearrangements in receptor ligand interactions. Science 1992;255(5050):1419–21.

[73] Smith SB, Finzi L, Bustamante C. Direct mechanical measurements of the elasticity of single DNA molecules by using magnetic beads. Science 1992;258(5085):1122–6.

[74] Ashkin A, Schutze K, Dziedzic JM, Euteneuer U, Schliwa M. Force generation of organelle transport measured in vivo by an infra red laser trap. Nature 1990;348(6299):346–8.

[75] Merkel R, Nassoy P, Leung A, Ritchie K, Evans E. Energy landscapes of receptor-ligand bonds explored with dynamic force spectroscopy. Nature 1999;397(6714):50–3.

[76] Ahimou F, Denis FA, Touhami A, Dufrene YF. Probing microbial cell surface charges by atomic force microscopy. Langmuir 2002;18:9937–41.

[77] Dupres V, Verbelen C, Dufrene YF. Probing molecular recognition sites on biosurfaces using AFM. Biomaterials 2006;28(15):2393–402.

[78] Lower SK, Hochella Jr MF, Beveridge TJ. Bacterial recognition of mineral surfaces: nanoscale interactions between Shewanella and alpha-FeOOH. Science 2001;292(5520): 1360–3.

[79] Allen J, Liu Y, Kim YL, Turzhitsky VM, Backman V, Ameer GA. Spectroscopic translation of cell–biomaterial interactions. Biomaterials 2007;28(2):162–74.

[80] Hench LL. Thompson Ian twenty first century challenges for biomaterials. J R Soc Interface 2010;7(Suppl. 4):S379–91.

[81] Notinger I, Hench LL. Raman micro spectroscopy: a non-invasive tool for studies of individual living cells in vitro. Expert Rev Med Rev 2006;3(2):215–34.

[82] Notinger I, Verrier S, Romanska H, Bishop AE, Polak JM, Hench LL. In situ characterization of living cells by Raman spectroscopy. Spectroscopy 2002;16:43–51.

[83] Schanen BC, Karatoki AS, Seal S, Drake III DR, Warren WL. Exposure to titanium dioxide nanomaterials provokes inflammation of an in vitro human immune construct. ACS Nano 2009;3(9):2523–32.

[84] Griffith LG. Polymeric biomaterials. Acta Mater 2000;48(1):263–77.

Chapter 5.2

Characterization of Bacteria–Biomaterial Interactions, from a Single Cell to Biofilms

Nehal I. Abu-Lail and Haluk Beyenal

Gene and Linda Voiland School of Chemical Engineering and Bioengineering, Washington State University, Pullman, WA, USA

Chapter Outline

Characterization of Biomaterials. http://dx.doi.org/10.1016/B978-0-12-415800-9.00006-1

5.2.1. INTRODUCTION

According to the Society for Biomaterials, a biomaterial is any natural or man-made material that comprises all or part of a living structure or a biomedical device, which performs, augments or replaces a natural function. Biomaterials can be used for therapeutic or diagnostic purposes [1,2]. Examples of biomaterials include orthopedic, cardiac and dental implants [3–5]; contact lenses [6]; biomedical devices [7]; urinary and dialysis catheters [8]; and fracture fixation devices [9]. Irrespective of type or function, all biomaterials are prone to bacterial infections [10]. The probability of occurrence of a biomaterial-centred bacterial infection (BCBI) depends on many factors, including the virulence of the bacteria present at the biomaterial site [11]; the quality of the sterilization techniques used in surgical procedures [12]; the administration of antibiotics before and after surgery [13]; the type of the biomaterial and whether it is implanted or not; the site of implantation [10]; and finally the age, health and immunity of the patient [3]. BCBIs can be treatable or severe, and they can be acute or chronic [14]. Severe infections are generally associated with indwelling devices and cardiac implants [15]. Such infections require surgical interventions to remove the infected biomaterial and implant a new one [15,16]; otherwise, they can cause the morbidity of local tissues or even the disability or mortality of patients [17]. In the United States, about half of the two million annual cases of nosocomial infections are associated with indwelling devices [15], with a fatality-to-case ratio that ranges between 5 and 60% [10]. Despite the advent of changes in health care, infection-control practices, antimicrobial use, modern standards in the control of sterility within the operating room environment and adequate protocols of peri-operative antibiotic prophylaxis [4], nosocomial infections still affect 10% of all inpatients, delay their discharge by 11 days [18], cause surgical interventions to cost two to eight times as much as when no infection is acquired and are still considered the major reason for the failure of implanted biomedical devices [19–21].

Bacteria are ubiquitous in hospital environments [22] as well as on the skin [23]. An analysis of skin bacterial flora indicates that both pathogenic and non-pathogenic bacterial species, representative of most bacterial genera, are present on the skin, with *Pseudomonas* having the largest share of 60% [23,24]. Similarly, an analysis of data from the National Nosocomial Infections Surveillance System on 410,000 infections which occurred from 1986 to 2003 revealed that *bacilli* were largely responsible for the four infections most frequently acquired in hospitals: pneumonia, surgical site infection (SSI), urinary tract infection (UTI), and bloodstream infection [25]. Note that both SSI and UTI represent BCBIs. Common *bacilli* bacteria include *Staphylococcus epidermidis*, *Staphylococcus aureus*, *Escherichia coli* and *Pseudomonas* [8].

Bacterial adhesion to biomaterial surfaces is the essential step in the pathogenesis of BCBIs [20,26]. Implanted biomaterials tend to potentiate bacteria on their surfaces so that normally friendly species or opportunistic organisms become virulent pathogens [27]. Virulence is also enhanced because both bacteria and biomaterials interfere with host-defence mechanisms [27]. Bacterial adhesion to biomaterials is a multi-step process that proceeds as follows [1,28]. First, a biomaterial implanted in the body is coated with a conditioning film of peptides [29] and a mixture of serum and plasma proteins. These include abundant proteins such as albumins [30] and proteins, which may serve as binding ligands to the receptors of colonizing bacteria, such as fibronectin (FN), fibrinogen and vitronectin [10,31]. Protein adsorption on biomaterials is mediated by the same types of nonspecific forces that mediate bacterial adhesion to abiotic surfaces, as will be discussed later [30]. Second, bacteria at the implantation site transport to the biomaterial by molecular diffusion, Brownian motion or convective transport [10,20]. Microorganisms in nature and disease are dependent on substratum attachment for optimal growth and development [27]. Once on a biomaterial surface, bacteria establish their initial attachment to it (Fig. 5.2.1a) *via* nonspecific and specific forces [21,32].

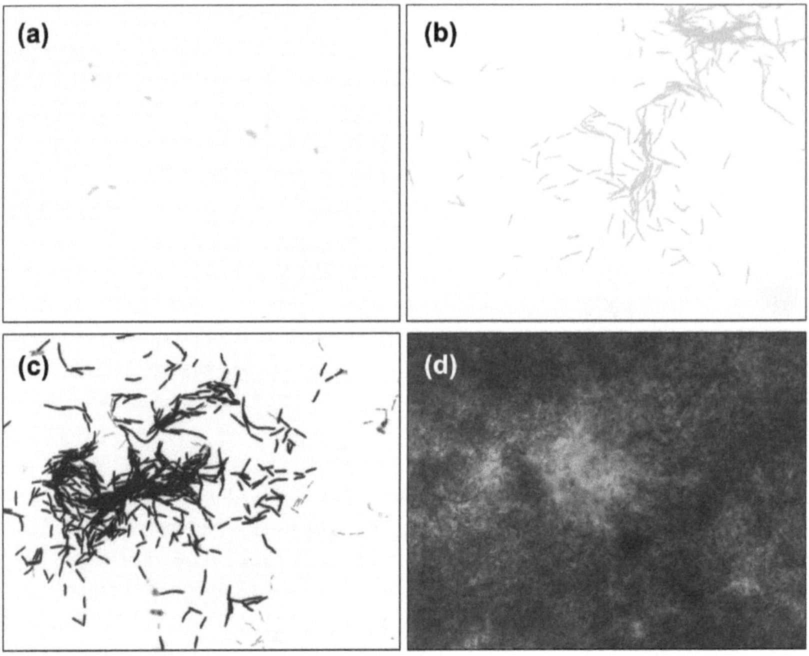

FIGURE 5.2.1 Example images showing the steps of bacterial adhesion to a biomaterial surface. (a) Single bacterial cells attached to a biomaterial surface. (b) The cells divide on the surface and start covering the surface. (c) The cells start to colonize the surface and grow as multi-layers. (d) The colonies mature, express EPS and develop a biofilm. The white areas in the images are voids, and the green areas show bacterial cells expressing green fluorescent protein.

The relative contributions of specific and nonspecific forces to bacterial interactions with biomaterials are dependent on the physiochemical surface properties of the biomaterial and the bacterial cells as well as on the associated flow and environmental conditions [20]. The initial nonspecific reversible unicellular attachment to the biomaterial surface is called the *docking stage* and is dominated by long-range interactions (>50 nm) such as dispersion, van der Waals, electrostatic and Lewis acid–base hydrophobic interactions [20,21,30,32]. Physiochemical properties that play important roles in nonspecific interactions include the wettabilities, roughnesses, and charges of the interacting surfaces [33]. In addition, the pH and ionic strength of the medium in which the bacteria–biomaterial interactions are taking place are important to consider. van der Waals forces will largely depend on the wettabilities of the two surfaces, the area of contact between them, and the hydrophobicity of the environmental medium [34]. Electrostatic forces largely depend on the charges of the two interacting surfaces, the separation distance and area of contact between them, and the ionic strength of the medium of interaction [35]. Finally, nonspecific forces are dependent on biomaterial roughness, structure and porosity. The porosity of a biomaterial can increase the surface area available for pathogens to attach to the biomaterial [17].

Once a bacterial cell is in close proximity to a biomaterial surface, irreversible short-range (<5 nm) specific forces play an important role in establishing bacterial adherence to the biomaterial [21]. These specific forces can be categorized as (1) chemical forces such as dipole–dipole (Keesom) interactions, dipole-induced dipole (Debye) interactions, ion–dipole interactions and hydrogen bonding [36] and (2) the biological ligand–receptor forces that can be established between bacterial cells and protein-conditioning films coating biomaterials. This stage is called the *locking stage* [21]. Biopolymeric molecules present on the bacterial surface, such as pili, fimbriae, lipopolysaccharides (LPS) and capsular polysaccharides, are thought to play a role in mediating the specific forces governing the short-range interactions between biomaterials and bacterial cells [37]. Initial bacterial adhesion to surfaces is thus mediated by microscopic, molecular interactions between stereochemically complementary surface structures arising over short distances [38].

Once bacterial cells have established their initial attachment to a biomaterial (Fig. 5.2.1a), cells start to divide and multiply on the biomaterial surface (Fig. 5.2.1b). After that, cells form multi-layers and develop colonies (Fig. 5.2.1c) [39]. Meanwhile, cells start to produce copious amounts of extracellular polymeric substances (EPS). EPS consist of a mixture of proteins, polysaccharides, nucleic acids and lipids and are known to play a critical role in promoting or inhibiting cell/solid surface adhesion by means of steric interactions [40]. A three-dimensional network of differentiated cells densely populating and impeded in EPS matrix is called a *biofilm* (Fig. 5.2.1d) [41]. Biofilms forming on biomaterials in the body are mostly composed of mixed bacterial cultures, in which the EPS produced by different cells get mixed together and develop a mixed-cell-line biofilm [10]. The activity and virulence of bacterial pathogens are correlated with their capacity for biofilm formation [39]. It was shown that 76% of 25 patients who developed prosthesis-related infections had bacteria growing in biofilms [28]. Finally, cells detach from biofilms [39]. The detachment starts with the biofilm development and continues as long as biofilm grows on the surface. The equilibrium between the attachment and the detachment of bacterial cells determines the pseudo-steady-state biofilm structure [42]. Biofilm formation on biomaterials is mediated by a number of factors, including biomaterial surface properties, shear stress, composition of nutrient solution, pH, and temperature [43]. Some of these parameters control the initial biofilm attachment, and some of them control the pseudo-steady-state biofilm structures [42,44,45].

The biofilm matrix is generally characterized by a very high resistance to rigorous antibiotic, detergent and biocidal treatments and resistance to stressful environmental conditions such as UV radiation, pH variations, osmotic shock, desiccation and flow conditions [10]. For example, compared to planktonic bacterial cultures, bacterial cells grown in biofilms display 10–1000 times higher antibiotic resistance [10]; doses of antibiotics 500–5000 times higher than those required to kill planktonic organisms suspended in blood, urine or other body fluids are needed to eradicate them [46]. Scanning electron microscopy (SEM) has shown that most catheters removed from patients diagnosed with a UTI are colonized by bacterial biofilms after only 7 days [37]. Biofilms persist in the urinary tract and on catheter surfaces because the microorganisms contained within a biofilm are well protected from urine flow and host-defence mechanisms [37]. Add to the resistance mechanisms described above the ease of horizontal gene transfer between resistant and non-resistant bacterial species, which makes it difficult to predict how biofilms will respond to environmental conditions [47]. Thus, biofilms are very difficult to eradicate using standard clinical procedures, and BCBIs often can only be eradicated by the removal of the infected implant.

Many approaches to minimizing BCBIs have been investigated. Research has been conducted to develop antimicrobial surface coatings that can prevent the initial bacterial colonization of a biomaterial, leading to minimized biofilm formation [39]. For example, coating stainless steel and titanium rods with the antiseptic dye gendine before implanting them into bovine bone soaked in donor calf serum *in vitro* prevented biofilms from forming for 2 weeks [30]. Other approaches include the development of anti-adhesive surfaces such as synthetic polymer coatings [48], delivering antibacterial agents to disrupt cell–cell communication and thus, prevent bacterial aggregation or biofilm formation, and killing bacteria directly (lysing the cell membranes) [49]. Clinically, determining the factors that affect the initial adhesion of bacteria to biomaterials, the precursor to infection, is essential to developing adhesion-resistant materials that are effective for either short- or long-term use [33]. Finally, because cohesive cell–cell strength is a primary factor affecting the balance between the growth and the detachment of cells in biofilms, the quantification of cell–cell adhesion in biofilms is essential to understanding, predicting, and modelling biofilm development [50].

Impeding biofilm formation will require a better understanding of how to stop the initial attachment of bacteria to biomaterials [37,51] as well as of how to quantify biofilms formed on biomaterials [52]. This review will start with introducing current methodologies used to quantify the initial adhesion of bacteria to surfaces and subsequent biofilm formation on such surfaces. A focus on how atomic force microscope (AFM) can be used to quantify the initial attachment of bacteria to biomaterials will follow. In addition, a discussion of how AFM can be used to explore the roles of various physical, topographic, chemical, biological and environmental factors in governing the initial adhesion of bacteria to biomaterials will be given. At present, the roles of various bacterial surface molecules and appendages in controlling bacterial adhesion is still poorly understood [37]. Improving our understanding of the molecular mechanisms that govern the initial bacterial adhesion to biomaterials is key to efforts aimed at the identification of mechanisms capable of preventing biofilm formation on biomaterials [43] or at the design of new strategies for preventing, detecting, or controlling BCBIs [53,54]. However, since it is very likely that biofilms will still form on biomaterials and we are far from preventing that from happening [27], a detailed section on how biofilms that have developed on biomaterials can be characterized and quantified will be given [50]. The differences among various microscopy techniques used to image biofilms will be highlighted. The types of stains that can be used to image various components of biofilms including cells and EPS will be discussed. Finally, an overview of how biofilm images can be analyzed to obtain structural parameters capable of describing biofilms will be given.

5.2.2. QUANTIFICATION OF BACTERIAL INTERACTIONS WITH SURFACES

5.2.2.1. Initial Bacterial Adhesion to Surfaces

To date, the molecular and physical interactions that govern initial bacterial adhesion to biomaterials have not been understood in detail. Although the molecular origin of specific and nonspecific forces is the same [30], it is often convenient to distinguish between the two when describing the initial bacterial adhesion to a given substratum. Initial bacterial adhesion to surfaces was previously commonly quantified using the Derjaguin–Landau–Verwey–Overbeek (DLVO) theory of colloid stability [55]. DLVO theory takes into account the roles of the van der Waals attractive and electrostatic repulsive force components in governing bacterial adhesion to a surface [34]. However, classical DLVO theory was often unable to explain bacterial adhesion to surfaces. Among the main shortcomings of the theory are assuming the bacterial cell as a rigid surface with a homogenous distribution of charge [34], neglecting the roughness of the bacterial surface that results from the presence of molecular appendages, neglecting the roughness of the biomaterial surface due to the presence of asperities on such surfaces [56] and ignoring the role of other forces, such as solvation forces, ligand–receptor forces, depletion forces and hydrophobic forces, in governing the initial events of bacterial adhesion to surfaces [56]. In biological aqueous systems, hydrophobic interactions are usually the largest long-range nonspecific forces and thus should not be ignored [57]. To overcome this last shortcoming, extended DLVO theory was developed. In this model, short-range acid–base and steric interactions are accounted for in addition to the van der Waals and electrostatic forces that were considered in the classical DLVO model [58]. Finally, soft-particle analysis of DLVO theory was developed by Ohshima to account for the softness of the bacterial cells as well as for the charge distribution on the bacterial surface [34,35,59]. Soft-particle analysis of DLVO theory was able to explain the nanoscale initial adhesion forces measured between *E. coli* JM109 or *Pseudomonas putida* KT2442 and a silicon nitride model surface in an array of environmental conditions [34,35].

In addition to the various DLVO models used to quantify initial bacterial adhesion to surfaces, bacterial adhesion can be quantified using a thermodynamic approach, first proposed by van Oss et al. [57]. This thermodynamic model treats bacterial cells and biomaterials as continuum surfaces. The wettability and the charge of the bacterial surface are often quantified using contact angle and electrophoresis measurements, respectively [30,60]. Counting bacterial cells attached to a surface when they are still individual cells has been used as a means of describing initial bacterial adhesion to a surface. However, this approach is qualitative, time-consuming and has low sensitivity [61]. A direct method of quantifying the initial adhesion of bacterial cells to solid surfaces including biomaterials is very important to designing biomaterials or coatings capable of inhibiting bacterial adhesion to their surfaces [36]. Bacterial adhesion strength can be evaluated through the measurement of the shear force required to detach bacteria from a biomaterial surface using flow chambers or spinning disks [16]. Similarly, adhesion forces can be calculated based on the surface tension forces required to detach cells from substrates through the passage of air bubbles [16]. Although these methods are quantitative, most of them suffer from limitations such as low sensitivity and precision, complexity, high cost, and lengthy procedures [16]. The atomic force microscope (AFM), invented in 1986, has been used extensively to quantify the initial adhesion of bacterial single cells to surfaces of interest such as biomaterials [30]. The use of AFM to quantify bacterial interactions with biomaterials will be detailed later in this review due to its enormous advantages [52,62].

5.2.2.2. Biofilm Formation on Surfaces

The study of biofilm development (Fig. 5.2.1b–d) requires an understanding of microbial growth and of biofilm structure. Although there are many techniques available for quantifying these, each has its limitations. Therefore, combined techniques are required when studying biofilms. Quantifying biofilms formed on surfaces is largely achieved via microscopic techniques [60]. A variety of microscopy techniques including optical microscopy, fluorescence microscopy, atomic force microscopy [61], confocal laser scanning microscopy (CLSM) [62], and environmental scanning electron microscopy (ESEM) [63] have been used to detect biofilms, image them and quantify their structures. In addition to imaging, spectroscopic techniques such as attenuated total reflectance (ATR)/ Fourier transform infrared (FTIR) laser spectrometry have been used to quantify the macroscale attachment of bacteria to surfaces [64]. Finally, substrate-adhered cells have been enumerated by direct counting methods including light microscopy, epifluorescence microscopy, and SEM and by indirect counting methods such as colony forming unit (CFU) plate counting, radiolabeling, spectrophotometry, and bioluminescence analysis [16]. The methodologies and stains required to obtain high-quality images of biofilms on biomaterials will be discussed later in this review. In addition, the software packages most commonly used to extract parameters of interest from such images will be compared.

5.2.3. ATOMIC FORCE MICROSCOPY (AFM) USE IN INVESTIGATIONS OF BIOLOGICAL SYSTEMS

Since its invention in 1986 [63], the use of AFM to investigate biological systems has been continuously progressing [64–67]. AFM has been used successfully at a fine resolution to image biological molecules including DNA [68], RNA [52,69], proteins [70,71] and lipopolysaccharides [72,73]. AFM has also been used to image both live microbial [74–76] and mammalian cells [77,78] in their native environments. In addition to imaging, AFM has been used to quantify interactions between various biomolecules, such as ligand–receptor interactions [79,80]; and forces in complementary DNA base pairings [81]; and interactions between cells and surfaces, such as those experienced in bacterial adhesion to inert and biotic surfaces [82–84]. In addition, AFM has been successfully used to quantify rates of protein folding and unfolding [85], antibody–antigen binding strength [86], mechanical properties of cells and biomacromolecules [87], bio-sensing [88] and to map proteins on surfaces of cells [89].

5.2.3.1. Components and Working Principle of AFM

Scanning probe microscopes (SPM) are classed in two categories, scanning tunnelling microscopes (STMs) and AFMs [90,91]. The STM is the ancestor of all SPMs [63]. It was invented in 1982 by Gerd Binning and Heinrich Rohrer at IBM Zurich [63]. Five years later, Binning and Rohrer were awarded the Nobel Prize in physics for their invention of STM. The usefulness of STM is limited to samples that are considered to be good electrical conductors [90]. Binning et al. invented AFM in 1986 [63] to obviate this limitation. With AFM, conductive and non-conductive surfaces including metals, polymers, ceramic materials and biological samples can be scanned [91].

As its name implies, AFM is capable of imaging surfaces with an atomic resolution while at the same time, measuring forces between two surfaces of interest [63]. The heart of the AFM is a tip attached to the end of a cantilever. To image a surface of interest, the tip scans the surface in the *XY* plane. Upon scanning, repulsive or attractive interaction forces between the tip and the surface scanned will cause the cantilever to deflect upwards or downwards, respectively [91]. The deflection of the cantilever (Z-plane) is monitored using a laser beam that is reflected off the surface of the cantilever onto a split quadratic photodiode detector [62]. By recording the deflection values over the *XY* plane, a three-dimensional image of the surface can be constructed.

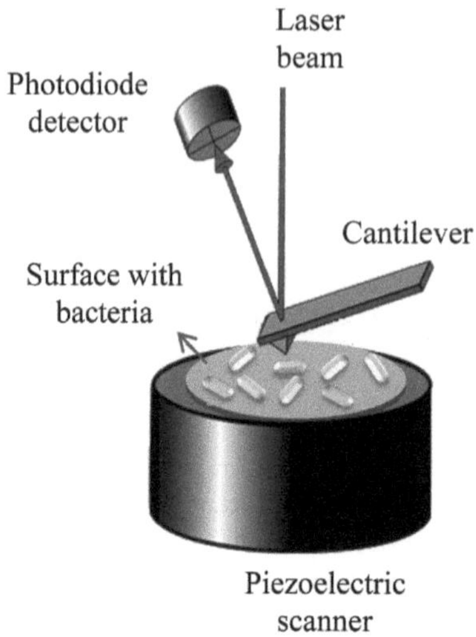

FIGURE 5.2.2 The main components of AFM.

By considering the cantilever as a spring with a known spring constant (k), deflection values (d) can be converted to force values (F) as a function of position on the sample using Hooke's law ($F = kd$) [92]. Figure 5.2.2 shows the main components of an AFM.

5.2.3.2. Operating Modes of AFM

AFM can operate in several modes, with contact and tapping modes being the most frequently used. In contact mode, a constant force is applied to the cantilever (force set point), thereby pushing the tip against the sample as it scans the surface [54,93]. Upon interaction, the force measured between the tip and the surface will be different from the force set point applied to the cantilever. To correct for this, a signal is sent from the proportional integral (PI) controller to the piezoelectric actuator placed within the scanner. Based on the signal, the scanner moves the sample up or down in the Z-direction to match the force applied to the cantilever [63]. A constant force implies a constant cantilever deflection. Although it is relatively easy to operate the AFM in contact mode, the force applied to the cantilever can be damaging to biological soft samples and even to some harder samples such as silicon [91]. To avoid such damages, tapping mode is generally used. In tapping mode, the cantilever oscillates over the sample of interest at a set point amplitude of oscillation. Because of the interactions with the surface, the amplitude of the oscillation of the cantilever varies from that of the set point. Again, the scanner is moved up or down to keep the cantilever oscillatory amplitude constant [94–96]. As can be understood from the way both contact and tapping modes operate, the feedback control system used in AFM is considered one of its most important features [97]. Both tapping and contact modes can be run in air or under liquid conditions [62,98].

5.2.3.3. Nomenclature of AFM

Depending on the mode being used or the application sought from the AFM experiment, AFM has been given many different names in the literature. For example, if the AFM cantilever is moved laterally, the AFM is called frictional or lateral force microscopy [99–101]. Similarly, if the cantilever is tagged by a certain chemical group, the AFM is called chemical force microscopy [102]. Other common names for AFM in the literature include single-molecule force spectroscopy [83], molecular recognition spectroscopy [103], and colloidal probe spectroscopy [104].

5.2.3.4. Advantages of AFM

The extensive use of AFM to investigate biological systems can be largely credited to five main advantages [74,103,105–107]. First, AFM studies of biological systems can be performed in their native environments in liquid or in air [106,108,109]. Studying cells and biomolecules in their native environments is critical to predicting their functional performance. Second, AFM requires minimal sample preparation or alteration for measurements. It requires no chemical fixation or dehydration of samples [110]. Third, AFM is characterized by high lateral and vertical resolutions as well as by a high signal-to-noise ratio [62]. With clean samples that lack large surface roughnesses, the resolution of AFM can be 0.1 nm in the *XY* plane and 0.01 nm in the *Z* plane [91]. Fourth, AFM allows for synergistic imaging and force measurements on a sample of interest to be preformed [111]. This dual functionality makes AFM studies unique. For example, with AFM, a bacterial cell can be imaged for its ultrastructure in three dimensions and simultaneously be characterized for its mechanical properties, the distribution of a certain protein on its surface, or the distribution of charge or hydrophobic domains on its surface [112,113]. Finally, AFM studies are quantitative. For example, AFM is capable of detecting piconewton (pN) forces between biomolecules to a very fine resolution [114,115]. By applying various theoretical models to AFM data, specific and nonspecific forces can be quantified and decoupled for their importance in dominating the interactions within biological systems [32,116].

5.2.3.5. AFM Imaging

AFM can be used to image surfaces in real time in air or liquid environments [106]. AFM imaging can be done in various modes including tapping, contact, phase, deflection and friction [117]. The choice of the mode to be used for imaging depends on the purpose of the measurement [118]. The quality of the images obtained depends on the size of the tip used, sample softness, medium in which the images were collected, fluctuation of solvation forces, modification of the tip by suspended contaminants in the medium, and stability of the sample being imaged as well as the types of forces that dominate the interactions between the tip and the surface being imaged [69,117,119,120]. Images can be used to characterize surface morphology and roughness [36,76,121], reveal details of surface ultrastructure [122], quantify cell dimensions [123] and monitor variations in the surface properties and morphology of cells as a function of variable environments [124]. Sometimes, images can be used to illustrate how surfaces interact. For example, images can be used to show how proteins change the topography of DNA as they interact [125].

5.2.3.6. AFM Force Measurements

A considerable amount of work has been done to investigate the roles of factors such as the physiochemical properties of bacteria and surfaces and environmental stresses in the initial nanoscale adhesion of bacteria to surfaces [33,35,38,61,83,126–131]. Measuring forces between bacteria and biomaterials in native environments represents a special case of quantifying nanoscale interactions between bacteria and surfaces [16,30,58]. Prior to force measurements, images of the biomaterial or the bacterial surface should be captured in the medium of interest and characterized [131] (Fig. 5.2.3A). In addition, the force constant of each cantilever should be determined. The cantilever spring constant can be determined using the Cleveland method [132] or from the power spectral density of the thermal noise fluctuations in the air or liquid [133]. For details on various methods used to calibrate AFM cantilevers, please refer to Emerson et al. [134]. Once the surface of interest has been located *via* topographical scanning, the oscillation of the cantilever is stopped, and extending (approach) and retracting deflection–displacement curves are measured between the bacterial surface biopolymers and the biomaterial in the medium of interest (Fig. 5.2.3C and D) [34]. The rate at which these curves are measured, the ramp size of the measurement and the number of points collected per curve should all be consistent between variable curves or treatments for purposes of data comparisons. The design of appropriate controls should always be considered to ensure that measurements are reproducible. For example, the equality of force measurements made by an SAM-coated cantilever on a bacteria-free area of a sample of interest before and after taking a force measurement on a bacterial cell will ensure that the cantilever properties were not altered by contact with the bacterial surface biopolymers.

In general, AFM measures forces between two surfaces in a cycle that consists of an approach curve recorded while the probe is approaching the surface (Fig. 5.2.3c) and a retraction curve recorded while the probe is moving away from the surface (Fig. 5.2.3d). Raw AFM piezo position–deflection data files (Fig. 5.2.3b) are then converted to distance–force files (Fig. 5.2.3c and d)

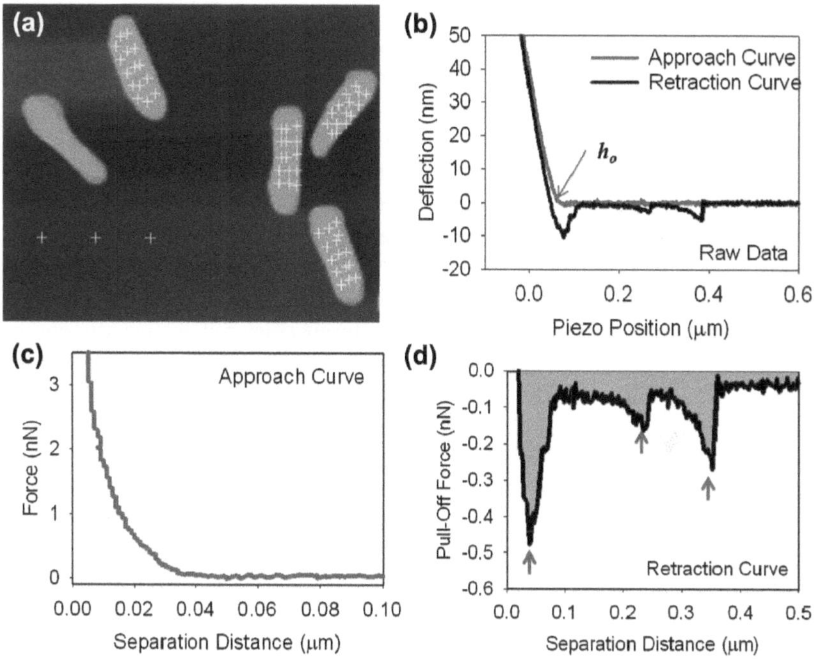

FIGURE 5.2.3 (a) Tapping-mode images of *Listeria monocytogenes* EGDe immobilized on gelatin-coated mica in water. The white cross-marks on the bacterial cells indicate locations of force measurements. The locations on the mica marked by cross-marks were used to check for tip contamination by bacterial molecules between force measurements. The image is 6 μm × 5 μm × 500 nm. (b) Raw data collected by AFM. The blue arrow indicates the location of the contact point between the tip and the bacterial biopolymers. (c) Converted approach data. (d) Converted retraction data. The red arrows point to adhesion events, and the grey-shaded area indicates the work of adhesion.

as described previously [135,136]. It is worth mentioning, though, that the precise determination of the contact point (h_0, Fig. 5.2.3B) between an AFM tip and a soft biological surface such as a bacterial biopolymer brush has proven to be a challenging task for all researchers working with soft samples [87]. This is mainly because the location of the contact point is largely affected by the force applied by the cantilever to the surface being studied, the heterogeneity in the mechanical properties of the surface indented, and the nature of the interactions between the tip and the surface of interest [137]. For statistical purposes, repetitive measurements should be performed using an array of cantilevers or biomaterial surfaces and using a sufficient number of bacterial cells cultivated from separate cultures [138]. In addition, force measurements should be done on variable points on the surface of interest. Volume–force imaging is one way in which the surface of interest is pixelated and measurements are done on the whole surface [139]. Another alternative relies on using the point-and-shoot feature available in AFM software to select force measurement locations on a surface of interest (Fig. 5.2.3a) [131]. Covering the whole surface is especially important when bacterial interactions with surfaces are being quantified. This is mainly due to the heterogeneity of the bacterial biopolymers that make up the bacterial surface and play an important role in the adhesion of bacteria to biomaterials [138]. If cellular probes were used in force measurements, the position and integrity of the immobilized cells should be checked using a light microscope after the experiments [140].

The force–distance curve measured during the approach between a bacterium and a surface such as a biomaterial (Fig. 5.2.3c) usually shows a repulsive barrier that has to be overcome prior to adhesion. Approach curves can be used to determine the steric repulsion forces present between bacteria and biomaterials [141,142], the thickness and the grafting density of the bacterial surface biopolymer brush [143], and the elasticity of the surface of interest [135,144]. Generally, if captured at the same ramp size and using the same rate and resolution, approach curves can be averaged because of the minimal variations observed between individual curves [34,35]. In comparison, retraction curves measured between bacteria and surfaces such as biomaterials (Fig. 5.2.3d) usually display individual or multiple peaks. The arrows shown in Fig. 5.2.3d indicate adhesion events. The pull-off force of each peak in the retraction curve is equivalent to the adhesion force and represents the sum of all interaction forces measured between the bacterial surface biopolymers and the biomaterial of interest. The baseline for each peak is usually taken at zero force. The user should define the resolution at which peaks can be distinguished from noise. This resolution will vary from one machine to another. When analyzed, retraction curves should be considered individually because of the expected heterogeneity in the measured adhesion forces [131,138]. Retraction curves are generally used to quantify bacterial adhesion strength in terms of forces and energies [131,138], elasticity of bacterial biopolymers [83,145], interactions specificity [146], and homogeneity of the bacterial surface biopolymers [83].

Because of the heterogeneity of the adhesion forces measured between bacteria and surfaces such as biomaterials, a variety of statistical tools have been implemented in the literature to describe adhesion data. A common method is to divide the adhesion forces into pins and plot the data in the form of a histogram [131]. Various dynamic peak function models, such as the Poisson [116], normal [35], Gamma [35], log-normal [131] and Wiebull [147] models, are then applied to the distribution of adhesion affinities to describe the heterogeneities of the adhesion data. Another approach depends on calculating the mean, mode, median and range of the adhesion data collected [131,147]. The median values of adhesion collected using various treatments of interest are generally compared using a statistical rank sum test [123,131].

5.2.4. USE OF AFM FOR QUANTIFICATION OF BACTERIA–BIOMATERIAL INTERACTIONS

As with any force measurements carried out using AFM, measurements of forces between bacteria and biomaterials require an appropriate experimental design, in this case, one that will mimic the interactions between biomaterials and bacteria in the most reasonable way. This implies that decisions need to be made with regard to whether the AFM cantilever will be used to represent the biomaterial or the bacterial cells [33], how bacteria will be immobilized during measurements [148,149] and the type of the cantilever to be used, in particular, its spring constant and whether it will be functionalized or not [150]. Quantifying the interaction forces between bacteria and biomaterials falls into three experimental designs in the literature. First, the biomaterials tested cannot be represented by the materials of existing cantilevers. Examples include stainless steel [151], polystyrene, Teflon and poly(methyl methacrylate) [152]. In this case, bacterial cells have to be immobilized on the cantilever, allowing for a greater variety of biomaterials to be tested for their interactions with bacteria [33] (Fig. 5.2.4). Cantilevers that are modified to have bacterial cells on them are called *cellular probes*. Using cellular probes allows for cell–cell interactions to be quantified as well [153]. Various methods have been discussed in the literature for preparing cellular probes;

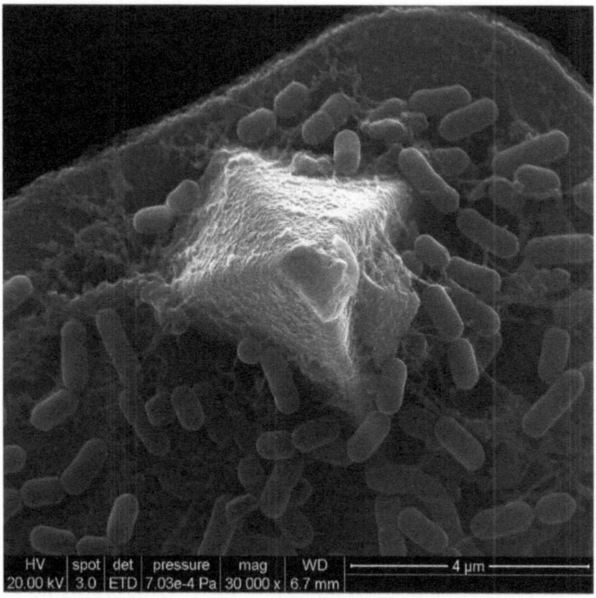

FIGURE 5.2.4 SEM image of an AFM cantilever modified with *Listeria monocytogenes* EGDe using a 0.1% w/v poly-L-lysine (PLL).

these will be discussed below. A cellular probe can have a single cell on it, in which case, AFM is called single-cell force spectroscopy [55], or it can have a biofilm on it [61].

Second, chemically functionalized probes to mimic biomaterials chemistry or coatings of biomaterials have been frequently employed in the literature to quantify bacteria–biomaterial interactions [30,38]. AFM that employs chemically functionalized cantilevers is referred to as *chemical force spectroscopy* in the literature [31]. In this case, bacterial cells are generally immobilized on surfaces or grown as biofilms on materials of interest. Bacteria can be immobilized on surfaces using covalent bonding [83], physical adsorption to gelatin-coated surfaces [108], or mechanical trapping of the cells in porous polycarbonate membranes with a pore size similar to that of the bacterial cell size [29]. Images of the bacterial cells on the surface are generally acquired to determine the locations of force measurements on the bacterial cell surface [131]. To vary the cantilever hydrophobicity or charge, cantilevers are generally coated with self-assembled monolayers (SAMs) of thiol derivatives [30]. To establish SAMs, AFM cantilevers are coated with a gold layer and a covalent bond is established between the gold and the thiol molecules being used [154]. Using this approach, cantilevers that have end functionalities of COO^-, OH, and CH_3 have been used in the literature [29]. AFM can also be coated with proteins to mimic biomaterials coated with a conditioning layer of proteins [155]. AFM can be used to probe the thickness of the SAMs or the protein coating employed on the cantilever. To do this, surfaces are prepared in parallel using the exact protocols used to prepare the cantilevers. The AFM scratch technique is then employed to estimate the height of the coating. Briefly, coatings are carefully removed from surfaces using a razor blade tip. The film thickness is then determined using cross-sectional analysis of the AFM height images taken at the boundary between the scratched and non-scratched regions as the difference in height between the region with film coating and that of the reference surface [87]. However, the section analysis feature used for analysing a film thickness cannot necessarily detect the bottom of the film since the measurement is based on topographical mapping. Furthermore, AFM height data have been questioned because of the height shrinkage effects on samples due to forces applied by the cantilever [155].

Finally, interaction forces between biomaterials and bacteria can be measured using lateral force spectroscopy [156]. This approach differs from the two approaches discussed above in that it relies on lateral movement of the AFM cantilever rather than the normal movement of the cantilever. In this approach, bacterial cells are generally grown on the biomaterial of interest [156]. Using the AFM tip, detachment of the biofilm from the biomaterial surface is established when the bending moment of the cantilever is sufficient to overcome the cell-surface adhesion force [16]. Although this approach is interesting, it has yet to be demonstrated on single cells rather than on biofilms.

5.2.4.1. Preparation of Cellular Probes

Razatos et al. were the first to modify AFM cantilevers with bacterial cells to quantify interactions between biomaterials and bacterial cells [152]. To immobilize bacteria on cantilevers, a pellet of bacterial cells grown to the phase of interest was manually transferred onto tips coated with polyethyleneimine. The pellet was further treated with an additional drop to 2.5% vol/vol glutaraldehyde, and samples were incubated at 4 °C for 1–2 h. The cantilevers were then rinsed in water and air-dried [152]. Later, Kang et al. developed a new methodology for modifying the coating and prepared AFM cantilevers with live individual cells [55]. To prepare the single-cell probes, the extreme end of a polydopamine-coated tipless cantilever was lowered towards a bacterial suspension spread on a silicon oxide wafer. A single bacterial cell was then held in contact with the polydopamine-coated cantilever for approximately 1 min to allow strong binding to take place [55]. The viability of the cells was confirmed *via* fluorescence and metabolic assays. When adhesion forces measured between a quartz surface and cellular probes with *Saccharomyces cerevisiae*, *Bacillus subtilis*, and *E. coli* prepared according to the gluteraldehyde-treated cells protocol pioneered by Razatos et al. and the bioinspired single-live-cell probes pioneered by Kang et al. in 10 mM TRIS buffer with pH 8.0 were compared, the results were significantly different [55]. The gluteraldehyde-treated cellular probes showed adhesive forces between the bacterial probes and the quartz sample, while the live-cell probes showed repulsive forces, suggesting that gluteraldehyde treatment of cells leads to the rearrangement of bacterial cell structure and thus, influences cell behavior significantly [55]. In addition to the two approaches discussed above, other chemical methodologies have been used in the literature to immobilize bacterial cells on AFM probes. These include the attachment of bacterial cells to AFM tips coated with positively charged poly-L-lysine (PLL) (0.1% w/v [30] or 0.01% w/v [157,158] in water) or to AFM tips coated with hexadecanethiol (HDT) (20 mM in ethanol) [33] and the immobilization of cells at the end of standard V-shaped AFM tipless cantilevers using a small amount of glue and a housed-in micromanipulator [140]. Irrespective of the approach used to coat cantilevers with cells, the presence of cells on probes is generally confirmed *via* SEM images [30]. For a list of studies that used cellular probes to quantify interactions between bacteria and biomaterials, please see the grey-shaded rows in Table 5.2.1.

5.2.5. EXAMPLES FROM THE LITERATURE OF AFM STUDIES OF BACTERIA–BIOMATERIAL INTERACTIONS

The development of bacterial-infection-resistant materials and treatments requires a detailed knowledge of the factors and forces involved in bacterial adhesion to biomaterial surfaces. Initial bacterial adhesion to biomaterial surfaces is governed by reversible physiochemical forces that include

TABLE 5.2.1 A Summary of AFM Studies that Quantified the Initial Adhesion of Bacteria to Biomaterial Surfaces. Grey-Filled Rows Indicate the Use of Cellular Probes in the Study

Model Bacteria	Biomaterial	AFM Conditions and Other Tools Used to Quantify Bacterial Adhesion and Properties [Reference]
5.1 The Role of Surface Coatings in the Strength of Attachment of Bacteria to Biomaterials		
Staphylococcus mutans with LT11 and without IB03987 antigen I/II	Salivary proteins on 0.06 N/m tip.	* $k = 0.06$ N/m, silicon nitride (Si_3N_4) cantilevers. * Adhesion maps, decay length, repulsion forces and adhesion histograms from forces measured at room temperature in adhesion buffer. * Comparison to parallel flow chamber experiments [148].
Streptococcus sanguinis and *S. mutans*	Stainless steel 316 used for orthodontic brackets with and without a salivary conditioning film.	* Ultrasharp cantilevers. * Adhesion force measurements at room temperature in adhesion buffer. * SEM [158].
Staphylococcus epidermidis	Fetal bovine serum (FBS) and fibronectin (FN), both adsorbed on a gold-coated slide.	* $k = 0.06$ N/m, Si_3N_4 cantilevers. * Force measurements in PBS. * QCM-D [155].
S. epidermidis	SAMs of isophthalic acid (IPA) or isophthalic acid with silver (IAG).	* $k = 0.05–0.019$ N/m, silicon tips. * Force measurements in 0.1 M MES buffer or PBS in the presence of FN or FBS, histograms of forces were shown. * Bacterial retention assays were done as well [30].

E. coli JM109	Silicone modified with various biopolymers and silanes, including: heparin, hyaluronan, and self-assembled octadecyltrichlorosilane (OTS), and fluoroalkylsilane (FAS). Mica was used as a control.	* $k = 0.13$ N/m. * Force measurements in 1 mM Tris solution at room temperature. * SEM and contact angles [38].
S. cerevisiae NCYC 1681 (Yeast)	Hydrophobic-coated mica with methyl silane, BSA-coated mica and hydrophilic mica.	* $K = 0.4$ N/m, V-shape tipless cantilevers. * Force measurements in 0.01 M NaCl at 25 °C, pH = 5. * Bacteria grown till exponential, stationary and death phase and effect of contact time on adhesion were studied too. * Zeta potential and BATH assays [140].
S. epidermidis negative and positive biofilm-forming strains	Hydrophilic and $HS(CH_2)_{17}CH_3$ hydrophobic-decorated Si_3N_4.	* $k = 0.02–0.025$ N/m, Si_3N_4 cantilevers. * Images of single bacterial cells, 16 × 16 force-volume mapping in MQ water and histograms of forces and WLC [29].
Isogenic uropathogenic *E. coli* strains with variable colanic acid expression	Hydrophobic *N*-octadecyltrichlorosilane (OTS) Glass-coated coverslips and hydrophilic glass cleaned substrates.	* $k = 0.06$ N/m, Si_3N_4 cantilevers. * Force measurements in phosphate-buffered saline (PBS, pH 7.4) or Tris buffer (pH 7.4). * Parallel-plate flow chamber experiments, zeta potential and contact angles [37].
E. coli D21	Hydrophilic glass or hydrophobic *N*-octadecyltrichlorosilane (OTS)-treated glass substrates coated with block copolymers, poly(ethylene glycol) (PEG)-lysine dendron or Pluronic F127 surfactant.	* $k = 0.06$ N/m, Si_3N_4. * Force measurements in 1 mM Tris buffer, pH 7.4, or in PBS, pH 7.5. * Contact angle, flow cell experiments and fluorescence microscopy [169].

(*Continued*)

TABLE 5.2.1 A Summary of AFM Studies that Quantified the Initial Adhesion of Bacteria to Biomaterial Surfaces. Grey-Filled Rows Indicate the Use of Cellular Probes in the Study—cont'd

Model Bacteria	Biomaterial	AFM Conditions and Other Tools Used to Quantify Bacterial Adhesion and Properties [Reference]
S. sobrinus, S. mutans and S. mitis BMS	Enamel particles, bare or with pellicle coating.	* $k = 0.06$ N/m with enamel particles attached to the probe. * Force measurements at room temperature in Millipure water [162].
5.2 The Role of Environmental Conditions and Antibiotics in the Attachment of Bacteria to Biomaterials		
S. epidermidis 35984	Silicone surface with loaded antibiotic rifampicin drug.	* $k = 11.5$ N/m, ultrasharp silicon non-contact 'Golden' tip. * AFM images of air-dried samples taken in tapping mode. * SEM and measurement of drug release [170].
E. coli strains with and without fimbriae	Silicon nitride in the presence of cranberry juice	* $k = 0.13 \pm 0.02$ N/m, Si_3N_4 tips. * Force measurements and histograms of adhesion [164].
5.3 The Role of the Roughness of Biomaterials in Bacteria-Biomaterial Interactions		
S. epidermidis	A total of six SAMs (IPA, IAG, PDT, PLL, BSA and HDT).	* $k = 0.042 \pm 0.001$ N/m ($n = 15$). * Force measurements in MES buffer. * Film thickness measurements were obtained from ellipsometry, roughness by AFM and contact angles [33].

S. cerevisiae and *Leuconostoc oenus*	Patterned polyurethane surface.	* $k = 0.06$ N/m, sharp Si_3N_4 (NP-S) tips. * Imaging of bacteria using tapping or contact mode in hexane or in water. * Frictional force microscopy. * QCM-D [165].
E. coli	Nanostructured substrates and flat gold and glass substrates.	* $k = 0.58$ N/m and a radius of 10 nm, ultrasharp Si_3N_4. * AFM images of biofilms in PBS buffer. * Genomics qPCR and proteomics (2D-DIGE), SEM and confocal laser microscopy [166].
Pseudomonas aeruginosa and *Staphylococcus aureus*	Hydrophobic titanium oxide surfaces with defined topography (smooth or with randomly spaced surface pits of 0.5 μm diameter).	* $k = 0.05–0.12$ N/m Si_3N_4 cantilevers. * Images of biofilms in contact mode under water. * Count of cells remaining after shear applied by successive scanning. * Contact angles, energy-dispersive X-ray spectroscopy (EDX) and live-dead staining [54].
5.4 Quantification of the Strength of Interactions Measured between Biomaterials and Bacterial Cells		
Pseudomonas aeruginosa, Staphylococci and *Serratia*	* Soft and rigid RGP contact lenses made of lotrafilcon A. * Corneas cut off intact pig eyes.	* $k = 0.04 =$ N/m, tipless V-shaped cantilevers (DNP-0). * Force measurements in 0.9% NaCl at room temperature. * Weibull analysis of adhesion forces. * Contact angles and interfacial free energies [157].
Escherichia coli	Mica, hydrophilic glass, hydrophobic glass, polystyrene, and Teflon.	* $k = 0.06$ N/m, Si_3N_4 cantilevers. * Force measurements in Tris buffer. * Contact angles, zeta potential and DLVO analysis [58].

(*Continued*)

TABLE 5.2.1 A Summary of AFM Studies that Quantified the Initial Adhesion of Bacteria to Biomaterial Surfaces. Grey-Filled Rows Indicate the Use of Cellular Probes in the Study—cont'd

Model Bacteria	Biomaterial	AFM Conditions and Other Tools Used to Quantify Bacterial Adhesion and Properties [Reference]
Uropathogen *Enterococcus faecalis*	Medical-grade polymers including polyurethane, polyamide, and poly(tetrafluoroethylene).	* $k = 0.06$ N/m. * Imaging of bacteria in water attached to surfaces and measurement of the lateral force required to detach cells from surfaces. * SEM and roughness of polymer surfaces by AFM [16].
Pseudomonas aeruginosa biofilm	Sheets of aluminium, steel, rubber, and polypropylene.	* $k = 2.0–2.5$ N/m, Si_3N_4 cantilevers. * Contact mode images in air; lateral force required to detach cells from surfaces and normal adhesion forces to bare silicon nitride were measured; and dimensions of cells in biofilms were calculated [156].
5.5 AFM Imaging of Biofilms and Characterization of their Extracellular Polymeric Substances		
St. guttiformis	Hydrophobic poly(*tert*-butylmethacrylate) (P*t*BMA) and hydrophilic polystyrene maleic acid (PSMA) surfaces.	* $k = 0.032$ or 42 N/m, Si_3N_4 cantilevers. * Height and friction imaging under ambient air conditions, 23 °C and 45% relative humidity. * Contact angles [117].

electrostatic, van der Waals, and hydrophobic interactions, followed by the establishment of irreversible interactions such as specific receptor–ligand-binding events [159]. The initial adhesion to surfaces is also affected by the composition of the biopolymers on the bacterial surface, the medium of interactions, the environmental conditions and the physiochemical characteristics of the biomaterial surface to which bacteria are attaching and whether it is coated by a conditioning film or not [160]. Although the importance of these factors in the initial bacterial adhesion to a surface is acknowledged, a fundamental understanding of how they affect the strength and the types of forces that govern the adhesion is still lacking [37]. Clinically, it is of interest to determine the factors affecting bacterial adhesion to biomaterials, the precursor to BCBIs [31]. It is also important to design bacterial-adhesion-resistant materials or coatings. In designing such materials, knowledge of surface texture, roughness, wettability and charge is important [33]. Below, studies in the literature which attempted to investigate various aspects of bacteria–biomaterial interactions are discussed and classified into five different categories (Table 5.2.1).

5.2.5.1. The Effect of Surface Coatings on the Strength of Attachment of Bacteria to Biomaterials

Immediately upon implantation, biomaterial substrates become coated with abundant proteins, such as albumin, in a nonspecific way. Soon after that, albumin is displaced by larger proteins, such as FN and fibrinogen [31]. Bacteria in the biomaterial environment can interact specifically with these proteins, accelerating the colonization and infection of biomaterials [161]. Until recently, the nonspecific adsorption of proteins on implanted biomaterial surfaces has been quantified using experimental techniques such as mass spectroscopy, X-ray photoelectron spectroscopy, surface plasmon resonance, and secondary-ion mass spectrometry [31]. Although these techniques can provide a wealth of information about the amount and types of proteins adsorbed on the biomaterials, they are incapable of providing information on the strength of the physiochemical interactions involved in the adsorption process, which render protein adsorption on biomaterials irreversible and uncontrollable [31]. In comparison, AFM studies of how proteins interact with biomaterials and with bacterial cells can improve our ability to design anti-fouling biomaterial surfaces. For example, interaction forces between salivary proteins and *Staphylococcus mutans* with (LT11) and without (IB03987) antigen I/II measured using AFM were correlated with the adhesion of the strains to saliva-coated glass in a parallel-plate flow chamber [148]. The results of this study indicated that the combined specific and nonspecific adhesion forces were significantly stronger for parent strain LT11 than for mutant strain IB03987 [148]. The same group decoupled the adhesion forces measured between a conductive stainless steel surface and *streptococci* both in the absence and in the presence of an adsorbed salivary conditioning film using a Poisson analysis of the adhesion forces. Their results indicated that the repulsive nonspecific Lifshitz–van der Waals forces dominated the interactions of *streptococci* with saliva-coated stainless steel, but, interestingly, attractive nonspecific forces were revealed on stainless steel in the absence of a salivary conditioning film [158]. In another study, AFM probes coated with *S. epidermidis* cells were used to quantify bacterial interaction forces with gold surfaces coated with fetal bovine serum (FBS) or FN [155]. These studies point to the importance of exploring the effects of protein conditioning layers on interactions measured between biomaterials and bacterial cells.

One approach to preventing BCBIs is to coat the biomaterial surface with a layer prohibitive of the adsorption of both conditioning proteins and bacterial cells. The use of SAMs, poly(ethylene glycol) (PEG), and biosurfactants as protein- and bacteria-resistant coatings will be discussed below. SAMs can function as nonfouling coatings by altering physiochemical properties of the biomaterial including the biomaterial wettability, charge and free surface energy [161]. To test the efficacy of SAMs in resisting bacterial and protein adsorption to their surfaces, Liu et al. investigated the strength of attachment of *S. epidermidis* immobilized on AFM cantilevers to gold-coated SAMs [30]. SAMs terminated with isophthalic acid (IPA) or isophthalic acid with silver (IAG) in the presence of FBS or FN were investigated. Their results indicated that *S. epidermidis* had weak nonspecific interactions with the SAMs when FBS was present, while strong specific interactions were observed when FN was present because of ligand–receptor interactions [30]. Similarly, Cao et al. studied the effects of various hydrophobic coatings including heparin, hyaluronan and self-assembled octadecyltrichlorosilane (OTS) and fluoroalkylsilane (FAS) layers on the ability of silicon to resist *E. coli* JM109 adhesion [38]. Silicone has been widely used as an implant due to its flexibility, fatigue resistance, high degree of chemical inertness, thermal stability and non-toxicity [38]. Their results showed that short-range repulsive forces were dominant between the FAS-coated silicon and *E. coli*, while long-range attractive forces dominated the interactions between heparin-coated silicon and *E. coli*. Their work indicated that hydrophobicity cannot be used as a criterion to predict bacterial adhesion. Rather, both the native properties of bacterial strains and the specific functionalities of biomaterial surfaces determine the strength of the interaction forces, and thus, the extent of adhesion [38]. Salerno et al. investigated the adhesion of four bacterial strains which differed in the thicknesses of their LPS layers to silica surfaces modified with alkylsilane layers with variable lengths using AFM [56]. Their results indicated that the adhesion forces decreased as the thickness of the LPS layer increased and as the solution Debye length became longer than the alkylsilane [56]. They concluded their study by emphasizing the importance of molecular details in predicting the initial adhesion of bacteria to surfaces. Bowen et al. quantified the adhesion of metabolically active *S. cerevisiae* cells to a hydrophilic mica surface as a control, a mica surface with a hydrophobic methyl silane coating and a mica surface coated with bovine serum albumin (BSA) in an aqueous 0.01 M NaCl environment. Their results indicated that hydrophobic surfaces adhered to the yeast cells more than the hydrophilic control [140]. The growth phase of the cells was found to affect the adhesion too [140]. In another example of how hydrophilic Si_3N_4 and $HS(CH_2)_{17}CH_3$, hydrophobic AFM cantilevers affecting the properties and interactions of four negative and positive biofilm-forming *S. epidermidis* strains in MQ water, Hu et al. showed that all strains displayed adhesion strengths similar to that of the hydrophilic silicon nitride cantilever, while adhesion strengths were significantly different when the hydrophobic probe was used. Adhesion forces were lower for the positive biofilm strains than for the negative biofilm-forming strains [29]. Hanna et al. investigated the role of physiochemical and specific binding interactions in the adhesion of colanic acid exopolysaccharide, isogenic, uropathogenic *E. coli* expressing mutant strains to various substrates using AFM. Their results indicated that electrostatic interactions were not solely responsible for the repulsive forces measured between the colonic acid mutant strains and the hydrophilic substrates. In addition, hydrophobic interactions were not found to play a significant role in the adhesion of the colanic-acid-mutant strains [37].

In addition to SAMs, PEG chains are generally immobilized on biomaterial surfaces in the form of a brush that extends from the biomaterial interface into the solution [51]. The efficacy of PEG brushes in decreasing bacterial adhesion to biomaterials was investigated by Razatos et al. AFM was used to quantify interaction forces between *E. coli* D21 bacteria and hydrophilic or hydrophobic glass substrates treated with *N*-octadecyltrichlorosilane (OTS) and coated with a block copolymer, (PEG)-lysine dendron or Pluronic F127 surfactant, respectively [51]. Their results indicated that polymeric brush layers have a combined effect of blocking the long-range attractive forces of interaction between bacteria and substrates as well as introducing repulsive steric effects [51].

Bacterial interactions with biomaterials have also been controlled using biosurfactant coatings. Van Hoogmoed et al. utilized AFM to quantify the interaction forces between enamel with or without pellicle and two strains of *S. mutans* [162]. Their results indicated that the use of *Streptococcus mitis* BMS biosurfactant decreased the attractive forces between the bacterial cells and the enamel materials and increased the repulsive forces upon approach [162].

5.2.5.2. The Role of Environmental Conditions and Antibiotics in the Attachment of Bacteria to Biomaterials

The administration of antibiotics to patients before and after biomaterial implantation is considered a gold standard in the medical community [15]. However, the ability of bacterial cells to develop resistance to antibiotics along with the use of the bulk of the antibiotic drugs to reduce early-onset bacterial adhesion limits the efficacy of antibiotics over the long term [4]. One approach to overcoming this is to coat biomaterials with antibiotics or to incorporate antibiotics into biomaterials such that a sustained release of antibiotics is controlled over a long period of time [163]. Once designed, the effects of antibiotic coatings on initial bacterial adhesion and further biofilm growth should be quantified [36].

AFM studies which addressed the effects of antibiotic coatings on biofilm formation, are very limited in the literature. Liang et al. used AFM and SEM to investigate the effect of the antibiotic rifampicin on the colonization and growth of *S. epidermidis* 35984 on the surface of silicone [36]. Rifampicin was cast into the silicon before it was cured. AFM images of biofilms which formed on silicon with and without rifampicin indicated that multi-layer biofilms were able to form on the silicon surface in the absence of rifampicin, in contrast to only sparse biofilm structures forming on rifampicin-loaded surfaces after incubation for the same duration [36].

Bacterial adhesion to surfaces can also be controlled by environmental media [34]. Cranberry juice has long been believed to be an alternative medicine for the treatment of UTIs. In one study, Liu et al. used AFM to explore the molecular scale effects of cranberry juice on the adhesion to silicon nitride of two *E. coli* strains, one of which expresses P-fimbriae and the other of which has no fimbriae [164]. They observed that the equilibrium length of the P-fimbriae on the *E. coli* decreased from 148 to 48 nm upon being exposed to cranberry juice [164].

5.2.5.3. The Role of the Roughness of Biomaterials in Bacteria–Biomaterial Interactions

In addition to covering biomaterials with coatings resistant to protein and bacterial cells, the texture and roughness of biomaterials can be controlled as a means of controlling bacteria–biomaterial interactions. Studies have indicated that irregularities of polymeric surfaces normally promote bacterial adhesion and biofilm accumulation [53]. Emerson et al. explored the relationships between surface wettability and root-mean-square roughness (Rq) on the adhesion of *S. epidermidis* to a variety of chemically and texturally distinct model substrata using AFM [33]. Their results indicated that substrate hydrophobicity alone was incapable of explaining the adhesion forces measured between the substrates and bacterial cells. Similarly, weak correlations between adhesion and Rq were observed for most surfaces. They concluded their study by stating that adhesion to biomaterials may be influenced by the area available for interactions at the macroscale. However, at the nanoscale, bacteria may not be able to conform to substrate asperities and ridges [33]. In another study, the effect of introducing patterns onto a polyurethane surface on the ability of *S. cerevisiae* and *Leuconostoc oenus* to adhere to and interact with it was studied using quartz crystal microbalance (QCM) and AFM [165] simultaneously. Their results indicated that the interactions of bacteria with yeast molecularly imprinted polymers (MIPs) were very weak. The weak interactions were influenced by the large dimensions of the yeast imprint compared to the bacterial cells. The size of the yeast MIPs exceeded the size of the bacteria by 3–5 times, thus rendering the yeast MIPs nearly flat to the bacteria [165]. Rizzello et al. showed that highly controlled nanostructured substrates can impact bacterial morphology and genomic and proteomic responses [166]. Using AFM imaging, the authors showed that type-1 fimbriae typically disappeared from *E. coli* adherent on nanostructured substrates, as opposed to bacteria adherent on reference glass or flat gold surfaces [166]. Finally, Whitehead et al. investigated the ease of removal of differently sized and shaped Gram-negative *Pseudomonas aeruginosa* (rods, 1 μm width × 3 μm length) and Gram-positive *S. aureus* (cocci, 1 μm diameter) from substrata with defined surface topographies and features [54]. Hydrophobic smooth or rough titanium oxide surfaces with 0.5 μm diameter randomly spaced surface pits were produced. Their results indicated that *S. aureus* were removed more easily from smooth surfaces than from rough surfaces. In contrast, *P. aeruginosa* cells were removed more easily from the 0.5 μm featured surfaces than from the smooth surfaces. They concluded the study by stating that both the type of the bacterial cells and the topography of the surfaces play roles in controlling bacterial adhesion to biomaterial surfaces [54].

5.2.5.4. Quantification of the Strength of Interactions between Biomaterials and Bacterial Cells

AFM is unique in its ability to quantify the interaction forces that govern the initial adhesion of bacteria to biomaterials [20]. For example, the adhesion forces of eight Gram-negative and Gram-positive *P. aeruginosa*, *Staphylococci* and *Serratia* bacterial strains to two types of contact lenses were quantified and compared to the adhesion forces measured on real corneal surfaces taken from pigs. The bacterial adhesion to actual corneas was larger than that measured on contact lenses [157]. In another example, AFM was used to quantify the adhesion forces between a cellular probe with *E. coli* cells and mica, hydrophilic glass, hydrophobic glass, polystyrene, and Teflon. Consistent with prior qualitative observations, the authors showed that bacterial adhesion is enhanced by the surface hydrophobicity of the substrate [58]. In a third study, AFM was used to quantify and compare the initial adhesion forces of the uropathogen *Enterococcus faecalis* with medical-grade polymers polyurethane (PU), polyamide (PA), and poly(tetrafluoroethylene) (PTFE) [16]. The detachment of individual live cells from the polymeric surfaces through the application of increasing force using unmodified cantilever tips was assessed. The results showed that the lateral force required to detach *E. faecalis* cells from polymeric substrates varied depending on the hydrophobic nature of the polymeric surface, with the highest forces measured on PU, followed by PA and then PTEE [16]. Finally, the formation of *P. aeruginosa* biofilms on a variety of biomaterial surfaces including sheets of aluminium, steel, rubber, and polypropylene was assessed using AFM [156]. AFM was used to investigate the morphology, transition, and adhesiveness of biofilms. Their results indicated that biofilms can remain on rough surfaces even when substrates are immersed in hot water [156].

5.2.5.5. AFM Imaging of Biofilms and Characterization of Their Extracellular Polymeric Substances

In addition to quantifying the forces that govern the interactions between bacteria and biomaterials, AFM can be used to image bacterial cells and their ultrastructure and to image biofilms forming on biomaterials. AFM can also be used to characterize the EPS secreted by bacteria and to quantify their role in the adhesion and aggregation of bacterial cells on solid surfaces. For example, AFM was used to study *Staleya guttiformis* biofilms forming on hydrophobic poly(*tert*-butylmethacrylate) (P*t*BMA) and hydrophilic polystyrene maleic acid (PSMA) surfaces [117]. AFM was also used to investigate the ultrastructure of these biofilms. AFM images indicated that bacterial attachment to the PSMA surface was negligible, while multi-layer biofilms formed on the P*t*BMA surface over the same period [117]. Noncontact-mode imaging of the multilayer biofilms that formed on the P*t*BMA revealed a capsular EPS surface having granular structures with lateral dimensions of 30–50 nm and a vertical roughness of 7–10 nm [117].

5.2.5.6. Additional Means by Which AFM can be Used to Quantify Bacteria–Biomaterial Interactions

In addition to the approaches discussed above, there are a few other means by which AFM can be used to investigate bacteria–biomaterial interactions. First, AFM can be a valuable tool for screening for anti-adhesion molecules that can be used to inhibit biofilm formation. In one study, the treatment of two biofilm-positive *S. epidermidis* strains with two chemical inhibitor compounds led to a loss of adhesion [29]. Second, AFM can be used to quantify the interactions between bacterial cells and mammalian cells present at the implantation site. For example, one way to minimize biofilm formation on BCBIs is achieved *in vivo* when white blood cells and phagocytes successfully fight infections. AFM was used successfully to study interactions between *Vibrio cholerae* bacterial cells and lipid bilayers containing a specific molecule of interest [167]. In a similar manner, bacteria immobilized on cellular probes can be used to quantify interactions with mammalian cells immobilized on surfaces in media of interest. Third, AFM imaging can be used to compare the structures of bacterial cells impeded in biofilms of various bacterial mutants. In one study, AFM tapping-mode images of biofilms of wild-type and quorum-sensing mutants of *Pseudomonas chlororaphis* O6 (PcO6) revealed elongated cell structures in both types of mutant biofilm cells. Images also indicated that the wild-type strain was capable of forming multi-layered biofilms under conditions in which the quorum-sensing mutant, deficient in the GacS sensor kinase, could not mature beyond a monolayer biofilm structure [168]. Fourth, the cohesiveness of young biofilms can be assessed using AFM by calculating the volume of biofilm displaced and the corresponding frictional energy dissipated as a function of biofilm depth when an AFM cantilever rasters laterally on the biofilm surface under an elevated load. The pioneers of this methodology showed that the cohesive energy increased with the depth of the biofilm for activated sludge taken from a metropolitan wastewater treatment plant, from 0.10 ± 0.07 to 2.05 ± 0.62 nJ/μm^3 [50]. The effects of various environmental conditions on biofilm cohesiveness can be investigated as well. Understanding the cohesive interactions in the biofilm matrix under a variety of conditions can lead to the design of new strategies for controlling biofilm development based on disrupting or protecting the matrix holding the biofilm together. Finally, AFM can be used to investigate how bacterial cells can adapt to various environments. Bacteria are 'naturally programmed' to respond to several stimuli by morphological adaptation [166].

5.2.6. CHARACTERIZATION OF BIOFILMS ON BIOMATERIAL SURFACES

Typically, optical microscopy is used to characterize biofilms on surfaces [44,169]. Optical microscopy is used to characterize (1) the initial attachment of cells, (2) the distribution of cells in EPS, (3) the distribution of the EPS and its components, and (4) the role of environmental conditions in biofilm structure [170]. To image biofilms, generally specific fluorophores are required [170]. Most of the time, cells expressing green fluorescent protein (GFP) are used to image cells within biofilms [171,172]. The biofilm images are further analyzed to quantify numerical parameters characterizing the size and shape of the microcolonies and EPS in the biofilm. These processes are generally performed automatically using software. Time-lapse digitized images of the biofilms are used to quantify the dynamics of microbial attachment to surfaces and biofilm growth on surfaces. These images are later used to characterize the rates of attachment of cells to biofilm and of detachment of cells from biofilm and to generate kinetic expressions quantifying various biofilm processes on surfaces.

Optical microscopy uses visible light to illuminate and image biofilms on biomaterial surfaces. This technique has two main limitations: (1) the resolution of the image is limited by the wavelength used, and (2) out-of-focus objects degrade biofilms that are thicker than a single layer. The theoretical resolution of a microscope is given by Eqn (5.2.1).

$$d = \lambda / 2N_a \tag{5.2.1}$$

where d is the distance between two objects that can be separated in the field of view, λ is the wavelength of the light, and N_a is the numerical aperture of the objective. Therefore, for single-cell imaging, an objective with a larger magnification and a larger numerical aperture is preferred. Equation (5.2.1) also explains why we need to use AFM for nanoscale observations. Simply, light microscopy does not provide the resolution needed for single-cell imaging.

Fluorescence microscopy is one of the preferred techniques for imaging living biofilms on biomaterial surfaces [173–175]. In fluorescence microscopy, the biofilm sample is excited and the emitted light is used for imaging. The emission light always has a higher wavelength than the excitation light [176]. Epifluorescence microscopy is a type of fluorescence microscopy, in which the illuminating light is delivered thorough the objective. This is an excellent technique when the biomaterials are opaque. For laser microscopes, the illuminating light is a laser light. The main advantage of a laser microscope is that the light can be delivered at a high intensity to a small area at a desired wavelength. Confocal scanning laser microscopy (CSLM) rejects out-of-focus objects and is generally used to generate crisp images of biofilms in two and three dimensions. To image biofilm components on biomaterial surfaces, generally the component of the biofilm to be imaged needs to be stained using fluorescent proteins. In this chapter, we refer to cells and EPS as biofilm components. Some general stains can be used to stain organic components in biofilms and can be used to image entire biofilms, such as fluorescein isothiocyanide (FITC) [177,178]. There are many different fluorescent dyes that can be used to stain biofilms, and their use is detailed in the literature [177,179–183].

One of the most commonly used types of stains in biofilm imaging is live/dead stains [180,182]. If the cell is dead and the cell membrane is compromised, the stain penetrates the cell and it is imaged as a dead cell. One example of this type of stain is a nucleic acid stain, propidium iodide, which is quite popular with biofilm researchers [180,182,184]. Other stains are used to stain all of the cells, live and dead. Thiazole orange is a permanent dye which enters all cells, live and dead. The combination of these two stains can be used to image live and dead cells. Since a red signal is stronger than a green signal, live cells fluoresce green and dead cells fluoresce red. This type of staining generates very valuable information about biofilms on biomaterial surfaces. For example, someone interested in developing a biomaterial with an antibiotic which is effective at killing attached cells can use live/dead staining to estimate the effectiveness of the biomaterial in this process [185]. However, if the goal is to image only attached cells, a bacterium which simply expresses a fluorescent protein can be used. An example of this is *P. aeruginosa*, which constitutively produces green fluorescent protein (GFP) and can be imaged easily [171,172].

Since EPS play a critical role in the structure of biofilms and their resistance to antibiotics, it is necessary to image them. However, imaging EPS is not as simple as imaging cells. This is because EPS are composed of many different types of organic molecules, such as polysaccharides, proteins, nucleic acids, and lipids. Currently, there is no general stain for EPS. The available stains target certain groups in the EPS. For example, there are fluorescent stains for some types of polysaccharides, but not for all those present in biofilms. Specifically, researchers have used Alcian blue to stain EPS to target anionic molecules only [186] and lipid fractions have been stained using Nile red [177,181,187]. We should note that the available stains for imaging EPS are lately increasing significantly. For example, F10318 FilmTracer™ SYPRO® Ruby from Invitrogen can stain most classes of proteins, including glycoproteins, phosphoproteins, lipoproteins, calcium-binding proteins, fibrillar proteins and other proteins that are difficult to stain [188]. It is best to check company catalogues to find the most suitable stain (such as www.invitrogen.com). The overall conclusion here is that components of biofilm can be stained individually or using multiple stains.

It is important to note that the imaging of biofilm components is limited by the availability of specific probes that can selectively target them and by the nature of biomaterials. Biofilms are usually imaged on smooth surfaces. However, there are many biomaterials which have porous surfaces because of their intended application. In this case, the quantification of biofilms becomes extremely difficult. For such porous biomaterials, biofilms can be tested on a smooth surface replica and the results can be later extrapolated to the porous surfaces.

5.2.7. QUANTIFYING BIOFILM STRUCTURE

Once a biofilm has been imaged, it is necessary to quantify the biofilm structure. In this chapter, biofilm structure refers to the morphology and distribution of the imaged components of the biofilms. Images of biofilm components are used to quantify parameters characterizing their structure. These parameters should be related to the experimental system being studied. Parameters describing biofilms can be quantified from two-dimensional or three-dimensional images. A two-dimensional image represents a single layer in a biofilm, most often a single layer in a CSLM image or a light microscopy image. Three-dimensional images are reconstructed from stacks of two-dimensional images taken using CSLM. Most commonly, two classes of parameters are computed from such images: (1) textural and (2) areal (for two-dimensional images)/volumetric (for three-dimensional images).

5.2.7.1. Textural Parameters

Textural parameters are used to describe repeating patterns of local variations in grey-level intensity [44,170]. Textural parameters are calculated from grey-level images. Textural parameters can be calculated for 2D or 3D images. They quantify biofilm structure by comparing the intensities, positions, and/or orientations of the pixels. Textural parameters are calculated in two or three dimensions using a grey-level co-occurrence matrix, which represents the distribution of changes in the grey-level values of neighbouring pixels in the calculated directions [44]. Examples of textural parameters are textural entropy (TE), homogeneity and heterogeneity.

5.2.7.2. Areal/Volumetric Parameters

Areal and volumetric parameters give information about imaged biofilm components [44]. They are calculated from thresholded 2D and 3D biofilm images, respectively. Areal/volumetric parameters are concerned with the size, shape and orientation of the imaged biofilm components. Each parameter measures a unique characteristic feature of an imaged component of a biofilm or the background in the image. Examples of areal parameters are the average run lengths (in the horizontal and vertical directions), average and maximum diffusion distances (MDDs), fractal dimension (FD), and areal porosity (AP). Examples of volumetric parameters are the average run lengths (in the *X*, *Y* and *Z* directions), aspect ratio, average and MDDs and FD.

5.2.7.3. Software Packages Used for Biofilm Image Analysis and Structure Quantification

There has been a significant amount of research effort dedicated to quantifying biofilm structure, and image analysis software packages have been developed to quantify biofilm structures from these images [189–192]. The most widely used software includes Amira (Visage Imaging, San Diego, California), *bio*-Image_L [193], COMSTAT [190], DAIME [189], ImageJ (National Institutes of Health), Imaris (Bitplane, Zurich, Switzerland), ISA [170], Matlab with Image Processing Toolbox (The MathWorks, Inc., Natick, Massachusetts), PHLIP [194] and Volocity (PerkinElmer, Waltham, Massachusetts). Our research group uses ISA, which was developed by our group using MATLAB®, which allows us for detailed custom analysis. As an example of biofilm structure quantification, Fig. 5.2.5 shows biofilm images taken (A) right after inoculation and (B) 1 day later. At the time of the inoculation, the AP was quantified to be around 0.996, showing a small amount of cell attachment. However, a day later, the areal porosity (AP) had decreased to 0.853, demonstrating biofilm growth. The increase in fractal dimension (FD) shows an increase in roughness at the edges. Since, initially, there were only single cells, the FD was close to one. A day later, the edges of the cluster had developed and FD had increased. The average diffusion distance (ADD) and maximum diffusion distance (MDD) increased with biofilm growth as expected. The horizontal (HRL) and vertical (VRL) run lengths were very close to each other, demonstrating that the average sizes in the horizontal and vertical directions were approximately the same. As expected, the perimeter (P) increased significantly because of the development of the biofilm cluster from a single

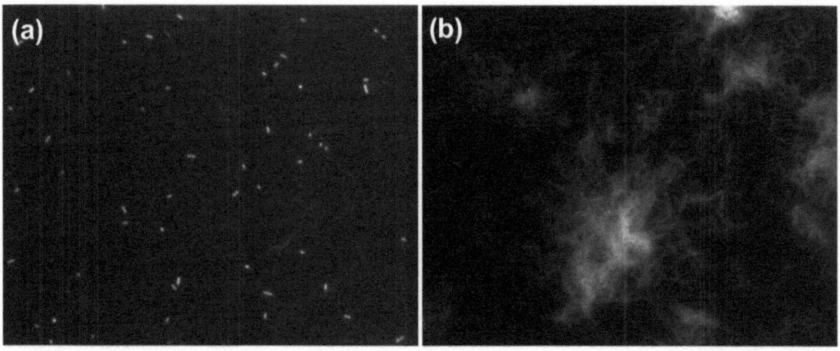

FIGURE 5.2.5 Two biofilm images consisting of individual cells. These were analyzed and the results are presented in Table 5.2.2.

cell to a cell cluster. The energy (E) and homogeneity increased with time. The textural entropy (TE) was around 5.7, demonstrating that the color variations in these images are almost identical. This is expected since the images consist of single cells expressing GFP.

There are numerous biofilm structural parameters presented in the literature [195–202]. Table 5.2.2 shows some of the parameters we use. However, generally, only a few of these parameters are used in research in the literature [203–205]. As an illustration, Table 5.2.3 shows a list of several peer-reviewed studies since 1990 that used two-dimensional structural parameters. On average, each study utilized four parameters out of a total of 26 different parameters available. In part, this trend is justified by the fact that only a handful of parameters computed from biofilm images, e.g. biofilm porosity [42,206–208], can be intuitively related to identifiable biological processes, and authors tend to select these parameters in their studies. The parameters most used in the literature are biovolume and AP. These parameters provide information about the volume of the biofilm or the surface coverage of the biofilm. If the researcher is interested in quantifying the amount of biofilm, these are the right parameters to use. However, if the researcher is interested in how cell colony dimensions change over time, different parameters, such as horizontal and vertical run lengths should be used [45,207].

TABLE 5.2.2 The Numerical Values of the Calculated Parameters of Images a and b, Shown in Figure 5.2.5

#	AP	FD	ADD	MDD	HRL	VRL	Perimeter	E	TE	H
A	0.996	1.111	2.05	6.32	8.47	7.70	1889	0.0047	5.777	0.337
B	0.853	1.31	20.41	111.51	27.60	27.17	18,302	0.01557	5.657	0.56

TABLE 5.2.3 List of Several Studies that Used 2D Structural Parameters to Quantify Biofilm images

Reference	Microorganism	2D Parameters
Kim et al. 2010	*Pseudomonas putida*	E, FD, H, L, SBC, SB, TE
Raajan et al. 2008	KP&P*	AP, ADD, AHRL, AVRL, FD, TE
Kim et al. 2006	Mixed species	AP
Lewandowski et al. 2004	KP&P	AP
Beyenal et al. 2004	KP&P	AP, ADD, AHRL, AVRL, E, FD, H, MDD, P, TE
Battin et al. 2003	Mixed species	AP, BF
Purevdorj et al. 2002	*Pseudomonas aeruginosa*	ASM, AP, ADD, AHRL, AVRL, FD, IDM, MDD
Lomander et al. 2002	*Escherichia coli*	BC, BPA, BPC, BPP
Jackson et al. 2001	KP&P	AP, ADD, FD, TE
Beyenal and Lewandowski 2001	*Desulfovibrio desulfuricans* *Pseudomonas fluorescens*	AP, FD
Yang et al. 2000	KP&P	ASM, AR, ADD, AHRL, AVRL, FD, IDM, MDD, TE
Lewandowski et al. 1999	KP&P	AP, FD, MDD, TE
Okabe et al. 1998	Mixed species	AP, MDM, SSA
Zahid and Ganczarczyk 1994a	Mixed species	AP, FD, MPD, P, PS, SA
Zahid and Ganczarczyk 1994b	Mixed species	AP, MPD
Obert et al. 1990	*Ashbya gossypii* *Streptomyces griseus*	BD, FD

The Parameters Used Include Angular Second Moment (ASM), Areal Porosity (AP), Average Diffusion Distance (ADD), Average Horizontal Run Length (AHRL), Average Vertical Run Length (AVRL), Biofilm Coverage (BC), Biofilm Diameter (BD), Biofilm Fragmentation (BF), Biofilm Patch Area (BPA), Biofilm Patch Circularity (BPC), Biofilm Patch Perimeter (BPP), Energy (E), Fractal Dimension (FD), Homogeneity (H), Inverse Difference Moment (IDM), Lacunarity (L), Maximum Diffusion Distance (MDD), Maximum Pore Dimension (MPD), Mean Diameter of Macropores (MDM), Perimeter (P), Pore Size (PS), Specific Biofilm Contacts (SBC), Specific Biofilms (SB), Specific Surface Area (SSA), Surface Area (SA) and Textual Entropy (TE).

* *KP&P= Klebsiella pneumoniae, Pseudomonas aeruginosa and Pseudomonas fluorescens.*

Reprinted from [202].

5.2.8. ANALYSIS OF BIOFILM IMAGES

There are software packages, developed by various research groups, for quantifying variable parameters from biofilm images. In some cases, the software installed to operate microscopes can perform some limited analysis of biofilm images. Although there are commercially available software packages for biofilm image analysis, they have limited use. Therefore, many research groups develop their own software. Table 5.2.4 lists the available software packages for quantifying parameters from biofilm images.

TABLE 5.2.4 Currently Available Software Packages for Quantifying Biofilm Structure

Parameters	ISA [41,44]	COMSTAT [192]	PHLIP [219]	DAIME [191]
Biovolume, and surface area	+	+	+	+
Volume to surface area ratio (or surface to volume ratio)	+	+	−	+
Area occupied by each layer	+	+	−	−
Porosity	+	+*	+*	−
Average diffusion distance, average maximum diffusion distance, and fractal dimension	+	Only 2D distribution	−	−
Average *X*, *Y*, and *Z* run lengths and aspect ratio in 3D	+	−	−	−
Spatial spreading and co-localization (2D and 3D)	−	−	+	−
Spatial arrangement and fluorescence intensity	−	−	−	+
Textural entropy, energy, and homogeneity	+	−	−	
Mean thickness, maximum thickness, roughness coefficient, volume of the microcolonies identified at the substratum	+	+	−	
Identification and areal distribution of microcolonies at the substratum	−	+	−	−

(+) Indicates an available option in the software and (−) indicates that the parameter cannot be computed by the software.

* *The parameter can be computed by the user from other parameters reported by the software.*

5.2.9. CONCLUSIONS

Quantifying bacterial interactions with biomaterials is critical for designing biomaterials and biomaterial coatings capable of preventing bacterial attachment to their surfaces. It is needed to reduce the chances of the bacterial contamination of implant materials during surgery. The reduction of the bacterial contamination of biomaterials is the first step towards the elimination of biomaterial-centered bacterial infections (BCBIs). Bacterial interactions with surfaces generally follow a multi-step approach, in which single cells initiate the attachment to the surface. This initial attachment is generally followed by colonization of the surface and biofilm formation. The eradication of biofilms requires detailed understanding of the various steps leading to biofilm formation, described in Fig. 5.2.1 in this review. The first part of the review focused on how AFM can be used to quantify the initial attachment of bacterial cells to biomaterials. An introduction to the main components of AFM, its working principle and modes of operation were given. The various methodologies that can be implemented in AFM to investigate bacteria–biomaterial interactions were discussed. Finally, examples from the literature were given in support of the various capabilities of AFM in the exploration and quantification of bacteria–biomaterial interactions. The second part of the review focused on the quantification of structures and components of biofilms grown on biomaterials. An introduction to the different types of microscopy techniques that can be used to image biofilms was given. Then, the various types of stains that can be used to image various components of biofilms were introduced. Biofilm images can be used to obtain a wealth of information on various structural parameters of the biofilm. The most commonly used parameters extracted from biofilm images were introduced. Finally, the most commonly available software packages used to extract such parameters from biofilm images were compared.

SOURCES FOR FURTHER INFORMATION AND ADVICE

Protocols for imaging biofilms and the computations we generally use to characterize biofilm structures are described in our book: Lewandowski, Z. and Beyenal, H. 2007. Fundamentals of Biofilm Research. CRC.

ACKNOWLEDGMENTS

Abu-Lail would like to thank the National Science Foundation (NSF, grant EEC-0823901), the National Institutes of Health (NIH, grant number 1R03AI077590-01A1) and the 3M Corporation for financial support of this work. Beyenal acknowledges NSF award #0954186.

REFERENCES

[1] Gristina AG. Biomaterial-centered infection: microbial adhesion versus tissue integration. Science 1987;237:1588–95.

[2] Larry LL, Hench L, Polak JM. Third-generation biomedical materials. Science 2002;295:1014–7.

[3] Busscher HJ, Rinastiti M, Siswomihardjo W, van der Mei HC. Biofilm formation on dental restorative and implant materials. J Dent Res 2010;89:657–65.

[4] Campoccia D, Montanaro L, Arciola CR. The significance of infection related to orthopedic devices and issues of antibiotic resistance. Biomaterials 2006;27:2331–9.

[5] Trainini JC, Cichero D, Lago N, Giordano R, de Paz J, Elencwajg B, et al. Autologous cellular cardiac-implant. Basic Appl Myol 2003;13:39–43.

[6] Russo A, Maconi G, Spinelli P, Di Felice G, Eboli M, Andreola S, et al. Effect of lifestyle, smoking, and diet on development of intestinal metaplasia in *H. pylori* positive subjects. Am J Gastroenterol 2001;96:1402–8.

[7] Chaterji S, Kwon K, Park K. Smart polymeric gels: redefining the limits of biomedical devices. Prog Polym Sci 2007;32:1083–122.

[8] Dasgupta MK. Exit-site and catheter-related infections in peritoneal dialysis: problems and progress. Nephrology 2000;5:17–25.

[9] Chao EY, Kasman RA, An KN. Rigidity and stress analyses of external fracture fixation devices: a theoretical approach. J Biomech 1982;15:971–83.

[10] Cheng G, Zhang Z, Chen SF, Bryers JD, Jiang SY. Inhibition of bacterial adhesion and biofilm formation on zwitterionic surfaces. Biomaterials 2007;28:4192–9.

[11] Sudagidan M, Erdem I, Cavusoglu C, Ciftcioglu M. Investigation of the surface properties of *Staphylococcus epidermidis* strains isolated from biomaterials. Mikrobiyol Bul 2010;44: 93–103.

[12] Cuperus PL, Vandermei HC, Reid G, Bruce AW, Khoury AE, Rouxhet PG, et al. The effect of serial passaging of *Lactobacilli* in liquid-medium on their physicochemical and structural surface characteristics. Cells Mater 1992;2:271–80.

[13] Costerton JW, Khoury AE, Ward KH, Anwar H. Practical measures to control device-related bacterial infections. Int J Artif Organs 1993;16:765–70.

[14] Dasse KA. Infection of percutaneous devices: prevention, monitoring, and treatment. J Biomed Mater Res A 2004;18:403–11.

[15] Darouiche RO. Current concepts: treatment of infections associated with surgical implants. Engl J Med 2004;350:1422–9.

[16] Senechal A, Carrigan SD, Tabrizian M. Probing surface adhesion forces of *Enterococcus faecalis* to medical-grade polymers using atomic force microscopy. Langmuir 2004;20: 4172–7.

[17] Homminga GN, Bulstra SK, Kuijer R, Vanderlinden AJ. Repair of sheep articular cartilage defects with a rabbit costal perichondrial graft. Acta Orthop Scand 1991;62: 415–8.

[18] Giacometti A, Cirioni O, Ghiselli R, Goffi L, Mocchegiani F, Riva A, et al. Efficacy of polycationic peptides in preventing vascular graft infection due to *Staphylococcus epidermidis*. J Antimicrob Chemother 2000;46:751–6.

[19] Sengupta S, Uemura S, Patwardhan S, Huber V, Grozema FC, Siebbeles LDA, et al. Columnar mesophases based on zinc chlorophyll derivatives functionalized with peripheral dendron wedges. Chem Eur J 2011;17:5299–309.

[20] Katsikogianni MG, Missirlis YF. Interactions of bacteria with specific biomaterial surface chemistries under flow conditions. Acta Biomater 2010;6:1107–18.

[21] Bushnak IA, Labeed FH, Sear RP, Keddie JL. Adhesion of microorganisms to bovine submaxillary mucin coatings: effect of coating deposition conditions. Biofouling 2010; 26:387–97.

[22] Getchell-White SI, Donowitz LG, Gröschel DHM. The inanimate environment of an intensive care unit as a potential source of nosocomial bacteria: evidence for long survival of *Acinetobacter calcoaceticus*. Infect Control Hosp Epidemiol 1989;10:402–7.
[23] Selwyn S, Ellis H. Skin bacteria and skin disinfection reconsidered. Br Med J 1972;1.
[24] Liston J. The occurrence and distribution of bacterial types on flatfish. J Gen Microbiol 1957;16:205–16.
[25] Gaynes R, Edwards JR, System NNIS. Overview of nosocomial infections caused by Gram-negative *Bacilli*. Clin Infect Dis 2005;41:848–54.
[26] An YH, Friedman RJ. Laboratory methods for studies of bacterial adhesion. J Microbiol Methods 1997;30:141–52.
[27] Gristina AG, Naylor P, Myrvik Q. Infections from biomaterials and implants: a race for the surface. Med Prog Technol 1988;14:205–24.
[28] Gristina AG, Costerton JW. Bacterial adherence to biomaterials and tissue: the significance of its role in clinical sepsis. J Bone Joint Surg Am 1985;67:264–73.
[29] Hu YF, Ulstrup J, Zhang JD, Molin S, Dupres V. Adhesive properties of *Staphylococcus epidermidis* probed by atomic force microscopy. Phys Chem Chem Phys 2011;13:9995–10003.
[30] Liu Y, Strauss J, Camesano TA. Adhesion forces between *Staphylococcus epidermidis* and surfaces bearing self-assembled monolayers in the presence of model proteins. Biomaterials 2008;29:4374–82.
[31] Wang MS, Palmer LB, Schwartz JD, Razatos A. Evaluating protein attraction and adhesion to biomaterials with the atomic force microscope. Langmuir 2004;20:7753–9.
[32] Abu-Lail NI, Camesano TA. Specific and nonspecific interaction forces between *Escherichia coli* and silicon nitride, determined by Poisson statistical analysis. Langmuir 2008; 24:4420.
[33] Emerson IV RJ, Bergstrom TS, Liu Y, Soto ER, Brown CA, McGimpsey WG, et al. Microscale correlation between surface chemistry, texture, and the adhesive strength of *Staphylococcus epidermidis*. Langmuir 2006;22:11311–21.
[34] Abu-Lail NI, Camesano TA. Role of ionic strength on the relationship of biopolymer conformation, DLVO contributions, and steric interactions to bioadhesion of *Pseudomonas putida* KT2442. Biomacromolecules 2003;4:1000–12.
[35] Abu-Lail NI, Camesano TA. Role of lipopolysaccharides in the adhesion, retention, and transport of *Escherichia coli* JM109. Environ Sci Technol 2003;37:2173–83.
[36] Liang X, Wang A, Cao T, Tang H, McAllister II JP, Salley SO, et al. Effect of cast molded rifampicin/silicone on *Staphylococcus epidermidis* biofilm formation. J Biomed Mater Res 2006;76A:580–8.
[37] Hanna A, Berg M, Stout V, Razatos A. Role of capsular colanic acid in adhesion of uropathogenic *Escherichia coli*. Appl Environ Microbiol 2003;69:4474–81.
[38] Cao T, Tang H, Liang X, Wang A, Auner GW, Salley SO, et al. Nanoscale investigation on adhesion of *E. coli* to surface modified silicone using atomic force microscopy. Biotechnol Bioeng 2006;94:167–76.
[39] Costa F, Carvalho IF, Montelaro RC, Gomes P, Martins MCL. Covalent immobilization of antimicrobial peptides (AMPs) onto biomaterial surfaces. Acta Biomater 2011;7:1431–40.
[40] Gattas-Asfura KM, Weisman E, Andreopoulos FM, Micic M, Muller B, Sirpal S, et al. Cinnamate-functionalized gelatin: synthesis and "Smart" hydrogel formation via photo-cross-linking. Biomacromolecules 2005;6:1503–9.
[41] Beyenal H, Donovan C, Lewandowski Z, Harkin G. Three-dimensional biofilm structure quantification. J Microbiol Methods 2004;59:395–413.

[42] Lewandowski Z, Beyenal H, Myers J, Stookey D. The effect of detachment on biofilm structure and activity: the oscillating pattern of biofilm accumulation. Water Sci Technol 2007;55:429–36.

[43] Lorite GS, Rodrigues CM, de Souza AA, Kranz C, Mizaikoff B, Cotta MA. The role of conditioning film formation and surface chemical changes on *Xylella fastidiosa* adhesion and biofilm evolution. J Colloid Interface Sci 2011;359:289–95.

[44] Lewandowski Z, Beyenal H, CRC Press; 2007.

[45] Lewandowski Z, Beyenal H. Biofilm monitoring: a perfect solution in search of a problem. Water Sci Technol 2003;47:9–18.

[46] Sjollema J, Sharma PK, Dijkstra RJB, van Damb GM, van der Mei HC, Engelsman AF, et al. The potential for bio-optical imaging of biomaterial-associated infection in vivo. Biomaterials 2010;31:1984–95.

[47] del Pozo JL, Patel R. The challenge of treating biofilm-associated bacterial infections. Clin Pharmacol Ther 2007;82:204–9.

[48] Alarcon CDL, Twaites B, Cunliffe D, Smith JR, Alexander C. Grafted thermo- and pH responsive co-polymers: surface-properties and bacterial adsorption. Int J Pharm 2005; 295:77–91.

[49] Qiu Y, Zhang N, An YH, Wen X. Biomaterial strategies to reduce implant-associated infections. Int J Artif Organs 2007;30:828–41.

[50] Ahimou F, Semmens MJ, Novak PJ, Haugstad G. Biofilm cohesiveness measurement using a novel atomic force microscopy methodology. Appl Environ Microbiol 2007;73:2897–904.

[51] Razatos A, Ong Y-L, Boulay F, Elbert DL, Hubbell JA, Sharma MM, et al. Force measurements between bacteria and poly(ethylene glycol)-coated surfaces. Langmuir 2000;16:9155–8.

[52] Hansma HG, Pietrasanta LI, Auerbach ID, Sorenson C, Golan R, Holden PA. Probing biopolymers with the atomic force microscope: a review. J Biomater Sci Polym Ed 2000;11:675–83.

[53] de Oliveira SS, Nascimento JDS, Povoa DC, de Araujo SA, Gamon MR, Bastos MCD. Genetic analysis of the bacteriocin-encoding plasmids pRJ6 and pRJ9 of *Staphylococcus aureus* by transposon mutagenesis and cloning of genes involved in bacteriocin production. J Appl Microbiol 1998;85:972–84.

[54] Whitehead KA, Rogers D, Colligon J, Wright C, Verran J. Use of the atomic force microscope to determine the effect of substratum surface topography on the ease of bacterial removal. Colloids Surf B Biointerfaces 2006;51:44–53.

[55] Kang S, Elimelech M. Bioinspired single bacterial cell force spectroscopy. Langmuir 2009;25:9656–9.

[56] Salerno MB, Logan BE, Velegol D. Importance of molecular details in predicting bacterial adhesion to hydrophobic surfaces. Langmuir 2004;20:10625–9.

[57] van Oss CJ, Giese RF. Role of the polar properties of water in separation methods. Sep Purif Rev 2011;40:163–208.

[58] Ong Y-L, Razatos A, Georgiou G, Sharma MM. Adhesion forces between *E. coli* bacteria and biomaterial surfaces. Langmuir 1999;15:2719–25.

[59] Ohshima H, Kondo T. Electrophoretic mobility and Donnan potential of a large colloidal particle with a surface charge layer. J Colloid Interface Sci 1987;116:305–11.

[60] Gordesli FP, Abu-Lail NI. The role of growth temperature in the adhesion and mechanics of pathogenic *L. Monocytogenes*: an AFM study. Langmuir 2012;28(2):1360–73.

[61] Razatos A, Ong YL, Sharma MM, Georgiou G. Evaluating the interaction of bacteria with biomaterials using atomic force microscopy. J Biomater Sci Polym Ed 1998;9:1361–73.

[62] Abu-Lail N, Camesano T. AFM and SMFM studies of biopolymers. In: Schwarz James A, Contescu Cristian I, Putyera Karol, editors. Dekker encyclopedia of nanoscience and nanotechnology. Taylor & Francis Group, LLC; 2004. p. 119–31.

[63] Ohnesorge FM, Horber JKH, Haberle W, Czerny CP, Smith DPE, Binnig G. AFM review study on pox viruses and living cells. Biophys J 1997;73:2183–94.

[64] Firtel M, Beveridge TJ. SPM in Microbiology. Micron 1995;26:347–62.

[65] Simon A, Durrieu MC. Strategies and results of atomic force microscopy in the study of cellular adhesion. Micron 2006;37:1–13.

[66] Dufrene YF. AFM, a powerful tool in microbiology. J Bacteriol 2002;184:5205–13.

[67] Casuso I, Rico F, Scheuring S. Biological AFM: where we come from – where we are – where we may go. J Mol Recognit 2011;24:406–13.

[68] Lyubchenko YL, Gall AA, Shlyakhtenko LS, Lindsay SM. AFM imaging of DNA and other biological molecules: use of silylated mica. Biophys J 1996;70. Wamg6-Wamg6.

[69] Fay MJ, Walter NG, Burke JM. Imaging of single hairpin ribozymes in solution by atomic force microscopy. RNA 2001;7:887–95.

[70] Allen MJ, Lee C, Lee JD, Pogany GC, Balooch M, Siekhaus WJ, et al. Atomic force microscopy of mammalian sperm chromatin. Chromosoma 1993;102:623–30.

[71] Assis OBG, Vieira DC, Bernardes R. AFM characterization of protein net formation on a fibrous medium. Braz J Chem Eng 2000;17:245–9.

[72] Abu-Lail NI, Camesano TA. Polysaccharide properties probed with atomic force microscopy. J Microsc Oxf 2003;212:217–38.

[73] Tasker S, Matthijs G, Davies MC, Roberts CJ, Schacht EH, Tendler SJB. Molecular resolution imaging of dextran monolayers immobilized on silica by atomic force microscopy. Langmuir 1996;12:6436–42.

[74] Beech IB, Smith JR, Steele AA, Penegar I, Campbell SA. The use of atomic force microscopy for studying interactions of bacterial biofilms with surfaces. Colloids Surf B Biointerfaces 2002;23:231–47.

[75] Bushell GR, Watson GS, Holt SA, Myhra S. Imaging and nano-dissection of tobacco mosaic-virus by atomic force microscopy. J Microsc Oxf 1995;180:174–81.

[76] Ahimou FO, Touhami A, Dufrene YF. Real-time imaging of the surface topography of living yeast cells by atomic force microscopy. Yeast 2003;20:25–30.

[77] Allen MJ, Bradbury EM, Balhorn R. The natural subcellular surface structure of the bovine sperm cell. J Struct Biol 1995;114:197–208.

[78] Kreplak L, Wang H, Aebi U, Kong XP. Atomic force microscopy of mammalian urothelial surface. J Mol Biol 2007;374:365–73.

[79] Lo YS, Huefner ND, Chan WS, Stevens F, Harris JM, Beebe TP. Specific interactions between biotin and avidin studied by atomic force microscopy using the Poisson statistical analysis method. Langmuir 1999;15:1373–82.

[80] Moy VT, Florin EL, Gaub HE. Adhesive forces between ligand and receptor measured by AFM. Colloids Surf A Physicochem Eng Asp 1994;93:343–8.

[81] Ke C, Humeniuk M, S-Gracz H, Marszalek PE. Direct measurements of base stacking interactions in DNA by single-molecule atomic force spectroscopy. Phys Rev Lett 2007;99.

[82] Atabek A, Camesano TA. Atomic force microscopy study of the effect of lipopolysaccharides and extracellular polymers on adhesion of *Pseudomonas aeruginosa.* J Bacteriol 2007;189:8503–9.

[83] Abu-Lail NI, Camesano TA. Elasticity of *Pseudomonas putida* KT2442 surface polymers probed with single-molecule force microscopy. Langmuir 2002;18:4071–81.

[84] He T, Shi ZL, Fang N, Neoh KG, Kang ET, Chan V. The effect of adhesive ligands on bacterial and fibroblast adhesions to surfaces. Biomaterials 2009;30:317–26.

[85] Ng SP, Rounsevell R, Randles LG, Steward A, Clarke J. Single molecule studies of protein folding by atomic force microscopy(AFM). Biophys J 2005;88:184A.

[86] Li LY, Chen SF, Oh SJ, Jiang SY. In situ single-molecule detection of antibody-antigen binding by tapping-mode atomic force microscopy. Anal Chem 2002;74:6017–22.

[87] Park BJ, Abu-Lail NI. Variations in the nanomechanical properties of virulent and avirulent *Listeria monocytogenes*. Soft Matter 2010;6:3898–909.

[88] Lam Y, Abu-Lail NI, Alam M, Zauscher M. Using Microcantilever deflection to detect HIV-1 envelope glycoprotein gp120. Nanomedicine: Nanotechnology, Biology and Medicine 2006;2:222–9.

[89] Dupres V, Alsteens D, Andre G, Verbelen C, Dufrene YF. Fishing single molecules on live cells. Nano Today 2009;4:262–8.

[90] Blanchard CR. Atomic force microscopy. Chem Educ 1996;1:1–8.

[91] Vilalta-Clemente A, Gloystein K. Principles of atomic force microscopy (AFM) based on the lecture of Prof. Nikos Frangis. Physics of Advanced Materials Winter School 1–10 2008.

[92] Lin DC, Dimitriadis EK, Horkay F. Robust strategies for automated AFM force curve analysis - I. Non-adhesive indentation of soft, inhomogeneous materials. J Biomech Eng 2007;129:430–40.

[93] Dorobantu LS, Gray MR. Application of atomic force microscopy in bacterial research. Scanning 2010;32:74–96.

[94] Bachmann RT, Edyvean RGJ. AFM study of the colonization of stainless steel by *Aquabacterium commune*. Int Biodeterior Biodegrad 2006;58:112–8.

[95] Camesano TA, Natan MJ, Logan BE. Observation of changes in bacterial cell morphology using tapping mode atomic force microscopy. Langmuir 2000;16:4563–72.

[96] Colton RJ, Baselt DR, Dufrene YF, Green JBD, Lee GU. Scanning probe microscopy. Curr Opin Chem Biol 1997;1:370–7.

[97] Binning G, Quate CF, Gerber C. Atomic force microscope. Phys Rev Lett 1986;6:930–3.

[98] Wright M, Revenko I. Tapping Mode AFM operation in fluid. Veeco Application Notes 2004.

[99] Choi HG, Lee IH, Kim YK, Lee WH, Choi JW. Lateral force microscopy investigation of bacteriorhodopsin adsorption onto mixed self-assembled monolayers. Mol Cryst Liq Cryst 2000;349:307–10.

[100] Leggett GJ, Brewer NJ, Chonga KSL. Friction force microscopy: towards quantitative analysis of molecular organization with nanometer spatial resolution. Phys Chem Chem Phys 2005;7:1107–20.

[101] Chang DP, Abu-Lail NI, Coles JM, Guilak F, Jay GD, Zauscher S. Friction force microscopy of lubricin and hyaluronic acid between hydrophobic and hydrophilic surfaces. Soft Matter 2009;5:3438–45.

[102] Brewer NJ, Leggett GJ. Chemical force microscopy of mixed self-assembled monolayers of alkanethiols on gold: Evidence for phase separation. Langmuir 2004;20:4109–15.

[103] Gaboriaud F, Dufrene YF. Atomic force microscopy of microbial cells: application to nanomechanical properties, surface forces and molecular recognition forces. Colloids Surf B Biointerfaces 2007;54:10–9.

[104] Zauscher S, Klingenberg DJ. Surface and friction forces between cellulose surfaces measured with colloidal probe microscopy. Nord Pulp Pap Res J 2000;15:459–68.

[105] Nunez ME, Martin MO, Chan PH, Duong LK, Sindhurakar AR, Spain EM. Atomic force microscopy of bacterial communities. Environ Microbiol 2005;397:256–68.

[106] Robichon D, Girard JC, Cenatiempo Y, Cavellier JF. Atomic force microscopy imaging of dried or living bacteria. C R Acad Sci III 1999;322:687–93.
[107] Alessandrini A, Facci P. AFM: a versatile tool in biophysics. Measurement Science and Technology 2005;16:R65–92.
[108] Doktycz MJ, Sullivan CJ, Hoyt PR, Pelletier DA, Wu S, Allison DP. AFM imaging of bacteria in liquid media immobilized on gelatin coated mica surfaces. Ultramicroscopy 2003;97:209–16.
[109] Ikai A. STM and AFM of bio/organic molecules and structures. Surf Sci Reports 1996;26:263–332.
[110] El Kirat K, Burton I, Dupres V, Dufrene YF. Sample preparation procedures for biological atomic force microscopy. J Microsc Oxf 2005;218:199–207.
[111] Hussain MA, Agnihotri A, Siedlecki CA. AFM imaging of ligand binding to platelet integrin alpha(IIb)beta(3) receptors reconstituted into planar lipid bilayers. Langmuir 2005;21:6979–86.
[112] Ahimou F, Denis FA, Touhami A, Dufrene YF. Probing microbial cell surface charges by AFM. Langmuir 2002;18:9937–41.
[113] Scheuring S, Dufrêne YF. Atomic force microscopy: probing the spatial organization, interactions and elasticity of microbial cell envelopes at molecular resolution. Mol Microbiol 2010;75:1327–36.
[114] Florin EL, Moy VT, Gaub HE. Adhesion forces between individual ligand-receptor pairs. Science 1994;264:415–7.
[115] Hinterdorfer P, Dufrene YF. Detection and localization of single molecular recognition events using atomic force microscopy. Nat Methods 2006;3:347–55.
[116] Abu-Lail NI, Camesano TA. Specific and nonspecific interaction forces between *Escherichia coli* and silicon nitride, determined by Poisson statistical analysis. Langmuir 2006;22:7296–301.
[117] Pham DK, Ivanova EP, Wright JP, Nicolau DV. In: SPIE-the international society for optical engineering; 2003. p. 151–9.
[118] Bolshakova AV, Kiselyova OI, Filonov AS, Frolova OY, Lyubchenko YL, Yaminsky IV. Comparative studies of bacteria with an atomic force microscopy operating in different modes. Ultramicroscopy 2001;86:121–8.
[119] Tsilimbaris MK, Lesniewska E, Lydataki S, Le Grimellec C, Goudonnet JP, Pallikaris IG. The use of atomic force microscopy for the observation of corneal epithelium surface. Investig Ophthalmol Vis Sci 2000;41:680–6.
[120] Wei ZQ, Wang C, Bai CL. Surface imaging of fragile materials with hydrophobic atomic force microscope tips. Surf Sci 2000;467:185–90.
[121] Arnold JW, Bailey GW. Surface finishes on stainless steel reduce bacterial attachment and early biofilm formation: scanning electron and atomic force microscopy study. Poult Sci 2000;79:1839–45.
[122] Sun W, Romagnoli JA, Palazoglu A, Stroeve P. Characterization of surface coats of bacterial spores with atomic force microscopy and wavelets. Ind Eng Chem Res 2011;50:2876–82.
[123] Park B-J, Abu-Lail NI. The role of the pH conditions of growth on the bioadhesion of individual and lawns of pathogenic *L. monocytogenes* cells. J Colloids Interface Sci 2011;358:611–20.
[124] Sahu K, Bansal H, Mukherjee C, Sharma M, Gupta PK. Atomic force microscopic study on morphological alterations induced by photodynamic action of Toluidine Blue O in *Staphylococcus aureus* and *Escherichia coli*. J Photochem Photobiol B Biol 2009;96:9–16.

[125] Poma A, Spano L, Pittaluga E, Tucci A, Palladino L, Limongi T. Interactions between saporin, a ribosome-inactivating protein, and DNA: a study by atomic force microscopy. J Microsc Oxf 2005;217:69–74.

[126] Boonaert CJP, Dufrene YF, Derclaye SR, Rouxhet PG. Adhesion of *Lactococcus lactis* to model substrata: direct study of the interface. Colloids Surf B Biointerfaces 2001;22: 171–82.

[127] Fang HHP, Chan KY, Xu LC. Quantification of bacterial adhesion forces using atomic force microscopy (AFM). J Microbiol Methods 2000;40:89–97.

[128] Jeyachandran YL, Venkatachalam S, Karunagaran B, Narayandass SK, Mangalaraj D, Bao CY, et al. Bacterial adhesion studies on titanium, titanium nitride and modified hydroxyapatite thin films. Mater Sci Eng C Biomim Supramol Syst 2007;27:35–41.

[129] Sheng XX, Ting YP, Pehkonen SO. Force measurements of bacterial adhesion on metals using a cell probe atomic force microscope. J Colloid Interface Sci 2007;310:661–9.

[130] Xu L, Logan BE. Analysis of bacterial adhesion using gradient force analysis method and colloid probe AFM. Langmuir 2004;20:8817–22.

[131] Park BJ, Haines T, Abu-Lail NI. A correlation between the virulence and the adhesion of *Listeria monocytogenes* to silicon nitride: an atomic force microscopy study. Colloids Surf B Biointerfaces 2009;73:237–43.

[132] Jingushi S, Scully SP, Joyce ME, Sugioka Y, Bolander ME. Transforming growth factor Beta-1 and fibroblast growth factors in rat growth plate. J Orthop Res 1995;13:761–8.

[133] Hutter JL, Bechhoefer J. Calibration of atomic force microscope tips. Rev Sci Instrum 1993;64:1868–73.

[134] Emerson RJ, Camesano TA. On the importance of precise calibration techniques for an atomic force microscope. Ultramicroscopy 2006;106:413–22.

[135] Abu-Lail NI, Camesano TA. The effect of solvent polarity on the molecular surface properties and adhesion of *Escherichia coli*. Colloids Surf B Biointerfaces 2006;51:62–70.

[136] Touhami A, Nysten B, Dufrene YF. Nanoscale mapping of the elasticity of microbial cells by AFM. Langmuir 2003;19:4539–43.

[137] Mahaffy RE, Park S, Gerde E, Kas J, Shih CK. Quantitative analysis of the viscoelastic properties of thin regions of fibroblasts using atomic force microscopy. Biophys J 2004;86:1777–93.

[138] Park BJ, Abu-lail NI. Heterogeneities in the adhesion energies measured between *Listeria* pathogenic and non-pathogenic species and silicon nitride probed using atomic force microscopy. Biofouling 2011;27:543–59.

[139] Dupres V, Menozzi FD, Locht C, Clare BH, Abbott NL, Cuenot S, et al. Nanoscale mapping and functional analysis of individual adhesins on living bacteria. Nat Methods 2005;2:515–20.

[140] Bowen WR, Lovitt RW, Wright CJ. Atomic force microscopy study of the adhesion of *Saccharomyces cerevisiae*. J Colloid Interface Sci 2001;237:54–61.

[141] Alexander SJ. Adsorption of chain molecules with a polar head a scaling description. J Phys II (Paris-France) 1977;38:983–7.

[142] de Gennes PG. Polymers at interface: a simplified view. Adv Colloid Interface Sci 1987;27:189–209.

[143] Butt HJ, Kappl M, Mueller H, Raiteri R, Meyer W, Ruhe J. Steric forces measured with the atomic force microscope at various temperatures. Langmuir 1999;15:2559–65.

[144] Dimitriadis EK, Horkay F, Maresca J, Kachar B, Chadwick RS. Determination of elastic moduli of thin layers of soft material using the atomic force microscope. Biophys J 2002;82:2798–810.

[145] Janshoff A, Neitzert M, Oberdorfer Y, Fuchs H. Force spectroscopy of molecular systems: single molecule spectroscopy of polymers and biomolecules. Angew Chem Int Ed 2000; 39:3213–37.

[146] Dupres V, Verbelen C, Raze D, Lafont F, Dufrene YF. Force spectroscopy of the interaction between *Mycobacterial* adhesins and heparan sulphate proteoglycan receptors. Chem phys chem 2009;10:1672–5.

[147] van der Mei HC, de Vries J, Busscher HJ. Weibull analyses of bacterial interaction forces measured using AFM. Colloids Surf B Biointerfaces 2010;78:372–5.

[148] Xu C-P, van de Belt-Gritter B, Dijkstra RJB, Norde W, van der Mei HC, Busscher HJ. Interaction forces between salivary proteins and *Streptococcus mutans* with and without antigen I/II. Langmuir 2007;23:9423–8.

[149] Meyer RL, Zhou XF, Tang LN, Arpanaei A, Kingshott P, Besenbacher F. Immobilization of living bacteria for AFM imaging under physiological conditions. Ultramicroscopy 2010;110:1349–57.

[150] Cidade GAG, Costa LT, Weissmuller G, Neto AJD, Roberty NC, de Moraes MB, et al. Atomic force microscopy as a tool for biomedical and biotechnological studies. Artif Organs 2003;27:447–51.

[151] Denes AR, Somers EB, Wong ACL, Denes F. 12-Crown-4-ether and tri(ethylene glycol) dimethyl-ether plasma-coated stainless steel surfaces and their ability to reduce bacterial biofilm deposition. J Appl Polym Sci 2001;81:3425–38.

[152] Razatos A, Ong YL, Sharma MM, Georgiou G. Molecular determinants of bacterial adhesion monitored by atomic force microscopy. Proc Natl Acad Sci U S A 1998;95:11059–64.

[153] Emerson RJ, Camesano TA. Nanoscale investigation of pathogenic microbial adhesion to a biomaterial. Appl Environ Microbiol 2004;70:6012–22.

[154] Chang D, Abu-Lail NI, Guilak F, Jay G, Zauscher S. Conformational mechanics, adsorption, and normal force interactions of lubricin and hyaluronic acid on model surfaces. Langmuir 2008;24:1183–93.

[155] Strauss J, Liu YT, Camesano TA. Bacterial adhesion to protein-coated surfaces: an AFM and QCM-D study. JOM 2009;61:71–4.

[156] Oh YJ, Lee NR, Jo W, Jung WK, Lim JS. Effects of substrates on biofilm formation observed by atomic force microscopy. Ultramicroscopy 2009;109:874–80.

[157] Qu W, Busscher HJ, Hooymans JMM, van der Mei HC. Surface thermodynamics and adhesion forces governing bacterial transmission in contact lens related microbial keratitis. J Colloid Interface Sci 2011;358:430–6.

[158] Mei L, van der Mei HC, Ren Y, Norde W, Busscher HJ. Poisson analysis of *Streptococcal* bond strengthening on stainless steel with and without a salivary conditioning film. Langmuir 2009;25:6227–31.

[159] Wright CJ, Armstrong I. The application of atomic force microscopy force measurements to the characterization of microbial surfaces. Surf Interface Anal 2006;38:1419–28.

[160] Abu-Lail NI. An atomic force microscopy look at the molecular world of living bacteria. In: Camesano TA, Mello CM, editors. Microbial surfaces: structure, interactions, and reactivity, ACS Symposium Series 984. Washington, DC: American Chemical Society; 2007. p. 133–62.

[161] Liu YT, Strauss J, Camesano TA. Thermodynamic investigation of *Staphylococcus epidermidis* interactions with protein-coated substrata. Langmuir 2007;23:7134–42.

[162] van Hoogmoed CG, Dijkstra RJ, van der Mei HC, Busscher HJ. Influence of biosurfactant on interactive forces between mutans Streptococci and enamel measured by AFM. J Dent Res 2006;85:54–8.

[163] Sarghini S, Paulussen S, Terryn H. Atmospheric pressure plasma technology: a straight forward deposition of antibacterial coatings. Plasma Process Polym 2011;8:59–69.

[164] Liu YT, Black MA, Caron L, Camesano TA. Role of cranberry juice on molecular-scale surface characteristics and adhesion behavior of *Escherichia coli*. Biotechnol Bioeng 2006;93:297–305.

[165] Hayden O, Bindeus R, Dickert FL. Combining atomic force microscope and quartz crystal microbalance studies for cell detection. Meas Sci Technol 2003;14:1876–81.

[166] Rizzello L, Sorce B, Sabella S, Vecchio G, Galeone A, Brunetti V, et al. Impact of nanoscale topography on genomics and proteomics of adherent bacteria. ACS Nano 2011;5:1865–76.

[167] Adams EL, Almagro-Moreno S, Boyd EF. An atomic force microscopy method for the detection of binding forces between bacteria and a lipid bilayer containing higher order gangliosides. J Microbiol Methods 2011;84:352–4.

[168] Anderson AJ, Britt DW, Johnson J, Narasimhan G, Rodriguez A. Physiochemical parameters influencing the formation of biofilms compared in mutant and wild type cells of *Pseudomonas chlororaphis* O6. Water Sci Technol 2005;52:21–5.

[169] Lewandowski Z, Beyenal H, editors. Methods for imaging and quantifying the structure of biofilms, in biofilms in the food and beverage industries. Woodhead Publishing; 2009.

[170] Beyenal H, Lewandowski Z, Harkin G. Quantifying biofilm structure: facts and fiction. Biofouling 2004;20:1–23.

[171] Klayman BJ, Klapper I, Stewart PS, Camper AK. Measurements of accumulation and displacement at the single cell cluster level in *Pseudomonas aeruginosa* biofilms. Environ Microbiol 2008;10:2344–54.

[172] Rani SA, Pitts B, Beyenal H, Veluchamy RA, Lewandowski Z, Davison WM, et al. Spatial patterns of DNA replication, protein synthesis, and oxygen concentration within bacterial biofilms reveal diverse physiological states. J Bacteriol 2007;189:4223–33.

[173] Bjorkoy A, Fiksdal L. Characterization of biofouling on hollow fiber membranes using confocal laser scanning microscopy and image analysis. Desalination 2009;245:474–84.

[174] Merod RT, Warren JE, McCaslin H, Wuertz S. Toward automated analysis of biofilm architecture: bias caused by extraneous confocal laser scanning microscopy images. Appl Environ Microbiol 2007;73:4922–30.

[175] Wagner M, Ivleva NP, Haisch C, Niessner R, Horn H. Combined use of confocal laser scanning microscopy (CLSM) and Raman microscopy (RM): investigations on EPS – Matrix. Water Res 2009;43:63–76.

[176] Sauders BD, Wiedmann M, Desjardins M, Fenlon C, Davenport N, Hibbs JR, et al. Recurrent *Listeria monocytogenes* infection: relapse or reinfection with a unique strain confirmed by molecular subtyping. Clin Infect Dis 2001;33:257–9.

[177] Larsen P, Olesen BH, Nielsen PH, Nielsen JL. Quantification of lipids and protein in thin biofilms by fluorescence staining. Biofouling 2008;24:241–50.

[178] Schmid T, Panne U, Haisch C, Niessner R. Biofilm monitoring by photoacoustic spectroscopy. Water Sci Technol 2003;47:25–9.

[179] Chen MY, Lee DJ, Tay JH, Show KY. Staining of extracellular polymeric substances and cells in bioaggregates. Appl Microbiol Biotechnol 2007;75:467–74.

[180] GrayMerod R, Hendrickx L, Mueller LN, Xavier JB, Wuertz S. Effect of nucleic acid stain Syto9 on nascent biofilm architecture of *Acinetobacter sp* BD413. Water Sci Technol 2005;52:195–202.

[181] Greenspan SL, Beck TJ, Resnick NM, Bhattacharya R, Parker RA. Effect of hormone replacement, alendronate, or combination therapy on hip structural geometry: a 3-year, double-blind, placebo-controlled clinical trial. J Bone Miner Res 2005;20:1525–32.

[182] Nancharaiah YV, Venuopalan VP, Wuertz S, Wilderer PA, Hausner M. Compatibility of the green fluorescent protein and a general nucleic acid stain for quantitative description of a *Pseudomonas putida* biofilm. J Microbiol Methods 2005;60:179–87.

[183] Priester JH, Horst AM, Van De Werfhorst LC, Saleta JL, Mertes LAK, Holden PA. Enhanced visualization of microbial biofilms by staining and environmental scanning electron microscopy. J Microbiol Methods 2007;68:577–87.

[184] McNamara CJ, Lemke MJ, Leff LG. Underestimation of bacterial numbers in starvation-survival mode using the nucleic acid stain DAPI. Arch Hydrobiol 2003;157:309–19.

[185] Kim J, Pitts B, Stewart PS, Camper A, Yoon J. Comparison of the antimicrobial effects of chlorine, silver ion, and tobramycin on biofilm. Antimicrob Agents Chemother 2008; 52:1446–53.

[186] Wetzel RG, Ward AK, Stock M. Effects of natural dissolved organic matter on mucilaginous matrices of biofilm communities. Arch Hydrobiol 1997;139:289–99.

[187] Fowler SD, Greenspan P. Application of Nile red, a fluorescent hydrophobic probe, for the detection of neutral lipid deposits in tissue sections: comparison with oil red O. J Histochem Cytochem 1985;33:833–6.

[188] Frank KL, Patel R. Poly-N-acetylglucosamine is not a major component of the extracellular matrix in biofilms formed by icaADBC-positive *Staphylococcus lugdunensis* isolates. Infect Immun 2007;75:4728–42.

[189] Daims H, Lucker S, Wagner M. Daime: a novel image analysis program for microbial ecology and biofilm research. Environ Microbiol 2006;8:200–13.

[190] Heydorn A, Nielsen AT, Hentzer M, Sternberg C, Givskov M, Ersboll BK, et al. Quantification of biofilm structures by the novel computer program COMSTAT. Microbiology Uk 2000;146:2395–407.

[191] Lewandowski Z, Beyenal H. Fundamentals of biofilm research. Boca Raton, FL: CRC Press – Taylor & Frances Group; 2007.

[192] Milferstedt K, Pons MN, Morgenroth E. Textural fingerprints: a comprehensive descriptor for biofilm structure development. Biotechnol Bioeng 2008;100:889–901.

[193] Zapico P, de Paz M, Medina M, Nunez M. The effect of homogenization of whole milk, skim milk and milk fat on nisin activity against *Listeria Innocua*. Int J Food Microbiol 1999;46:151–7.

[194] Mueller LN, de Brouwer JFC, Almeida JS, Stal LJ, Xavier JB. Analysis of a marine phototrophic biofilm by confocal laser scanning microscopy using the new image quantification software PHLIP. BMC Ecol 2006;6.

[195] Daims H, Wagner M. Quantification of uncultured microorganisms by fluorescence microscopy and digital image analysis. Appl Microbiol Biotechnol 2007;75:237–48.

[196] Milferstedt K, Pos MN, Morgenroth E. Analyzing characteristic length scales in biofilm structures. Biotechnol Bioeng 2009;102:368–79.

[197] Ponsa P, Amante B, Roman JA, Oliver S, Diaz M, Vives J. Higher education challenges: introduction of active methodologies in engineering curricula. I J Eng Educ 2009;25: 799–813.

[198] Xavier JB, White DC, Almeida JS. Automated biofilm morphology quantification from confocal laser scanning microscopy imaging. Water Sci Technol 2003;47:31–7.

[199] Lowe DR, Byerly GR. Ironstone bodies of the Barberton greenstone belt, South Africa: products of a Cenozoic hydrological system, not Archean hydrothermal vents!. Geol Soc Am Bull 2007;119:65–87.

[200] Daims H, Wagner M. In situ techniques and digital image analysis methods for quantitative spatial localization patterns of nitrifiers and other microorganisms in biofilm and flocs,

in methods in Enzymology. Pt B,496. In: Klotz MG, Stein LY, editors. Research on nitrification and related processes, vol. 46; 2011. p. 185–215.
[201] Haagensen JAJ, Regenberg B, Sternberg C. Advanced microscopy of microbial cells. In: Muller S, Bley T, editors. High resolution microbial single cell analytics, vol. 124; 2011. p. 21–54.
[202] Renslow R, Lewandowski Z, Beyenal H. Biofilm image reconstruction for assessing structural parameters. Biotechnol Bioeng 2011;108:1383–94.
[203] de Carvalho C, da Fonseca MMR. Assessment of three-dimensional biofilm structure using an optical microscope. Biotechniques 2007;42:616–20.
[204] Luef B, Neu TR, Zweimuller I, Peduzzi P. Structure and composition of aggregates in two large European rivers, based on confocal laser scanning microscopy and image and statistical analyses. Appl Environ Microbiol 2009;75:5952–62.
[205] Rodriguez SJ, Bishop PL. Three-dimensional quantification of soil biofilms using image analysis. Environ Eng Sci 2007;24:96–103.
[206] Okabe S, Kuroda H, Watanabe Y. Significance of biofilm structure on transport of inert particulates into biofilms. Water Sci Technol 1998;38:163–70.
[207] Jackson G, Beyenal H, Rees WM, Lewandowski Z. Growing reproducible biofilms with respect to structure and viable cell counts. J Microbiol Methods 2001;47:1–10.
[208] Xavier J d B, Picioreanu C, van Loosdrecht MCM. A general description of detachment for multidimensional modelling of biofilms. Biotechnol Bioeng 2005;91:651–69.

Chapter 6

In Vivo Characterization of Biomaterials

Samit K. Nandi* and Subhasish Biswas†

*Department of Veterinary Surgery and Radiology, West Bengal University of Animal & Fishery Sciences, Kolkata, India, †Department of Livestock Products Technology, West Bengal University of Animal & Fishery Sciences, Kolkata, India

Chapter Outline

Characterization of Biomaterials. http://dx.doi.org/10.1016/B978-0-12-415800-9.00007-3

6.1. INTRODUCTION

Biomaterials are naturally occurring items intended to interface with living cells to evaluate, treat, augment or replace any tissue, organ or function of living systems. Besides the above widespread utility, the use of these materials are ever expanding covering wide range of area starting from medical use to the use in the emerging fields like food engineering and nanotechnology. Whatever may be the use and application, the essential prerequisite to entertain a material as a biomaterial is it should be biocompatible and biologically sustainable in order to get guarantee over the objective for which it is being used. Biocompatibility of a material is very host specific and is dependent on its end use application. There is a wide range of materials envisaged within the list of materials as biomaterials. It starts with metals, ceramics, glasses, polymers, etc. The present inclusion of number of items such as some derivatives of marine animals, insects, and some plant-origin materials like herbs are also expanding the list of biomaterials. The list cannot be completed as the scope and research on these aspects of utilizing the biomaterials have been very extensively investigating for medical and related applications. The identification of synthesizing or modifying polymers is making this area very interesting where different angels of science are working together for the benefit of minimizing human suffering to have better life.

The development of the application of biomaterial dates back thousands of years where the inclusion of process of development in polymeric biomaterials could be considered as a remarkable process. The application of synthetic polymers in medicine is, however, a recent aspect of study. The highest development in this area could be achieved in 1940s during the Second World War and it was for the need of curing orthopaedic problems besides other health abnormalities starting from vascular graft, breast and facial prosthesis, blood–device interfaces, etc. In spite of all these developments during the later half of the twentieth century, scientist attempts for developing novel polymeric materials with better biocompatibility and mechanical properties suitable for the need of biomedical application. The progress was more potent when it has been clubbed with the biotechnology and pharmaceutical science where development of materials could be started and to be preferred for *in vivo* application. The serious questions on biostability, long-term biocompatibility, and biodegradability had also been suitably redressed with the views of emerging issues like tissue engineering and drug delivery system. This shift in biomaterial research is due to the constant observation and a material use on the patients. The *in vivo* observations have given the thought for identifying metabolic transformation within the system during the process of convalescence leading to get objective notes for further improvement on material selection and in minimizing adverse body reaction through metabolic outputs. The origin of bacterial polyesters and polyamides as a very suitable material qualifying a number of desirable characteristics like non-toxic, biocompatible, biodegradable material found a variety of applications in biomedical research.

The system interaction between cellular microenvironment and the rest of the body needs to be understood when the material to be placed *in vivo* for any purpose. Sometimes, it is also compared with the observations of performance differences in engineered tissues both *in vivo* and *in vitro*. The question of tissue signals generated, metabolic product accumulation, and the hormone appearances should be adequately mimicked at a micro-bioreactor level at the site of application of the materials to have a meaningful prediction to the total success of the item to be placed for the purpose. Some scientists are in opinion that to get a total satisfaction on viability and functionality in question of vascularization and its development at the site of application of biomaterial, a toxicity study could be undertaken using cell culture analog concept at molecular level.

In controlling microenvironments of the tissue systems, besides other important aspects of biomimetic are important elements considered to be emphasized to understand the processes that permitting the tissue response towards a particular biomaterial. It is an inherent concept in natural processes that has lead to a broad spectrum of new process and materials modelled on mimetic system. The other fields like in pollution prevention, the fundamentals of biomimicry could be successfully acceptable and used. The key properties on which the living system relies include multifunctionality, hierarchical organization, adaptability, reliability, self-regulation and reparability. All these properties could be incorporated into a system that mimics living processes, other way this can be called as *intelligence*. Molecular design has permitted the fabrication of items that can mimic active transport of ions and non-invasive sensors have also been designed to monitor *in vivo* glucose concentration and carrier molecules have been developed for controlling the release of pharmaceuticals. These successes are often accomplished through multidisciplinary approach where morphology and properties development and molecular engineering of the thin films and biotechnology and engineering biosciences are the major organizational themes. The morphology, properties development and molecular engineering part are two important aspects of biomimetics. Polymer blends, new materials which are introduced as biomaterials, have created a rapid development in biomimetrics. The phase behavior, phase architecture and morphology and interfacial properties of polymer blends are the unique aspects in morphology and property development of the materials. These are very useful for developing artificial organs, the cardiovascular items, orthopaedics, dentistry and drug delivery system. There may have some biological deficiency particularly in synthetic polymers where enhancement of mechanical characteristics on interpenetrating network is a matter of great importance. Preparations fabricated from copolymerization of biopolymers like fibrin, collagen and hyaluronic acid with synthetic polymers like polyurethane, polyacrylic acid and polyvinyl alcohol are now developed, having adequate morphological properties for development of adequate need of application considering biomimetics. The question of molecular engineering is focused for thin film design and to develop fundamental understanding of organic thin film composites and molecular design of membrane system. Therefore, all these in together should be viewed for better understanding of biomimetics and its interaction with the tissue system when applied *in vivo*.

6.2. IDEAL CHARACTERISTICS OF BIOMATERIAL FOR *IN VIVO* APPLICATION

- Composition and mechanical properties should be equivalent to that of the bone.
- Should have a high resistance to wear and a modulus of elasticity similar to that of the bone to minimize bone resorption around the implant.
- Adequate biocompatibility (immunologically compatible), tailorable biodegradability, ability to initiate osteogenesis; in short, the material should closely mimic the natural bone.
- Do not elicit undesirable side effects including infection and immune response.
- Adequate porosity for sustained delivery of drugs, cell seeding, migration and distribution since these features help in directing tissue formation and function. Porosity is also responsible for the diffusion of oxygen and nutrients in the three-dimensional (3D) structure of the scaffold.
- Once applied in living system, it should not be removed by second surgery upon completion of task.
- Should have good osseointegration properties, i.e. chemical (bonding osteogenesis) or physicochemical (connective tissue osteogenesis) bond between implant and bone.
- Should have excellent corrosion resistance in the physiologic milieu.

6.3. ANIMAL MODEL IN ORTHOPAEDIC SURGERY

All the animal experiments are to be carried out according to the Laws of animal welfare and prior approval by the Institutional animal Ethics Committee.

The experimental animals are to be kept in stalls/cages prior to surgery. Food has to be withdrawn 36 h, and water 6 h prior to anaesthesia. A pre-anaesthetic examination is to be performed including haematology and a chemscreen consisting of liver enzymes, urea, creatinine, protein, albumin, sodium and potassium. After recovery of the animals from anaesthesia, they should be initially kept in small groups in stables/respective cages. After 10 days, the skin sutures would be removed and the animal's, especially small ruminants like sheep, goat, etc., would be allowed to join the flock on the fields. They should be checked daily for lameness and additional clinical problems. Food and water are to be given ad libitum.

Surgery: Animals are to be sedated with standard anaesthetic protocol and to be received as pre- and post-operative prophylaxis 30,000 IU penicillin/kg body weight (BW) (Hoechst AG, Germany) and 6 mg gentamicin/kg BW intravenously twice a day. In addition, small ruminants are to be received subcutaneously 500 units of equine tetanus serum as a single application however; this tetanus injection is not required for laboratory animals. Analgesia is to be maintained through injection of 0.01 mg buprenorphine/kg BW i.v. peri-operatively and additionally 4 mg carprofen/kg BW i.v. (Rimadyl®, Pfizer Inc., NY, USA) post-operatively for 3 days. The area in the region of the incisions is to be clipped and disinfected in the standard manner. The animals are to be placed in left or right lateral recumbency on the operating table with the limbs placed in a horizontal position. The approach to the bone should always be from the lateral side if operation is carried out involving femur or humerus bone. Otherwise, medial approach is better especially in radius–ulna and tibia bone. After standard surgical approaches in the specified location, drilling of bone is to be done and immediately after drilling, the holes are to be filled with test biomaterials. After filling the drill holes, the different wound layers are to be closed separately with non-resorbable suture material as per standard technique. Immediately after surgery, the exact locations of the material-filled holes could be determined using radiographs. Time point of sacrifice depends upon the material to be tested.

In addition to drill holes defect, critical size bone defect can be made by removing a piece of whole thickness of cortex by micromotor dental drill measuring little more than the size of the implant. Cold, sterile normal saline is to be poured continuously over the site of the bone defect during drilling to prevent excessive production of heat during surgery. The area is to be flushed with few drops of gentamicin solution and then, the respective materials would be placed within the created bed. The implants are to be secured by suturing the periosteum, muscle, and subcutaneous tissue using chromic catgut. The wound have to be protected with a protective dressing and bandages until removal of skin sutures (Fig. 6.1).

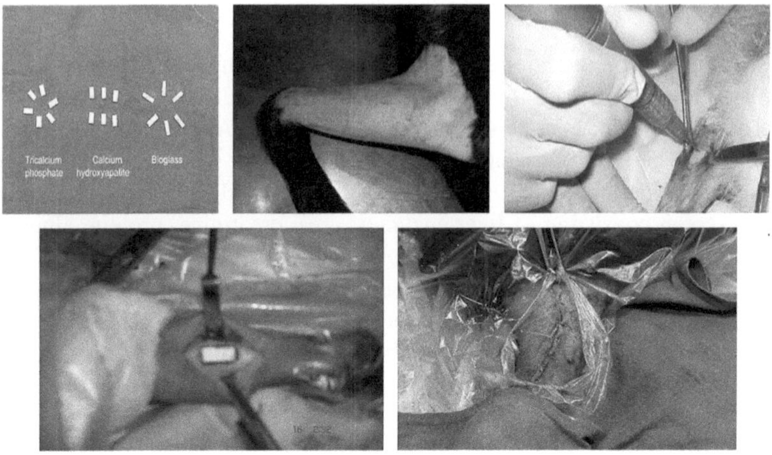

FIGURE 6.1 Rectangular blocks of bioceramics sample along with surgical operations for placement of bioceramics blocks in radius bone of goat.

6.3.1. Animal Model in Dental Surgery for Characterization of Biomaterials

The development and modification of dental implants are continuous process for many years in an effort to create an optimal interaction between the body and the implanted material. Periodontitis is one of the common problems in human teeth where the breakdown of the tooth-supporting structures until tooth loosening is the ultimate sequel. Scaling and root-planning together with rigorous oral hygiene are the general therapeutic strategies to arrest the progression of the disease. Appropriate experimental animal models are needed for testing, validating and subsequent development of newer therapeutic means using biomaterials. In this context, various species of experimental animal models can be used for testing of varieties of biomaterials. Although number and size of implants to be tested will influence directly the species of animal chosen for a study, laboratory animals can be suitably used for such situation.

Experimental periodontal defects in animal model may be of acute defect model, the chronic defect model and combination of both. Acute model consists of creation of surgically induced defect by removing all the periodontal components like bone, cementum and periodontal ligament along with control defect. In the chronic model, defects can be created by placing orthodontic elastics, silk sutures or ligatures around teeth for a period of 12–20 weeks. In the combined acute/chronic model, the defects are surgically created followed by placing of ligatures to make sure about calculus accumulation and to check spontaneous regeneration of the defects. The alveolar crest of the ramus bone can be utilized for creation of periodontal and osseous defects. The *in vivo* characterization parameters of dental material include histology, radiography, scanning electron microscopy (SEM), immunohistochemistry, etc.

6.4. ANIMAL MODELS IN SPINAL SURGERY AND CHARACTERIZATION OF BIOMATERIALS

The pathological condition of spine such as spondylosis, scoliosis, spondylolisthesis, tumour, infection, post-traumatic fracture and instability may all lead to an abnormal correlation between adjacent vertebral structures. Accordingly, the aim of fusion models is to stabilize the spinal column by removing intervertebral articulations and positioning the segments in proper, mechanically advantageous alignment which is a challenging job to the spinal surgeons. Clinical trials usually are not sufficient to get optimal outcome due to uncontrolled variables of the host that include systemic factors such as age, steroid use, nutritional depletion, osteoporosis and nicotine abuse, including idiopathic variations in their inherent ability that have an impact on bone healing [1]. Therefore, data resulting from clinical trials are insufficient to interpret for the use of tissue-engineered bone constructs, which needs lot of pertinent information.

In spinal fusion surgery, animal models have been used to assess biologic factors since 1913. These models have been extensively used to establish proof of concept, possibility and effectiveness of tissue-engineered constructs ranging from bone grafts, bone graft substitutes, growth factors and, more recently, gene therapy-based constructs. These validated models provide relatively reproducible and quantifiable data that are otherwise impracticable to obtain from human clinical trials, cadaver studies or computer simulations. Moreover, in animals, the local and biologic environment are subjected to minimal variation and surgical techniques are controlled in a better way keeping local conditions more consistent [2,3].

Historically, the rabbit model has been used initially for testing of biologics in connection to spine. Small animals like mouse and rat were avoided owing to their small anatomy [4]. Larger animals provide better surgical simulation of the human, but are more expensive, and more time is required to arrive at conclusion. Small animal models are now gaining importance for investigational studies since they are less expensive and heal rapidly, which shortens time to outcome. Of late, the rat posterolateral intertransverse fusion model has been used for initial testing of biologics and scaffolds in the spine [5]. Mouse model is less frequently used [6]. Transgenic small animals are also now promising model for the testing of biologic factors in many different conditions [7,8].

In spine models, fusion rate and the amount and quality of bone formation (BF) are the key measures and can be determined by radiographs, computed tomography (CT) and manual assessment of the excised spine. BF can be objectively quantified with CT [9,10]. Strength and union correlate with the histological cellular matrix composition of the fusion mass [11]. Immunohistochemistry may further illustrate cell phenotypes [4]. CT definitions of trabecular number, connectivity, and mineral density are compared to test mechanical properties, modulus, strength, and fatigue resistance [12]. Further, biologic outcome may be simply evaluated with quantitative and qualitative histologic evaluation and mechanical integrity of the fusion construct using validated destructive and non-destructive mechanical testing protocols.

Fusion area	Animal species	Time point of evaluation	Characterization parameters	Investigator
Anterior cervical spinal fusion	Sheep	12 weeks	• Radiological examination—before and after surgery (day 0), at 1, 2, 4, 8 and 12 weeks. • The disc space height, intervertebral angle and lordosis angle at the same time points. • Histomorphologic and histomorphometric analysis at the end of the study to determine the new bone formation. • Polychrome sequential labelling to determine the new bone formation. • CT scans to determine bone mineral density, bone mineral content and bony callus volume.	Kandziora et al. [13]
Anterior cervical spinal fusion and decompression	Goat	12 weeks	• Histological examination • Biomechanical stiffness testing • Fluorochrome analysis to assess the rate of bone growth	Zdeblick et al. [14–17]
Anterior lumber spinal fusion	Sheep Monkey	6 months	• Histological examination • Radiological examination • CT scan findings	Sandhu et al. [18] Boden et al. [19]
Posterior lumber spinal fusion	Rat	4 weeks	• Histological examination • Biomechanical stiffness testing • Radiological examination	Boden et al. [20] Louis-Ugbo et al. [21]
	Rabbit	5 weeks		
	Canine	3 months		Sandhu et al. [22]
	Sheep	6 months	In addition to above, X-ray absorptiometry to determine mineral densities.	Blattert et al. [23]

6.5. CHARACTERIZATION PARAMETERS

6.5.1. Biochemical Investigation

Bone healing is a local process that has an effect on systemic mineral homeostasis. The latter involves vitamins, hormones, enzymes and other factors. Bone healing is mainly monitored by physical and serial radiographic examinations of the fracture site. Apart from the radiological survey of bone callus formation, some serum biochemical markers are successfully used for this purpose [24,25]. This complex is characterized by a close inter-relationship of bone resorption on one hand and BF on the other, biochemical markers that include both processes together with the primary macroelements that build the inorganic bone matrix.

Some biochemical markers of BF, such as serum alkaline phosphate (ALP) [26], serum calcium [27] and serum inorganic phosphorus [28] activity, may be useful in evaluating the progress of healing. In an experimental procedure of biomaterial implanted bone, these activities can be monitored on day 0, 3, 7, 15, 30, 60, 90 post-operatively in a 3 month study. In general, there is a significant increase of serum calcium and serum phosphorus during early stages of fracture healing [29,30]. Similarly, during the fracture healing, the total serum ALP activity increased significantly as early as the 10th day and by day 30, returned to initial values [24]. The increase in ALP activity may be suggestive of an increase in osteoblastic proliferation at and around the implant during initial phase of fracture healing and also related to abundant fibrous tissue formation [31–33]. ALP was reported to release inorganic phosphate from phosphoric esters with consequent precipitation of a compound of calcium and phosphorus in the matrix of fracture site [34]. Besides, the concentration of bone-specific ALP assay can be determined using an Alkphase B™ kit (Metra Biosystems Inc., Mountain view, CA) and absorbance data can be analyzed using EIA smart software.

Apart from ALP activity, osteocalcin could be an earlier marker to quantify the osteogenic responses [35]. The ALP that localizes on the cell membrane of the osteoblasts and osteocalcin is produced exclusively by osteoblasts, making these osteoblastic markers. The concentration of osteocalcin can be measured using a Novocalcin™ kit (Metra Biosystems Inc., Mountain view, CA) and absorbance data can be analyzed using EIA smart software.

Besides, the level of tartrate-resistant acid phosphatase (TRAP) activity in post-implanted serum of animal can be measured using an acid phosphatase reagent kit. This activity is also helpful to determine the amount of osteoclasts within the explants from the reconstructed defects. TRAP staining can be performed in deparaffinised and rehydrated paraffin sections as per the protocol of Manufacturer (Sigma, Deisenhofer, Germany).

TRAP Staining Procedure:

- Preheat to 37 °C two Coplin jars with 50 ml of Buffer I solution each.
- Take one Coplin jar and add 0.5 ml of Buffer II solution, add slides and incubate at 37 °C for 45 min.
- A few minutes before the time is up, mix 1 ml of Buffer III solution and 1 ml of Buffer IV solution for 30 s and let it sit for 2 min.
- Add this mixed solution to the other Coplin jar of 50 ml Buffer I solution, mix and add the slides without rinsing.
- Incubate at room temperature for ~5 min.
- Rinse, counterstain with haematoxylin for 40 s and put slides in 0.05% ammonia water.
- Dehydrate through alcohol, clear in xylene and cover slip.

Osteoclasts————————— Bright red
Background ——————— Green

Buffer I solution: Sodium acetate (anhydrous)—9.2 g; L-(+)tartaric acid—11.4 g; distilled water—950 ml; glacial acetic acid—2.8 ml; dissolve and adjust pH to 4.7–5.0 with 5 M sodium hydroxide; bring the total volume to 1 L.
Buffer II solution: Naphthol AS-BI phosphate (store at −20 °C)—0.1 g; ethylene glycol monoethyl ether—5 ml.
Buffer III solution: Sodium nitrite—1 g; distilled water—20 ml.
Buffer IV solution: Pararosaniline chloride—1 g; 2 N HCL—20 ml; heat to 60 °C but do not boil and filter through kimwipe.

6.5.2. Radiological Investigation

Radiography is one of the vital tools in characterization of *in vivo* orthopaedic biomaterials both before and after implantation and useful to assess the union at the host bone–implant interfaces during the follow-up [36,37]. Radiographic evaluation is generally focused on the continuity between host bone and implant, and the change of implant radiopacity. The radiographs are rather low in resolution level and some areas close to the implant can be difficult to access when the complete host bone is imaged. A distinct radiolucent zone at the interface between implants and the host bone is generally seen on the immediate post-operative radiographs. The absence of this radiolucent zone is considered to be an indication of union between the implant and the host bone. With time, the radiolucent site gradually decreased and the radiolucent fracture line disappeared. In resorbable implants, the density of bone decreased more with the passage of time and bone remodelling [38]. In plain X-ray, calcium phosphate biomaterials like beta-tricalcium phosphate and hydroxyapatite are radio-opaque. Most BF occurs beneath the biomaterials because of osteoconductive effect. Plain X-ray is usually used for monitoring any dislocation of biomaterial *in vivo*. In spinal

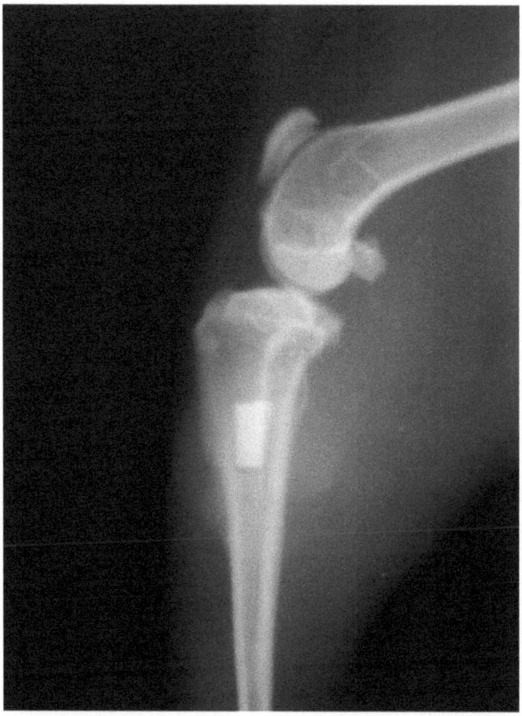

FIGURE 6.2 Radiographic image showing presence of rectangular implant in critical sized defect of proximal part of tibia bone in animal.

application of biomaterial, high-resolution radiography is used for precision measurements of fusion mass and fusion rate of fusion complex after sample harvesting.

Lateral and anterioposterior (A/P) radiographs of implanted limb are generally taken before and immediately after surgery as well as 2, 4, 6, 8, 12, 16 weeks if the study protocol is of 4 months. The radiographic factors are 14 mAs; 50 Kvp and 90 cm FFD for rabbit model (Fig. 6.2). Two persons independently and blindly can evaluate all the radiographs with regard to the presence of osseous union at each host bone–implant interface as indicated by obliteration of the transverse radiolucent line between the implant and the host bone that was observed immediately post-operatively, the presence of osseous callus overlying the implant, and the presence of periosteal new bone spanning the whole defect from the distal to the proximal host bone–implant interface. The time at which the callus around the implant was the thickest should also be noted. Calipers can be used to measure the thickness of the callus on the medial aspect of the implant in three locations: the proximal and distal host bone–implant interfaces and the middle portion of the implant. The mean thickness of the callus is to be calculated at 4, 8, 12, and 16 weeks.

6.5.2.1. Some Key Points to be Considered During Interpretation of Radiograph

- Immediately post-operatively, the implants of calcium phosphate have a granular appearance because the pores were empty.
- With the progression of new BF, the appearance of the implant became smoother and more radio-opaque.
- A distinct radiolucent zone at the interface between the implant and the host bone can be visible on the immediate post-operative radiographs.
- The absence of this radiolucent transverse zone is considered to be an indication of union between the implant and the host bone.
- Stabilization of fracture by increasingly radiodense bone within the implant or surrounding the implant, or both.
- More degree of radiopacity of implants, more progression of formation of bone.

Bone union (BU), callus formation and bone resorption can be estimated from the radiographs. BU can be defined as either disappearance or partial bridging of the gap between bone and the implant. The scoring system proposed by Johnson et al. (1996) can be used for evaluation of BU in which proximal union is graded as 0–3 and distal union as 0–3 and accordingly, the highest possible score for BU is 6 [39]. BF can also be scored, the maximum score being 4. The combined score (BU + BF) refers to the sum score for BU and BF, the maximum score being 10. The radiograms can be digitized with a ccd camera and the area of callus and can be evaluated from the lateral view radiograms using a digital image analysis system. Bridging of the fracture callus has been described by Schmidmaier et al. [40], using the scoring system by two individual observers: (a) complete bridging (four cortices bridged), (b) incomplete bridging (one to three cortices bridged) and (c) no bridging (no cortex bridged).

6.5.3. X-ray Computed Microtomography (Micro-CT) Analysis

Conventional histology is the relatively easiest methodologies for qualitative analysis of new BF in the implants. Conversely, 3D structure data and quantitative analysis are difficult to obtain. A computer-based method for the automated quantification of bone tissue in 2D histological sections of decalcified specimens has been described [41]. Even though, a 3D structure of the new bone could be derived by analysis of serial sections of the implant. This approach is not useful since the histological treatment of the sample by decalcification undergoes before analysis and the technique does not permit determination of the volume of the remaining scaffold.

Micro-CT is identified to be a unique technique for the non-invasive, non-destructive 3D characterization of materials in medicine, material science and biology. It is a 3D radiographic imaging technique, used in medical and industrial applications. Unlike such systems, which typically have a maximum spatial resolution of about 1 mm^3, micro-CT is capable of achieving a spatial resolution of the order of 1 μm^3. In particular, X-ray micro-CT associated with X-ray synchrotron radiation is a possible experimental technique to explore BF in porous scaffolds [42–45]. The technique is similar to conventional CT, but several advantages come from the use of synchrotron radiation [43]. Synchrotron radiation provides the option to keeping selected X-rays with a small energy bandwidth from the wide and continuous energy spectrum, maintaining at the same time the photon flux rate high enough for efficient imaging. High-spatial resolution images are generated from 10 μm to 1 μm with high signal-to-noise ratio [43,44] and finally, 3D images of the sample can be obtained [42]. This methodology offers the possibility of carrying out a complete experimental plan for investigation of the influence of numerous parameters such as pore size and spatial distribution with regard to the 3D imaging of newly formed bone and of the scaffold especially in bioceramic, together with quantitative information. This also allows establishing the efficiency of BF in scaffolds with different microstructural parameters or under different circumstances such as variable sources of osteogenic cells, increased or decreased mechanical loading, etc. Ultimately, new scaffolds can be tested in large animal models before their routine clinical application [46]. Micro-CT study is a recent imaging technique and has great potential in characterization of biomaterials in bone tissue engineering [47].

6.5.3.1. Methodologies for Micro-CT Analysis

The bone specimens consisting of healed test bone and adjacent host bone are to be collected from the animals at different time points post-operatively. Sample preparation includes fixing in 5% E. M. grade glutaraldehyde phosphate solution for 24–48 h and subsequently washing twice for 30 min initially with phosphate buffered saline (PBS) (pH 7.4) followed by distilled water. The sample is to be dehydrated thereafter in serial concentration of ethanol (30–100%) for 30 min each. Specimens will be first dried and then fixed with hexamethyldisilazane (HMDS) for 20 min. All evaluations are to be done on a micro-CT 50 imaging system (Scanco Medical, Brüttiselen, Switzerland) using the following parameters: resolution: 10 μm; energy: 55 kVp; intensity: 145 μA; frame averaging: 2.

6.5.4. Tissue–Materials Interactions—Histology and Histomorphometric Analysis

6.5.4.1. Histology with Decalcified Bone Sections

Histological observations provide more detailed knowledge about the cellular events during incorporation of different types of implants. For this study, the implanted bone/implant is to be collected from the animals at each time point after sacrifice. The section is to be cut (3–4 mm thick) including both normal and implanted area of bone. The bone piece is to be washed thoroughly with

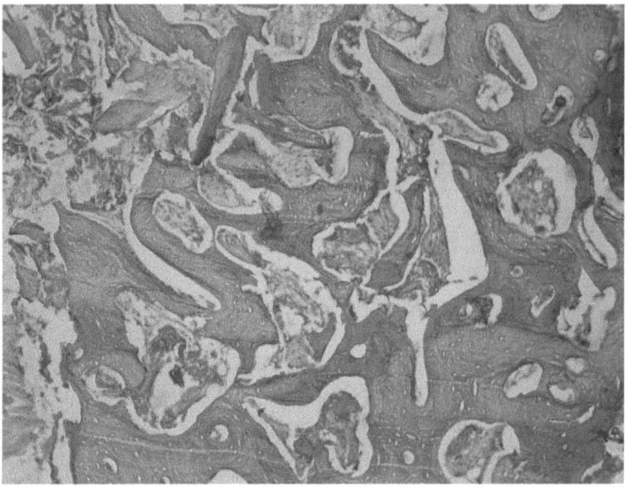

FIGURE 6.3 Histological images of decalcified implant-bone after sacrificing of animal.

normal saline and fixed in 10% neutral formaldehyde. Subsequently, bones are to be decalcified in 10% EDTA or Goodling and Stewart's fluid containing formic acid 15 ml, formalin 5 ml and distilled water 80 ml solution or 0.1 N hydrochloric acid and it should be stirred daily and changed once in 3 days. The section is to be checked regularly for the status of decalcification. They are considered as completely decalcified when sections become flexible, transparent and easily penetrable by pin. The decalcified tissues are embedded in paraffin and 4–6 μm sections are to be cut using a microtome. The sections are to be mounted on a slide, stained using haematoxylin and eosin and finally viewed using light microscope equipped with a digital camera connected to a computer to observe for status of the bone implants and cellular response of host bone to the implants [48] (Fig. 6.3).

6.5.4.2. Histology with Undecalcified Bone Sections

Histology with undecalcified bone sections proved to be suitable to assess bone healing as well as the behavior of biodegradable materials. By sectioning the bone samples in midlength and perpendicular to the original defect, the stage and ratio of new BF versus material resorption can be optimally visualized and calculated at its most remote location. Since any biomaterial used in bone finally will be replaced by creeping substitution, it may be carefully understood that if the centre and middle of the original biomaterial are replaced with new bone, the rest of the original bone defect has already been replaced and is already undergoing bone remodelling. Serial sections can be prepared throughout the entire original defect if found necessary although the extensive workload and expenses associated with preparing sections of non-decalcified bone may prove inhibitive.

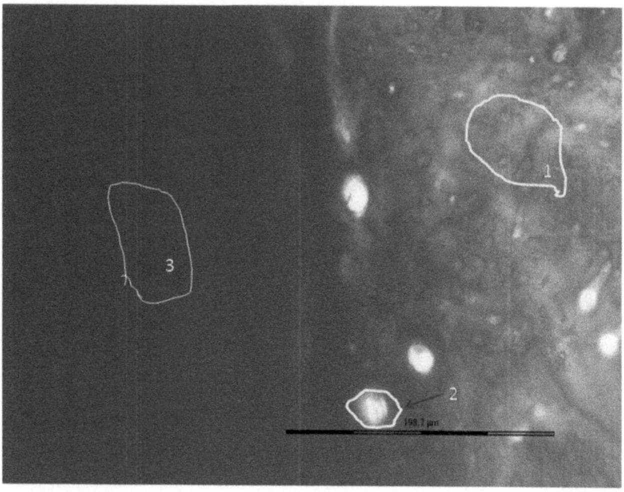

FIGURE 6.4 Undecalcified image of bone-metal interface 1. Mineralized bone, 2. New bone, 3. Metal surface.

For undecalcified histological analysis, bone–implant specimens are to be fixed in 10% buffered formalin solution and dehydrated in a graduated ethanol (70%, 95%, and 100%) series. After embedding samples in Spurr's resin, each undecalcified implant block is to be sectioned perpendicular to the implant surface using a low-speed diamond blade. After polishing, the sections are to be stained by Masson Goldner's trichrome stain and observe under light microscope (Olympus BH-2, Olympus America Inc., USA) (Fig. 6.4).

6.5.4.2.1. Procedure for Fixing and Embedding Bone Tissue with SPURRs Resin

- Fixation: 10% buffered formalin for 60 h.
- Dehydration: 70% ethanol for 60 min, 95% ethanol for 6 h, 100% ethanol twice for 6 h each. Then ethanol:acetone (1:1) for 6 h, 100% acetone twice for 6 h Each.
- Infiltration: acetone:SPURRs (1:1) thrice for 6–8 h each.
- 100% SPURRs thrice for 6–8 h each.
- Embedding: 100% SPURRs at 70 °C for overnight.

[Note: Before changing to 100% SPURRs, leave the glass vial open with acetone:SPURRs (1:1) overnight. This will evaporate the acetone and have a better infiltration of SPURRs (make sure you put enough acetone:SPURRs (1:1) so that when the acetone evaporates, the sample gets fully covered with SPURRs).]

6.5.4.2.2. Staining Resin Sections of Bone [Goldner Trichrome (Modification)]

- Sections to distilled water twice for 15 min each.
- Place sections in Weigert's haematoxylin for 20 min and then wash in water.
- Differentiate with 0.5% acid alcohol followed by wash in water for 20 min.
- Ponceau/acid fuchsin/azophloxine for 5 min and then rinse in 1% acetic acid for 10 s.
- Phosphomolybdic acid/orange G for 20 min and then rinse in 1% acetic acid for 10 s.
- Light green for 5 min, rinse in water and blot dry.
- Rinse in alcohol twice and wash in methyl cyclohexane twice.
- Finally mount in Picomount.

Observations: Nuclei—blue/black; mineralized bone/muscle—green; osteoid/collagen—red.

Composition of solution as above
Weigert's haematoxylin:
Solution A: Haematoxylin 10 g, distilled water 1000 ml—Ripen for at least 2 weeks before use.
Solution B: Ferric chloride (hydrated) 11.6 g, distilled water 1000 ml, 2% hydrochloric acid 10 ml.
For use, mix equal points of A and B immediately before required. Do not keep working solution pre-made.
Ponceau de Xylidine/Acid Fuchsin: Ponceau de xylidine 1.5 g, acid fuchsin 0.5 g, acetic acid (conc) 2 ml, distilled water 98 ml.
Azophloxine: Azophloxine 0.5 g, acetic acid (conc) 0.6 ml, distilled water 99.4 ml.
Ponceau/Acid Fuchsin/AzophloxinE (working solution): Ponceau de Xylidine/acid fuchsin 12 ml, azophloxine 8 ml, 0.2% acetic acid 80 ml. Re-use and keep made up.
Phosphomolybdic Acid/Orange G: Phosphomolybolic acid 6 g, Orange G 4 g, distilled water 1000 ml.
Light green: Light green 2 g, acetic acid (conc) 2 ml, distilled water 1000 ml.

Toluidine blue: Sections to distilled water twice for 15 min, toluidine blue/EDTA for 1 h, wash in distilled water, blot dry (37 °C overnight or 60 °C 8 h), rinse in methyl cyclohexane and finally mount in Picomount.

Observations: Nuclei—blue; mineralized bone—light purple; osteoid colorless—pale blue; mineralized front—light blue; reversal line—dark blue/purple.

6.5.4.3. *Histomorphometry*

Histomorphometry of the bone samples can be carried out using an automatic software program with the appropriately established 'macro' that allows the calculation to measure the area and percentage of bone, fibrous tissue and remnants of biomaterials within the original defect. Histomorphometry can be performed on the entire cancellous bone area including the transitional zone but excluding the endocortical perimeter. For these measurements (MEG), a Zeiss Axiophot microscope (Carl Zeiss, Thornwood, NY, USA) and semiautomatic image analysis software (KS400, Kontron Elektronik, Munich, Germany) can be used.

The primary two-dimensional measures include total tissue area, cancellous bone area, bone perimeter, osteoid perimeter, osteoblast perimeter, osteoid area, osteoclast perimeter, osteoclast number, osteoid width, wall width, double-label perimeter, single-label perimeter, and inter-label width. Necessary magnifications required are (a) tissue and bone areas at 156×, (b) surface-based parameters at 400×, (c) osteoid, wall, and inter-label widths at 100 μm intervals at 625× and is to be corrected for obliquity. All structural indices are to be calculated according to the modified American Society for Bone and Mineral Research (ASBMR) histomorphometry nomenclature [49].

Structural parameters resulting from the primary measurements include cancellous bone volume (BV/TV; percentage of tissue area), trabecular thickness (Tb.Th), and trabecular number (Tb.N). Static parameters of formation and resorption resulting from the primary measurements include osteoid volume (OV/BV; % of bone area), osteoid surface (OS/BS), osteoblast surface (Ob.S/BS), and osteoclast surface (Oc.S/BS) measured as a percentage of the bone surface. The number of osteoclasts (N.Oc/T.Ar) can be measured per millimetre square tissue area. Osteoclasts may be defined as multinucleated, light-blue-stained cells containing foamy cytoplasm lying close to an eroded lacuna or on the bone surface.

Dynamic parameters resulting from the primary measurements are as follows. Mineralizing surface (MS/BS) calculated as the percentage of bone surface with double plus 0.5 single labels. The surface-based bone formation rate (BFR/BS; mm^3/mm^2/year) can be calculated as [(MAR × 0.365) × MS/BS]/100, where the mineral apposition rate (MAR) can be determined by dividing the inter-label width by the interval labelling time. BFR can also be referenced to bone volume (BFR/BV) and tissue volume (BFR/TV) and expressed as percent per year. The activation frequency of new remodelling units can be calculated as BFR/BS divided by mean wall thickness (wall width/wall number).

Histomorphometry is also an important tool for determining the host's vascularization using the section stained with CD-31 for both the centre and margin of the implantation bed. The biomaterial's total area as well as the vessel numbers is marked using the *NIS-Elements annotations and measurements* tool and the total number of vessels on each slide as well as their mean diameter are determined. The number of vessels/millimetre square is to be considered from the total vessel number found in the biomaterial and the total biomaterial area (in mm^2). The vascularization percent can be determined from the area of all vessels with regard to the total implantation area. The distribution of the vessels within the implantation bed can be measured by relating the amount of the different vessel diameter fractions to the total amount of the vessels.

6.5.5. Histochemistry

The histological slides can be used to observe connective tissue ingrowth within the scaffold as well as to stain for bone-specific matrix deposition using histochemical staining with Movat's pentachrome and Sirius red stains as per protocols of Movat and Sweat et al. [50,51]. These stains are frequently used to identify reticular and collagen fibres as well as bone or osteoid matrix. Sirius red stains give collagen red, whereas Movat's pentachrome stains make bone tissue green, collagen yellow, and osteoid red.

6.5.6. Immunohistochemistry

During fracture healing, the main and preliminary process is the cell proliferation to build up the fracture callus composed of different tissue types. Cell proliferation can be detected by the antibody against the proliferating cell nuclear antigen (PCNA), which has been recognized as an auxiliary protein to DNA polymerase delta and expressed from the early G1 phase to the S phase, and declines through the G2 and M phases [52].

For conducting immunohistochemical study, the scaffold together with surrounding tissues will be taken after sacrificing the animal followed by fixing in 4% paraformaldehyde and decalcification in 10% EDTA. The samples are then dehydrated in graded alcohol and embedded in paraffin. Tissue blocks are sectioned at 5 μm thickness and used for immunohistochemical staining either with the monoclonal antibody against PCNA (PC10, 1:800, Oncogene, Germany) or with the monoclonal antibody against BrdU (1:1000, Sigma, Germany). A biotinylated rat absorbed mouse secondary antibody may be used to identify the primary antibody. The avidin–biotin complex detection system (ABC method) together with ALP and New Fuchsin as chromogen may be used. The sections/slices are then counterstained with methyl green to detect cell nuclei and cartilage. The negative control may be made by omitting the primary antibody. Finally, the morphology and the proliferating cells in the callus are to be analyzed with an image analysis system (Zeiss KS 400, Germany). The detailed staining procedure has been available in www.abcam.com/technical.

6.5.7. Fluorochrome Labelling Study

The procedure of bone remodelling is initiated by osteoblasts, which are bone-forming cells that also stimulate the differentiation of osteoclasts for resorption of bone prior to its new deposition and mineralization [53,54]. Numbers and activities of osteoblasts and osteoclasts are firmly controlled by extracellular cytokine levels, to which these cells also contribute, and by matrix receptor–ligand interactions, such that a balance of the two cell types is normally maintained. To understand the mechanisms of unusual bone remodelling, a number of different fluorescent stains have been developed to detect and quantify bone mineralization. Among the most successful of these are divalent cation chelating agents such as calcein, DCAF (2,4-bis[N,N′-di(carboxymethyl)–aminomethyl] fluorescein) [55–57] tetracycline and its derivatives [56,58,59], alizarin red derivatives [56,60], xylenol orange [56,61] and bone-targeting drugs such as pamidronate [62]. These agents while incorporation during periods of remodelling directly bind to areas undergoing calcification at the bone/osteoid (unmineralized bone) interface.

Tetracycline fluorescence labelling of bone has been used for decades to analyze bone remodelling as well as regeneration processes but usually a qualitative measure. It is non-toxic to both the patient and osteoblasts, is inexpensive, can be taken orally easily, has a short plasma half-life, is reliably incorporated into host bone, and is readily visible with fluorescent microscopy. Oxytetracycline follows the ionized calcium and the fixation by a process of adsorption which is restricted to areas where active deposition of mineralized tissue is taking place [63–65]. The labelled new bone and old bone emitted bright golden yellow and dark sea green fluorescence, respectively. The methodology is useful in assessing the amount of new bone formation and fracture healing when observing under UV light [66–70]. These parts were of value in studying gross bone architecture and histological mapping of new BF using fluorescent bone markers [71].

Tetracycline double labelling of bone is used to calculate kinetic data on bone turnover. Tetracycline binds to newly formed bone at the bone/osteoid (demineralized bone) interface where it shows as a linear fluorescence. If a second dose is given 11–14 days after the first dose, the amount of bone formed during that interval can be calculated by measuring the distance between the two fluorescent labels. Although double labelling of bone is necessary to calculate kinetic indices of bone turnover, important information can be obtained even if patients have not been labelled with tetracycline. Several different types of tetracycline may be used; each will fluoresce a different color. While it is preferable to label patients with two different types of tetracycline, the same brand may be used for both labels.

6.5.7.1. *Methodologies for Tetracycline Labelling Study*

Tetracycline, i.e. oxytetracycline dihydrate at a dose rate 25 mg/kg BW is to be given 3 weeks before the sacrifice of animal and sometimes another injection may be given 10 days before of first injection for double toning of bone. The implanted segments of the bone are collected and transverse section (2–3 mm) thickness including the implanted area is cut with the help of hacksaw. Undecalcified ground sections are to be prepared as described by Nandi et al. [68–70]. The sections are then ground to 20 μm thickness using different grade sand papers. Final grinding is done over the bone under moderate pressure using slow circular motions. During grinding, ground sections are examined repeatedly under microscope (10×) for transparency of the sections and presence of structural details of the bone. The specimens are kept wet by dipping in water

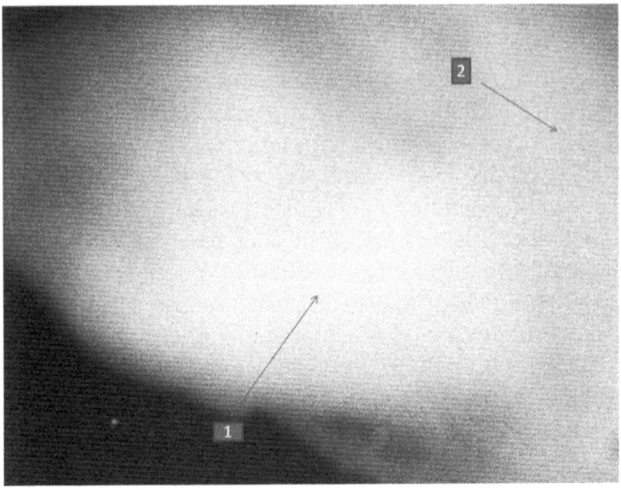

FIGURE 6.5 Fluorochrome labeling picture of undecalcified bone sample showing new bone formation. 1. New bone, 2. Old bone.

during the entire procedure. The ground-undecalcified sections are observed under ultraviolet light (Leitz Orthoplan Universal) Widefield Microscope with incidental light (Excitation filter, BP-400 range) for tetracycline labelling to find out the amount and source of newly formed bone (Fig. 6.5).

a. Working protocol of dose (mg/kg body weight) and preferred route of administration of different fluorochrome in different species of animals

- Emission spectra: Tetracycline—yellow range; Xylenol orange—orange color; Calcein green—green color; Calcein blue—blue color; Alizarin red—red color.

Type of animal	Tetracycline	Xylenol orange	Calcein green	Calcein blue	Alizarin red/ Complexone	Route of administration (Preferred)
Mouse	20	80	10		30	Upto the size of rabbits-subcutaneous (S.C.)
Rat	25	90	10	30	30	
Guinea pig	–		20		100	
Rabbits	25	90	10		30	
Cats	30		30		–	Large animals-slow intravenous (I.V.) or intramuscular (I.M)
Dogs	20	80	10		–	
Pigs	25		15		–	
Goat/Sheep	20	80	10	30	25	

b. Solvents

- Tetracycline powder—Physiologic saline at a concentration of 10–25 mg/ml (unstable and should not be stored long times)
- Alizarin red/Complexone—Distilled water or 2% $NaHCO_3$ at a concentration of 15–30 mg/ml
- Calcein green—2% $NaHCO_3$ at a concentration of 10–20 mg/ml
- Xylenol orange—2% $NaHCO_3$ at a concentration of 20 g/L

c. Standard time protocol of administration of fluorochrome (multiple labels) in some animal species

- Mouse: Calcein green-2 weeks, alizarin red-3 weeks, xylenol orange-5 weeks.
- Rat: Calcein green-2 weeks, xylenol orange-4 weeks.
- Rabbit: Calcein green-3 weeks, Alizarin red-4 weeks, oxytetracycline two labels at an interval of 10 days and last injection 3 weeks before sacrifice.
- Dog: Calcein green-3 weeks, oxytetracycline-5 weeks, xylenol orange-7 weeks.
- Goat: Calcein green-3 weeks, oxytetracycline-5 weeks, xylenol orange-7 weeks.

Some important points to be considered during fluorochrome labelling study

- Marker interval (labels interval) should be 2–3 weeks apart.
- Tetracycline is hepatotoxic in rats and hence unsafe in that species.
- Administration through intravenous route should be slow to prevent the risk of severe hypocalcaemia.
- Longer interval will increase the likelihood for label escape and shorter intervals will make discrimination between the labels technically more difficult even at high magnification.
- Multiple labels with short intervals have the risk of inhibiting bone mineralization.

6.5.8. Angiography

The periosteal and surrounding soft tissue vessels play a vital role in the early stage of fracture healing [72]. The periosteal blood supply is developed from the surrounding soft tissue immediately following disruption of nutrient arteries in long bone fractures [73]. These new blood vessels added in the construction of early periosteal callus [74] and extraosseous callus, which eventually united the fracture. An alteration in the blood supply caused by fracture has been reported to be the local enhancement of three primary components of the different vascular system, i.e. primary nutrient artery, metaphyseal arteries and periosteal arteries [75]. A rapid growth of developing pattern of blood vessels was correlated with the organization and classification of the callus.

The progress of new blood vessels during fracture healing passes through five stages: hyperaemia, proliferation of periosteal and medullary vessels, differentiation of proliferated vessels and their organization [76].

Osteomedullography and angiography have been used in recent past in veterinary orthopaedics for evaluation of allogenic bone grafts [77], non-union fracture [78], ceramic and bioglass implants [68–70,79] and diagnosis of bone diseases [80]. Revascularization of bone grafts is an essential prerequisite in their incorporation in the host bed [81]. Besides, the development of new blood vessels to support rapid vascularization of biomaterials and tissue engineering constructs remains a difficult challenge. However, nature of graft/implants may influence their revascularization and subsequent osteogenesis and bone healing. Hence, the need of assessing the acceptance of bone graft/implant becomes paramount necessity, which can be accomplished by osteomedullography, angiography and micro-CT-based angiography technique.

6.5.8.1. Technique of Angiography

Angiography is performed by standard anaesthesia and surgical procedure. As for example of angiography in radius and tibia bone, the corresponding radial/tibial arteries are located and exteriorized and catheterized using polyethylene catheters. The catheter is then pushed downwards into the artery and a tight ligature was applied around the catheter, so as to prevent any leakage or backflow of the blood/contrast materials from the artery. An elastic tourniquet is applied at the proximal part to prevent retrograde flow of the contrast medium. A syringe containing 15 ml sodium iothalmate (Conray® 420) is connected to the catheter. The contrast material is infused with a regular gentle digital pressure and radiographs are taken. The catheter is removed and the puncture of the artery is sutured with the help of 4-0 chromic catgut. The tourniquet is thereafter released. Both ligature and clamp placed earlier in the artery are removed and the skin wound is closed. For better visualization of the arteries, lead oxide suspension (20% W/V) can be used after euthanizing the animal at the end of the experiment and radiographs can be taken as usual. The angiograms are helpful as a tool for assessing vascular response of the host bone and surrounding tissue in the implanted area and visualization of the implant.

Sodium iothalmate, as a contrast media, has been successfully used by various scientists for the visualization of different vascular patterns [77,82]. Since the contrast material recommended for angiography procedures is drained out very quickly, it is difficult to visualize minute vascular branches. An immediate radiographic exposure is necessary with Conray-420 to minimize the venous drainage when the procedure is delayed. Lead oxide soap suspension (20%), as a contrast media, is found satisfactory for the visualization of major arteries including minute vascular branches and may be used for angiographic evaluation of fracture healing [79,83]. This material is toxic, cannot be drained out by the venous system and as such, and the animal is to be sacrificed before performing angiography.

6.5.8.2. *Micro-CT-Based Angiography Technique*

The vascular regrowth at the defect area can be investigated post-surgery by micro-CT-based angiography technique [84]. The micro-CT-based technique can help to quantitatively assess the vascularity in the early stages of bone regeneration [85]. This technique involves induction of standard anaesthesia followed by introducing a 25-gauge catheter into the abdominal aorta and injection of 250 units (0.25 mL of 1000 units/mL) heparin. The hind limb vasculature of animal is to be cleared with PBS followed by fixation with 10% neutral buffered formalin and clearance again with PBS using a peristaltic pump (Masterflex, Cole–Parmer). The animal is to be euthanized by an overdose of anaesthesia before the formalin perfusion. A radio-opaque, lead chromate-based contrast agent (Flow Tech, Carver, MA) may then be injected and allow to polymerize for at least 2 h. The bone-like femur (if implanted in this bone) along with its musculature is to be excised carefully, fixed in 10% neutral buffered formalin for 48 h, and decalcified for 2 weeks using a formic acid-based solution (Cal-Ex II, Fisher Scientific). The decalcified bone samples then are to be rinsed in PBS and stored in 10% neutral buffered formalin until imaging. The sample can be imaged in a micro-CT system (Viva-CT, Scanco Medical) at a 21.5 μm voxel size. Two volumes of interest (VOIs) are then be defined to evaluate the vessels inside the defect only and inside including directly neighbouring to the defect periphery.

The percent vascularization (PV) within each VOI was calculated as follows:

$$\%\ \text{vascularization} = \frac{\text{vessel volume } (\mu\text{m}^3)}{\text{VOI volume } (\mu\text{m}^3)} \times 100\%$$

$$\text{Standardized }\%\text{ vascularization} = \frac{\text{vessel volume } (\mu\text{m}^3)}{\text{volume } (\mu\text{m}^3) - \text{average volume of per scaffold } (\mu\text{m}^3)} \times 100\%$$

6.5.8.3. *Techniques of Osteomedullography*

Osteomedullography (intraosseous phlebography) of the implanted animal can be performed under standard anaesthesia. One small hole is to be drilled at the distal metaphysis using a 14-gauge bone marrow biopsy needle into the medullary cavity of the test bone. The correct positioning of the needle in the medullary cavity can be assessed by the presence of blood on aspiration in a glass syringe attached to the needle. The heparinized normal saline is then infused for flushing the medullary cavity. Ten millilitres of contrast media like sodium iothalmate (Conray-420) is to be rapidly injected into the medullary cavity and lateral radiographs are to be taken during and immediately after injection. Osteomedullograms are observed for the venous drainage at the grafted/implanted area.

6.5.8.4. *Intravital Microscopy Imaging*

Intravital fluorescence video microscopy is another technique for observing vascularization that can be performed at early days of implantation, i.e. on 7, 14, and 28 days using a vertical illumination fluorescence microscope unit (Olympus Schweiz AG, Volketswil, Switzerland) equipped with ×-2.5, ×-20 and ×-40 objectives and a fluorescence filter for green light (excitation wavelength: 460–480 nm, emission wavelength: 495–540 nm). The microcirculation is to be visualized in regions of interest by intravenous injection of 50 μl of the high-molecular weight plasma marker fluorescein isothiocyanate (FITC)-labelled dextran (FD 2000, Sigma Aldrich; Buchs, Switzerland; 2.5% concentrated solution in 0.9% saline, molecular weight 2000 kD) in anaesthetized animal. Vessels can be recorded with a digital microscope camera directly before and 1 min post-injection at ×-40 magnification. Mice were kept under anaesthesia (see above) during microscopy. To terminate anaesthesia, an antidote mixture containing atipamezole hydrochloridum (1.25 mg/kg BW), sarmazenilum (0.5 mg/kg BW) and naloxone hydrochloridum (0.6 mg/kg BW) was injected subcutaneously.

6.5.8.5. *Analysis of Blood Vessel Ingrowth (Angiogenesis)*

Tissue sections are immunostained for von Willebrand factor (vWF), a protein present in large quantities in subendothelial matrices such as blood vessel basement membranes. Imaging is to be carried out by means of a Nikon Eclipse E800 light microscope and a Spot RT digital camera (Diagnostic Instruments, Sterling Heights, MI, USA). Blood vessels, designated by vWF staining, are counted manually at 100× magnification in the total scaffold, and normalized to tissue area with the use of Image Pro Plus Software (Media Cybernetics, Carlsbad, CA, USA). Details of Staining Protocol and Kit components have been available in www. chemicon.com.

6.5.9. Transmission Electron Microscopy (TEM) Study

TEM study is a recent advancement for *in vivo* characterization of biomaterials to study the interfaces between the materials and newly formed bone. From structural and morphological view point, the investigation also deals whether a newly formed apatite layer or an electron-dense layer is present at the interface between bone and implants. The morphology and organization of the individual phases will also be examined at low resolution, contrary to the study of their interfaces at and between the colonies by a high-resolution technique (HRTEM) involving lattice plain imaging [86].

6.5.9.1. Technique of TEM Study

The technique involves using undecalcified sectioning rather decalcified one since decalcification eliminates the spatial relation between organic substances and inorganic material.

The specimens consist of both test bone from fracture healing site and adjacent old bone at the end of the experiment followed by fixation in 10% buffer formaldehyde solution. The samples are to be dehydrated in a series of graded alcohol solutions followed by embedding in hydroxyethyl-methacrylate resin. Thin sections of undecalcified samples of bone–implant interface are to be prepared with a Reichert ultramicrotome fitted with a diamond knife. The sectioning is to be done transversely to the long axis of bone and finally will be placed on uncoated TEM copper grids of 400 mesh for better support.

High-resolution TEM (HRTEM) using a JEOL Jem 2010 fitted with a LaB6 filament at 200 keV will be used for observing half of the as-cut thin sections directly. The other half will be stained and examined in a conventional transmission electron microscope, a JEOL Jem 100 CX, at 100 keV. The staining of the thin sections supported on the TEM grids will be done by osmium tetroxide vapour for 30 min in an airtight dish located in a fume cupboard. During microscopy, various parts of the interfaces will be examined in bright field at low- and high-resolution levels including lattice plain imaging [87].

6.5.10. Densitometry

Between the years 1981 and 2010, there has been ever-increasing interest within the orthopaedic community in the non-invasive measurement of bone mineral content (BMC) of various regions of skeleton. However, densitometry technique can be applied either in axial or appendicular skeleton. This medical bone density (or bone mineral density (BMD)) is measured by a procedure called densitometry, which would be computed as mass per volume and normally referring to the amount of mineral matter per square centimetre of bones. The procedure is frequently performed in the radiology departments of hospitals or clinics for detection of osteoporosis and fracture risk. The technique is painless and non-invasive with low radiation exposure. Bone densitometry has a wide range of application in recent day orthopaedic practices for detection of BMD as much as lower levels when compared to conventional radiography.

Dual energy X-ray absorptiometry (DEXA) is a fast, non-invasive technique that enables highly accurate measurement of BMD [88]. In DEXA, the radiation dose is low enough to allow BMC and BMD measurements in different skeletal sites and in longitudinal studies with a high-resolution, single-beam, small animal scan (QDR-2000, Hologic, Inc., Waltham, MA, USA). Individualized regions of interest (R1) should be constructed over the surgical site of each slice of the interest bone. Corresponding regions will be constructed on each slice of the interest bone using the 'compare' feature of the software. The ratio to be calculated for each region of interest (R1) is as follows:

$$\frac{\text{BMD (or BMC) of a given region of the bone}}{\text{BMD (or BMC) of the corresponding area of the bone}}$$

6.5.11. Biomechanical Testing

Bone graft substitutes should offer optimal biomechanical strength of bone that are under high weight-bearing loads, or that are subjected to high-torsion forces. The biomechanical strength is an outcome of a complex interaction between the bone and the bone graft substitute material. In an ideal situation, a bone substitute material should offer the same biomechanical strength as the bone being replaced. However, *in vivo* interactions like integration, incorporation and bioresorption play a significant role in biomechanical behavior of an implanted bone substitute. In general, human cortical bone has a compressive strength of 130–290 MPa and a tensile strength of 90–190 MPa, while the compressive strength of cancellous bone is of 2–38 MPa [89]. A very few bone substitute materials offer a suitable biomechanical strength like that of cortical bone. The compressive strengths of calcium phosphate grafts are equivalent to those of cancellous bone although have inadequate resistance to tensile and shear forces. Calcium sulphate provides only least structural support. The compressive strength of bioactive glass is of 91–197 MPa but its tensile strength is not that of cortical bone [90]. As a result, most of the bone substitutes available are inappropriate for grafting of major cortical bone loss/defects without additional support.

Histological and mechanical assessment is the main tool for *in vivo* analysis of orthopaedic and dental biomaterials [91]. Several methods exist for quantitative assessment of implant integration and one of the preferred options is based on the measurement of contact percentage between implant material and host tissue. The frequently used method for quantitative evaluation of the bond strength of a healed implant–bone interface is the push-out test, which also allows for evaluating interface shear strength [92,93]. During the testing, the most relevant parameters to be considered are (a) clearance of the hole in the support jig, (b) Young's modulus of the implant, (c) cortical thickness, and (d) implant diameter. Implant fixation can be a result of bonding with bone or of mechanical interlocking [94]. The shear strength of the bone–implant interface generally increased over time due to bone ingrowth into the implant.

6.5.11.1. Protocol of Push-Out Test of in vivo Bone Sample

Push-out test is to be carried out at the end of the experiment after sacrificing the animal and harvesting of interest bone. After removal of soft tissue, the

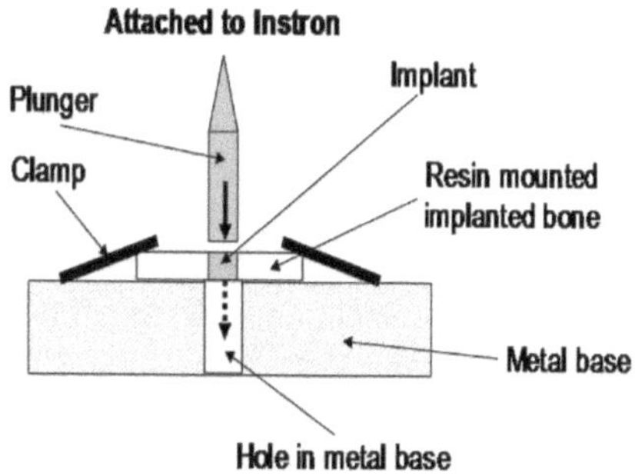

FIGURE 6.6 Schematic representation of the protocol of the push-out test performed.

implanted bone is to be refrigerated in gauze, dampened with physiological saline. The proximal and distal metaphysis are to be excised, thus leaving the diaphyseal segment containing the implants (if implants are placed in diaphysis). Each implant is to be pushed out of the bone with a plunger of rectangular/circular cross-section depending upon the design of the implant. The samples are to be placed on a support jig, with a hole larger in diameter than the implant perimeter. For the push-out test, an Instron 5500R UTM may be used at a constant displacement speed of 0.5 mm/min. An illustration of the push-out test set-up including test and specimen geometries is shown in Fig. 6.6. The end of the plunger is to be brought into close alignment with the model intramedullary end of the implant. Load will be applied until the bone–implant interface ruptured or there was compressive collapse of the implant as observed from the peak on the load–deformation curve. To calculate the contact area for each specimen, the thickness of the cortical bone around the implant is to be measured at four sites. Their mean is supposed to approximate the actual cortical thickness. The load at which an implant is detached from the bone or at which the bone will break is designated as the 'failure load'. The product of the height and circumference of the implant dissected from the segment is measured as the contact area between the implant and the surrounding bone. The failure load divided by the contact area is to be calculated to estimate the bonding strength between the bone and the implant. Average of the results is to be taken for calculation. Care should be taken to ensure that the line of action of the loading force is collinear with the long axis of each implant. Bone overgrowth at the outward, muscular site of the implant is to be removed by polishing, which in turn will improve the accuracy of alignment during the test. The instrument is to be calibrated before each test.

The 'interfacial shear strength' is calculated using the following formula:

$$F = \frac{S}{\Pi \times d \times t}$$

where $S =$ interfacial shear strength, $F =$ force applied on implant, $d =$ implant diameter, $t =$ cortical thickness.

Torque test is the alternative way for biomechanical testing of bones. To start with the testing, the bones are to be thawed at room temperature. During the testing, the bones are to be kept moistened to avoid the potential effect of drying [95] followed by embedding in moulds with two-component fibreglass resin using a torsional shaft of 8 cm. After hardening of the resin, the bones are to be placed in the torque machine and torsionally loaded at a constant angular speed of 6.5°/s until failure [96]. The testing parameters include maximal torque capacity (MTC), maximal angular deformation (MA), bone torsional stiffness (BS) and maximal absorbed energy (MAE).

It has been recommended that the torque test is more dependent on the bone–implant interface while the push- and pull-out test is more dependent on the surrounding bone support [97]. A limitation of the push- and pull-out tests is that most of them are to be carried out after animal sacrifice and even after tissue fixation creating doubts as to the effects of post-retrieval procedures on the 'true' values of the anchorage of the implant in the tissue.

6.5.12. Scanning Electron Microscopic Study

The importance of the interactions between biological components and the surface of implanted materials has made considerable dimensions on the development of tools for surface modification and analysis. A variety of methods are designed for surface analysis but with their pros and cons. Successful osseointegration leading to bone–implant fusion depends on various conditions during peri-implantation bone healing. There is an irreparable damage of bone–implant interface if implant is not well tolerated by the body and that might have to be removed [98], whereas a biocompatible implant exhibits good adjacent osseous growth and successful bone–implant fusion. SEM was used for evaluating both the implant surface and the interaction with bone tissue, where detectors for secondary electrons (SE) and back-scattered electrons (BSE) were used, respectively.

6.5.12.1. Protocol for SEM Study

The bone specimens consisting of healed test bone and adjacent host bone are to be collected from the animals at the end of the experiment and samples are to be prepared following the procedure of Staniforth et al. [99]. They are fixed in 5% electron microscopy grade glutaraldehyde phosphate solution for 24–48 h. The fixed samples are to be subsequently washed twice for 30 min initially with PBS (pH 7.4) followed by distilled water. The specimens are to be dehydrated thereafter in serial concentration of ethanol (30%–100%) for 30 min each followed by drying with HMDS for 20 min. To obtain electromagnetic signals, the samples are to be golden ion sputtered on Jeol ion sputter Model JFC 1100 at 7–10 mA and 1–2 KV for 5 min. Finally, they are to be examined in SEM for the study of their different surfaces to understand the orientation and distribution of newly formed

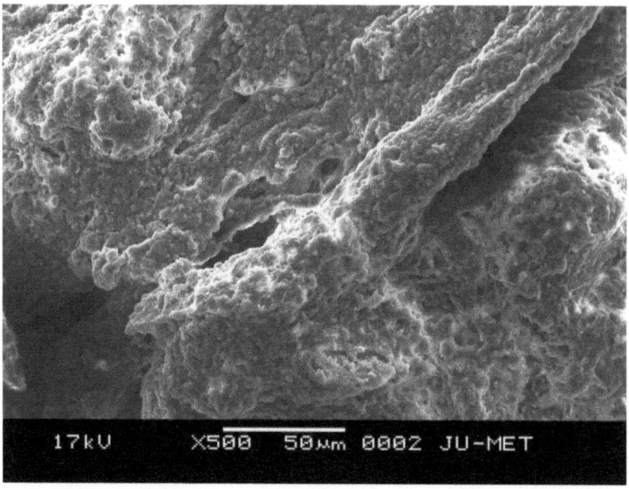

FIGURE 6.7 Scanning electron microscopic picture of biomaterial implanted healed bone.

osseous tissues and distribution/absorption of biomaterials at the implanted site. The evaluation can be carried out in SEM using magnification levels between 100 and 20,000 times (Fig. 6.7).

6.5.13. Resonance Frequency Analysis (RFA)

Widespread use of RFA in clinical surgery has been started as a tool for measurement of implant primary stability, consequently as an indicator for the possibilities to perform immediate and early functional loading of the implants. During the healing period, the changes in implant stability can be measured by RFA as per the method described by Meredith et al. [100]. This could provide quantitative data on the bone–implant interface stiffness. The process involves the attachment of a small transducer with the implant immediately after insertion and immediately before sacrifice of the animal. Oscillations are induced by a piezo electric element and the responding resonance frequency is recorded. The measurements can be performed parallel and perpendicular to the long axis of the interest bone. The resonance frequency can also be determined from the marginal bone height, stiffness of the bone and the length of the transducer [101]. As the stability increases with the passage of time, the remodelling of the bone–implant interface also became better [100,102].

6.5.14. Ultrasound Elasticity Imaging (UEI)

Non-invasively, monitoring the extent of scaffold degradation and tissue development is the current research interest of the researchers which will greatly help to evaluate regenerative therapies. It not only provides valuable information to measure matrix development, such as collagen, elastin and hard tissues, but also helps in determining the load-bearing capability of the scaffold/tissue construct. Moreover, to optimize therapies, it is of imperative necessity to understand relative rates of tissue regeneration and scaffold degradation for the specimen as a function of time. Currently, available analysis methods for quantifying tissue formation include primarily histology and direct mechanical measurements. Although these methods provide valuable information, they involve animal sacrifice and scaffolds destruction, hence prohibiting reuse of samples after characterization.

UEI has been emerged as a valuable non-invasive tool to investigate mechanical characteristics of biological tissue [103,104]. It estimates tissue motion based on speckle tracking between two frames before and after deformation of the tissue. Here, generally, tissue is to be pushed with an ultrasound transducer itself and recording the tissue deformation with real-time ultrasound images [105]. This comprises a static deformation with an external force. The procedure involves the measurement of weight and strain of each sample before and after degradation, the coefficient of variation (COV = standard deviation/mean) and finally based on the direct stress–strain measurement curve, the elastic modulus can be calculated [106].

6.6. BIODISTRIBUTION STUDIES

The fundamental of biodistribution is to develop a successful targeted nanodelivery system to understand the factors affecting biodistribution in experimental and material design aspects. This is very pertaining to understand due to the fact that at present, there is no standardized practice for the evaluation of the performance on targeted delivery system. Besides, it also causes difficulties to compare the similar studies and their results thereafter. The related factors of biodistribution are the following things like the initial dose [107,108], methods of administration [109], experimental model [110] and the experimental time range. This targeted nanodelivery system is a continuous process and by means of developing standardized practice, the effect of experimental conditions on results will speak on its final success for an established system. In spite of the fact that there are no standardized procedure in measuring biodistribution, it is under the basis of commonalities and tend that are in practice for its measurement. The use of autoradiography for identifying nanoparticles biodistribution is one of the most common studies where the measurement of biodistribution can be possible [111]. The success of autoradiography in measuring biodistribution is due to the ability and easiness for quantitative measurement of biodistributable materials even when it is very small concentration. Even though, there are a number of drawbacks of radiography like it requires tissue excision and there must have the safety consideration in practicing this technique. In spite of these constraints, there are some emerging techniques for quantifying biodistribution system which are CT, fluorescence imaging, inductively coupled atomic emission spectroscopy, inductively coupled plasma-mass spectrometry, micro-positron emission tomography, MRI imaging, radiography, etc. However, to have a better understanding of the target site with improved higher accumulation of targeting ligands must have to be assessed to know the biodistribution for systems on which the proposed work has been undertaken. It is evident that higher concentration of nanodelivery system at the target site allows a greater availability of cell uptake. This underscores that the physical properties affecting biodistribution are equally important for active and passive targeting. The shape, size, surface charge and mechanical properties can be selected based on the effects of biodistribution. The biodistribution of nanoparticles is size dependent as shown by two recent studies on gold nanoparticles and PEGylated gold nanoparticles [112,113]. While studying on nanoparticles accumulation *in vivo*, the accumulation primarily in the liver, spleen and blood is measured by the concentration (ng/g tissue) and percent of initial doses.

One of the important aspects of shape of the nanoparticles *in vivo* is having a very effective performance to predict the success of the item as a biodistributer which is only beginning to be explored [114,115]. Park et al. evaluated the magnetic iron oxide nanoworms, elongated iron oxide nanoparticles and compared them to spherical iron oxide nanoparticles to observe the biodistribution ability of the items [116]. The study is significant to know overall biodistribution of the two systems and their site of accumulation which can be graded as the highest in spleen followed by liver and kidney with a note that the circulation time in both the systems lies between the half lives of 15 and 18 h.

While studying the importance of size of the bio-nanoparticles to affect biodistribution, the determination of 'flexibility or shape adaptability' of the particles is a matter of concern [117]. The cross-linked nanoparticles with the core composition of either glassy polystyrene or fluid-like polymethacrylate have been studied by Sun et al., and concluded that polystyrene cores (18 ± 3 nm) had a greater retention in the blood compared to small particles 24 ± 3 nm. Here also, the accumulation of biomaterials of smaller size could be observed in the different visceral organs (in liver, lungs and spleen).

Surface properties such as surface charge and PEGylation also affect the biodistribution of nanoparticles. Generally, presence of neutral hydrophilic polymers at the surface of the nanoparticles could neutralize the charges at the surface and this is also associated with improved blood retention time and decreased opsonisation by RES which gives rise to a term 'stealth properties' [118]. However, addition of PEG, on the other hand, could increases the surface charge of the PCL nanoparticles and there would be a significant improvement of blood retention time with a drop in nanoparticles accumulation in liver and lungs.

The composition of polymer is a determining factor affecting biodistribution and this has been noted by making experiment in a rabbit model with poloxamer 407 coating on polystyrene and PLGA nanoparticles were found to be most effective in clearing of the biomaterial from the blood and the liver than that of other similar biomaterials with different coating like poloxamine 904 and poloxamine 908.

The last but not the least, it is the chains and end-group functionalities of the nanoparticles that determine and play a very important role in biodistribution of biomaterials. All these together are important in lowering systemic toxicity which is a goal to develop a successful targeted delivery system. The only fraction of the total dose administered remained accumulated at the targeted site and the distribution of the biomaterial at different tissues and organs and its clearance rate through the system of the experimental unit is also a matter of extreme importance.

6.7. *IN VIVO* CHARACTERIZATION OF BIOMATERIALS IN SOFT TISSUE SURGERY

Acute and chronic wounds caused by burns, congenital cutaneous anomalies, cutaneous ulcers, trauma, and surgery pose significant challenge to surgeons in curing such maladies. The situations may aggravate in patients suffering from diabetes, paralysis, renal failure, malnutrition, etc., where wound healing is essential and challenging [119]. Autologous full or split thickness skin grafts are being considered the 'gold standard' for clinical situations requiring skin replacement, yet morbidity from grafting may crop up at both the recipient and donor sites [120]. Additionally, there may be insufficient areas of undamaged donor skin available to attain adequate coverage in burns patients with a large total body surface area. To meet the needs, researchers have attempted to create bio-artificial, cellular-based skin substitutes that mimic the properties of autologous skin grafts and to provide sufficient mechanical support. As a result, there is paramount necessity to culture cellular components on appropriate biomaterial matrices or scaffolds to produce natural multi-layered design of skin vis-à-vis management of such situation.

6.7.1. Histological Analysis

The histological study is carried out to assess the evidence of progressive healing in terms of inflammatory cell proliferation, fibroplasias, epithelisation and neo-vascularisation.

The skin including the wound/granulated tissues collected at different intervals is to be transferred to 10% neutral buffered formalin for 24 h at 4 °C. The formalin-fixed tissues are to be dehydrated through grades of alcohol, cleared in xylene and then embedded in paraffin wax (58–60° m.p.). The moulds are to be labelled and stored until use. Five to seven micrometres deparaffinised sections are stained with haematoxylin following counterstained with eosin. Masson's trichrome staining is to be done to observe collagen deposition in the granulated tissue. The sections are to be made perpendicular to the anterior–posterior axis and perpendicular to the surface of the wound with positioning on a glass slide, and finally stained with haematoxylin–eosin (HE) reagent [120].

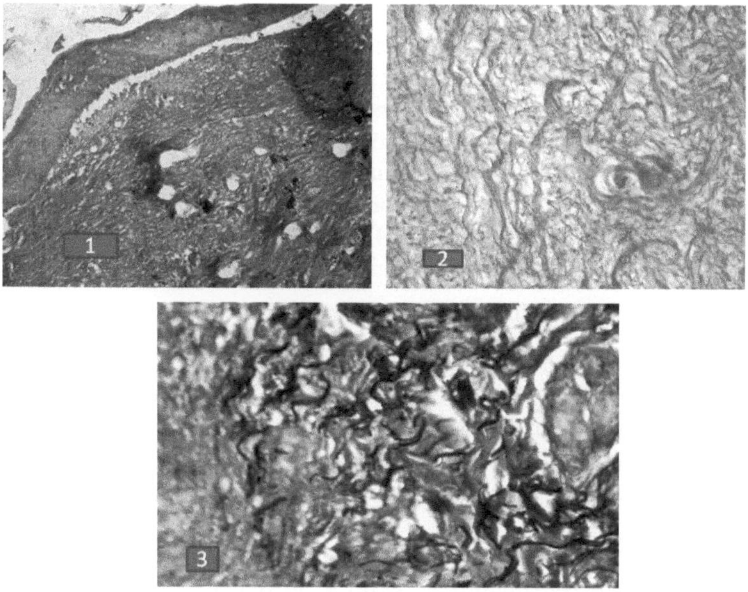

FIGURE 6.8 Histochemical picture of healed biomaterial implanted wound 1. Collagen, 2. Reticulin, 3. Elastin.

6.7.2. Histochemical Analysis

This study demonstrates the presence and quantifies the intensity of collagen, elastin and reticulin fibre during different phases of reparative process. The presence of the above was also visually quantified on the same scale as described for histomorphological study. The following criteria were adopted while visually quantifying the histomorphological and histochemical parameters as per method described by Kumar [121] (Fig. 6.8).

6.7.3. Matrix Metalloproteinase (MMP) and MMP9 Assay

MMPs are a family of at least 24 Zn^{2+}-dependent endopeptidases comprising secreted MMP and membrane-type MMP (MT-MMP). These enzymes have been concerned in wound repair by their ability to remodel extracellular matrix components. Indiscriminate pharmacologic inhibition of MMP also blocks epithelial resurfacing, emphasizing the physiological role of one or more MMP in wound healing [122–124].

6.7.3.1. Technique

The granulation tissues can be utilized for analysis of MMP8 and MMP9 by immunoblot analysis [125]. At first, the tissue proteins are to be processed through electrophoretic separation by sodium dodecyl sulphate-polyacrylamide gel electrophoresis and consequently, the proteins are to be electrotransferred onto a nitrocellulose membrane. The membrane is then to be blocked in blocking buffer containing 0.05% Tween-20 and 5% bovine serum albumin for 4 h, followed by incubation in rat anti-MMP8 and anti-MMP9 primary antibodies. After three washings with Tris-buffered saline containing 0.5% Tween-20, the nitrocellulose membrane is to be incubated in ALP-conjugated secondary antibody followed by washing with buffer, and the bands are visualized using (nitro blue tetrazolium/5-bromo-4-chloro-3-indolyl phosphate) (NBT/BCIP) as a substrate.

6.7.4. Transmission Electron Microscopy

The tissue specimens collected from wound site are to be fixed in 3% glutaraldehyde, buffered with 0.1 M sodium carbonate, pH 8.0, containing 0.2 M sucrose and final fixation is to be performed with 1% osmium tetroxide in the same buffer for 2 h at 4 °C. After dehydration of tissue sample in ethanol, embedding is to be done in a mixture based on Epon 812. Ultra-thin sections are to be prepared with a Reichert Ultracut S ultratome (Leica, Nussolch, Germany) and stained with 2% aqueous uranyl acetate and lead acetate. Then the samples are to be examined in a JEOL JEM-1010 electron microscope (JOEL, Tokyo, Japan) at 60 kV and photographed.

6.7.5. Scanning Electron Microscopy

SEM is one of the best modalities to obtain 3D imaging of the healing tissue. The procedure starts with the osmic maceration of tissues by removing cytosol and helps in visualization of cytoplasmic ultrastructure. The samples are then to be fixed with 0.25% glutaraldehyde and 0.25% paraformaldehyde in 0.1 M nacacodylate buffer (pH 7.2) for 20 min at room temperature. After washing in 0.1 M cacodylate buffer (pH 7.2), the samples are to be post-fixed in a solution of 1% osmium tetroxide and potassium ferrocyanide for 2 h. Each tissue sample is then to be cut into slices about 0.2 mm thick and post-fixed in 1% osmium tetroxide and 1.25% potassium ferrocyanide for 1 h. Slices are to be washed in PBS (pH 7.2) and then immersed in 0.1% osmium tetroxide in PBS for 48 h. The dehydration of slices include with an increasing concentration of ethanol, subjected to critical point drying with CO_2. Dried slices are to be mounted on stubs, gold coated with a Sputter K250 coater, and then to be observed with an SEM-FEG XL 30 microscope (Philips, Eindhoven, The Netherlands).

6.7.6. Immunohistochemistry

The tissue sections from the wound site can be used to analyze the presence of infiltrated macrophages. The procedure involves the use of freshly excised granulation tissues followed by embedding in embedding medium (OCT, Tissuetek; Sakura Finetek Inc., Torrance, CA) and frozen. Serial transverse sections around the granulation tissues are to be prepared at 10 μm using a cryostat (JUNG CM 3000; Leica Microsystems GmbH, Wetzlas, Germany). The cryostat sections are to be selected on poly-L-lysine coated slides and fixed with 2% paraformaldehyde. Blocking solution containing 2% goat serum and 3% bovine serum albumin in Dulbecco's modified PBS can be utilized for blocking non-specific sites for 10 min at 20 °C. Finally, the sections are to be treated/stained with FITC-conjugated mouse anti-rat CD11b (Integrin αM chain, Mac-1 α chain) monoclonal antibody (BD Biosciences, San Jose, CA), mounted and observe under confocal microscope (Leica TCS SP5).

6.7.7. Biochemical Analysis

The attainment of wound strength is made by cross-linking of collagen and remodelling of connective tissue. Hydroxylation of proline and lysine in ribosomes occurs during collagen chain synthesis, where ascorbic acid plays significant role as a co-factor in converting proline residues into hydroxyproline [126]. Hexosamine acts as a ground substratum for the synthesis of ECM. Both the amount of hydroxyproline and hexosamine can be estimated from 5 mg of defatted, dry granulation tissue [127,128].

Further, in order to characterize the cytosolic content of fibroblasts and to evaluate the mitogenic activity of these cells, proteoglycans can be estimated as per the protocol of Ruggeri et al. [129]. The technique involves the fixation of tissue samples for 2–4 h at 4 °C in 3% glutaraldehyde (in PBS 0.1 M, pH 7.4) and then reduced into small blocks of about 100 μm. Pre-incubation of the sample is prerequisite for 1 h at room temperature in 25 mM acetate buffer, pH 5.8, containing $MgCl_2$ at the same concentration used in the subsequent staining solution followed by further incubation in the staining solution (0.05% w/v Alcian Blue 8 GX in 25 mM acetate buffer and $MgCl_2$ at final concentrations of 0.05, 0.3, 0.8, 1, 1.2 M) for 7 h. After rinsing for 1 h in $MgCl_2$–acetate buffer, specimens are to be placed first in 0.01 N HCL for 4 h, then in distilled water for 1 h, and finally in PBS for 30 min. The specimens are lastly dehydrated in an ethanol series followed by inclusion in epoxy resin.

6.7.8. Apoptosis Assay

Various methods to identify apoptotic cells have been recognized based on the morphological and biochemical events of apoptosis. Biochemical techniques may be easier methods for detection, but often fail to differentiate between apoptotic and necrotic pathways. However, morphological and biochemical methods can provide meaningful information about apoptosis.

The procedure involves deparaffinisation of paraformaldehyde-fixed tissue sections followed by dehydration in an ethanol series and finally washing three times in PBS. Apoptosis detection can be made by the TUNEL assay using the Apoptosis Detection System-Fluorescein (Promega, Madison, WI, U.S.A.).

Nuclear staining in the procedure can be done by 4′,6-Diamidino-2-phenylindole dihydrochloride.

6.7.9. Mechanical Testing

The mechanical properties of tissue engineering scaffolds are of paramount importance for performing their adequate function *in vivo*. Characterization of the mechanical properties of the scaffolds can be carried out by uniaxial tensile testing [130]. In-house designed bone-shaped die and a cutting board can be used for harvesting samples. Uniaxial tensile tests are to be carried out in a mechanical tester (Tytron 250, MTS Systems Corp., Eden Prairie, MN). After careful harvesting, the samples are to be loaded to failure at a constant displacement rate of 0.2 mm s^{-1}. Engineering stresses are to be calculated by dividing the recorded loads by the cross-sectional area of the samples and the thickness of the samples. Final tensile strength (UTS) and failure strain are indicated by the peak stress and maximum deformation withstand by the samples prior to failure. Tensile modulus is the slope of the linear portion of the stress–strain curve between 25% and 80% of the UTS of the sample. Stress–strain curves are to be plotted and analyzed. Results are expressed as means ± standard deviation.

6.8. SUMMARY

A new wave of biomaterials-based approaches assures to promote regeneration of damaged or diseased tissue. Understanding *in vivo* cell/tissue/medical implant interactions is crucial in medical device design. The choice of models and metrics influences the success or failure of a therapy, with the choices including evaluations of impacts on quality of life and economics of intervention. Tracking tissue injury, scaffold degradation, and tissue regeneration non-invasively is also of great significance for regenerative medicine. Traditional methods to evaluate the biological response to implants, material degradation, and tissue regeneration have mostly relied on histology, where biomaterials or devices are implanted into animals that are sacrificed at different points. New measurement tools, engineering methods, design principles, non-invasive and real-time imaging assays are urgently needed to shift the field forward. Recent developments in the design of biomaterials, molecular imaging probes, and label-free imaging techniques allow the monitoring of dynamic cellular/tissue/material response in live animals.

REFERENCES

[1] Sandhu HS, Boden SD. Biologic enhancement of spinal fusion. Orthop Clin North Am 1998;29:621–31.

[2] Ludwig SC, Boden SD. Osteoinductive bone graft substitutes for spinal fusion. Orthop Clin North Am 1999;30:635–45.

[3] Schimandle JH, Boden SD. The use of animal models to study spinal fusion. Spine 1994;19:1998–2006.

[4] Muschler GF, Raut VP, Patterson TE, Wenke JC, Hollinger JO. The design and use of animal models for translational research in bone tissue engineering and regenerative medicine. Tissue Eng Part B Rev 2010;16:123–45.

[5] Lu SS, Zhang X, Soo C, Hsu T, Napoli A, Aghaloo T, et al. The osteoinductive properties of Nell-1 in a rat spinal fusion model. Spine 2007;7:50–60.

[6] Rao RD, Bagaria VB, Cooley BC. Posterolateral intertransverse lumbar fusion in a mouse model: surgical anatomy and operative technique. Spine 2007;7:61–7.

[7] Peterson B, Iglesias R, Zhang J, Wang JC, Lieberman JR. Genetically modified human derived bone marrow cells for posterolateral lumbar spine fusion in athymic rats: beyond conventional autologous bone grafting. Spine 2005;30:283–9.

[8] Lu J, Bhargav D, Wei AQ, Diwan A. Posterolateral intertransverse spinal fusion possible in osteoporotic rats with BMP-7 in a higher dose delivered on a composite carrier. Spine 2008;33:242–9.

[9] Carlisle E, Fischgrund JS. Bone morphogenetic proteins for spinal fusion. Spine 2005;5:240S–9S.

[10] Barck KH, Lee WP, Diehl LJ, Ross J, Gribling P, Zhag Y, et al. Quantification of cortical bone loss and repair for therapeutic evaluation in collagen-induced arthritis, by micro-computed tomography and automated image analysis. Arthritis Rheum 2004;50:3377–86.

[11] Farey ID, McAfee PC, Gurr KR, Randolph MA. Quantitative histologic study of the influence of spinal instrumentation on lumbar fusions: a canine model. J Orthopaedic Res 1989;7:709–22.

[12] Ding M, Cheng L, Bollen P, Schwarz P, Overgaard S. Glucocorticoid induced osteopenia in cancellous bone of sheep: validation of large animal model for spine fusion and biomaterial research. Spine 2010;35:363–70.

[13] Kandizora F, Schmidmaier G, Schollmeier G, Bail H, Pflugmacher R, Gorke T, et al. IGF-I and TGF-beta1 application by a poly-(D, L-lactide)-coated cage promotes intervertebral bone matrix formation in the sheep cervical spine. Spine 2002;27(16):1710–23.

[14] Zdeblick TA, Wilson D, Cooke ME, Kunz DN, McCabe R, Ulm MJ, et al. Anterior cervical discectomy and fusion. A comparison of techniques in an animal model. Spine 1992;17(Suppl. 10):S418–26.

[15] Zdeblick TA, Cooke ME, Wilson D, Kunz DN, McCabe R. Anterior cervical discectomy, fusion, and plating. A comparative animal study. Spine 1993;18(14):1974–83.

[16] Zdeblick TA, Cooke ME, Kunz DN, Wilson D, McCabe RP. Anterior cervical discectomy and fusion using a porous hydroxyapatite bone graft substitute. Spine 1994;19(20):2348–57.

[17] Zdeblick TA, Ghanayem AJ, Rapoff AJ, Swain C, Bassett T, Cooke ME, et al. Cervical interbody fusion cages. Spine 1998;23(7):758–65.

[18] Sandhu HS, Toth JM, Diwan AD, Seim HB, Kanim LE, Kabo JM, et al. Histologic evaluation of the efficacy of rhBMP-2 compared with autograft bone in sheep spinal anterior interbody fusion. Spine 2002;27(6):567–75.

[19] Boden SD, Martin GJ, Horton WC, Truss TL, Sandhu HS. Laparoscopic anterior spinal arthrodesis with rhBMP-2 in a titanium interbody threaded cage. J Spinal Disord 1998;11(2):95–101.

[20] Boden SD, Titus L, Hair G, Liu Y, Viggeswarapu M, Nanes MS, et al. spine fusion by local gene therapy with a cDNA encoding a novel osteoinductive protein (LMP-1). Spine 1998;23(23):2486–92.

[21] Louis-Ugbo J, Kim HS, Boden SD, Mayr MT, Li RC, Seeherman H, et al. Retention of 125I-labeled recombinant human bone morphogenetic protein-2 by biphasic calcium phosphate or a composite sponge in a rabbit posterolateral spine arthrodesis model. J Orthop Res 2002;20(5):1050–9.
[22] Sandhu HS, Kanim LE, Kabo JM, Toth JM, Zeegen EN, Liu D, et al. Effective doses of recombinant human bone morphogenetic protein-2 in experimental spinal fusion. Spine 1996;21(18):2115–22.
[23] Blattert TR, Delling G, Dalal PS, Toth CA, Balling H, Weckbach A. Successful transpedicular lumbar interbody fusion by means of a composite of osteogenic protein-1 (rhBMP-7) and hydroxyapatite carrier. Spine 2002;27(23):2697–705.
[24] Komnenou A, Karayannopoulou M, Polizopoulou ZS, Constantinidis TC, Dessiris A. Correlation of serum alkaline phosphatase activity with the healing process of the long bone fractures in dogs. Vet Clin Pathol 2005;34(1):35–8.
[25] Paskalev M, Krastev S, Filipov J. Changes in some serum bone markers after experimental fracture and intramedullary osteomyelitis and osteosynthesis in dogs. Trakia J Sci 2005;3(5):46–50.
[26] Kind PRN, King EJ. Alkaline phosphatase activity determination in serum. J Clin Pathol 1954;7:322–6.
[27] Trinder P. Calorimetric micro-determination of calcium in serum. Analyst 1960;85:889–94.
[28] Gomori G. A modification of calorimetric phosphorus determination. J Lab Clin Med 1942;27:955–8.
[29] Suryawanshi SB, Maiti SK, Varshney VP, Singh GR. Effect of anabolic hormones on haemato-biochemical changes in fracture healing in dogs. Indian J Anim Sci 1999;69(6):404–6.
[30] Soliman FA, Hassan SYS. Serum calcium and Phosphorus in rabbits during fracture healing with reference to parathyroid activity. Nature 1964;204:693–4.
[31] Pandey SK, Udupa KN. Effect of anabolic hormone on certain metabolic response after fracture in dogs. Indian Vet J 1981;58:37–41.
[32] Shifrin LZ. Correlation of serum alkaline phosphatase with bone formation rates. Clin Orthop Relat Res 1970;70:212–5.
[33] Aithal HP, Mogha IV, Singh GR, Swarup D. Serum acid and alkaline phosphatase activities following bone grafting in goats. Indian J Vet Surg 1995;16:47–9.
[34] Volpin G, Rees JA, Ali SY, Bently G. Distribution of alkaline phosphatase activity in experimentally produced callus in rats. J Bone Joint Surg 1986;68:629–34.
[35] Herrmann M, Klitscher D, Georg T, Frank J, Marzi L, Herrman W. Different kinetics of bone markers in normal and delayed fracture healing of long bones. Clin Chem 2002;48:2263–6.
[36] Arinzeh TL, Peter SJ, Archambault MP, van Den Bos C, Gordon S, Kraus K, et al. Allogenic mesenchymal stem cells regenerate bone in a critical-sized canine segmental defect. J Bone Joint Surg Am 2003;85:1927–35.
[37] Bruder SP, Kraus KH, Goldberg VM, Kadiyala S. The effects of implants loaded with autologous mesenchymal stem cells on the healing of canine segmental bone defects. J Bone Joint Surg Am 1998;80:985–6.
[38] Van, Hemert WL, Willems K, Anderson PG, van, Heerwarden RJ, Wymenga AB. Tricalcium phosphate granules or ridge wedge performs in open wedge high tibial osteotomy: a radiological study with a new evaluation system. Knee 2004;11:451–6.
[39] Johnson KD, Frierson KE, Keller TS, Cook C, Scheinberg R, Zerwekh J, et al. Porous ceramics as bone graft substitutes in long bone defects: a biomechanical, histological, and radiographic analysis. J Orthop Res 1996;14(3):351–69.

[40] Schmidmaier G, Wildemann B, Heeger J, Gabelein T, Flyvbjerg A, Bail HJ, et al. Improvement of fracture healing by systemic administration of growth hormone and local application of Insulin-like growth factor-1 and transforming growth factor- β1. Bone 2002;31(1):165–72.

[41] Martin I, Mastrogiacomo M, De Leo G, Muraglia A, Beltrame F, Cancedda R, et al. Fluorescence microscopy imaging of bone for automated histomorphometry. Tissue Eng 2002;8:847.

[42] Salomé M, Peyrin F, Cloetens P, Baruchel J, Spanne P, Suortti P, et al. Assessment of bone micro-architecture using 3D computed microtomography. ESRF Newsl 1997;1:26–8.

[43] Salomé M, Peyrin F, Cloetens P, Odet C, Laval-Jeantet AM, Baruchel J, et al. A synchrotron radiation microtomography system for the analysis of trabecular bone samples. Med Phys 1999;26:2194–204.

[44] Peyrin F, Salomé M, Cloetens P, Laval-Jeantet AM, Ritman E, Rügsegger P. Micro-CT examinations of trabecular bone samples at different resolutions: 14, 7 and 2 micron level. Technol Health Care 1998;6:391–401.

[45] Weiss P, Obadia L, Magne D, Bourges X, Rau C, Weitkamp T, et al. Synchrotron X-ray microtomography (on a micron scale) provides three-dimensional imaging representation of bone ingrowth in calcium phosphate biomaterials. Biomaterials 2003;24:4591–601.

[46] Mastrogiacomo M, Muraglia A, Komlev V, Peyrin F, Rustichelli F, Crovace A, et al. Tissue engineering of bone: search for a better scaffold. Orthod Craniofac Res 2005;8:277–84.

[47] Renghini C, Komlev V, Fiori F, Verne E, Baino F, Vitale-Brovarone C. Micro-CT studies on 3-D bioactive glass–ceramic scaffolds for bone regeneration. Acta Biomater 2009;5: 1328–37.

[48] Ghosh SK, Nandi SK, Kundu B, Dutta S, De DK, Roy SK, et al. In vivo response of porous hydroxyapatite and ß-tricalcium phosphate prepared by aqueous solution combustion method and comparison with bioglass scaffolds. J Biomed Mater Res Appl Biomater 2008;86:217–27.

[49] Parfitt M, Drezner MK, Glorieux FH, Kanis JA, Malluche H, Meunier PJ. Bone histomorphometry: standardization of nomenclature, symbols, and units. J Bone Miner Res 1987;2:595–610.

[50] Movat HZ. Demonstration of all connective tissue elements in a single section; pentachrome stains. AMA Arch Pathol 1955;60(3):289–95.

[51] Sweat F, Puchtler H, Rosenthal SI. Sirius red F3ba as a stain for connective tissue. Arch Pathol 1964;78:69–72.

[52] Bolton WE, Freeman JW, Mikulka WR, Healy CG, Schmittling RJ, Kenyon NS. Expression of proliferation-associated antigens (PCNA, P.120, p.145) during the reentry of G0 cells into the cell cycle. Cytometry 1994;17:66–74.

[53] Allen MJ. Biochemical markers of bone metabolism in animals: uses and limitations. Vet Clin Pathol 2003;32:101–13.

[54] Zaidi M. Skeletal remodeling in health and disease. Nat Med 2007;13:791–801.

[55] Bernhard JM, Blanks JK, Hintz CJ, Chandler GT. Use of the fluorescent calcite marker calcein to label foraminiferal tests. J Foraminiferal Res 2004;34:96–101.

[56] Pautke C, Vogt S, Tischer T, Wexel G, Deppe H, Milz S, et al. Polychrome labeling of bone with seven different fluorochromes: enhancing fluorochrome discrimination by spectral image analysis. Bone 2005;37:441–5.

[57] Suzuki HK, Mathews A. Two-color fluorescent labeling of mineralizing tissues with tetracycline and 2, 4-bis[N, N′-di-(carbomethyl)aminomethyl] fluorescein. Stain Technol 1966;41:57–60.

[58] Dhem A, Piret N, Fortunati D. Tetracyclines, doxycycline, and calcified tissues, Scand. J Infect Dis Suppl 1976:42–6.

[59] Milch RA, Rall DP, Tobie JE. Fluorescence of tetracycline antibiotics in bone. J Bone Joint Surg Am A 1958;40:897–910.

[60] Rahn BA, Perren SM. Alizarin complexone–fluorochrome for bone and dentine labeling. Experientia 1972;28:180.

[61] Rahn BA, Perren SM. Xylenol orange, a fluorochrome useful in polychrome sequential labeling of calcifying tissues. Stain Technol 1971;46:125–9.

[62] Zaheer A, Lenkinski RE, Mahmood A, Jones AG, Cantley LC, Frangioni JV. In vivo near-infrared fluorescence imaging of osteoblastic activity. Nat Biotechnol 2001;19:1148–54.

[63] Gibson CJ, Thronton VF, Brown WAB. Incorporation of tetracycline into impeded and unimpeded mandibular incisors of the mouse. Calcif Tissue Res 1978;26:29–31.

[64] Dahners LE, Bos GD. Fluorescent tetracycline labeling as an aid to debridement of necrotic bone in the treatment of chronic osteomyelitis. J Orthop Trauma 2002;16:345–6.

[65] Kleihues P, Sobin LH. World Health Organization classification of tumors. Cancer 2000;88:2887.

[66] Frost HM. Preparation of undecalcified bone sections by rapid manual method. Stain Technol 1958;33:273–7.

[67] Maiti SK, Singh GR. Different types of bone grafts and ceramic implants in goats. A triple fluorochrome labeling study. Indian J Anim Sci 1995;65:140–3.

[68] Nandi SK, Kundu B, Ghosh SK, De DK, Basu D. Efficacy of nano-hydroxyapatite prepared by an aqueous solution combustion technique in healing bone defects of goat. J Vet Sci 2008;9(2):183–91.

[69] Nandi SK, Ghosh SK, Kundu B, De DK, Basu D. Evaluation of new porous β-tri-calcium phosphate ceramic as bone substitute in goat model. Small Ruminant Res 2008;75(2–3):144–53.

[70] Nandi SK, Kundu B, Datta S, De DK, Basu D. The repair of segmental bone defects with porous bioglass: an experimental study in goat. Res Vet Sci 2009;86(1):162–73.

[71] Tam CS, Anderson W. Tetracycline labelling of bone in vivo. Calcif Tissue Int 1980;30:121–5.

[72] Wray JB, Lynch CJ. The vascular response to fracture of tibia in rat. J Bone Joint Surg 1959;4:1143–8.

[73] Macnab I, Dettass WG. The role of periosteal blood supply in the healing of fractures of the tibia. Clin Orthop Relat Res 1974;105:27–33.

[74] Gothman L. Vascular reaction in experimental fractures. Microangiographic and radioisotopic studies. Acta Chir Scand 1961;312(Suppl.):9–16.

[75] Schelling SH. Secondary (classical) bone healing. Semin Vet Med Surg (Small Animals) 1991;6:16–20.

[76] Koekenberg LTC. Vascularization in the healing of fractures. Springfield: Charles C. Thomas; 1963.

[77] Singh AP, Nigam JM, Singh GR. Allogenic bone transplantation in bovines. An angiographic study. Indian J Anim Sci 1978;48:374–80.

[78] Punto L, Puranen J, Mokka R. Osteomedullography in tibial shaft fracture of the dog and pig. J Am Vet Radiol Sci 1977;18:102.

[79] Maiti SK, Singh GR. Vascularization of composite bone grafts and ceramic implants in goats. Small Ruminant Res 1996;20:171–6.

[80] Steinbach HL, Jergesen F, Geilfiltan RS, Petarakis ML. Osseous phlebography. Surg Gynaecol Obstet 1957;104:215–26.

[81] Deleu J, Trueta. Vascularization of bone grafts in the anterior chamber of eye. J Bone Joint Surg 1965;47-B(2):319–29.
[82] Varshney AC, Singh H, Kumar A. Angiographic studies on experimental osteomyelitis in dog. Indian J Vet Surg 1994;15(2):74–7.
[83] Rao RLN, Singh AP, Nigam JM, Chawla SK. Response of intraosseous venous circulation of metacarpal fracture healing in buffalo-calves. Indian J Anim Sci 1985;55:632.
[84] Kolambkar YM, Dupont KM, Boerckel JD, Huebsch N, Mooney DJ, Hutmacher DW, et al. An alginate-based hybrid system for growth factor delivery in the functional repair of large bone defects. Biomaterials 2011;32(1):65–74.
[85] Duvall CL, Taylor WR, Weiss D, Wojtowicz AM, Guldberg RE. Impaired angiogenesis, early callus formation, and late stage remodeling in fracture healing of osteopontin-deficient mice. J Bone Miner Res 2007;22:286–97.
[86] De Aza PN, Luklinska ZB, Anseau MR, Guitian F, De Aza S. Electron microscopy of interfaces in a Wollastonite-tricalcium phosphate bioeutectic. J Microsc 1998;189(2): 145–53.
[87] De Aza PN, Luklinska ZB, Anseau MR, Guitian F, De Aza S. Transmission electron microscopy of the interface between bone and pseudowollastonite implant. J Microsc 2001;201(1):33–43.
[88] Devlin H, Horner K, Ledgerton DA. Comparison of maxillary and mandibular bone mineral densities. J Prosthet Dent 1998;79:323–7.
[89] An YH, Draughn RA. Mechanical testing of bone and bone – implant interface. Tulsa, OK: CRC Press LLC; 1999.
[90] Gheduzzi S, Webb JJ, Miles AW. Mechanical characterization of three percutaneous vertebroplasty biomaterials. J Mater Sci Mater Med 2006;17:421–6.
[91] Hamson KR, Toth JM, Stiehl JB, Lynch KL. Preliminary experience with a novel model assessing in vivo mechanical strength of bone grafts and substitute materials. Calcif Tissue Int 1995;57:64–8.
[92] Thomas KA, Kay JF, Cook SD, Jarcho M. The effect of surface macrotexture and hydroxylapatite coating on the mechanical strengths and histologic profiles of titanium implant materials. Biomed Mater Res 1987;21:1395–414.
[93] Verheyen CCPM, Dhert WJA, Braak LH, de K. Groot, Push-out tests evaluated by finite element analysis, Trans Soc Biomuter 1991;17:216.
[94] Dhert WJ, Verheyen CC, Braak LH, de, Wijn JR, Klein CP, de, Groot K. A finite element analysis of the push-out test: influence of test conditions. J Biomed Mater Res 1992;26:119–30.
[95] Turner CH, Burr DB. Basic biomechanical measurements of bone: a tutorial. Bone 1993;14:595–608.
[96] Jämsä T, Jalovaara P. A cost-effective, accurate machine for testing torsional strength of sheep long bones. Med Eng Phys 1996;18:433–5.
[97] Branemark R, Skalak R. An in-vivo method for biomechanical characterization of bone-anchored implants. Med Eng Phys 1998;20:216–9.
[98] Harris WH. The problem is osteolysis. Clin Orthop Relat Res 1995;311:46–53.
[99] Staniforth M, Goldstein J, Newbury DE, Lyman CE, Echlin P, Lifshin E, et al. Scanning electron microscopy and x-ray microanalysis. 3rd ed. New York Boston, Dordrecht, London, Moscow: Kluwer Academic/Plenum Publishers; 2002.
[100] Meredith N, Book K, Friberg B, Jemt T, Sennerby L. Resonance frequency measurements of implant stability in vivo. A cross-sectional and longitudinal study of resonance frequency measurements on implants in the edentulous and partially dentate maxilla. Clin Oral Implant Res 1997;8(3):226–33.

[101] Meredith N. Assessment of implant stability as a prognostic determinant. Int J Prosthodont 1998;11:491–501.

[102] Huang HM, Cheng KY, Chen CF, Ou KL, Li CT, Lee SY. Design of a stability-detecting device for dental implants. Proc Inst Mech Eng [H] 2005;219:203–11.

[103] Lerner RM, Parker KJ, Holen J, Gramiak R, Waag RC. Sonoelasticity: medical elasticity images de rived from ultra sound signals in mechanically vibrated targets. Acoust Imaging 1988;16:317–27.

[104] Hall TJ, Zhu Y, Spalding CS. In vivo real-time free-hand palpation imaging. Ultrasound Med Biol 2003;29:427–35.

[105] Ophir J, Céspedes I, Ponnekanti H, Yazdi Y, Li X. Elastography: a quantitative method for imaging the elasticity of biological tissues. Ultrason Imaging 1991;13:111–34.

[106] Kim K, Jeong CG, Hollister SJ. Non-invasive monitoring of tissue scaffold degradation using ultrasound elasticity imaging. Acta Biomater 2008;4:783–90.

[107] Bae Y, Nishiyama N, Kataoka K. In vivo antitumor activity of the folate-conjugated pH-Sensitive polymeric micelle selectively releasing adriamycin in the intracellular acidic compartments. Bioconjug Chem 2007;18:1131–9.

[108] Akiyama Y, Mori T, Katayama Y, Niidome T. The effects of PEG grafting level and injection dose on gold nanorod biodistribution in the tumor-bearing mice. J Control Release 2009;139:81–4.

[109] Sun W, Zou W, Huang G, Li A, Zhang N. Pharmacokinetics and targeting property of TFu-loaded liposomes with different sizes after intravenous and oral administration. J Drug Target 2008;16:357–65.

[110] Rossin R, Pan D, Qi K, Turner JL, Sun X, Wooley KL, et al. 64Cu-labeled folate-conjugated shell cross-linked nanoparticles for tumor imaging and radiotherapy: synthesis, radiolabeling, and biologic evaluation. J Nucl Med 2005;46:1210–8.

[111] Owens DE, Peppas NA. Opsonization, biodistribution, and pharmacokinetics of polymeric nanoparticles. Int J Pharm 2006;307:93–102.

[112] Zhang G, Yang Z, Lu W, Zhang R, Huang Q, Tian M, et al. Influence of anchoring ligands and particle size on the colloidal stability and in vivo biodistribution of polyethylene glycol-coated gold nanoparticles in tumor-xenografted mice. Biomaterials 2009;30:1928–36.

[113] Jong De WH, Hagens WI, Krystek P, Burger MC, Sips AJ, Geertsma RE. Particle size-dependent organ distribution of gold nanoparticles after intravenous administration. Biomaterials 2008;29:1912–9.

[114] Glangchai LC, Caldorera-Moore M, Shi L, Roy K. Nanoimprint lithography based fabrication of shape-specific, enzymatically-triggered smart nanoparticles. J Control Release 2008;125:263–72.

[115] Gratton SE, Ropp PA, Pohlhaus PD, Luft JC, Madden VJ, Napier ME, et al. The effect of particle design on cellular internalization pathways. Proc Natl Acad Sci U S A 2008;105: 11613–8.

[116] Park JH, von Maltzahn G, Zhang L, Schwartz MP, Ruoslahti E, Bhatia SN, et al. Magnetic iron oxide nanoworms for tumor targeting and imaging. Adv Mater 2008;20:1630–5.

[117] Sun X, Rossin R, Turner JL, Becker ML, Joralemon MJ, Welch MJ, et al. An assessment of the effects of shell cross-linked nanoparticle size, core composition, and surface PEGylation on in vivo biodistribution. Biomacromolecules 2005;6:2541–54.

[118] Avgoustakis K, Beletsi A, Panagi Z, Klepetsanis P, Livaniou E, Evangelatos G, et al. Effect of copolymer composition on the physicochemical characteristics, in vitro stability, and biodistribution of PLGA-mPEG nanoparticles. Int J Pharm 2003;259:115–27.

[119] Boyce ST. Warden GD: principles and practices for treatment of cutaneous wounds with cultured skin substitutes. Am J Surg 2002;183:445–56.

[120] McManus GA, Mowry RW. Staining methods: histological and histochemical. New York: Harper and Row; 1964. p. 8.

[121] Kumar A. Veterinary surgical techniques. New Delhi: Vikas Publishing House; 1996. pp. 151–165.

[122] Lund LR, Rømer J, Bugge TH, Nielsen BS, Frandsen TL, Degen JL, et al. Functional overlap between two classes of matrix-degrading proteases in wound healing. EMBO J 1999;18:4645–56.

[123] Ågren MS, Mirastschijski U, Karlsmark T, Saarialho-Kere UK. Topical synthetic inhibitor of matrix metalloproteinases delays epidermal regeneration of human wounds. Exp Dermatol 2001;10:337–48.

[124] Mirastschijski U, Impola U, Karsdal MA, Saarialho-Kere U, Ågren MS. Matrix metalloproteinase inhibitor BB-3103 unlike the serine proteinase inhibitor aprotinin abrogates epidermal healing of human skin wounds ex vivo. J Invest Dermatol 2002;118:55–64.

[125] Towbin H, Staehelin T, Gordon J. Electrophoretic transfer of proteins from polyacrylamide gels to nitrocellulose sheets: procedure and some applications. Proc Natl Acad Sci U S A 1979;76:4350–4.

[126] Phillips C, Wenstrup RJ. Biosynthetic and genetic disorders of collagen. In: Cohen K, Diegelman RF, Winblad WJ, editors. Wound healing, biochemical and clinical aspects. Philadelphia: Saunders; 1992. p. 252–61.

[127] Neuman RE, Logan MA. The determination of hydroxyproline. J Biol Chem 1950;184: 299–306.

[128] Adamsons RJ, Musco F, Enquist IF. The relationship of collagen content to wound strength in normal and scorbutic animals. Surg Gynaecol Obstet 1964;119:323–9.

[129] Ruggeri A, Dell'Orbo C, Quacci D. Electron microscopic visualization of proteoglycans with Alcian Blue. Histochem J 1975;7:187–97.

[130] Bouhout S, Perron E, Gauvin R, Bernard G, Ouellet G, Cattan V, et al. In vitro reconstruction of an autologous, watertight, and resistant vesical equivalent. Tissue Eng Part A 2010;16:1539–48.

Chapter 7.1

Structural and Biological Characterization of Scaffolds

Julia Will, Rainer Detsch and Aldo R. Boccaccini
Institute of Biomaterials, University of Erlangen-Nuremberg, Erlangen, Germany

Chapter Outline

7.1.1. INTRODUCTION

The most important function of a bone tissue engineering (TE) scaffold is its role as a template that allows cells to attach, proliferate, differentiate and organize into normal, healthy bone as the scaffold degrades. Figure 7.1.1 illustrates the most important factors involved in the design of TE scaffolds and their interdependencies, according to Guarino et al. [1]. Depending on the final application, scaffold requirements include matching the structural and mechanical properties with those of the recipient tissue and optimization of the micro-environment to support cell integration, adhesion and growth, issues that have become known as structural and surface compatibility of biomaterials [2].

Scaffolds have to fulfil many requirements, such as osteoconductivity [1,3,4], appropriate rate of biodegradation, interconnected porosity and pore diameters in a wide range (e.g. 10–500 μm) [1,3–6], microporosity [5,7–9],

Characterization of Biomaterials. http://dx.doi.org/10.1016/B978-0-12-415800-9.00008-5

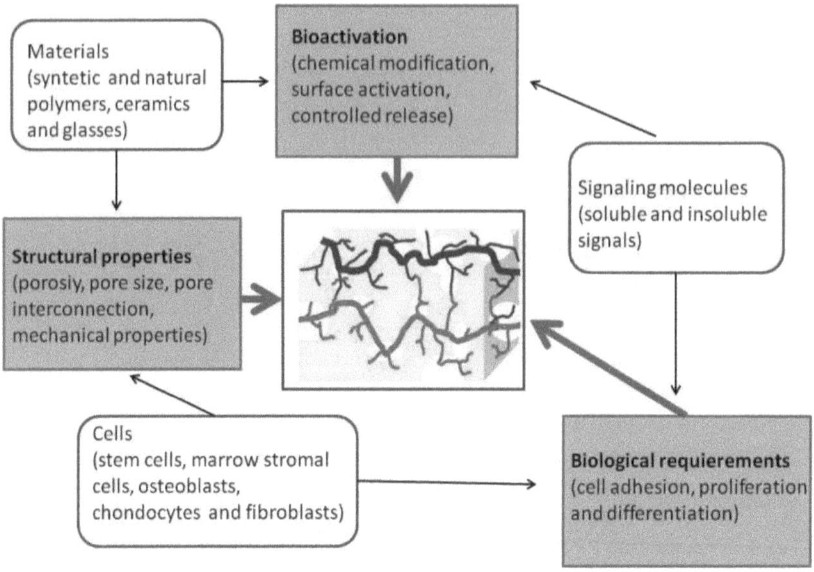

FIGURE 7.1.1 Schematic diagram of key factors involved in the design of optimal scaffolds for bone tissue engineering. *Modified after Ref. [1].*

suitable mechanical strength and structural integrity [7]. Therefore, the complete characterization of scaffold structure and properties is essentially the first step in the process of developing successful bone engineering scaffolds.

Characterization of scaffolds is far from trivial as they are inherently complex structures due to their porous nature [8]. An impressive number of suitable methods are available for scaffold characterization: multi-scale characterization methods have to be used, however, not all methods will be relevant or applicable in every case. One of the problems in choosing adequate characterization techniques is the identification of the key parameters required to adequately characterize scaffolds. Porosity is one such property, discussed widely in the literature [5,7,9–13]. Porosity is a measure of the void space in a material that can be determined from the ratio of the void volume to the bulk material volume [14,15]. Porosity is known to play a role in determining cell seeding efficiency in addition to the diffusion properties and mechanical strength of a scaffold [5,10]. However, it is not a unique parameter (i.e. structurally different materials can have identical porosities) and it cannot be used on its own to sufficiently characterize scaffolds. Rather than basing the scaffold characterization on one parameter, such as porosity, it has been suggested that a multi-parameter approach is needed [14]. Porosity, pore size, and permeability are thus interrelated architectural properties that have been shown to influence both fluid flow through the scaffold and scaffold mechanical properties. Candidate parameters to be determined include the pore-size distribution, pore interconnectivity, tortuosity, total pore surface, and permeability [9,12,14,16–18].

Numerous studies demonstrate that pore size, interconnectivity and porosity affect bone tissue regeneration, and these three design features appear to be the most important structural variables in an initial scaffold screening algorithm as developed by Cleyenbreugel et al. [19] for bone TE applications. Clearly, besides scaffold morphology and mechanical and surface properties, biological characterization of scaffolds by suitable cell culture methods is also required. In the following sections, a summary of characterization methods for scaffolds is presented.

7.1.2. CHARACTERIZATION OF SCAFFOLDS MORPHOLOGY AND POROSITY

Methods to visualize and subsequently quantify scaffold structures, e.g. scanning electron microscopy (SEM) [20–22], micro-computed tomography (μ-CT) [11,16,19,23–25], and confocal laser scanning microscopy [26] are acquiring increasing significance. μ-CT, in particular, has been widely used as an essential scaffold characterization tool with the potential to significantly enhance the understanding of pore structures in porous material in the millimetre to micrometre range [11,27]. This information can be applied to gain knowledge on the correlation between pore structure and mechanical properties and to improve the understanding of the relationship between pore-size distribution and permeability [28]. As compared to porosimetry methods, μ-CT can access both connected and isolated pores enabling the total void volume to be determined [12]. μ-CT is also capable of non-destructive assessment of mineralization processes within 3D scaffolds *in vitro* [13].

Porosity being the percentage of void space in a solid [29] is a morphological property independent of the material [5]. Porosity assessment via porosimetry is based on the study of the flow of gases or liquids (or both), across a porous structure. This method, therefore, is only suitable for the detection of open pores that allow fluid transport. Consequently, standard porosimetry methods cannot be used to assess total pore volume. Liquid intrusion methods (e.g. mercury intrusion porosimetry) are based on the pressurized penetration of a liquid into a porous structure and are capable of determining the total pore volume exposed to the outside of a structure. As with flow porosimetry, closed pores are hidden from the test. Other strategies used for scaffold characterization include gas pycnometry [10,30], gravimetry [30,31], gas adsorption [10], Archimedes principle [22], and molecular diffusion studies [14]. It is recognized that part of the difficulty in identifying a suitable measurement strategy is the fact that no single investigative technique is able to fully characterize the porous nature of scaffolds [14] if they exhibit porosity in different scales; e.g. in some cases from nano- to macroporosity.

7.1.3. PERMEABILITY

Successful bone TE depends on the scaffold's ability to allow nutrient diffusion to and waste removal from the regeneration site, therefore, permeability is a key parameter for the design of scaffolds. Permeability is directly related to the degree of pore interconnectivity [18]. Studies have found that cell growth into a scaffold depends on how well nutrients can permeate through the porous structure during the cell migration process [17,32,33].

Several permeability measurement systems have been developed for determining the permeability of scaffolds. Gravity-induced systems have been applied [33–36] using oil and water as fluid, respectively, or compressed air [37,38]. Swider et al. [39] used a high-resolution MRI methodology for characterizing the permeability and fluid velocity while Maxwell and Wei [38] used dry air as the fluid medium allowing rapid measurement operations. The permeability of a commercial bioceramic scaffold has been evaluated through numerical modelling by Sanz-Herrera et al. [40].

The physical principle used for the measurement of permeability is based on the measurement of the pressure drop caused by the introduction of the scaffolds in a fluid (e.g. water or cell culture medium). Thereby, the scaffold permeability can be easily determined by applying simple mathematical relations. Based on the relationship between the pressure drop gradient and the fluid flow velocity [18] given by Darcy's law, the intrinsic permeability k [m^2] can be obtained as the linear relationship between flow rate and pressure gradient by the following equation [41]:

$$Q = \frac{-kA}{\mu}\frac{(P_b - P_a)}{L}$$

where Q is the flow rate (m^3/s), A is the cross-sectional area (m^2), $P_b - P_a$ is the drop pressure between two points spaced L, μ is the dynamic fluid viscosity (Pa s), and L is the scaffold thickness.

The effects that scaffold permeability has on bone tissue regeneration have not been studied in depth using accurately controlled porous architectures with reproducibly designed permeability [42]. However there have been few studies that directly examined construct permeability and its effect on bone regeneration [42,43].

7.1.4. MECHANICAL CHARACTERIZATION OF SCAFFOLDS

Clearly, understanding the correlation between pore structure, porosity, and scaffold mechanical properties is crucial in the process of optimization of scaffold architecture [44–48]. In most of the cases, compressive mechanical testing is used to measure the mechanical strength of a scaffold. It can be considered that scaffolds for bone regeneration should have a minimum compressive strength of ~2 MPa and a minimum modulus of ~50 MPa, which are at the low range of properties for trabecular bone [42]. Studies usually report the measurement of compression strength via uniaxial testing. Specimens are compressed between two fixed steel plates at a rate of typically 1.0 mm/min. Compressive strength may be calculated as the load carried at the 0.2% offset point divided by the original scaffold cross-sectional area [42]. Thus, the compression strength test, being comprehensively and successfully employed to characterize the mechanical properties of porous ceramics [49], is the most applied method on hydroxyapatite, bioactive glass and composite scaffolds. The area under the stress–displacement curve obtained in a compression strength test is usually considered to access the work of fracture, a measure of the toughness of the scaffold [50].

7.1.5. BIOLOGICAL CHARACTERIZATION OF SCAFFOLDS

Scaffolds for bone TE need to be characterized *in vitro* by cell culture methods before *in vivo* and clinical studies take place. The *in vitro* investigations can be divided into two levels. The first level involves analyzing cyto and structural compatibility of the scaffold materials with selected cell lines. The second level considers the in-time interactions of specific, relevant cells and scaffolds, in particular, cell attachment, cell proliferation and especially osteogenic cell differentiation should be investigated. These detailed aspects of cell–scaffold interaction can be tested in the laboratory using primary cell culture.

7.1.5.1. First Level: Cyto and Structural Compatibility

Besides choosing the right sterilization method, the first question regarding application of scaffolds for bone TE is the cyto compatibility of the newly developed material. Evaluating biomaterial scaffolds requires information on how specific parameters such as material composition and topography will influence cell adhesion, viability, proliferation, necrosis and apoptosis. Testing cell interactions with scaffolds requires selection of parameters such as type of cells, duration of the exposure and method of evaluation. In screening tests, the effect of biomaterials (dense discs or porous scaffolds) on the functions of cell types is investigated by continuous cell lines. Different cell types are available for measuring cell behavior on biomaterials for bone regeneration, for example, MC3T3-E1 (mouse calvaria osteoblast-like cells), SaOs-2 and MG-63 (both human osteosarcoma cell lines) [51,52]. During various incubation periods of 24 h up to 3 days, essential parameters, like cell morphology, are determined using light or scanning electron microscope imaging. The number of attached cells (by intracellular *LDH* activity or *BrdU* assay) and the cell viability (by mitochondrial activity, MTT or WST-1) are analyzed.

Scaffolds are designed for homing cells and they should mimic the target tissue architecture [53]. Therefore, it is very important to understand cell ingrowth, cell orientation and nutrition supply of these cells inside a porous construct. Engineering of cell-scaffold constructs is limited by gradients in tissue quality from the periphery towards the centre, including inhomogeneous cellular proliferation and differentiation from outer parts of the scaffolds towards the centre. It is well known that high porosity, permeability, appropriate pore size and required surface properties allow cell adhesion, differentiation and proliferation [54]. These tissue quality gradients mostly depend on nutrient and oxygen supply within the culture.

Therefore, a variety of dynamic bioreactor concepts have been developed for *in vitro* characterization of scaffolds, for example, spinner flasks, rotating wall vessel constructs and perfusion bioreactors, which mimic the native microenvironment in bone tissue [55]. To evaluate the role of nutrient diffusion and consumption within cell-seeded scaffolds (among other things), a glucose diffusion model was developed [32]. For example, metabolic activity of cultured cells in a 3D construct can be measured by the resazurin assay. The *Alamar Blue* assay, another test suitable in cell biology, incorporates a fluorometric/colorimetric growth indicator based on detection of metabolic activity. A widespread and straightforward assay for monitoring cell behavior is the *Live/Dead* staining by *calcein AM* and ethidium homodimer (Fig. 7.1.2).

These cyto and structural compatibility tests are recommended for scaffold materials. Standardized protocols are employed and quantitative and comparable data can be generated. They allow for a fast assessment of the biomaterials, and, due to their sensitivity, potentially toxic materials can be discarded prior to animal testing.

For quantitative analysis, each experiment should be repeated three times with minimal four replicates. Differences between the mean values can be simply evaluated by different statistical approaches, like student-*t* test or

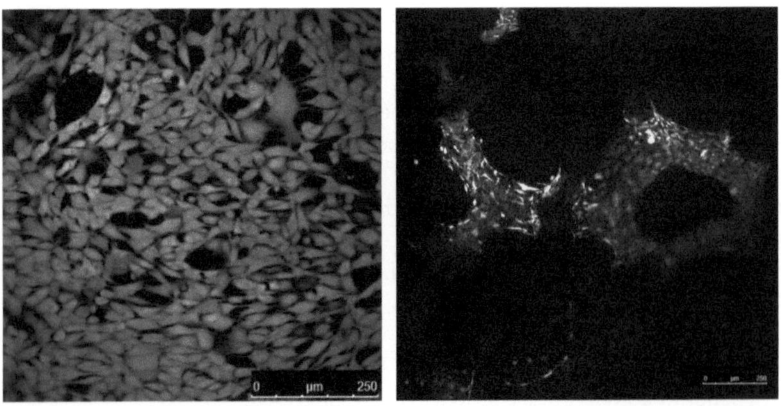

FIGURE 7.1.2 Fluorescence images of Live/Dead stained MG-63 osteoblast-like cells cultured on a dense disc (left) and on a three-dimensional bioactive glass scaffold (right). (Unpublished results, Institute of Biomaterials, University of Erlangen-Nuremberg).

one-way analysis of variance. The results are presented as average and mean ± standard deviation. Differences can be considered significant if their p-values are <0.05 (*p < 0.05) and highly significant if their p-values were <0.01 (**p < 0.01). Using representative microscope images, the morphology of cells on different surfaces should be examined by analyzing over 60–80 cells per sample. Mineralized area of a scaffold, for example, can be assigned by different colors to the respective grey values of SEM images using suitable imaging analysis software. Afterwards the area of interest (in percentage) can be calculated.

7.1.5.2. Second Level: Interaction and Osteogenic Differentiation

In the ideal scenario, scaffolds for bone regeneration should be custom made to the defect, utilizing suitable materials of optimal chemical composition. The effect of scaffold properties (topography, porosity, osteoconductivity, etc.) on specific cell functions must be studied, employing cells of the same type of those which will be in contact with the material *in vivo*. In this context, bone marrow from calvarium, periosteum, or trabecular bone have been claimed to be the most promising sources for mesenchymal stem cells (MSCs), which have a high proliferative ability and great capacity for differentiation [56,57]. Moreover, adipose derived cells and stem cells obtained from dental deciduous dental pulp also show osteogenic potential [58]. First, the differentiation of relatively undifferentiated MSCs along different lineages (for example, osteoblasts, chondrocytes and adipocytes, see Fig. 7.1.3) should be proved.

Typical signalling pathways relevant to osteoblast functions should be studied. In long-term cultivation periods *in vitro*, bone formation/mineralization should be finally checked. For imaging and visualizing the status of cell behavior in contact with the biomaterial, phase contrast microscopy, fluorescent microscopy, confocal laser scanning microscopy, electron microscopy, and

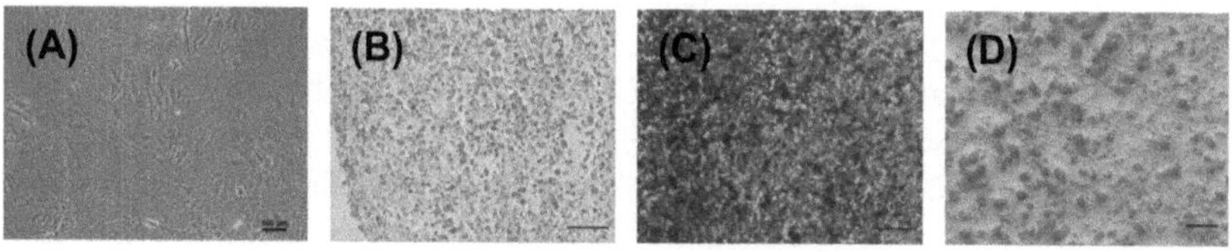

FIGURE 7.1.3 Light-microscope images of bone marrow stem cells cultured with complete medium showing the spindle-shaped morphology typical of undifferentiated MSC (A). Osteogenic differentiation of MSC is demonstrated by formation of calcium-hydroxyapatite-positive areas stained in grey by von Kossa staining (B), chondrogenic differentiation is shown by Safranin O staining (C) and adipogenic differentiation is demonstrated by intracellular lipid vacuoles stained in red by oil red O (D). (Unpublished results, Institute for Biomaterials, University of Erlangen-Nuremberg).

optical coherence tomography have been developed [59,60]. Besides the mentioned test assays and techniques, reverse transcription polymerase chain reaction for determining specific different RNA molecules within a cell or tissue as a measure of gene expression is a useful tool in this field of *in vitro* characterization [61]. Important factors for osteogenic differentiation are MSX-2, DLX-5, Runx-2 or osterix and alkaline phosphatase, collagen type 1, osteocalcin or bone sialoprotein, which represent typical markers to identify osteoblasts [62].

As mentioned above, challenges in bone scaffolding are the realization of a three-dimensional interconnecting porosity, a high total porosity and the optimization of pore geometries with an osteoconductive or even an osteoinductive surface for cell attachment, cell development and bone in-growth. Strategies to investigate cell–scaffold interactions are, in addition to the mentioned methods, the application of different drug delivery systems, which can be designed for (osteo-) inductive properties. Another challenge for future research is exploring different co-culture systems, like osteoblast–osteoclast for bone remodelling or osteoblast–endothelial cells for vascularization studies, in the context of scaffold characterization [54,63]. Additionally, tissue engineers employ computational modelling to understand the relationship between cellular response on scaffold properties, such as porosity, pore geometry as well as physical forces such as fluid flow, stresses and strains (within bioreactors) [64,65].

7.1.6. SUMMARY

Generally speaking, there is a trade-off among the requirements for an optimal scaffold, e.g. scaffold architectures designed to maximize nutrient diffusion typically result in decreased scaffold mechanical strength [42].

In vitro, lower porosity stimulates osteogenesis by suppressing cell proliferation and forcing cell aggregation. In contrast, *in vivo*, higher porosity and pore size result in greater bone in-growth, a conclusion that is supported by the absence of reports showing enhanced osteogenic response for scaffolds with low porosity. However, this trend results in diminished mechanical properties, thereby setting an upper functional limit for pore size and porosity. Thus, a balance must be reached depending on the kinetics of bone repair, rate of degradation of the scaffold material [5].

A smart combination of different characterization methods to access scaffold morphology, porosity, permeability, mechanical properties and *in vitro* cell behavior is thus required for developing optimized scaffold structures for successful bone regeneration.

REFERENCES

[1] Guarino V, Causa F, Ambrosio L. Bioactive scaffolds for bone and ligament tissue. Expert Rev Med Devices 2007;4:405–18.

[2] Ramakrishna S, Mayer J, Wintermantel E, Leong KW. Biomedical applications of polymer-composite materials: a review. Composites Sci Technol 2001;61:1189–224.
[3] Hutmacher DW, Schantz JT, Lam CXF, Tan KC, Lim TC. State of the art and future directions of scaffold-based bone engineering from a biomaterials perspective. J Tissue Eng Regen Med 2007;1:245–60.
[4] Rezwan K, Chen QZ, Blaker JJ, Boccaccini AR. Biodegradable and bioactive porous polymer/inorganic composite scaffolds for bone tissue engineering. Biomaterials 2006;27:3413–31.
[5] Karageorgiou V, Kaplan D. Porosity of 3D biomaterial scaffolds and osteogenesis. Biomaterials 2005;26:5474–91.
[6] Tabata Y. Biomaterial technology for tissue engineering applications. J Royal Soc Interface 2009;6:S311–24.
[7] Hutmacher DW. Scaffolds in tissue engineering bone and cartilage. Biomaterials 2006;21:2529–40.
[8] Mather LM, Morgan PS, White JL, Tai H, Kockenberger W, Howdle SM, et al. Imagebased characterization of foamed polymeric tissue scaffolds. Biomed Mat Res A 2008;3:1–11.
[9] Jones JR, Poologasundarampillai G, Atwood R, Bernard D, Lee P. Non-destructive quatntitatve 3D analysis for the optimisation of tissue scaffolds. Biomaterials 2007;28:1404–13.
[10] Ho ST, Hutmacher DW. A comparison of micro CT with other techniques used in the characterization of scaffolds. Biomaterials 2006;27:1362–76.
[11] Lin A, Barrow T, Cartmell S, Guldberg R. Microarchicteural and mechanical characterization of oriented porous polymer scaffolds. Biomaterials 2003;24:481–9.
[12] Moore M, Jabbari E, Ritman E, Lu L, Currier B, Windebank AJ, et al. Quantitatvie analysis of interconnectivity of porous biodegradable scaffolds with micro-computed tomography. J Biomed Mat Res A 2004;71:258–67.
[13] Cartmell S, Huynh K, Lin A, Nagarja S, Guldberg R. Quantitative microcomjputed tomography analysis of mineralization within three-dimensional scaffolds in vitro. J Biomed Mat Res A 2004;69:97–104.
[14] Tomlins PE, Grant P, Vadgama P, James S, Mikhalovsky S. Characterisation of polymeric tissue scaffolds. Meas Good Pract Guide 2006;89.
[15] Mas Estelles J, Krakovsky I, Rodriguez Hrnandex J, Piotrowska A, Monleon Pradas M. Mechanical properties of porous crosslinked poly(ethyl-acrylate) for tissue engineering. J Mater Sci 2007;42:9629–35.
[16] Rajagopalan S, Yaszemsiki M, Robb R. Evaluation of thresholding techniques for segmenting scaffold images in tissue engineering. Proc SPIE, medical imaging. In: Image processing 5370. 2004. p. 1456–65.
[17] Li S, De Wijn JR, Li J, Layrolle P, De Groot K. Macroporous biphasic calcium phosphate scaffold with high permeability/porosity ratio. Tissue Eng 2003;9:535–48.
[18] Ochoa I, Sanz-Herrera JA, Garcıa-Aznar M, Doblare M, Yunos DA, Boccaccini AR. Permeability evaluation of 45S5 Bioglasss-based scaffolds for bone tissue engineering. J Biomech 2009;42:257–60.
[19] Cleyenbreugel TV, Schrooten J, Oosterwyck HV, Sloten JV. Micro-CT-based screening of biomechanical and structural properties of bone tissue engineering scaffolds. Med Biol Eng Comput 2006;44:517.
[20] Oh SH, Park IK, Kim JM, Lee JH. In vitro and in vivo characteristics of PCL scaffolds with pore size gradient fabrication by a centrifugation method. Biomaterials 2007;28:1664–71.
[21] Maquet V, Blacher S, Pirard R, Pirard JP, Jerome R. Characterization of porous polylactide foams by image analysis and impedance spectroscopy. Langmuir 2000;16:10463–70.

[22] Grant PV, Vaz CM, Tomlins PE, Mikhalovska L, Mikhalovsky S, James S, et al. Physical characterisation of a polycaprolactone tissue scaffold. Blitz JP, Gun'ko JM, editors. Surface chemistry in biomedical and enventiromental science. DEPC-MPR-027; 2005. p. 1–45.

[23] Brunke O, Odenbach S, Beckmann F. Quantitative methods for the analysis of synchrotron-μCT datasets of metallic foams. Eur Phys J Appl Phys 2005;29:73–81.

[24] Cooper DML, Turinsky AL, Sensen CW, Hallgrimsson B. Quantitative 3D analysis of the canal network in cortical bone by micro-computed tomography. Anat Rec 2003;274B: 169–79.

[25] Otsuki B, Takemoto M, Fujibayashi S, Neo M, Kokubo T, Nakamura T. Pore throat size and connectivity determine bone and tissue ingrowth into porous implants: three-dimensional micro-CT based structural analyses of porous bioactive titanium implants. Biomaterials 2006;27:5892–900.

[26] Tjia JS, Moghe PV. Analysis of 3-D microstructure of porous poly(lactide–glycolide) matricies using confocal microscopy. Biomed Mat Res A 1998;43:291–9.

[27] Darling AL, Sun W. Microtomographic characterization of precision extruded poly-ε-caprolactone scaffolds. Biomed Mat Res B 2004;70:311–7.

[28] Landis EN, Petrell AL, Nagy EN. Examination of pore structure and durability issues using three dimensional image analysis of microtomographic data. Austin, TX: 14th ASCE Engineering Mechanics Conf; 2000.

[29] Leon CA. New perspectives in mercury porosimetry. Adv Colloid Interface Sci 1998;76–77:341–72.

[30] Barry H, Silva M, Cartnekk S, Guldberg R, Scotchford C, Howdle SM. Porous methacrylate tissue engineering scaffolds: using carbon dioxide to control porosity and interconnectivity. J Mat Sc 2006;41:4197–204.

[31] Maspero FA, Ruffieux K, Muller B, Wintermantel E. Resorbable defect analog PLGA scaffolds using CO_2 as solvent: structural characterization. J Biomed Mater Res 2002;62:89–98.

[32] Botchwey EA, Dupree MA, Pollack SR, Levine EM, Laurencin CT. Tissue engineered bone: measurement of nutrient transport in three-dimensional matrices. J Biomed Mater Res A 2003;67:357–67.

[33] Grimm MJ, Williams JL. Measurements of permeability in human calcaneal trabecular bone. J Biomech 1997;30:743–5.

[34] Haugen H, Will J, Köhler A, Hopfner U, Aigner J, Wintermantel E. Ceramic TiO_2-foams: characterisation of a potential scaffolds. J Eur Chem Soc 2004;24:661–8.

[35] Kohles SS, Roberts JB, Upton ML, Wilson CG, Bonassar LJ, Schlichting AL. Direct perfusion measurements of cancellous bone anisotropic permeability. J Biomech 2001;34:1197–202.

[36] Shimko DA, Nauman EA. Development and characterization of a porous poly(methyl methracrylate) scaffold with controllable modulus and permeability. J Biomed Mat Res B 2007;80:360–9.

[37] Chor MV, Li W. A permeability measurement system for tissue engineering scaffolds. Meas Sci Technol 2007;18:208–16.

[38] Maxwell VC, Wei L. Permeability measurement system for tissue engineering scaffolds. Meas Sci Technol 2007;18:208–16.

[39] Swider P, Conroy M, Pedrono A, Ambard D, Mantell S, et al. Use of high-resolution MRI for investigation of fluid flow and global permeability in a material with interconnected porosity. J Biomech 2007;40:2112–8.

[40] Sanz-Herrera JA, Kasper C, van Griensven M, Garcia-Aznar JM, Ochoa I, Doblare M. Mechanical and flow characterization of Sponceram carriers: Evaluation by homogenization theory and experimental validation. J Biomed Mater Res B: Appl Biomater 2008;87:42–8.

[41] Darcy H. Les Fontaines publiques de la ville de Dijon, Paris; 1856.
[42] Mitsak AG, Kemppainen JM, Matthew T, Harris T, Hollister SJ. Effect of polycaprolactone scaffold permeability on bone regeneration in vivo. Tiss Eng A 2011;17.
[43] Jeong C, Zhang H, Hollister SJ. Three-dimensional poly(1,8-octanediol-co-citrate) scaffold pore shape and permeability effects on sub-cutaneous in vivo chondrogenesis using primary chondrocytes. Acta Biomat 2011;7:505–14.
[44] Wei G, Ma PX. Structure and properties of nano-hydroxyapatite/polymer composite scaffolds for bone tissue engineering. Biomaterials 2004;26:4749–57.
[45] Yarlagadda P, Chandrasekharan M, JYM C. Recent Advances and current developments in tissue scaffolding. Biomed Mater Eng 2005;15:159–77.
[46] Yang S, Leong KF, Du Z, Chua CK. The design of scaffolds for use in tissue engineering. Part I. Traditional factors. Tissue Eng 2001;7:679–89.
[47] Yang S, Leong KF, Du Z, Chua CK. The design of scaffolds for use in tissue engineering. Part II. Rapid prototyping techniques. Tissue Eng 2002;8:1–11.
[48] Schumacher M, Deisinger U, Detsch R, Ziegler G. Indirect rapid prototyping of biphasic calcium phosphate scaffolds as bone substitutes: influence of phase composition, macroporosity and pore geometry on mechanical properties. J Mater Sci Mater Med 2010;21:3119–27.
[49] Ramay HR, Zhang M. Preparation of porous hydroxyapatite scaffolds by combination of the gel-casting and polymer sponge methods. Biomaterials 2003;24:3293–302.
[50] Chen QZ, Boccaccini AR. Poly(D, L-lactic acid) coated 45S5 Bioglass (R)-based scaffolds: Processing and characterization. J Biomed Mater Res Part A 2006;77A:445–57.
[51] Kartsogiannis V, Ng KW. Cell lines and primary cell cultures in the study of bone cell biology. Mol Cell Endocrinol 2004;30:79–102.
[52] Cenni E, Ciapetti G, Granchi D, Arciola C, Savarino L, Stea S, et al. Established cell lines and primary cultures in testing medical devices in vitro. Toxicol in vitro 1999;13(4–5).
[53] Roldán J, Chang E, Kelantan M, Jazayeri L, Deisinger U, Detsch R. Quantifying migration and polarization of murine mesenchymal stem cells on different bone substitutes by confocal laser scanning microscopy. J Craniomaxillofac Surg 2010;38:580–8.
[54] Detsch R, Uhl F, Deisinger U, Ziegler G. 3D-Cultivation of bone marrow stromal cells on hydroxyapatite scaffolds fabricated by dispense-plotting and negative mould technique. J Mater Sci Mater Med 2008;19:1491–6.
[55] Andrew B, Yeatts JP. Bone tissue engineering bioreactors: dynamic culture and the influence of shear stress. Bone 2011;48:171–81.
[56] Griffin M, Iqbal S, Bayat A. Exploring the application of mesenchymal stem cells in bone repair and regeneration. J Bone Joint Surg Br 2011;93:427–34.
[57] Jones E, Yang X. Mesenchymal stem cells and bone regeneration: current status. Injury 2011;42:562–8.
[58] Bajada S, Mazakova I, Richardson J, Ashammakhi N. Updates on stem cells and their applications in regenerative medicine. Tissue Eng Regen Med 2008;2:169–83.
[59] Smith L, Smallwood R, Macneil S. A comparison of imaging methodologies for 3D tissue engineering. Microsc Res Tech 2010;73:1123–33.
[60] Robb R. 3-D visualization in biomedical applications. Annu Rev Biomed Eng 1999;1: 377–99.
[61] Frank O, Heim M, Jakob M, Barbero A, Schäfer D, Bendik I, et al. Real-time quantitative RT-PCR analysis of human bone marrow stromal cells during osteogenic differentiation in vitro. J Cell Biochem 2002;85:737–46.
[62] Marie P, Fromigué O. Osteogenic differentiation of human marrow-derived mesenchymal stem cells. Regen Med 2006;1:539–48.

[63] Unger R, Halstenberg S, Sartoris A, Kirkpatrick C. Human endothelial and osteoblast co-cultures on 3D biomaterials. Methods Mol Biol 2011;695:229–41.

[64] Azuaje F. Computational discrete models of tissue growth and regeneration. Brief Bioinform 2011;12:64–77.

[65] Sengers B, Taylor M, Please C, Oreffo R. Computational modelling of cell spreading and tissue regeneration in porous scaffolds. Biomaterials 2007;28:1926–40.

Chapter 7.2

Mechanical Properties of Bioceramic Coatings on Medical Implants

Mangal Roy, Amit Bandyopadhyay and Susmita Bose
W. M. Keck Biomedical Materials Research Lab, School of Mechanical and Materials Engineering, Washington State University, Pullman, WA, USA

Chapter Outline

7.2.1. INTRODUCTION

Musculoskeletal disorders and related bone deficiencies are among the most important human health conditions that exist today due to increasing median age of our population. Because of numerous and complex functions of bone, there are many disorders, such as osteoarthritis, rheumatoid arthritis, osteoporosis, and bone cancer, that arise with time. In due course, these disorders turn so painful that clinical care by a physician or other healthcare professional becomes essential. In often cases, replacement or restoration of the damaged bone is done through an implant. It is projected that the number of total hip arthroplasties would increase from 240,000 to 570,000 in years from 2005 to 2030 [1]. Increased active lifestyle and sports-related injury lead to the requirement of a hip or knee replacements at an early stage of life, which can be as low as 40–45 years. Therefore, the new generation of implants needs to remain active for longer period of time. Coatings

Characterization of Biomaterials. http://dx.doi.org/10.1016/B978-0-12-415800-9.00009-7

on medical devices can modify surface properties to enhance biological response without compromising physical and mechanical properties of the actual implant. In often cases, it is required to achieve certain surface properties in a medical device to match the application requirement such as hydrophobicity, osteo-conductivity, and high wear resistance.

Load-bearing orthopaedic implants are primarily constructed from metals. However, a major problem associated with these metallic implants is their long-term fixation in bone. Therefore, the concept of applying bioactive hydroxy-apatite (HA) coating, a calcium phosphate ceramic composition that is stable at room temperature, on metallic implants as a surface modification technique was developed, where the HA-coated implant combined the good strength and ductility of the metal and the excellent biocompatibility and bioactivity of the HA. The advantages of HA coating on metallic implant include (1) rapid fixation and stronger bonding between the host bone and the implant that results in faster healing and (2) increased uniform bone in growth and/or outgrowth at the bone–implant interface [2]. A variety of surface coating technologies are developed to deposit HA on Ti-based alloys for load-bearing implants. Among them, plasma spraying is regarded as the most efficient and economical technique to prepare calcium ceramic phosphate coatings for orthopaedic and dental applications. Coatings of these bioceramics on Ti-based alloys also provide the appropriate surface chemistry for tissue compatibility without altering the bulk mechanical properties of material [3]. Despite the recognized biocompatibility of HA coatings with bone, controversies with respect to the application on titanium for dental and orthopaedic implants still exist. The usefulness of a coated implant can only be realized if the coating is physically and mechanically stable *in vivo*.

Coatings on medical devices are not only used to improve or modulate the biological properties, but also can improve certain mechanical properties of metallic implants and one such important property is wear. Wear happens between two articulating surfaces such as in hip (acetabulum cavity and femoral head), knee and shoulder joints. Most of these joint replacements are done through metallic implants although some ceramic counterparts have also been used. In often cases, improvements in wear properties of these metallic implants are necessary to enhance the lifetime and their performance. Reduced wear also ensures lower metal ion release in body to restrict any possible toxicological side effects [4].

In line of the complex performances of coatings on medical devices, it is vitally important to understand the mechanical properties of coatings in order to evaluate and estimate the long-term *in vivo* performance. This chapter is focused on various mechanical characterizations needed to understand and measure coating integrity and its performance.

7.2.2. COATING MICROSTRUCTURE

Surface of a biomaterial is the first interaction point when placed in a biological environment. Therefore, surface features of coatings play a significant role in determining not only the initial biological responses but also its long-term performance. The first line of coating characterization can be performed with bare eye. The coating on biomedical devices should be checked for uniformity and desired surface coverage of the substrate. The coating surface can also be checked microscopically at low magnification such as at 10× for pinholes, cracking, chipping, and unevenness [5]. Necessary modification in the process parameters should be adopted if any one of these defects is noticed in the coating microscopically and/or macroscopically. If the coating contains designed porosity, it should be microscopically measured and appropriately reported. The interconnectivity of the porosity should not only be measured on the surface, it should also be studied across the coating thickness. To observe pores within the coating, cross-sectional samples of the coating–substrate unit should be prepared. If the coating is fragile in nature (like most calcium phosphate ceramic coatings), the sample can be embedded in an epoxy resin and cured for 24 h. The sample can be sectioned using a low-speed diamond saw (to minimize vibrational damage) across coating thickness. The sectioned sample can be polished and observed under microscope to measure coating thickness and its uniformity. The sectioned samples can also be used for studying pores and micro-cracks within coating, the coating–substrate delamination and the grain sizes [6]. Surface topography has been reported to play a major role in the bioactivity of biomaterials [7]. Therefore, it is also important to determine the surface roughness of the coatings [8]. The widely used plasma sprayed coatings generally contain both micro- as well as macro-roughness. Surface profilometer can accurately measure micro/macro-roughness of coatings.

7.2.3. WEAR PROPERTIES

The wear of a coating can be determined by a sliding pin on disk or ball on disk apparatus. The experiment will also provide information about the coefficient of friction. When a pin or ball of specific dimension is hold against the coating surface with specific amount of load while the test machine moves either the coating substrate or the counterpart (pin or ball) [9–11], the friction coefficient and the wear damage can be accurately measured. Both pin on disk and ball on disk test mode can be used without any significant limitations on either test. The selection of test method is decided by the application of the material being tested. The test material is selected in such a way that it can withstand the load applied for the duration of the test without any significant mechanical damage. It is also important to know the coating thickness to apply a suitable load. The process parameters, as mentioned below, are selected in such a way that the coating does not crack or spall from the substrate during the test. Care should be taken to ensure that the pin/ball does not touch the coating substrate anytime during the test. It is generally recommended that the surface be grounded to a roughness of 0.8 μm (32 μin) before starting the test [12]. There are several parameters that can influence the wear test results.

Pin/ball material: The choice of pin/ball material is dictated by the practical application and the sample/material to be tested. For example, femoral head and acetabular cup in a hip replacement are often made of Co–Cr alloy and, therefore, its wear properties should be tested against similar material. The ball/pin material can be selected to be harder than actual material to ensure sufficient wear resistance of the material being tested. For example, hardened chrome steel ball (100Cr6, 58–63 HRC) and Si_3N_4 ball have been used to study the wear properties of SiC coating on Ti [13]. Hardened chrome steel ball (100Cr6, 58–63 HRC) is also used to determine the wear properties of tantalum (Ta) materials and Ta coatings on Ti substrate [14]. However, care should be taken in selecting the counter pin/ball material. The pin/ball material should not be extremely harder or softer than the coating. For example, a hardened chrome steel ball should not be used to determine the wear properties of HA coatings. Ultrahigh molecular weight polyethylene can be used to determine the HA coatings wear rate. Although these tests can offer some information about the wear resistance of a biomaterial, further device level characterization is also needed if that material is selected for a specific device. In this chapter describes some of those device level characterizations detail.

Load: Wear properties of coatings are evaluated at a predetermined load. Load is selected depending on the hardness of the coating and pin/ball. The load is applied normal to the surface of the coatings and the force is expressed in Newton. In often cases, more than one load is used to verify the results.

Speed: Wear of any material depends on the relative speed of the articulating surface. To determine wear of the coatings, specific relative sliding speed between the coating surface and pin/ball are used and is expressed in metres per second. Multiple speeds can be used to study the effects of speed on wear properties of coatings.

Distance: Total distance of articulation plays a significant role on the wear properties of coatings. Tests are performed at a predetermined total sliding distance of pin/ball on coating surfaces and are expressed in metres. In most cases, early wear takes place up to 200–500 m distance followed by a steady-state wear regime. Therefore, longer distance wear tests are recommended to fully understand the wear mechanisms and the non-linear wear degradation of materials.

Temperature: Temperature has a profound effect on wear properties of coatings. Wear tends to increase as the temperature of the test environment or wearing contact points increases. However, most of the wear tests of medical coatings are performed at or around 37 °C to understand the material's performance at body temperature.

Atmosphere: Wear test of medical coatings are generally carried out under fully immersed condition in simulated body fluids or with the fluids that will come in contact with coatings.

The wear test is performed with a wear tester, which is capable of controlling the parameters mentioned above. At a setting of desired parameters, the coating samples should be fixed to the base of the tester before starting the experiment. In case an experiment stops for some reason, it should not be restarted. A new contact position between coating and pin/ball should be established before repeating the experiment. The wear rate is calculated as follows:

$$\text{Coating volume loss, mm}^3 = \pi\,(\text{wear track radius, mm})\,(\text{track width, mm})^3/6\,(\text{sphere radius, mm})$$

$$\text{Wear rate (mm}^3\text{/Nm)} = (\text{Coating volume loss, mm}^3)/[\text{load (N)} \times \text{distance (m)}].$$

7.2.4. BOND STRENGTH OF COATINGS

The adhesive bond strength is one of the decisive factors for the *in vivo* application of bioceramic coatings. Higher adhesive bond strength not only ensures long-term stability of the coating *in vivo* but also prevents osteolysis, a bone resorption process generally triggered by wear debris. There are three primary coating failure mechanisms. If coating–substrate bonding is stronger than the interlayer adhesion of the coating, it will fail within the coating when subjected to a tensile loading and some part of the coating will remain to the substrate. This kind of coating failure is called *cohesive mode of failure*. On the other hand, if the coating is weakly bonded to the substrate, the delamination will happen at the coating–substrate interface and the failure mode is known as *adhesive failure*. In practical cases, coatings show both adhesive and cohesive failures and this is referred to as *mixed mode of failure*. Depending on coating thickness and coating technique, there are two primary methods of determining the bond strength of coatings.

7.2.4.1. Thermal Spray Coating

This test method determines the adhesive or cohesive strength of thermal sprayed ceramic and metallic coatings [15,16]. The test method is only applicable to coatings having a thickness >0.015 in or 0.38 mm [17]. The basic principle of this characterization is bonding the coating surface with another uncoated substrate with a bonding agent and application of a load normal to the coating surface to measure the failure load [6,18]. The recommended coating thickness is designed to prevent the bonding agent penetration through the coating and to the substrate. If the coating thickness is too low, then there is a possibility that the adhesive can penetrate the coating and bond the coating substrate with the counter substrate. In that case, the determined bond strength will be erroneous and represents the adhesive strength of the cured adhesive agent and not the coating. A schematic of the experimental set-up is shown in Fig. 7.2.1. The bonding agent is of special interest here. Although there is no standard strength value requirement for the bonding agent, it is recommended that the adhesive and/or cohesive strength of the bonding agent should be

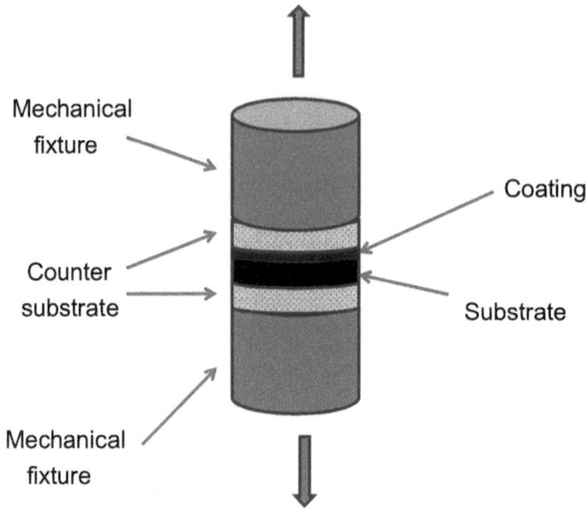

FIGURE 7.2.1 Schematic of adhesive testing set-up.

minimum than that required for the coatings being tested. The bonding agent should be fluid enough to cover the entire coating surface and bond to the counter substrate, however, should be viscous enough not to penetrate the recommended coating thickness (0.015 in or 0.38 mm).

The coating should be prepared on a flat metallic substrate of 0.9–1.0 in (23–25 mm) diameter. Special care should be given to prepare a coating of 0.015 in or 0.38 mm thick. A post-processing grinding and polishing can also be performed to prepare smooth matching surface. In case of post-processing, higher initial coating thickness should be achieved to accommodate for material removal during polishing. The counter substrate of the same diameter of the coating substrate should be sand blasted and attached to the surface of the coating using adhesive. Excessive bonding agent should be wiped off from the side of the coating. The substrate and counter substrate fixture need to hold together until the adhesive agent cures. Depending on the adhesive agent, heating the fixture at 120 °C for 2 h in an oven ensures proper curing of the adhesive. After curing, attach the fixture in a universal testing machine where normal tensile load can be applied and recorded. The fixtures should be subjected to a tensile test at a constant cross-head speed of 0.0013 cm/s until failure. The adhesive bond strength can be calculated as failure load/sample area (*F*/*A*).

$$S = F/A$$

where

S = adhesion or cohesion strength,

F = maximum load to failure, and

A = cross-sectional area.

At least five specimens need to be used for accuracy of the test. The bonding strength is reported as mean ± error.

7.2.4.2. General Calcium Phosphate Coatings

This test method applies to any thin or bulk calcium phosphate coatings on dense metal substrate. A brief description of the characterization technique has been mentioned in this chapter Orthopaedic Device Characterization. The characterization technique can also be used for porous metallic coating on dense substrates. Similar to thermal spray coatings, a polymeric adhesive is used to bond together the coating being tested and a counter substrate [19,20]. For very strong metallic coatings, the sample and counter substrate can be sintered together. A bulk tensile strength of 34.5 MPa (5000 psi) is the primary requirement of the adhesive. For a porous coating, extreme care should be taken to prevent the adhesive to touch the substrate through the porous network. According to the ASTM F1147 standard, the test specimen should be comprised a coating area of 5.07 cm^2 (0.78 in^2) and can be prepared by any coating techniques [21]. Samples of different surface area can also be used but should be reported in relation to 5.07 cm^2 (0.78 in^2). The adhesive should be applied on the coating surface followed by the matching counter substrate. The fixture should be held together by a load (resulting stress 0.138 MPa (20 psi)) and subjected to heating 176 °C (350 °F) for 2–3 h. Excess adhesive should be removed before proceeding further. The fixtures need to be attached to a gripping device that can apply a normal load to the coating surface. The tensile test is performed at a constant cross-head speed of 0.25 cm/min (0.10 in/min) until coating separation. The adhesive strength (MPa or psi) is calculated as

$$S = F/A$$

where

S = adhesion or cohesion strength,
F = maximum load to failure, and
A = cross-sectional area.

At least five specimens need to be used for accuracy of the test. The bonding strength is reported as mean $\pm$ error.

7.2.5. FATIGUE PROPERTIES OF COATINGS

In often cases, a freshly prepared coating is flawless or generally contains fewer defects. Therefore, defects like cracking or spalling cannot be readily determined in these cases. Moreover, generally used plasma spray coating method generates residual stress at the substrate–coating interface, which can degrade the bonding (both adhesion and cohesion) strength of the coatings. The residual stress and different substrate preparation techniques (grit blasting) can also lower the fatigue properties of substrate metals [22]. Therefore, fatigue property evaluation is an important tool to determine the long-term stability of coated implants without doing a long-term implant evaluation.

The fatigue properties of coatings can be determined in two different modes: shear fatigue and bending fatigue. The shear mode determines the adhesive/cohesive strength of coatings in fatigue condition, whereas the bending mode determines the effect of coating preparation on the fatigue properties of substrate. In addition, the bending fatigue test can provide information about cracking and spalling resistance of coatings.

The shear fatigue test specimens are prepared by coating an area of 19.05 mm (0.75 in) diameter [23]. A brief description of the characterization technique has been mentioned in this chapter Orthopaedic Device Characterization. An uncoated matching substrate of the same dimension should be bonded to the coatings using a polymeric adhesive. The adhesive can be a polymeric film (FM1000, 3M-weld 2214) or in bulk form (epoxy), however, should have a minimum shear strength of 34.5 MPa (5000 psi). The adhesive should be cured by heating the fixtures at 176 °C (350 °F) between 2 and 3 h with a load (resulting stress 0.138 MPa (20 psi)). Excess adhesive should be removed before proceeding further. The specimen assembly needs to be attached to a mechanical fixture that can apply shear load through the centre of the grip assembly. The test should run at a specified frequency (generally 50 Hz and should not exceed 170 Hz) and low load. The experiment is terminated when the coating separates from the substrate or when 10^7 cycles of loading has reached. The results should be reported as number of cycles before failure if failure has occurred. After experiment is complete, the coatings should be characterized for surface and internal cracking or spalling. For bending fatigue strength experiments, normally tangentially blending fillets of the substrate materials are used. Other bending test specimens of different geometry can also be used. The coating should be applied to the reduced section only. The specimen should be mounted in a rotating beam test machine, where desired load can be applied. Uniaxial fatigue bending test should be performed with fully reversible loading and unloading set-up and with constant amplitude. The test should run at a specified frequency (generally 50 Hz and should not exceed 170 Hz) and at a load of 620 MPa [22]. The experiment is terminated when the coating separates from the substrate or when 10^7 cycles of loading has reached. After experiment is complete, the coatings should be characterized for surface and internal cracking or spalling. Failure life of the substrate should be reported in number of cycles.

7.2.6. SHEAR TESTING OF COATINGS

Coatings on medical devices do not always experience normal tensile force (perpendicular to the coating surface). In practical applications, the coatings face a shear force parallel to the coating. Therefore, for the long-term stability of coated implants, it is important to know the shear strength of coatings. The following experimental technique can be used to determine the shear adhesive/cohesive strength of ceramic (prepared by any technique) or metallic coatings on dense meal substrates [24–26]. Similar to bond strength determination techniques, the coated surface is bonded to an uncoated matching counter substrate using a polymeric adhesive and the fixture is subjected to parallel shear until failure. The adhesive can be a polymeric film (FM1000, 3M-weld 2214) or in bulk form (epoxy), however, minimum shear strength should be 34.5 MPa (5000 psi) [27]. A brief description of the characterization technique has been mentioned in this chapter Orthopaedic Device Characterization. For porous coatings, care should be taken to prevent the adhesive from reaching the substrate and can be achieved by using highly viscous adhesive. For metallic coatings, the coated samples are bonded to the counter substrate using diffusion bonding or sintering. The coating should be prepared on a surface area of 2.84 cm^2 (0.44 in^2). If samples of different dimensions are used, the results should be reported in relation to 2.84 cm^2 (0.44 in^2). After preparing the coating–counter substrate arrangement (polymer adhesive for ceramic and sintering/diffusion bonding for metallic coatings), the fixture is held together by a load (resulting stress 0.138 MPa (20 psi)) and subjected to heating 176 °C (350 °F) between 2 and 3 h. Excess adhesive should be removed before proceeding further. The fixtures are held together with a gripping device that can apply a normal load to the coating surface. The shear test is performed at a constant cross-head speed of 0.25 cm/min (0.10 in/min) until coating separation. The adhesive strength (MPa or psi) is calculated as

$$S = F/A$$

where

S = adhesion or cohesion strength,

F = maximum load to failure, and

A = cross-sectional area.

At least five specimens need to be used for accuracy of the test. The shear strength is reported as mean $\pm$ error.

7.2.7. SUMMARY

This short chapter deals with mechanical properties of bioceramic coatings with special emphasis towards load bearing implants. We have discussed techniques to evaluate coating microstructures, wear properties, and bond strength under different loading conditions. Proper precaution should also be taken to generate reproducible and reliable properties for ceramic coatings, since large variation in data is quite common in these samples. It is important to understand that specific test method will not only depend on application, but also will be dictated by coating thickness, coating-substrate material combination and their manufacturing technique.

REFERENCES

[1] Huo MH, Dumont GD, Knight JR, Mont MA. What's new in total hip arthroplasty. J Bone Joint Surg Am 2011;93:1944–50. http://dx.doi.org/10.2106/JBJS.K.00656.

[2] Dumbleton J, Manley MT. Hydroxyapatite-coated prostheses in total hip and knee arthroplasty. J Bone Joint Surg Am 2004;86-A:2526–40.

[3] Narayanan R, Seshadri SK, Kwon TY, Kim KH. Calcium phosphate-based coatings on titanium and its alloys. J Biomed Mater Res B Appl Biomater 2008;85B:279–99. http://dx.doi.org/10.1002/jbm.b.30932.

[4] Dittrick S, Balla VK, Bose S, Bandyopadhyay A. In vitro wear rate and Co ion release of compositionally and structurally graded CoCrMo-Ti6Al4V structures. Mater Sci Eng C Mater Biol Appl 2011a;31:809–14. http://dx.doi.org/10.1016/j.msec.2010.07.009.

[5] ASTM Standard F1609–08. Standard specification for calcium phosphate coatings for implantable materials. PA: ASTM International, www.astm.org; 2008.

[6] Roy M, Bandyopadhyay A, Bose S. Induction plasma sprayed nano hydroxyapatite coatings on titanium for orthopaedic and dental implants. Surf Coat Technol 2011a;205:2785–92. http://dx.doi.org/10.1016/j.surfcoat.2010.10.042.

[7] Zinger O, Anselme K, Denzer A, Habersetzer P, Wieland M, Jeanfils J, et al. Time-dependent morphology and adhesion of osteoblastic cells on titanium model surfaces featuring scale-resolved topography. Biomaterials 2004;25:2695–711. http://dx.doi.org/10.1016/j.biomaterials.2003.09.111.

[8] Wang G, Liu X, Zreiqat H, Ding C. Enhanced effects of nano-scale topography on the bioactivity and osteoblast behaviors of micron rough ZrO_2 coatings. Colloids Surf B Biointerfaces 2011;86:267–74. http://dx.doi.org/10.1016/j.colsurfb.2011.04.006.

[9] Lee C-K. Fabrication, characterization and wear corrosion testing of bioactive hydroxyapatite/nano-TiO_2 composite coatings on anodic Ti–6Al–4V substrate for biomedical applications. Mater Sci Eng B 2012;177:810–8. http://dx.doi.org/10.1016/j.mseb.2012.03.034.

[10] Liu F, Xu JL, Yu DZ, Wang FP, Zhao LC. Wear resistance of micro-arc oxidation coatings on biomedical NiTi alloy. J Alloys Compd 2009;487:391–4. http://dx.doi.org/10.1016/j.jallcom.2009.07.145.

[11] Sathish S, Geetha M, Aruna ST, Balaji N, Rajam KS, Asokamani R. Sliding wear behavior of plasma sprayed nanoceramic coatings for biomedical applications. Wear 2011;271:934–41. http://dx.doi.org/10.1016/j.wear.2011.03.023.

[12] ASTM Standard G99–05. Standard test method for wear testing with a pin-on-disk apparatus. PA: ASTM International, www.astm.org; 2010.

[13] Das M, Bysakh S, Basu D, Sampath Kumar TS, Balla VK, Bose S, et al. Microstructure, mechanical and wear properties of laser processed SiC particle reinforced coatings on titanium. Surf Coat Technol 2011;205:4366–73. http://dx.doi.org/10.1016/j.surfcoat.2011.03.027.

[14] Dittrick S, Balla VK, Bose S, Bandyopadhyay A. Wear performance of laser processed tantalum coatings. Mater Sci Eng C 2011b;31:1832–5. http://dx.doi.org/10.1016/j.msec.2011.08.017.

[15] Eriksson R, Brodin H, Johansson S, Östergren L, Li X- H. Influence of isothermal and cyclic heat treatments on the adhesion of plasma sprayed thermal barrier coatings. Surf Coat Technol 2011;205:5422–9. http://dx.doi.org/10.1016/j.surfcoat.2011.06.007.

[16] Lima RS, Marple BR. From APS to HVOF spraying of conventional and nanostructured titania feedstock powders: a study on the enhancement of the mechanical properties. Surf Coat Technol 2006;200:3428–37. http://dx.doi.org/10.1016/j.surfcoat.2004.10.137.

[17] ASTM Standard C633–01. Standard test method for adhesion or cohesion strength of thermal spray coatings. ASTM International, www.astm.org; 2008.

[18] Roy M, Bandyopadhyay A, Bose S. Induction plasma sprayed Sr and Mg doped nano hydroxyapatite coatings on Ti for bone implant. J Biomed Mater Res B Appl Biomater 2011b;99B:258–65. http://dx.doi.org/10.1002/jbm.b.31893.

[19] Hadad M, Hockauf M, Meyer LW, Marot G, Lesage J, Hitzek R, et al. Adhesion evaluation of multilayered based WC–Co–Cr thermally sprayed coatings. Surf Coat Technol 2008;202:4399–405. http://dx.doi.org/10.1016/j.surfcoat.2008.04.016.

[20] Zavgorodniy AV, Mason RS, LeGeros RZ, Rohanizadeh R. Adhesion of a chemically deposited monetite coating to a Ti substrate. Surf Coat Technol 2012;206:4433–8. http://dx.doi.org/10.1016/j.surfcoat.2012.04.090.

[21] ASTM Standard F1147–05. Standard test method for tension testing of calcium phosphate and metallic coatings. PA: ASTM International, www.astm.org; 2011.

[22] Lynn AK, DuQuesnay DL. Hydroxyapatite-coated Ti–6Al–4V: part 1: the effect of coating thickness on mechanical fatigue behaviour. Biomaterials 2002;23:1937–46. http://dx.doi.org/10.1016/S0142-9612(01)00321-0.

[23] ASTM Standard F1160–05. Standard test method for shear and bending fatigue testing of calcium phosphate and metallic medical and composite calcium phosphate/metallic coatings. PA: ASTM International, www.astm.org; 2011.

[24] Albayrak O, El-Atwani O, Altintas S. Hydroxyapatite coating on titanium substrate by electrophoretic deposition method: effects of titanium dioxide inner layer on adhesion strength and hydroxyapatite decomposition. Surf Coat Technol 2008;202:2482–7. http://dx.doi.org/10.1016/j.surfcoat.2007.09.031.

[25] Pei X, Wang J, Wan Q, Kang L, Xiao M, Bao H. Functionally graded carbon nanotubes/hydroxyapatite composite coating by laser cladding. Surf Coat Technol 2011;205:4380–7. http://dx.doi.org/10.1016/j.surfcoat.2011.03.036.

[26] Wang J, Huang C, Wan Q, Chen Y, Chao Y. Characterization of fluoridated hydroxyapatite/zirconia nano-composite coating deposited by a modified electrocodeposition technique. Surf Coat Technol 2010;204:2576–82. http://dx.doi.org/10.1016/j.surfcoat.2010.01.042.

[27] ASTM Standard F1044–05. Standard test method for shear testing of calcium phosphate coatings and metallic coatings. PA: ASTM International, www.astm.org; 2011.

Chapter 8

Characterization of Orthopaedic Devices

Imran Khan, Malcolm Naylor and Gautam Gupta
Biomet Inc, Warsaw, IN, USA

Chapter Outline

8.1. MECHANICAL TESTING OF ORTHOPAEDIC DEVICES

8.1.1. Introduction

Preclinical mechanical testing allows the performance and compatibility of medical devices to be assessed *in vitro* by simulating physiological loading,

Characterization of Biomaterials. http://dx.doi.org/10.1016/B978-0-12-415800-9.00010-3

motion and in some cases a local biological environment, in a controlled and reproducible manner. In addition to ensuring new product designs meet the essential regulatory requirements, device testing allows the quality of products to be verified as part of quality control and ensures that patient safety is not compromised due to any changes in raw material suppliers and manufacturing methods. In addition to fulfilling regulatory requirements, reference is frequently made to mechanical test data in device manufacturers' marketing literature to support the efficacy of their products.

8.1.2. The Regulatory Importance of Preclinical Mechanical Testing

The design process for medical devices is highly regulated to ensure the safety of patients and healthcare professionals.

In the USA, the safety of medical devices is regulated by the Food and Drug Administration (FDA) [1]. There are two major processes by which medical devices requiring FDA review come to the American market. A medical device that may be considered substantially equivalent to another legally US marketed device must undergo a Premarket Notification (510(k)) submission. The general non-clinical testing requirements for such devices are described in FDA Regulation 21 CFR 807, Subpart E [2]. Medical devices that are considered to be in the highest risk category, Class III, typically go through a more stringent Premarket Approval (PMA) process before they can be placed on the market. Regulations applicable for PMA devices are described in FDA Regulation 21 CFR Part 814 [3].

In the European Union (EU), conformance to the requirements of the Medical Devices Directive (MDD) 93/42/EEC is necessary for a device manufacturer to achieve a CE mark (Conformité Européene) and to legally place a medical device on the market [4]. If a device falls within the scope of the MDD, then a Notified Body (an organization that has been accredited by a Member State to assess whether a product meets the MDD essential requirements) must be involved in the conformity assessment process. Examples of European Notified Bodies include: BSI, TÜV Nord Cert GmbH and G-Med [5].

To show compliance with the regulatory requirements of the FDA and the MDD, companies must have an effective quality management system in place to ensure that the Design Control Process is managed and controlled in a systematic and repeatable manner [1,6,7]. They must demonstrate certain mandatory requirements such as the establishment of an intended use for the device, a design plan and periodic design reviews through the entire design process. Furthermore, confirmation that design outputs meet design inputs is carried out through design verification ('did we design the device right?'), design validation ('did we design the right device?'), design specification transfer for manufacture. The documentation associated with the entire process collectively makes up a Design Dossier (also known as a Design History File or Technical File). The Design Dossier is maintained post-product release to include subsequent changes to the product and relevant post-market surveillance data [1,4,6]. Mechanical testing forms an important part of the design verification and validation of the product to ensure that risks identified as part of this process have been eliminated or reduced, so that the safety of patients or healthcare professionals is not compromised.

8.1.3. Test Standards

Testing of medical devices is normally carried out according to international standards, such as American Society for Testing and Materials (ASTM) International [8] and International Organization for Standardization (ISO) [9] standards, or regional device standards, such as the British Standards Institution (BSI) [10] and Deutsches Institut für Normung [11], where they are available.

Device standards will typically define the type of device to be tested and the laboratory-based performance requirements and reference criteria that it must meet. General design features are indicated with schematic drawings, reference to materials and separate test standards may also be indicated.

While there is no legal obligation to adhere to standards, regulatory authorities will normally expect adherence to available standards. Most standards have been agreed by consensus by a number of standards bodies made up of representatives from academia, industry, government and private organizations. They therefore represent a method of testing that is acceptable to most people working in the field and provide an agreed platform for the assessment of safety, reliability and performance of orthopaedic devices [8–11].

In the absence of a suitable standard test standard, an appropriate 'in-house' test standard may be used. In such cases, there must be clear documentation of the rationale and justification for the suitability of the testing method and parameters employed.

Tables 8.1 and 8.2 list the most commonly used ISO and ASTM orthopaedic device standards, including test standards.

8.1.4. Examples of Standard Test Methods for Evaluation of Orthopaedic Devices

Mechanical testing is conducted according to the intended function of the implant. The most widely used tests for orthopaedic devices are static tests and fatigue endurance tests (Tables 8.1 and 8.2). Static tests involve investigation of the strength or deformation characteristics of a device and its constituent parts when subjected to non-cyclic tensile, shear, bending, compressive or torsional loading, depending on the expected loading scenario *in vivo*. Since most orthopaedic devices are subject to repeated loading and unloading, cyclic

TABLE 8.1 List of Arthroplasty Medical Device Standards Developed by ASTM [8]

Designation	Title
F1223 - 08	Standard Test Method for Determination of Total Knee Replacement Constraint
F1357 - 09	Standard Specification for Articulating Total Wrist Implants
F1378 - 12	Standard Specification for Shoulder Prostheses
F1440 - 92(2008)	Standard Practice for Cyclic Fatigue Testing of Metallic Stemmed Hip Arthroplasty Femoral Components Without Torsion
F1672 - 95(2011)	Standard Specification for Resurfacing Patellar Prosthesis
F1714 - 96(2008)	Standard Guide for Gravimetric Wear Assessment of Prosthetic Hip Designs in Simulator Devices
F1781 - 03(2009)	Standard Specification for Elastomeric Flexible Hinge Finger Total Joint Implants
F1800 - 07	Standard Test Method for Cyclic Fatigue Testing of Metal Tibial Tray Components of Total Knee Joint Replacements
F1814 - 97a(2009)	Standard Guide for Evaluating Modular Hip and Knee Joint Components
F1820 - 97(2009)	Standard Test Method for Determining the Axial Disassembly Force of a Modular Acetabular Device
F1829 - 98(2009)	Standard Test Method for Static Evaluation of the Glenoid Locking Mechanism in Shear
F2009 - 00(2011)	Standard Test Method for Determining the Axial Disassembly Force of Taper Connections of Modular Prostheses
F2025 - 06	Standard Practice for Gravimetric Measurement of Polymeric Components for Wear Assessment
F2028 - 08	Standard Test Methods for Dynamic Evaluation of Glenoid Loosening or Disassociation
F2033 - 12	Standard Specification for Total Hip-Joint Prosthesis and Hip Endoprosthesis Bearing Surfaces Made of Metallic, Ceramic, and Polymeric Materials
F2068 - 09	Standard Specification for Femoral Prostheses – Metallic Implants
F2083 - 11	Standard Specification for Total Knee Prosthesis
F2091 - 01(2012)	Standard Specification for Acetabular Prostheses

TABLE 8.1 List of Arthroplasty Medical Device Standards Developed by ASTM [8]—cont'd

Designation	Title
F2345 - 03(2008)	Standard Test Methods for Determination of Static and Cyclic Fatigue Strength of Ceramic Modular Femoral Heads
F2385 - 04(2010)	Standard Test Method for Determining Femoral Head Penetration into Acetabular Components of Total Hip Replacement using Clinical Radiographs
F2580 - 09	Standard Test Method for Evaluation of Modular Connection of Proximally Fixed Femoral Hip Prosthesis
F2582 - 08	Standard Test Method for Impingement of Acetabular Prostheses
F2665 - 09	Standard Specification for Total Ankle Replacement Prosthesis
F2722 - 08	Standard Test Method for Evaluating Mobile Bearing Knee Tibial Baseplate Rotational Stops
F2723 - 08	Standard Test Method for Evaluating Mobile Bearing Knee Tibial Baseplate/Bearing Resistance to Dynamic Disassociation
F2724 - 08	Standard Test Method for Evaluating Mobile Bearing Knee Dislocation
F2777 - 10	Standard Test Method for Evaluating Knee Bearing (Tibial Insert) Endurance and Deformation Under High Flexion

fatigue testing may be considered more appropriate. In this case, the devices are tested using cyclic loading under tension, compression, bending, shear or torsion, again depending on the *in vivo* loading conditions that they are trying to replicate. While some standards provide test parameters such as load, numbers of cycles and frequency, other standards are vaguer and leave it up to the user to define certain test parameters. Many standards also require that the 'worst case' designs are investigated. Note that the concept of testing the 'worst case' scenario is discussed later in section Identifying the 'Worst Case'.

In the following sub-sections, some examples of mechanical tests carried out for evaluation of orthopaedic devices are described.

8.1.4.1. Hip Stem Fatigue Test

In service, the taper and neck region of a hip femoral stem experiences high loads and stresses. Every step that a patient takes represents a single

TABLE 8.2 ISO Bone and Joint Replacement Device Standards [12]

Designation	Description
ISO 7206-1:2008	Implants for surgery — Partial and total hip-joint prostheses — Part 1: Classification and designation of dimensions
ISO 7206-2:2011	Implants for surgery — Partial and total hip-joint prostheses — Part 2: Articulating surfaces made of metallic, ceramic and plastics materials
ISO 7206-4:2010	Implants for surgery — Partial and total hip-joint prostheses — Part 4: Determination of endurance properties and performance of stemmed femoral components
ISO 7206-6:1992	Implants for surgery — Partial and total hip-joint prostheses — Part 6: Determination of endurance properties of head and neck region of stemmed femoral components
ISO 7206-10:2003	Implants for surgery — Partial and total hip-joint prostheses — Part 10: Determination of resistance to static load of modular femoral heads
ISO 7207-1:2007	Implants for surgery — Components for partial and total knee-joint prostheses — Part 1: Classification, definitions and designation of dimensions
ISO 7207-2:2011	Implants for surgery — Components for partial and total knee-joint prostheses — Part 2: Articulating surfaces made of metal, ceramic and plastics materials
ISO 14242-1:2012	Implants for surgery — Wear of total hip-joint prostheses — Part 1: Loading and displacement parameters for wear-testing machines and corresponding environmental conditions for test
ISO 14242-2:2000	Implants for surgery — Wear of total hip-joint prostheses — Part 2: Methods of measurement
ISO 14242-3:2009	Implants for surgery — Wear of total hip-joint prostheses — Part 3: Loading and displacement parameters for orbital bearing type wear-testing machines and corresponding environmental conditions for test
ISO 14243-1:2009	Implants for surgery — Wear of total knee-joint prostheses — Part 1: Loading and displacement parameters for wear-testing machines with load control and corresponding environmental conditions for test
ISO 14243-2:2009	Implants for surgery — Wear of total knee-joint prostheses — Part 2: Methods of measurement

TABLE 8.2 ISO Bone and Joint Replacement Device Standards [12]—cont'd

Designation	Description
ISO 14243-3:2004	Implants for surgery – Wear of total knee-joint prostheses – Part 3: Loading and displacement parameters for wear-testing machines with displacement control and corresponding environmental conditions for test
ISO 14243-3:2004/ Cor 1:2006	Implants for surgery – Wear of total knee-joint prostheses – Part 3: Loading and displacement parameters for wear-testing machines with displacement control and corresponding environmental conditions for test
ISO 14879-1:2000	Implants for surgery – Total knee-joint prostheses – Part 1: Determination of endurance properties of knee tibial trays
ISO 17853:2011	Wear of implant materials – Polymer and metal wear particles – Isolation and characterization
ISO 21534:2007	Non-active surgical implants – Joint replacement implants – Particular requirements
ISO 21535:2007	Non-active surgical implants – Joint replacement implants – Specific requirements for hip-joint replacement implants
ISO 21536:2007	Non-active surgical implants – Joint replacement implants – Specific requirements for knee-joint replacement implants

loading–unloading cycle; these single events accumulate to several millions of cycles over the lifetime of the prosthesis, depending on the activity of the patient. Pedometer studies have shown that on average, hip and knee patients take approximately 1 million steps per annum, with more active patients taking more than three times this value [13]. The average peak loads in the hip joint during normal walking can exceed three times body weight, with activities like stair climbing exerting even greater forces [14]. For this reason, when any modifications are made to the neck geometry of a hip stem it is important to examine high cycle fatigue as part of the design validation.

The fatigue performance of the head and neck portion of a stemmed femoral hip component may be determined by testing in accordance *with ASTM F2068–09 'Standard Specification for Femoral Prostheses – Metallic Implants'* [15].

Here the stem is embedded in a potting medium up to the resection level of the femur, according to the recommendations of the manufacturer. A representative batch of six stems must withstand cyclic loading with a minimum load of 534 N (120 lb) and a maximum load of 5340 N (1200 lb) (representing approximately seven times body weight for an 80 kg person) for 10 million cycles at a test frequency of between 1 and 30 Hz. Stems with the 'worst case' stem and off-set head combination, i.e. those with the greatest lever arm, should be tested to provide the 'worst case' stress conditions within the product range. The test may be carried out dry or in saline to accelerate any corrosion or fatigue processes.

Figure 8.1 shows a femoral stem undergoing fatigue testing according to the ASTM F2068–09 [15] method. Here the largest head size and off-set is used to represent a 'worst case'.

8.1.4.2. Knee Tibial Fatigue Test

Fatigue fracture of knee tibial trays is a failure mechanism observed in Total Knee Replacements (TKR) [16–18]. It may be caused by loss of underlying bone support resulting from host biological responses, such as wear particle-induced osteolysis. Due to the reduced mechanical support and cyclic loading induced by normal walking, fatigue cracks may be generated, leading ultimately to fracture of the component. Other factors such as the high-temperature processing used to attach a porous coating to the device and design factors such as reduced substrate thickness have also been implicated as contributing factors leading to fatigue failure of tibial trays [19].

ASTM F1800–07 '*Standard Test Method for Cyclic Fatigue Testing of Metal Tibial Tray Components of Total Knee Joint Replacements*' [20] provides test parameters for determining and validating the fatigue properties of different tibial tray designs under cyclic loading. In this test, one condyle of the tibial tray is unsupported to represent poor cement or bone support *in vivo*.

FIGURE 8.1 Femoral hip stem neck fatigue testing setup according to *ASTM F2068 – 09* [15]. (Image courtesy of Biomet).

The tibial tray is mounted as a cantilever beam where one condyle is supported, either using a clamp or by fixing it using bone cement or another high strength epoxy (Fig. 8.2). Cyclic load is then applied to the unsupported condyle via a spherical indenter and a stiff, creep resistant spacer at a frequency of between 1 and 30 Hz, until either 10 million cycles have been performed or failure occurs. A recommended load is not specified in the standard and therefore values are typically adopted from the literature or other standards and can range from 500 N to greater than 2 kN [21]. Upon completion of the test, components are examined for any evidence of permanent damage such as cracking.

8.1.4.3. Glenoid Shear Test

Liner dissociation has been observed as a cause of failure of some designs of metal-backed glenoid components within shoulder joint replacements. This dissociation has been attributed to the failure of the snap-fit locking mechanism that provides mechanical attachment of the liner to the metallic backing [22,23].

In order to assess the integrity of the locking mechanism of a design of glenoid component, the static shear disassembly force of modular glenoid components used in shoulder prostheses may be evaluated using *ASTM F1829–98*

FIGURE 8.2 Tibial tray fatigue test setup according to *ASTM F1800 – 07* [20]. (Image courtesy of Biomet).

'Standard Test Method for Static Evaluation of the Glenoid Locking Mechanism in Shear' [24].

The test allows examination of the effects of materials, manufacturing, and design variables on the performance of metal-backed glenoid prostheses locking mechanisms when under static shear loading. The test is not designed to simulate *in vivo* loading but rather to allow comparison between different designs of metal-backed glenoid locking mechanism.

The test setup involves attaching the glenoid assembly (consisting of a polyethylene articular insert and a metal backing) to a fixture, such that load may be applied in an inferior-to-superior direction. A vertical load is applied to the articular insert with a blunt-edged loading applicator under a constant displacement rate (e.g. 25.4 mm/min) (Fig. 8.3). Testing is terminated either when the insert disengages from the metal backing, the disengagement force peaks and then progressively decreases, or the insert undergoes gross deformation without dislocation. The maximum load to failure is recorded.

8.1.4.4. Four-Point Bending Fatigue Testing of Intramedullary Fixation Devices

Intramedullary fixation devices (IMFDs), such as nails and rods, are used for the fixation of various bones and are typically used in the femur, tibia, humerus, radius and ulna. An IMFD subjected to cyclical loading can fail due to fatigue if stresses on the device exceed its endurance limit. The factors that influence

FIGURE 8.3 Glenoid shear test setup according to *ASTM F1829-98(2009)* [24]. (Image courtesy of Biomet).

such a failure are the stability of the fracture, a non-union or delayed union, and the nature of the rehabilitation programme [25,26].

In order to investigate the fatigue properties of IMFDs, testing may be carried out according to *ASTM F1264–03 'Standard Specification and Test Methods for Intramedullary Fixation Devices'* [27]. This standard defines performance criteria and methods for measurement of performance-related mechanical characteristics of IMFDs and their fixation to bone. It includes a test method for performing cyclic-bending fatigue using a four-point bending test setup.

The test utilizes a two-part fixture (top and bottom) capable of applying a uniform bending moment to the central portion of an IMFD. The IMFD specimen is supported by two outer support rollers, and the moment is applied through two inner loading rollers (Fig. 8.4).

The load level is determined through a number of possible methods, such as testing at several load levels to characterize the general fatigue behavior of an IMFD over a range of loads or stresses to generate an M-N or S-N curve, or determining the fatigue strength over 1 million cycles using an accepted method like the 'up-and-down' method [28].

Fatigue life can be determined at a specified maximum bending moment or over a predetermined number of cycles.

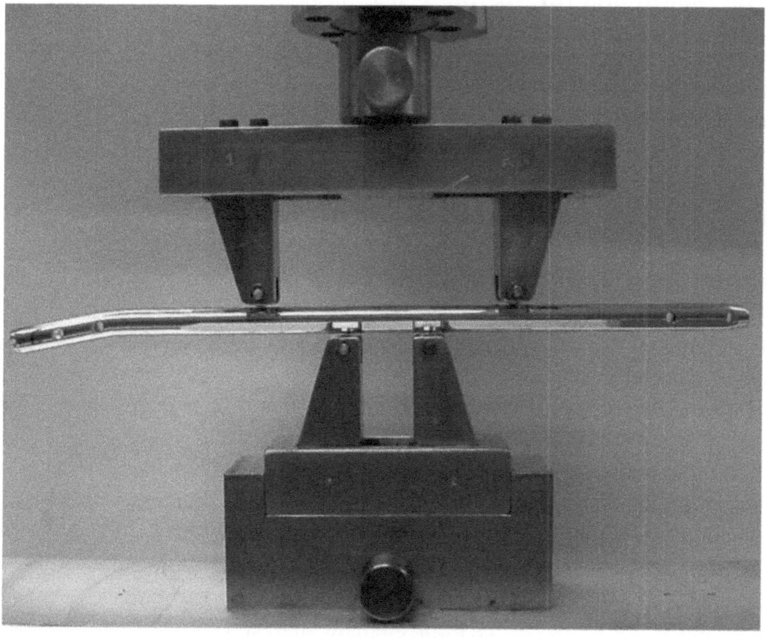

FIGURE 8.4 Four-point bend fatigue test of a femoral nail carried out according to *ASTM F1264 – 03* [27]. (Image courtesy of Biomet).

8.1.5. Evolution of Standard Test Methods

All standards undergo periodic review and as part of this process are able to evolve to address new devices, existing ambiguities or other issues encountered by users. An example is that of revision hip stems. *ISO 7206-4: 2002, 'Implants for surgery – Partial and total hip-joint prostheses – Part 4: Determination of endurance properties of stemmed femoral components'* [29] described a potting level for long revision stems that resulted in stresses that were far more severe than would be expected, even under extreme conditions *in vivo*. No load was specified, but a load of 2300 N specified in '*ISO 7206-8:1995 Implants for surgery – Partial and total hip joint prostheses – Part 8: Endurance performance of stemmed femoral components with application of torsion*' [30] was adopted by many users. This led to some clinically successful stem designs being unable to pass this test. A revision of this standard, ISO 7206-4: 2010 [31], changed the potting level and defined a reduced load requirement of 1200 N to make the test less severe (Fig. 8.5). Other test factors were also more clearly defined during this review. Table 8.3 shows the key differences between the two versions of ISO 7206-4.

8.1.6. Identifying the 'Worst Case'

The testing of every combination of a device design with all of its individual constituents can result in hundreds of combinations. A typical example would be a revision hip stem with multiple modular head and neck options/sizes. Since it is not practical from a time or cost perspective to test every single combination, often a 'worst case' example is identified to allow approval of an entire size range of components.

For some designs of implants, the 'worst case' may be obvious. For example, a femoral head with the largest off-set of +12 mm (largest lever arm) on the smallest femoral stem (smallest section) represents an obvious 'worst case' in terms of the resulting stresses in the neck of the femoral stem. For other designs, especially where a test involves an assembly, it may be more difficult to determine the exact combination of components that provides the harshest stress conditions.

The 'worst case' may differ depending on the type of mechanical test being considered. Therefore, for one type of test it may be the largest component in a particular size range and for another it may the smallest.

While documented experience from previous testing on a similar device or complaints data pointing to a higher risk of failure for a particular size may be a useful guide, for a new device finite element analysis (FEA) can be a valuable tool to determine the 'worst case' components or combination of components to test.

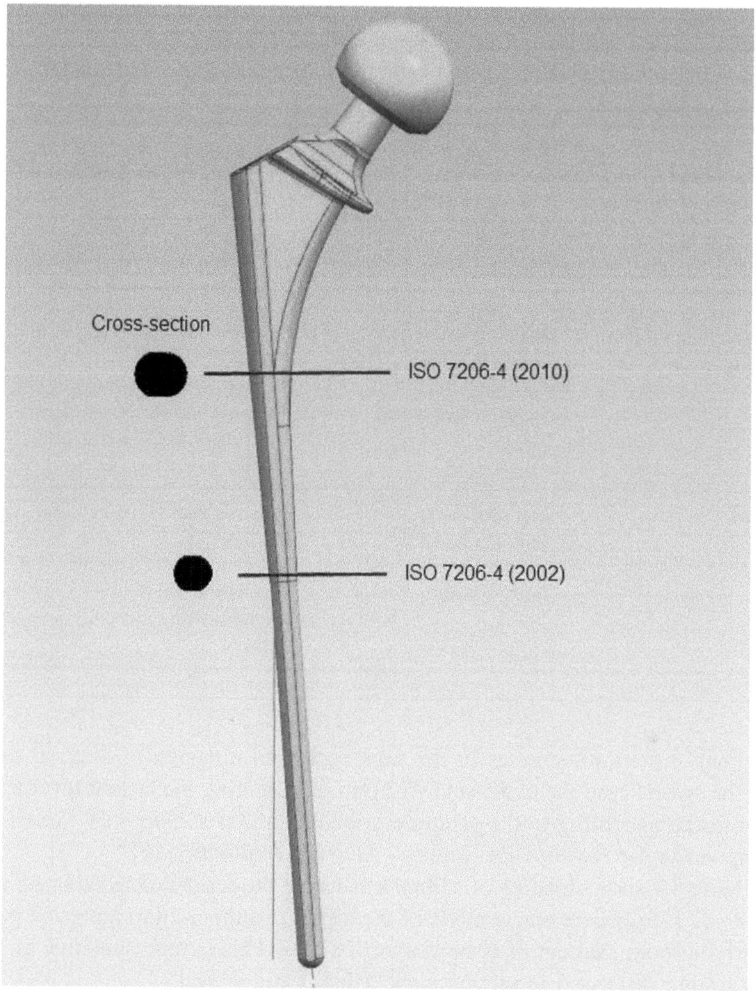

FIGURE 8.5 Potting levels according to ISO 7206-4: (2002) [29] and ISO 7206-4: 2010 [30] and the differences in cross-sectional area at the potting height for a stem with CT = 230 mm.

FEA is used widely in medical device development to verify if a design will have sufficient strength to withstand the loading conditions within the human body. The use of FEA provides a number of advantages for product development: improved product reliability, faster product development, examination of 'what if' design scenarios, reduced level/costs of physical testing, assistance with development of testing protocols, and identification of 'worst case' examples to test. Figures 8.6 and 8.7 show examples of FEA simulations used to investigate stress distributions in a hip stem and knee tibial tray, respectively.

TABLE 8.3 Comparison of ISO 7206-4: 2002 and ISO 7206-4: 2010 (CT = distance from the centre of the femoral head to the most distal point of the stem)

	ISO 7206-4: 2002 [29]	ISO 7206-4: 2010 [31]
Number of samples	Not specified	6
Number of cycles	5×10^6	5×10^6
Fluid test medium	–	Saline (0.9 g/L NaCl), 37 °C
Embedding medium distance (D)	For stem with CT < 200 mm: 0.4 × CT For stem with CT > 200 mm: CT-100	For stem with CT < 120 mm: 0.66 × CT For stem with 120 < CT < 250: 80 mm For stem with CT > 250 mm: CT-100 mm
Load	Not specified (2300 N typically used based on ISO 7206-8)	For stem with CT < 120 mm: 1200 N For stem with 120 mm < CT < 250 mm: 2300 N For stem with CT > 250 mm: 1200 N

Figure 8.6 shows stresses in the neck region of a design of femoral stem with the 'worst case' head off-set (+12 mm representing the largest lever arm) when tested according to the setup described in *ASTM F2068 – 09 'Standard Specification for Femoral Prostheses – Metallic Implants'* [15].

Figure 8.7 shows loading of a tibial tray as per the setup described in ASTM F1800–07 [20], where one condyle of the tray is unsupported to represent poor underlying bone cement or bone support *in vivo*. This is representative of the TKR testing discussed in section Knee Tibial Fatigue Test.

8.2. TRIBOLOGICAL TESTING OF JOINT IMPLANTS

8.2.1. Introduction

Orthopaedic joint replacement involves replacing the natural synovial joint and articular cartilage with synthetic materials that must provide extreme durability under complex loading conditions and variable lubrication conditions, depending on the condition of the joint capsule after surgery and during the lifetime of the implant.

Joint wear simulator testing is therefore an important part of the design verification activities for orthopaedic devices. Two common objectives for simulator testing are (a) to compare the wear rates for a new design with

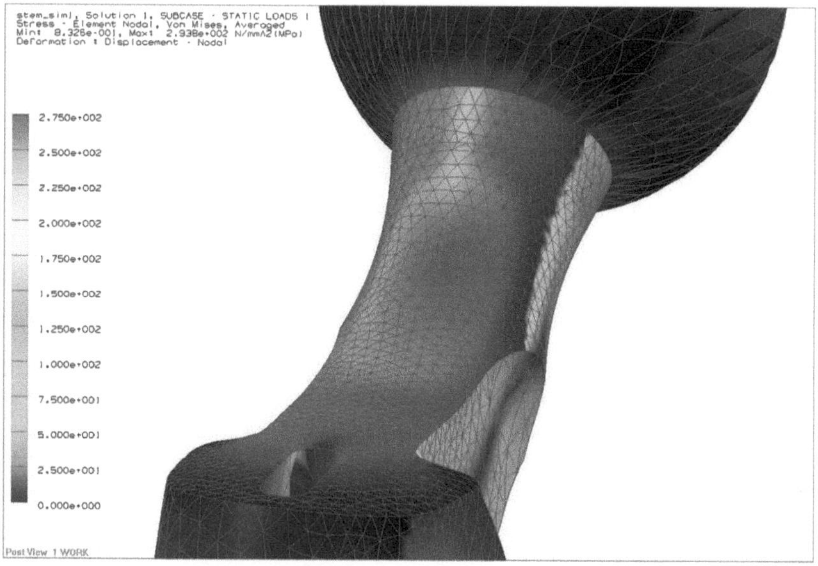

FIGURE 8.6 FEA simulation displaying stresses in the neck region of a design of a femoral stem with a +12 mm head when tested according to the setup described in *ASTM F2068 – 09* [15]. (Image courtesy of Biomet).

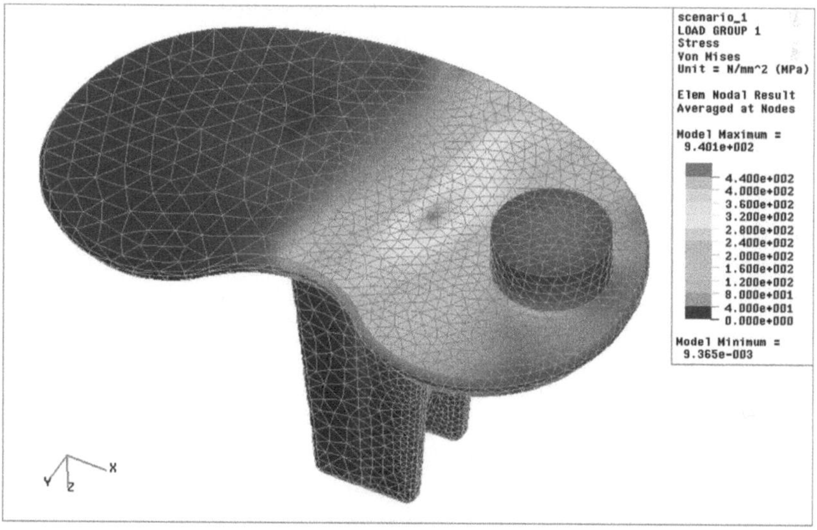

FIGURE 8.7 FEA simulation displaying stresses in a tibial tray according to the loading setup described in ASTM F1800-07 [20], where one condyle is unsupported to represent poor underlying bone cement or bone. (Image courtesy of Biomet).

a predicate design or (b) to compare the wear rates of two or more bearing materials.

Simulator test methods have evolved over several decades from simpler tests such as pin-on-plate (e.g. ASTM F732 [32]) to more realistic tests that replicate the forces and motions experienced during walking or other gait cycles and, importantly, allow the testing of actual components. Tests are designed to accelerate wear by simulating extreme conditions of load and motion while still producing clinically relevant wear modes and wear rates.

8.2.2. Test Standards

Wear simulator test standards have been developed for hip, knee and spinal joints by the Implants for Surgery technical committee TC150 of ISO (Table 8.4). Test standards have not yet been developed for other joints such as upper extremities or foot and ankle.

8.2.3. Hip Simulator Testing

Part 1 of ISO 14242 [33] defines the test conditions to be used for anatomical hip simulators. Samples are mounted in an anatomical orientation using methods comparable with the *in vivo* fixation (Fig. 8.8). The acetabular cup is mounted at an inclination of 30° to the horizontal. Pre-test measurements are not described in the standard, but typically include weight measurements, cup and head radius, form error and surface finish. Normally, the surfaces are photographed optically and by scanning electron microscopy in the unworn pre-test condition. Samples that are likely to absorb water from the test serum (e.g. many polymers) should be pre-soaked until the mass of the samples has equilibrated.

The lubricating test fluid is calf serum, diluted with deionized water to a protein concentration of 30 g dm^{-3} and filtered through a 2 micron filter. Optionally, antimicrobial reagents such as sodium azide may be used. For example, ASTM F732 recommends a concentration of 0.2–0.3% sodium azide plus 20 mM of ethylene-diaminetetraacetic acid (EDTA) for tests with ultra-high molecular weight polyethylene (UHMWPE) in order to bind calcium in solution and minimize precipitation of calcium phosphate onto the bearing surfaces. It should be noted that the use of any such additives is likely to affect the measured wear rates. During the test, the fluid is maintained at 37 °C.

The load cycle specified by ISO 14242 is a 3-kN double-peak curve based on an idealized walking gait cycle (Fig. 8.9).

The applied motions combine flexion–extension, internal–external rotation and abduction–adduction cycles, as illustrated in Fig. 8.10. The motions applied by hip simulators are designed to induce a crossing-path component to the relative motion of the bearing surfaces. This has been found to be essential in simulating the *in vivo* wear of UHMWPE [34].

TABLE 8.4 ISO Wear-Testing Standards

Designation	Description
ISO 14242-1	Implants for surgery — Wear of total hip-joint prostheses — Part 1: Loading and displacement parameters for wear-testing machines and corresponding environmental conditions for test
ISO 14242-2	Implants for surgery — Wear of total hip-joint prostheses — Part 2: Methods of measurement
ISO 14242-3	Implants for surgery — Wear of total hip-joint prostheses — Part 3: Loading and displacement parameters for orbital bearing type wear-testing machines and corresponding environmental conditions for test
ISO 14243-1	Implants for surgery — Wear of total knee-joint prostheses — Part 1: Loading and displacement parameters for wear-testing machines with load control and corresponding environmental conditions for test
ISO 14243-2	Implants for surgery — Wear of total knee-joint prostheses — Part 2: Methods of measurement
ISO 14243-3	Implants for surgery — Wear of total knee-joint prostheses — Part 3: Loading and displacement parameters for wear-testing machines with displacement control and corresponding environmental conditions for test
ISO 18192-1	Implants for surgery — Wear of total intervertebral spinal-disc prostheses — Part 1: Loading and displacement parameters for wear testing and corresponding environmental conditions for test
ISO 18192-2	Implants for surgery — Wear of total intervertebral spinal-disc prostheses — Part 2: Nucleus replacements

For samples that absorb water from the test serum, it is common to include load soak samples with the test samples. Many simulators provide active load soak stations that apply the specified load cycle but with no motion. This allows for the correction of the measured mass loss values on the wear stations by adding the mass gain experienced by the load soak samples.

Testing is conducted at a frequency of 1 Hz to a minimum of 5 million cycles. The test is stopped for weight measurements and serum changeout at least at half a million cycles, 1 million cycles and at least every million cycles thereafter until the test is terminated. Measurement methods are specified in ISO 14242-2 [35].

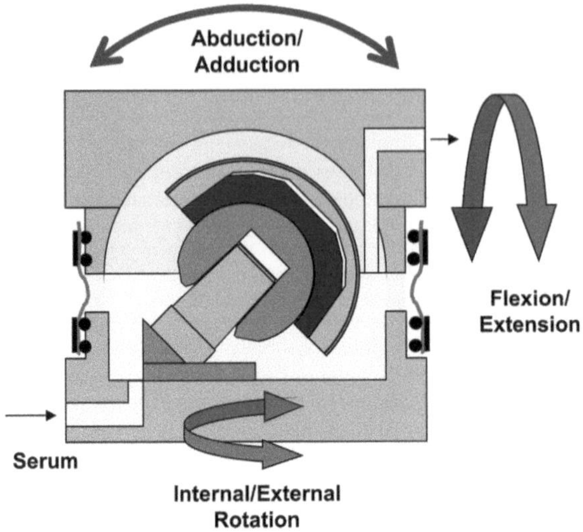

FIGURE 8.8 A typical sample mounting arrangement for ISO 14242-1 hip simulator tests.

An alternative to the anatomical hip simulator test is the orbital bearing type wear tester (Fig. 8.11), in which samples are mounted on a rotating inclined block such that the baseplate and chamber provide a ±23° biaxial rocking motion. The development of the orbital bearing test for testing UHMWPE is

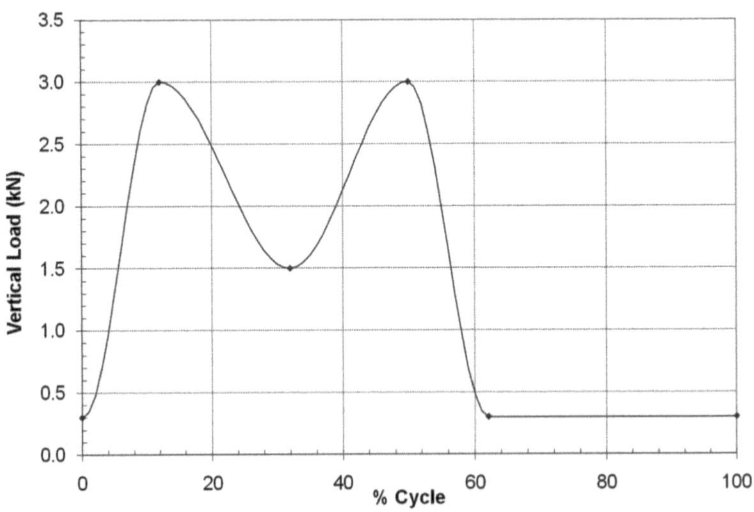

FIGURE 8.9 Representation of vertical load versus percentage of test cycle for ISO 14242 hip simulator testing.

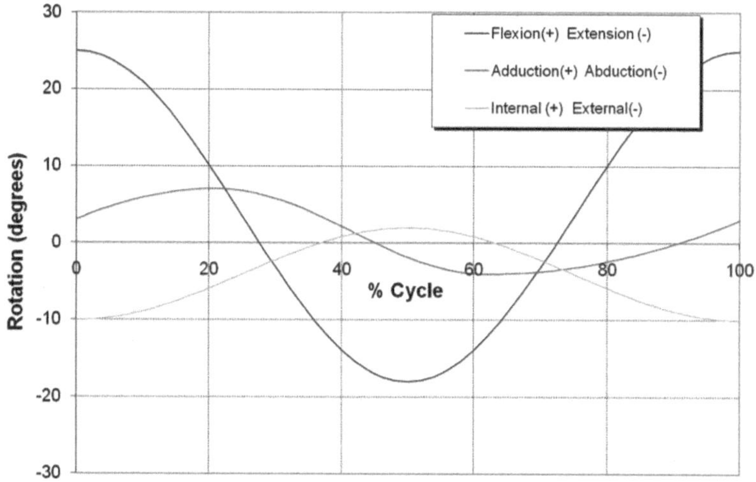

FIGURE 8.10 Representation of motions specified for ISO 14242-1 hip simulator testing.

described in Essner and Wang [36]. Test conditions are specified in ISO 14242-3 [37]. The loading cycle is the same as for ISO 14242-1.

It should be recognized that there are many factors that can affect clinical wear performance and it is difficult to incorporate all of these factors into a test.

FIGURE 8.11 Orbital bearing wear tester.

For example, Bowsher et al. [38] studied the effects of gait cycle on wear of metal on metal bearings, concluding that wear rates under simulated conditions of fast jogging were 5–12 times higher than for the normal walking gait. Williams et al. [39] found that cup inclination and microlateralization of the head both increased the wear rates of metal on metal bearings. Distraction of the head from the cup during the swing phase of the gait cycle has been found to increase wear rates of ceramic on ceramic bearings and reproduce clinically observed wear patterns [40]. The protein composition of lubricating serum has been found to affect wear rates of UHMWPE and PTFE acetabular components in complex ways [41].

Thus, hip simulation is a continually evolving technology. Much progress has been made in standardizing test methods but continuing research is constantly uncovering new factors that may need to be included in test design.

8.2.4. Knee Simulator Testing

The knee is a more complex joint than the hip because it has less conformity and greater laxity. The knee joint provides flexion–extension motion (like a mechanical hinge), with moderate axial (lift-off) motion or medial–lateral displacement or varus–valgus (abduction–adduction) motion. However, the knee is relatively more lax on anterior–posterior translation and internal–external rotation. The forces at the knee include muscle forces that react to ground reaction forces and produce resultant forces that can rise to many times body weight.

As for the hip, knee simulator testing has evolved over several decades [e.g. Ref. 42]. Generic tests such as pin-on-disk have been used, primarily as screening tests during the development of new materials. Anatomical knee simulator tests (Figs 8.12 and 8.13) are described in ISO 14243. Load-controlled testing is described in part 1 of the standard [43] and displacement-controlled testing in part 3 [44]. Measurement methods are described in ISO 14243 part 2 [45]. The rationale for conducting force- versus displacement-controlled testing has been discussed by Haider [42]. In general, the force-controlled method is considered to be more clinically relevant in order to simulate the changes in kinematics that may occur as the components wear.

In force-controlled testing, where the external anterior-posterior (AP) force and internal-external (IE) torque are controlled, the system requires additional simulation of the ligament and other soft tissue constraints that would occur *in vivo*. Thus, ISO 14243-1 includes suitable values of AP restraint and tibial rotation restraint stiffness for the case of posterior cruciate ligament (PCL)-sacrificing and PCL-retaining designs [43].

As for the hip simulator, testing is conducted at a frequency of 1 Hz to a minimum of 5 million cycles. The test is stopped for weight measurements and serum changeout at least at half a million cycles, 1 million cycles and at least every million cycles thereafter until the test is terminated. Measurement

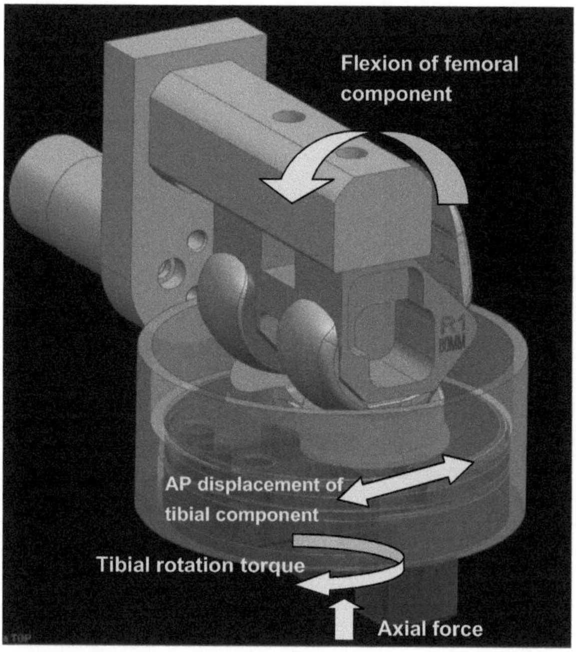

FIGURE 8.12 Example of an anatomical knee simulator test configuration.

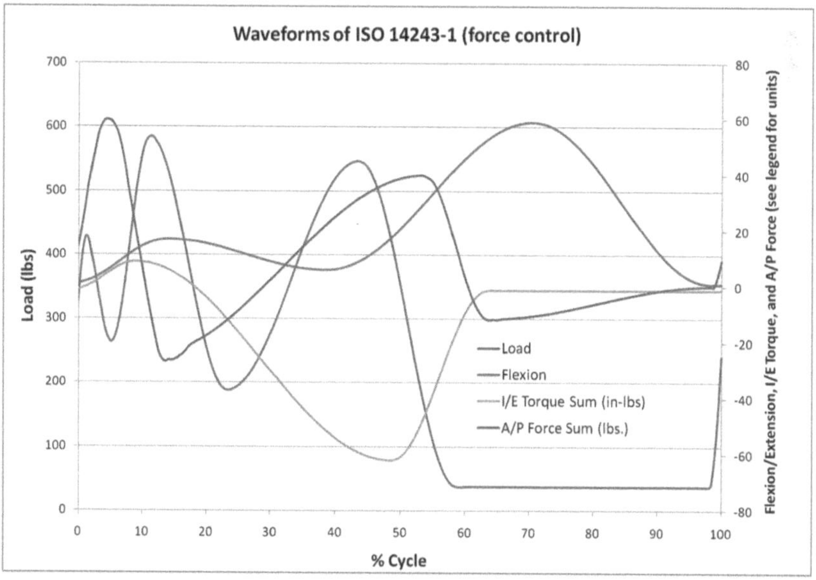

FIGURE 8.13 Representations of loading and motion waveforms specified for ISO 14243-1 knee simulator testing.

methods are specified in ISO 14243-2 [45]. For bearing materials that absorb water from the serum, the same provisions for pre-soaking the bearings and providing load soak corrections are applied.

8.2.5. Debris Analysis

Although most total joint prostheses remain stable for many years, it is generally accepted that loosening of the implant and local bone resorption is a potential result of biological reactions to particulate debris generated in the joint [e.g. Ref. 46]. Particle size, volume, morphology and composition can all affect biological response [e.g. Ref. 47]. Therefore, the characterization of debris from hip and knee simulator testing has become an important part of design verification activities.

Table 8.5 summarizes the test standards that are used in debris characterization. The process generally includes collection and preservation of test serum, digestion of proteins and lipids and filtration or centrifugation to extract the particles onto a suitable medium for microscopy. Typically, analysis is done in a high-resolution scanning electron microscope or a transmission electron microscope. Image analysis is conducted to determine the size and shape distribution of the particles. ASTM F1877 includes a classification scheme for description of particle morphology with reference micrographs [48].

Table 8.6 summarizes some debris size results from various hip simulator studies published in the literature.

8.2.6. Friction Testing

Friction measurements are often part of the tribological characterization of a joint design. The consequences of high frictional torque on the bearing surfaces can include high shear stress on the bone fixation surfaces, possibly leading to loss of fixation, or high torque on the modular junctions, such as the head–stem taper connection, possibly leading to increased micromotion and taper fretting.

TABLE 8.5 Debris Analysis Standards

Designation	Description
ASTM F561	Standard Practice for Retrieval and Analysis of Medical Devices, and Associated Tissues and Fluids
ASTM F1877	Standard Practice for Characterization of Particles
ISO 17853	Wear of implant materials – Polymer and metal wear particles – Isolation and characterization

TABLE 8.6 Literature Values of Debris Size from Various Hip Simulator Studies

	Hip Simulator Debris Size, nm			
	Max	Min	Mean	References
Metal on UHMWPE	400	50	150	[49]
Metal on highly crosslinked UHMWPE	200	50	110	
High carbon metal on metal, run-in	38	28	33	[50]
High carbon metal on metal, steady-state	37	32	34.5	
Ceramic on ceramic, no separation	20	4	9.2	[51]
Ceramic on ceramic, separation low range	35	3	6.4	
Ceramic on ceramic, separation high range	1000	200	400	

Static and kinematic friction coefficients for the natural synovial joints have been summarized in Ref. [52], with typical values in the range 0.003–0.010. Friction coefficients for various artificial hip joint material combinations are often significantly higher than these values, ranging from 0.002 to 0.3 in one study [53], depending on the articulating materials and the lubricant used for testing.

In general, friction testing has not yet been standardized to a great extent. Methods include pendulum testing [e.g. Ref. 54], simple flexion–extension test rigs [e.g. Ref. 53] and multi-station hip simulator testing [e.g. Ref. 55].

8.3. METALLIC COATINGS FOR ORTHOPAEDIC DEVICES

8.3.1. Introduction

Orthopaedic devices used in load-bearing applications are primarily metallic implants made out of biocompatible materials like Ti-alloys (e.g. ASTM F136 [56]), or CoCrMo alloys (e.g. ASTM F75 [57] or ASTM F1537 [58]). To enhance the integration of these metallic implants with the host bone, the surfaces of these devices are often coated with metallic coatings that facilitate bone apposition onto or ingrowth into the device, and thus enhance the long-term fixation of the device. The coatings can be generated by different processing methods like thermal plasma spraying or sintering of beads or diffusion bonding of thin wire fragments to create a wire mesh coating.

The metallic coatings are characterized per US FDA guidance documents for metallic coatings for orthopaedic devices – 'Guidance Document for Testing Orthopaedic Implants with Modified Metallic Surfaces Apposing Bone or Bone Cement' [59] and 'Guidance for Industry on the Testing of Metallic Plasma Sprayed Coatings on Orthopaedic Implants to Support Reconsideration of Post-market Surveillance Requirements' [60].

8.3.2. Tensile Adhesion Strength

The tensile adhesion test can be used to evaluate the bond strength between the coating and the substrate. The bond strength between the coating and the substrate can be characterized per ASTM F1147 [61]. The test utilizes standard 1″ diameter cylindrical metallic test buttons that are glued to the coating surface using a polymeric adhesive like FM1000. The face of the test buttons where the glue is applied are roughened with a 16-grit blast operation to ensure a better adhesion of the test button to the coated substrate. The test assembly consists of test buttons aligned normal to the coating-substrate interface of the device. The assembly is then pulled normal to the plane of the coating until there is a cohesive or adhesive failure of the coating, and the maximum load is recorded to calculate the adhesion strength of the coating to the substrate.

Per the FDA guidance documents for orthopaedic devices, the static tensile adhesion strength between the coating and the substrate should exceed 20 MPa [59,60].

8.3.3. Shear Strength

The static shear strength test can be used to evaluate the load it takes to shear the coating from the substrate. The static shear strength between the coating and the substrate can be characterized per ASTM F1044 [62]. The test utilizes standard 1″ diameter cylindrical metallic test buttons that are glued to the coating surface using a polymeric adhesive like FM1000. The face of the test buttons where the glue is applied are roughened with a 16-grit blast operation to ensure a better adhesion of the test button to the coated substrate. The test assembly consists of test buttons aligned normal to the coating-substrate interface of the device. The shear load is then applied parallel to the plane of the coating until there is a cohesive or adhesive failure of the coating, and the maximum load is recorded to calculate the shear strength of the coating.

Per the FDA guidance documents for orthopaedic devices, the static shear adhesion strength between the coating and the substrate should exceed 20 MPa [59,60].

8.3.4. Shear Fatigue

The shear fatigue strength of surface coatings can be characterized per ASTM F1160 [63]. The test assembly can be created similar to the setup described above for static shear testing. The assembly is then loaded under fatigue at stresses that are determined by the design requirements of the device to establish the endurance limit of the coating.

Per the FDA guidance documents for orthopaedic devices, the shear fatigue strength of the coating must be tested to at least 10^7 cycles [59,60].

8.3.5. Abrasion Resistance

The coatings on medical devices can subjected to abrasive forces that can cause particulate shedding from the coating surface. This could occur during the insertion of the device or can be caused from the micromotion of the device after implantation. The particulate shedding from the coating can cause third-body wear or can cause osteolysis of the tissue surrounding the implant. These phenomena can lead to implant loosening which can lead to implant failure and revision surgeries. As a result, it is of importance to evaluate the resistance of coatings to particle shedding. The abrasion test method per ASTM F1978 [64] can be used to evaluate the susceptibility of coatings to particle shedding. The test can be conducted with the Taber Abraser Model 5150 or equivalent and requires disc-shaped substrate with a flat coating surface that is abraded using a calibrated H-22 wheel. The rolling and rubbing of the wheel on the surface of the coating surface causes wear which is recorded as the cumulative mass loss of the coating. The initial weight of the sample is recorded, and the coating surface is abraded for 2, 5, 10 and 100 cumulative cycles under standard conditions specified in the ASTM standard. After each of the aforementioned cycle step, the samples are cleaned in fresh saline solution using ultrasonic cleaner and then completely dried per the conditions specified in the ASTM standard. The final weight of the sample after 100 cumulative cycles is recorded and the difference in initial and final sample weight is used to calculate the particulate shedding from the coating.

Per the FDA guidance documents for orthopaedic devices, the average cumulative weight loss of the coating after 100 abrasion cycles should be less than 65 mg [59,60].

8.3.6. Porosity and Pore size and Thickness

The surface coatings on orthopaedic devices are designed to allow space for bone to grow into and provide the necessary fixation of the implant to the host bone. The porosity, pore size and thickness of the surface coatings for devices can be characterized per ASTM F1854 [65].

For measuring the thickness, the samples are sectioned transverse to the coating-implant interface as showed in Fig. 8.14. The samples are then embedded in metallographic mounts, polished and photographed. The field of analysis is subsequently selected such that it is entirely contained between the top surface of the coating and the coating-substrate interface. For each sample,

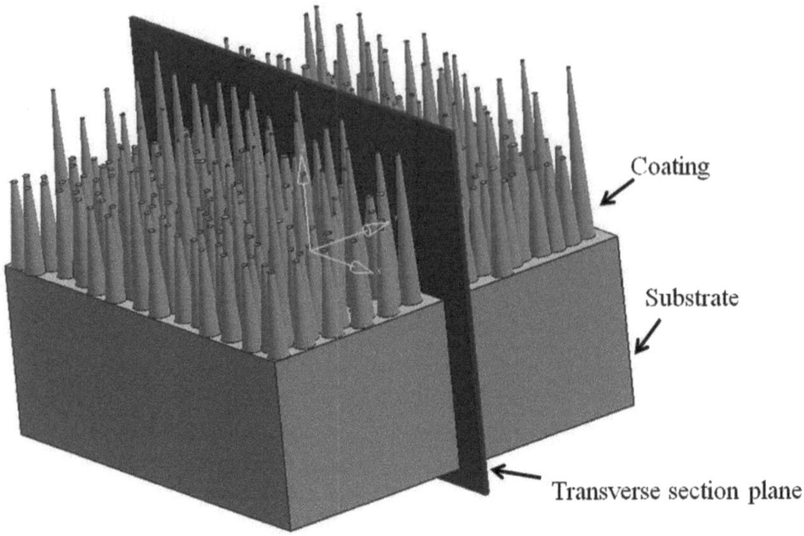

FIGURE 8.14 Schematic illustrating the use of transverse working surfaces for analysing the thickness of porous coatings per ASTM F1854.

multiple non-overlapping images should be acquired such that a continuous linear distance of at least 10 mm of porous surface is covered in the analysis. An array or grid is laid over the sample images such that grid is perpendicular to the plane of the coating and consists of parallel lines that are separated by no more than 100 microns. The analysis should be conducted with at least 10 consecutive gridlines for each image. The thickness of the coating is determined by measuring the intercepts between the substrate interface and top of the coating. The average of all the measurements represents the mean coating thickness for that working surface.

The pore size of the coating is characterized by embedding the samples such that the coating-substrate interface is parallel to the face of the mount that is polished, as showed in Fig. 8.15. The samples are polished to expose the coating surface, and images of the porous architecture are acquired. This coating depth level is referenced as 'zero' level. The sample is then ground and polished at incremental depths (100 microns or less) into the coating thickness and towards the substrate interface. The sample should be evaluated such that at least half the distance between the coating surface and the substrate interface is analyzed. At each depth level, images illustrating the porous architecture of the coating are acquired. An array or grid mesh is laid over the images such that the grid lines are no more than 100 microns apart. The intercepts of the grid lines and the coating are measured to determine the average pore size of the coating. The images should be taken from non-overlapping areas and should represent at least 5 mm^2 of total area of the coating.

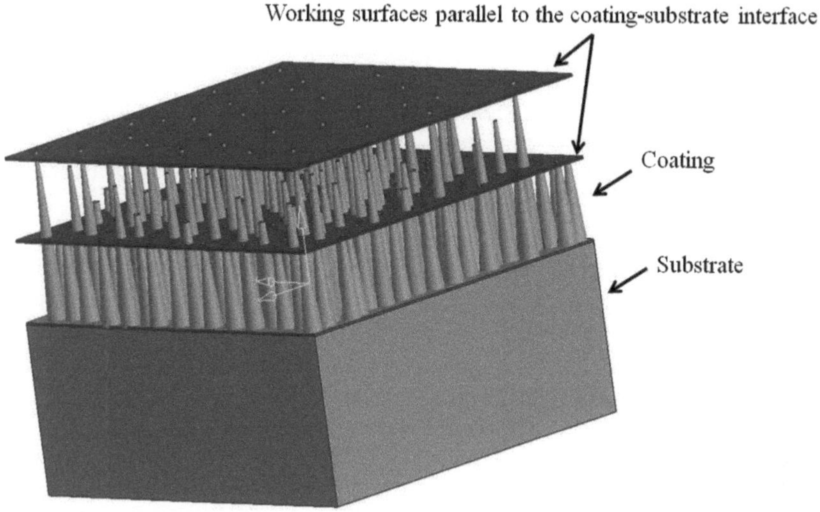

FIGURE 8.15 Schematic illustrating the use of parallel working surfaces for analysing the porosity characteristics of porous coatings per ASTM F1854.

As the coatings on medical devices are typically thin and have limited volume for intrusion, methods such as mercury porosimetry have limited applicability for analysing the porosity and pore size for such coatings. Image analysis is often the more accepted method and is better recognized and accepted by regulatory bodies like FDA and other agencies worldwide.

For porosity calculations using the image method, the samples can be analyzed either as shown in Fig. 8.14 or Fig. 8.15. If the parallel working surface system is utilized (as shown in Fig. 8.15), the working surfaces for measuring the porosity are prepared in accordance with the method described above for measuring pore size. The porosity from the working surfaces at each level is measured as the ratio of the area of the grid that falls within the porous space to the total grid area for that working surface. The overall porosity of the coating is calculated by the average of the porosity from each working surface. Again, as indicated in the pore size section above, the working surfaces should be selected to cover at least half the thickness of the coating between the coating surface and substrate interface.

If the transverse section method (shown in Fig. 8.14) is selected, the porosity is determined by a field that is entirely contained between the outer surface of the coating and the coating-substrate interface. The tallest peak of the coating is identified and a tangential line is drawn touching the peak such that it is also parallel to the coating-implant interface. If a grid is used to calculate the porosity, then an array of at least 100 regularly spaced points separated by less than 50 microns should be superimposed on the images.

The porosity is then calculated by the number of points of the grid that fall in the porous region to the total number of points in the grid. Volume percentage voids should be measured over a working surface area of at least 15 mm^2, with no overlapping areas. The height of the arrays should be at least half the distance from the coating surface to the substrate interface. It is generally recommended to include as much of the coating thickness as possible to obtain a true representation of the coating porosity.

The minimum pore size is considered to be around 100 μm for supporting bone ingrowth. However, pore sizes of over 300 μm are generally recommended to enhance bone formation and formation of capillaries for nutrient supply [66,67]. For orthopaedic devices, per the definition of porous coating in 21CFR888.3358 [68], FDA requires that the pore size should be between 100 and 1000 μm, the porosity be between 30% and 70% and the thickness of the coating be between 500 and 1500 μm.

8.3.7. Interconnectivity

It is essential for porous coatings to contain interconnected porous architecture to support bone ingrowth and to maintain constant blood and nutrient supply for new bone formation in the pore-pockets. It is generally recommended that a majority of the pores are interconnected to optimize conditions for osteoconduction and differentiation [69,70]. Additionally, it has been observed that interconnected channels below a certain critical size inhibit bone formation. However, channels above a certain size do not necessarily provide an additional benefit to promote new tissue formation for colonization of the entire volume of the implant [69].

Interconnectivity of the pores in a porous coating can be assessed by image analysis using optical microscopy or scanning electron microscopy.

8.3.8. Roughness

Medical devices with high surface roughness may provide an enhanced initial stability during and immediately post-implantation by providing a better scratch fit in contact with host bone tissue. As a result, the coatings for orthopaedic devices are generally engineered with higher surface roughness.

The surface roughness of the coatings is calculated and expressed as the arithmetic mean roughness or R_a. A profilometer can be used to determine the surface roughness for coatings that have uniform roughness over the entire surface or if the substrate geometry is not complex. It involves physically moving the profilometer over the peaks and valleys of the coating to capture the roughness. However, coatings on most orthopaedic devices contain high surface roughness and are applied to intricate shapes to match the anatomical geometry. Thus, it is not feasible to perform roughness measurements using physical profilometry for such coatings. In such cases, surface roughness can be estimated using optical profilometry techniques. The optical method involves using high precision microscopes to scan the samples to generate quantitative three-dimensional plots and images of the coating surface. To calculate surface roughness over large areas, the coating surface can be scanned at different locations and the images from these areas can be 'stitched' together to generate the average surface roughness representing that entire surface.

The FDA guidance document requires device manufacturers to provide roughness information for porous coatings on orthopaedic devices [59].

REFERENCES

[1] FDA 21 CFR 820 Quality System Regulation. Rockville, Maryland, USA: Food and Drug Administration; 2006.

[2] FDA 21 CFR 807 Establishment Registration and Device Listing for Manufacturers and Initial Importers of Devices. Rockville, Maryland, USA: Food and Drug Administration; 2011.

[3] FDA 21 CFR 814 Premarket Approval of Medical Devices. Rockville, Maryland, USA: Food and Drug Administration, http://www.accessdata.fda.gov/scripts/cdrh/cfdocs/cfcfr/CFRSearch.cfm?CFRPart=814; 1986.

[4] Council Directive 93/42/EEC of 14 June 1993 concerning medical devices. Off J Eur Commun 1993;vol. L169:1–43.

[5] List of bodies notified under directive 93/42/EEC. Off J Eur Commun 2012.

[6] BS EN ISO 13485: 2003 medical devices – quality management systems-requirements for regulatory purposes. London: British Standards Institute; 2003.

[7] PD CEN ISO/TR 14969: 2005 medical devices – quality management systems – guidance on the application of ISO 13485:2003. London: British Standards Institute; 2005.

[8] ASTM International, Standards World wide. West Conshohocken, Pennsylvania, USA: ASTM International. <http://www.astm.org>.

[9] Geneva, Switzerland: ISO – International Organization for Standardization. <http://www.iso.org>.

[10] London, UK: BSI Worldwide. <http://www.bsigroup.com/en/standards-and-publications>.

[11] Berlin-Tiergarten, Germany: Deutsches Institut für Normung e.V. (DIN). <http://www.din.de>.

[12] TC 150/SC 4 Bone and joint replacements. London, UK: BSI.

[13] Schmalzried TP, Szuszczewicz ES, Northfield MR, Akizuki KH, Frankel RE, Belcher G, et al. Quantitative assessment of walking activity after total hip or knee replacement. J Bone Jt Surg Am 1998;80:54–9.

[14] Bergmann G, Deuretzbacher G, Heller M, Graichen F, Rohlmann A, Strauss J, et al. Hip contact forces and gait patterns from routine activities. J Biomech 2001 Jul;34(7):859–71.

[15] ASTM F2068 – 09 Standard specification for femoral prostheses – metallic implants. West Conshohocken, Pennsylvania, USA: ASTM International.

[16] Morrey BF, Chao EY. Fracture of the porous-coated metal tray of a biologically fixed knee prosthesis. Report of a case. Clin Orthop Relat Res 1988 Mar;228:182–9.

[17] Jeremy L, David Stulberg S. In: Esakul KA, editor. Handbook of case Histories in failure analysis, vol. 2. ASM International; 1992.

[18] Abernethy PJ, Robinson CM, Fowler RM. Fracture of the metal tibial tray after Kinematic total knee replacement. A common cause of early aseptic failure. J Bone Jt Surg Br 1996 Mar;78(2):220–5.

[19] Cook SD, Thomas KA. Fatigue failure of noncemented porous-coated implants. A retrieval study. J Bone Jt Surg Br 1991 Jan;73(1):20–4.
[20] ASTM F1800 – 07 standard test method for cyclic fatigue testing of metal tibial tray components of total knee joint replacements. West Conshohocken, Pennsylvania, USA: ASTM International.
[21] Ahir SP, Blunn GW, Haider H, Walker PS. Evaluation of a testing method for the fatigue performance of total knee tibial trays. J Biomech 1999 Oct;32(10):1049–57.
[22] Mohammed K. Polyethylene and metal back glenoids in conventional total shoulder arthroplasty: New Zealand joint replacement registry experience. J Bone Jt Surg Br 2012;94. B no. SUPP XXI 56.
[23] Driessnack RP, Ferlic DC, Wiedel JD. Dissociation of the glenoid component in the Macnab/English total shoulder arthroplasty. J Arthroplasty 1990 Mar;5(1):15–8.
[24] ASTM F1829-98. Standard test method for static evaluation of the glenoid locking mechanism in shear. West Conshohocken, Pennsylvania, USA: ASTM International; 2009.
[25] Bucholz RW, Ross SE, Lawrence KL. Fatigue fracture of the interlocking nail in the treatment of fractures of the distal part of the femoral shaft. J Bone Jt Surg Am 1987 Dec;69(9):1391–9.
[26] Alvarez DB, Aparicio JP, Fernández EL, Múgica IG, Batalla DN, Jiménez JP. Implant breakage, a rare complication with the Gamma nail. A review of 843 fractures of the proximal femur treated with a Gamma nail. Acta Orthop Belg 2004 Oct;70(5):435–43.
[27] ASTM F1264 – 03 standard specification and test methods for intramedullary fixation devices. West Conshohocken, Pennsylvania, USA: ASTM International.
[28] Little RE. Tables for estimating Median fatigue limits, STP 731. Philadelphia: American Society for Testing and Materials 1981;176.
[29] ISO 7206–4. Implants for surgery – partial and total hip joint prostheses – Part 4: determination of endurance properties of stemmed femoral components. London, UK: BSI Worldwide; 2002.
[30] ISO 7206–8. RSS Implants for surgery – partial and total hip joint prostheses – Part 8: endurance performance of stemmed femoral components with application of torsion. London, UK: BSI Worldwide; 1995.
[31] ISO 7206–4. Implants for surgery – partial and total hip joint prostheses – Part 4: determination of endurance properties of stemmed femoral components. London, UK: BSI Worldwide; 2010.
[32] ASTM F732, standard test method for wear testing of polymeric materials used in total joint prostheses.
[33] ISO 14242 Part 1, implants for surgery – wear of total hip-joint prostheses – Part 1: loading and displacement parameters for wear-testing machines and corresponding environmental conditions for test.
[34] Wang A, Essner A, Cooper J. The clinical relevance of hip simulator testing of high performance implants. Semin Athroplasty 2006;17:49–55.
[35] ISO 14242 Part 2, implants for surgery – wear of total hip-joint prostheses – Part 2: methods of measurement.
[36] Essner A, Wang A. Tribological assessment of UHMWPE in the hip. In: Kurtz SM, editor. UHMWPE biomaterials handbook. 2nd ed. Elsevier; 2009. p. 369–79.
[37] ISO 14242 Part 3, implants for surgery – wear of total hip-joint prostheses – Part 3: loading and displacement parameters for orbital bearing type wear testing machines and corresponding environmental conditions for test.

[38] Bowsher JG, Nevelos J, Williams PA, Shelton JC. 'Severe' wear challenge to 'as-cast' and 'double heat-treated' large-diameter metal-on-metal hip bearings. Proc Inst Mech Eng H 2006;220:135–43.

[39] Williams S, Leslie I, Isaac G, Jin Z, Ingham E, Fisher J. Tribology and wear of metal-on-metal hip prostheses: influence of cup angle and head position. J Bone Jt Surg Am 2008 Aug;90(Suppl. 3):111–7.

[40] Nevelos J, Ingham E, Doyle C, Streicher R, Nevelos A, Walter W, et al. Microseparation of the centers of alumina-alumina artificial hip joints during simulator testing produces clinically relevant wear rates and patterns. J Arthroplasty 2000;15(6):793–5.

[41] Wang A, Essner A, Schmidig G. The effects of lubricant composition on in vitro wear testing of polymeric acetabular components. J Biomed Mater Res B Appl Biomater 2004 Jan 15;68(1):45–52.

[42] Haider H. Tribological assessment of UHMWPE in the knee. In: Kurtz SM, editor. UHMWPE biomaterials handbook. 2nd ed. Elsevier; 2009. p. 381–408.

[43] ISO 14243 Part 1, implants for surgery – wear of total knee-joint prostheses – Part 1: loading and displacement parameters for wear-testing machines with load control and corresponding environmental conditions for test.

[44] ISO 14243 Part 3, implants for surgery – wear of total knee-joint prostheses – Part 3: loading and displacement parameters for wear-testing machines with displacement control and corresponding environmental conditions for test.

[45] ISO 14243 Part 2, implants for surgery – wear of total knee-joint prostheses – Part 2: methods of measurement.

[46] Amstutz HC, Campbell P, Kossovsky N, Clarke IC. Mechanism and clinical significance of wear debris-induced osteolysis. Clin Orthop Relat Res 1992, March;1(276):7–18.

[47] Matthews JB, Besong AA, Green TR, Stone MH, Wroblewski BM, Fisher J, et al. Evaluation of the response of primary human peripheral blood mononuclear phagocytes to challenge with in vitro generated clinically relevant UHMWPE particles of known size and dose. J Biomed Mater Res 2000, November 1;52(2):296–307.

[48] ASTM F1877. Standard practice for characterization of particles. J ASTM Int 2010.

[49] Ries MD, Scott ML, Jani S. Relationship between gravimetric wear and particle generation in hip simulators: conventional compared with crosslinked polyethylene. J Bone Jt Surg Am 2001;83(Suppl. 2 Pt 2):116–22.

[50] Firkins PJ, Tipper JL, Saadatzadeh MR, Ingham E, Stone MH, Farrar R, et al. Quantitative analysis of wear and wear debris from metal-on-metal hip prostheses tested in a physiological hip joint simulator. Bio-medical Mater Eng 2001;11:143–57.

[51] Fisher J, Nevelos J, Hatton A, Tipper J, Ingham E, Doyle C, et al. Wear debris generation in alumina–alumina total hip joints, an in vivo and in vitro comparison, 47th Annual Meeting ORS, San Francisco, 6, 2001.

[52] Coefficients of Friction of Human Joints; The Physics Factbook™ edited by Glenn Elert – written by his students, Available at http://hypertextbook.com/facts/2007/ConnieQiu.shtml.

[53] Scholes SC, Unsworth A. Comparison of friction and lubrication of different hip prostheses. Proc Instn Mech Engrs Part H 2000;214:49–57.

[54] Shen M.C, Rieker C.B, Gnepf P, Liebentritt G, Schon R. Effect of clearance on frictional torque characteristics of metal-on-metal THA, 51st Annual Meeting of the Orthopedic Research Society, 2005, Poster 1196.

[55] Weisenburger J, Haider H, Naylor M, Schroeder D, White B, Unsworth A, Garvin K. Annual Meeting of the Orthopedic Research Society, 2012, Poster presentation #2074: 'Friction of a variety of total hip replacements'.

[56] ASTM F136 – 11: Standard Specification for Wrought Titanium-6 Aluminum-4 Vanadium ELI (Extra Low Interstitial) Alloy for Surgical Implant Applications (UNS R56401). West Conshohocken, Pennsylvania, USA: ASTM International.

[57] ASTM F75 – 12: Standard Specification for Cobalt-28 Chromium-6 Molybdenum Alloy Castings and Casting Alloy for Surgical Implants (UNS R30075). West Conshohocken, Pennsylvania, USA: ASTM International.

[58] ASTM F1537 – 11: Standard Specification for Wrought Cobalt-28 Chromium-6 Molybdenum Alloys for Surgical Implants (UNS R31537, UNS R31538, and UNS R31539). West Conshohocken, Pennsylvania, USA: ASTM International.

[59] Guidance document for testing orthopedic implants with modified metallic surfaces apposing bone or bone cement. Rockville, Maryland, USA: U.S. Food and Drug Administration; April 28, 1994.

[60] Guidance for industry on the testing of metallic plasma sprayed coatings on orthopedic implants to support reconsideration of postmarket surveillance requirements. Rockville, Maryland, USA: U.S. Food and Drug Administration; Feb 2, 2000.

[61] ASTM F1147-05. Standard test method for tension testing of calcium phosphate and metallic coatings. West Conshohocken, Pennsylvania, USA: ASTM International; 2011.

[62] ASTM F1044-05. Standard test method for shear testing of calcium phosphate coatings and metallic coatings. West Conshohocken, Pennsylvania, USA: ASTM International; 2011.

[63] ASTM F1160-05. Standard test method for shear and bending fatigue testing of calcium phosphate and metallic medical and composite calcium phosphate/metallic coatings. West Conshohocken, Pennsylvania, USA: ASTM International; 2011.

[64] ASTM F1978 – 12: Standard Test Method for Measuring Abrasion Resistance of Metallic Thermal Spray Coatings by using the Taber Abraser. Rockville, Maryland, USA: U.S. Food and Drug Administration.

[65] ASTM F1854 – 09: Standard test Method for stereological evaluation of porous coatings on medical implants. U.S. Food and Drug Administration, Rockville, Maryland, USA.

[66] Bobyn JD, Pilliar RM, Cameron HU, Weatherly GC. The optimum pore size for the fixation of porous-surfaced metal implants by the ingrowth of bone. Clin Orthop 1980;150:263–70.

[67] Karageogiou V, Kaplan D. Porosity of 3D biomaterial scaffolds and osteogenesis. Biomaterials 2005;26:5474–91.

[68] 21 CFR888.3358; Code of Federal Regulations: Title 21, Volume 8, April 1, 2011.

[69] Baril E, Lefebvre LP, Hacking SA. Direct visualization and quantification of bone growth into porous titanium implants using micro computed tomography. J Mater Sci Mater Med 2011 May;22(5):1321–32.

[70] Otsuki B, Takemoto M, Fujibayashi S, Neo M, Kokubo T, Nakamura T. Pore throat size and connectivity determine bone and tissue ingrowth into porous implants: three-dimensional micro-CT based structural analyses of porous bioactive titanium implants. Biomaterials 2006 Dec;27(35):5892–900.

Chapter 9

Characterization of Cardiovascular Implantable Devices

Ming H. Wu and Hengchu Cao
Edwards Lifesciences LLC, One Edwards Way Irvine, CA, USA

Chapter Outline

9.1. CARDIOVASCULAR SYSTEM

The cardiovascular system consists of the heart and the vasculature. The heart is a muscular pump providing energy that drives the blood flow through the circulatory system [1]. As shown in Fig. 9.1, the heart consists of four chambers – left atrium, left ventricle, right atrium and right ventricle. The left chambers and the right chambers are divided by the septum. Blood that flows through these chambers is regulated by four cardiac valves. In the left side of the heart, the oxygenated blood returning from the pulmonary

Characterization of Biomaterials. http://dx.doi.org/10.1016/B978-0-12-415800-9.00011-5

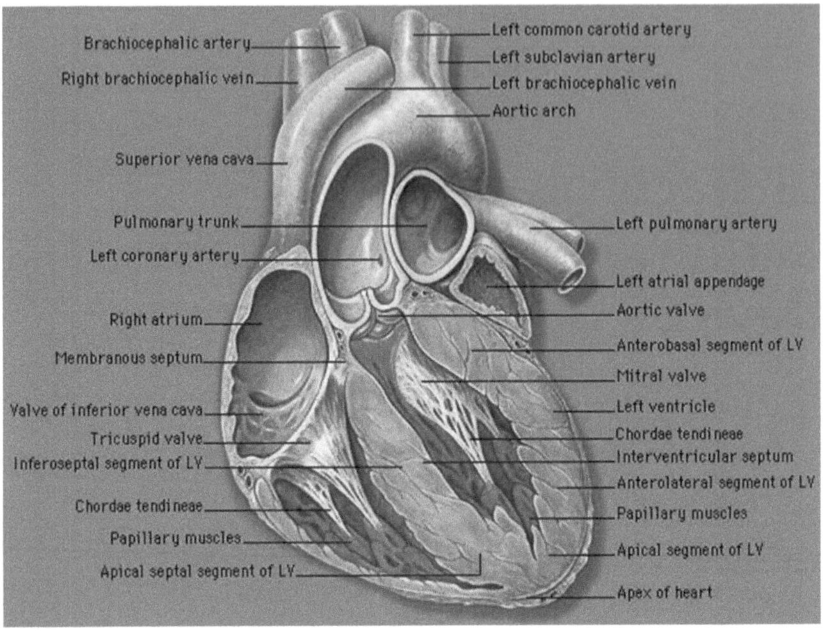

FIGURE 9.1 Anatomy of the heart. *Source: Peter J. Lynch, Yale University. Reprinted with permission.*

circulation flows through the pulmonary veins and fills the left atrium. During the diastolic cardiac cycle, the left ventricle relaxes and the blood in the left atrium goes through the mitral valve and rapidly fills the left ventricle. During the following systolic cardiac cycle, the left ventricle contracts and the oxygen-rich blood is ejected through the aortic valve into the aorta. The circulatory system is illustrated in Fig. 9.2. Oxygen-rich blood travels through the arterial blood vessels and capillaries of the systemic circulation supplying oxygen and nutrients to vital organs, muscular tissues and skeleton while collecting carbon dioxide along the circulation path. The deoxygenated blood returns through the venous vessels into the vena cava (systemic veins) and back to the right atrium. In the right side of the heart, the returned blood flows through the tricuspid valve and fills the right ventricle during the diastolic cycle. During the systole, the right ventricle contracts and ejects the blood in the right ventricle through the pulmonary valve into the pulmonary arteries. The deoxygenated blood now flows back into the pulmonary circulation to be re-oxygenated.

Oxygenation and nutrients for the heart are supplied by the coronary circulation to the heart muscle (myocardium). When the aortic valve closes during diastole, the arterial pressure in the ascending aorta provides the driving force to perfuse the coronary arteries. The oxygen-depleted blood returns

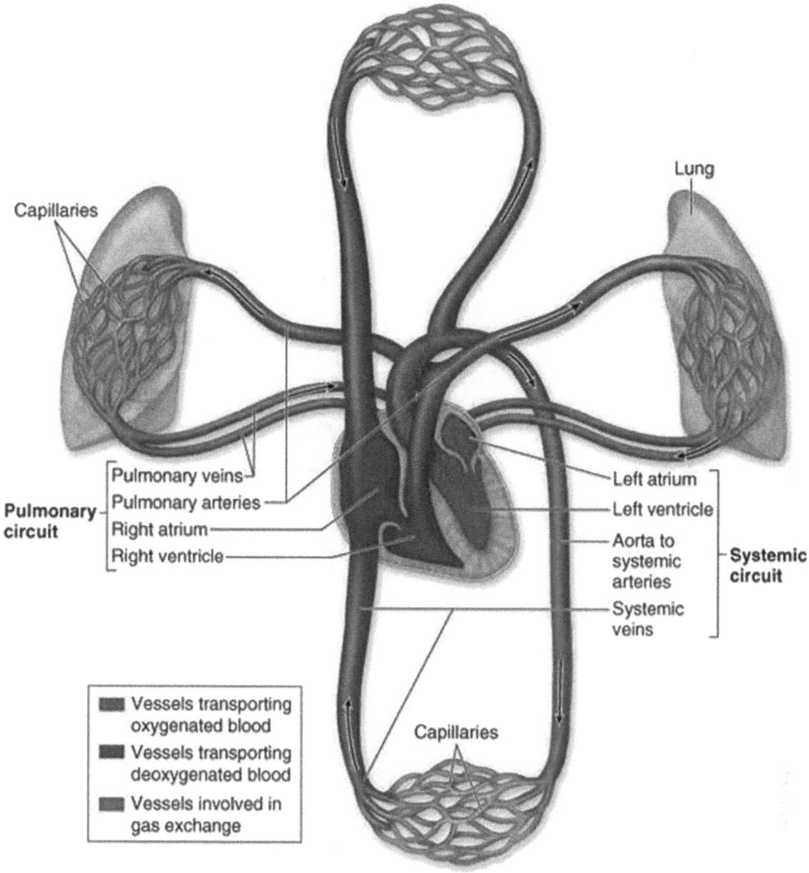

FIGURE 9.2 Systemic and pulmonary circulatory systems. *Source: Reprinted with permission of The McGraw-Hill Companies.*

though the coronary veins and sinus back to the right atrium. The coronary circulation system is illustrated in Fig. 9.3.

The cardiac rhythm of contraction and relaxation is regulated by the electrical impulse and conduction system in the heart (Fig. 9.4). At the beginning of the systolic cardiac cycle, the pace making function of the sinoatrial node (SA node) emits an electrical impulse, which is conducted through the atria to the atrioventricular node (AV node), causing the atrial to contract. The impulse is further transmitted through the specialized conduction system of HIS bundle, bundle branches and the Purkinje fibres into the ventricular myocardium, causing the ventricles to contract. The rhythm of this impulse generation and conducting pathway control the sequence and the effectiveness of a healthy cardiac function [2].

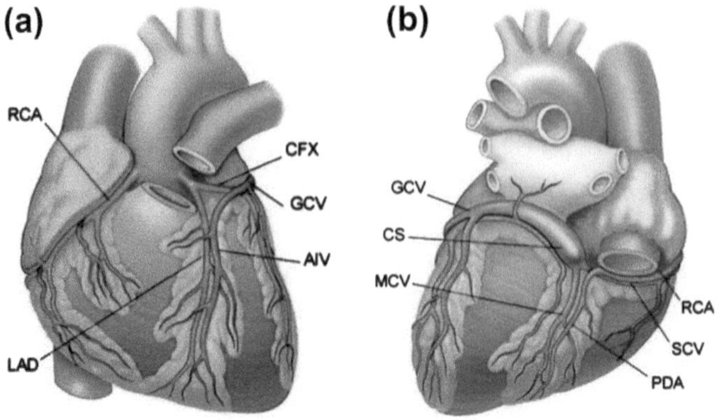

Anterior (A) and posterior (B) views of epicardial coronary circulation. LAD indicates left anterior descending coronary artery; AIV, anterior interventricular vein; CFX, circumflex coronary artery; RCA, right coronary artery; GCV, great cardiac vein; PDA, posterior descending artery; CS, coronary sinus; MCV, middle coronary vein; and SCV, small coronary vein.

FIGURE 9.3 Illustration of coronary circulation. *Source: Adapted with permission from: Percutaneous In Situ Coronary Venous Arterialization, Stephen N. Oesterle, MD; Nicolaus Reifart, MD; Eugen Hauptmann, MD; Motoya Hayase, MD; Alan C. Yeung, MD; Circulation (2001) 103, Page 2540.*

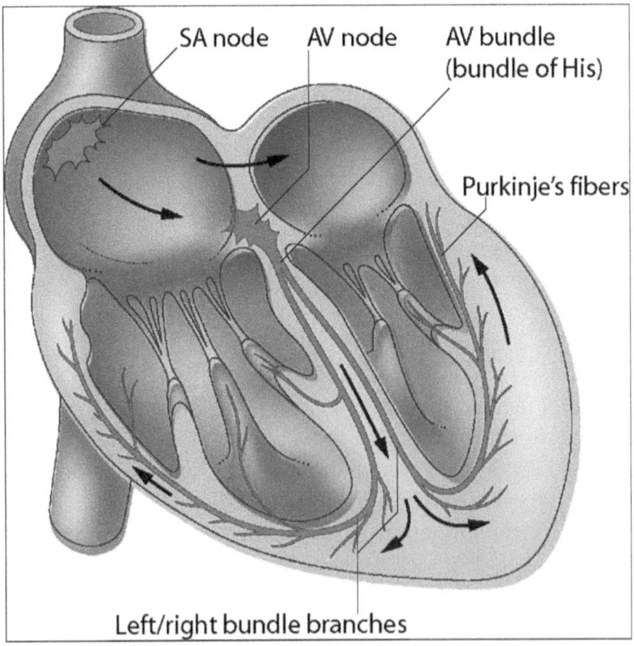

Conduction system of the heart. By permission from Cunningham JG, Textbook of Veterinary Physiology, Saunders, 2002

FIGURE 9.4 Cardiac conducting system. *Source: Reprinted with permission from Cunningham JG, Textbook of Veterinary Physiology, Saunders, 2002, Imprint of Elsevier.*

9.2. TYPES OF CARDIOVASCULAR IMPLANTABLE DEVICES

Recent advances in medical device industry have resulted in many life-saving cardiovascular implantable devices, ranging from stents and heart valve prosthesis to cardioversion defibrillators and left ventricular assist devices. It is the intent of this chapter to keep the content concise by focussing the scope on the following specific devices. It should be recognized that the general methodology for evaluating and understanding these cardiovascular devices can be extended to other devices currently available or still under development.

9.2.1. Vena Cava Filters

Pulmonary embolism is a critical, sometimes fatal, condition where blood clots developed in deep veins (deep vein thrombosis); most often in legs and occasionally in arms, travel into the pulmonary circulation and embolize the blood flow. Anti-coagulation treatments of heparin and warfarin are standard therapies for this condition. For patients who have high risks of deep vein thrombosis but are contraindicated to anti-coagulation treatment or to whom the anti-coagulation treatment fails to be effective, vena cava filters are placed into the inferior vena cava (IVC) to capture thrombolytic emboli for preventing recurrent pulmonary embolism. These devices are introduced via catheterization accesses through femoral, jugular, sub-clavian, or antecubital vein. Most of the vena cava filters are designed as permanent implants. Retrievable temporary implants are also available, which can be removed up to several months after implantation. Examples of permanent IVC filters include Greenfield filter (titanium, Boston Scientific), Simon filter (Nitinol, C.R. Bard), TrapEase (Nitinol, Cordis), Bird's Nest filter (316L stainless steel, Cook) and Venatech LP filter (phynox, B. Braun). Retrievable IVC filters include OptEase (Nitinol, Cordis), Celect (conichrome, Cook), Günther Tulip (conichrome, Cook), Option (Nitinol, Rex Medical) and G2 X (Nitinol, C.R. Bard). Photographs of Simon filter and Celect filter are shown in Fig. 9.5.

9.2.2. Septal Closure Devices

Septal defects are congenital heart anomalies resulting from incomplete embryological development of the septum, the muscular wall separating the right and left sides of the heart. The defect leaves a hole in the septum and allows blood flow between the right and the left chambers. Septal defect can occur at various locations in both atrial and ventricular septa. One example of atrial septal defect is patent foramen ovale (PFO), which occurs when the foetus foramen ovale fails to close entirely after birth. Ventricular septal defects (VSD) are found most often in the membranous and muscular areas of the

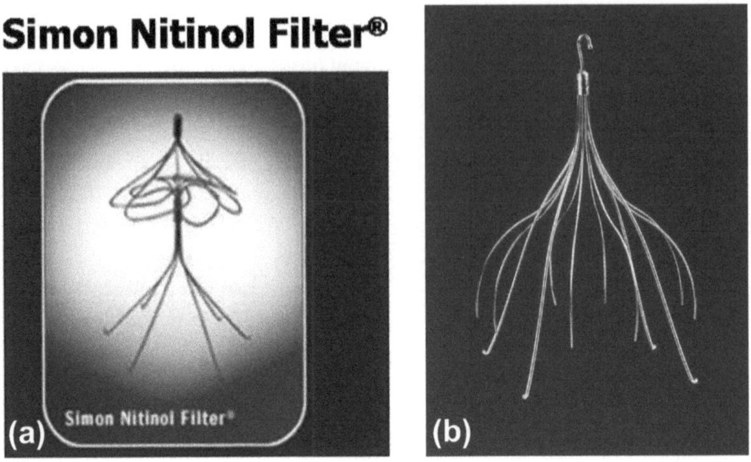

FIGURE 9.5 Vena cava filters: (a) Bard Simon filter. *Source: C.R. Bard Inc. Reprinted with permission* and (b) Cook Celect filter. *Source: Cook Medical. Reprinted with permission.*

ventricular septum. VSD in adults are uncommon but can occur as a complication of heart attacks.

Small septal defects may not cause any symptom. Large septal defects, however, can lead to a shunt of oxygenated blood flow from the left to the right atrium or ventricle and creates a cardiac insufficiency. PFO is an exception where the shunting flows from right to left due to its flap-valve-like structure. Because of the high pressure in the left side of the heart, shunts of septal defects inevitably lead to pressure overload in the right heart. Early symptoms of septal defect may include hypoxaemia (insufficient oxygen supply), palpitations, syncope, and reduced exercise tolerance. More severe problems such as arrhythmias, strokes and pulmonary hypertension can progressively develop. When left untreated, the defect may cause elevated pressure of the right-sided heart chambers to a point that the shunt reverses to flow from the right to the left. The condition is known as 'Eisenmenger syndrome' and can cause heart failure and significant mortality.

Children and adults who are symptomatic due to septal defects generally require treatment to close the defects. Patients with large septal defects are operated on conventional open heart surgical procedure where the defect is closed using surgical patches. For patients indicated for transcatheter septal closure procedures, current available devices include Amplatzer Septal Occluder (Nitinol, AGA Medical Corporation), STARflex (MP35N, NMT Medical, Inc.) and the Helex device (Nitinol, WL Gore & Associates, Inc.). All of these transcatheter septal closure devices are of self-expanding designs and exemplary photographs of STARFlex and Amplatzer are shown in Fig. 9.6.

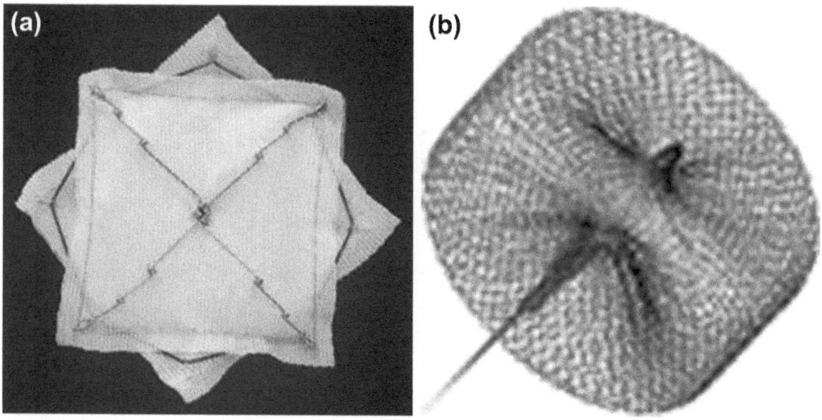

FIGURE 9.6 Septal defect closure devices: (a) NMT STARTFlex and (b) AGA Amplatzer. *Source: St. Jude Medical Inc. Reprinted with permission.*

9.2.3. Atrial Appendage Closure Devices

Atrial fibrillation (AF) is the most common cardiac arrhythmia, where abnormal electrical impulses in the atrium disrupt the regular heart beat and prevent the heart from pumping blood efficiently. Disorganized heart beats can lead to stagnant blood flow in the left atrium and the left atrial appendage (LAA), resulting in increased propensity for thrombus (blood clot) formation in these cardiac chambers. Patients in AF are at a greater risk for stroke and heart attack as the thrombus can break off and migrate into the cerebral and coronary circulations causing distant vessel embolization. Anti-coagulants such as warfarin are the standard therapy for reducing stroke risk in these patients. Recently percutaneous LAA closure devices, such as the self-expanding Nitinol WATCHMAN device by Boston Scientific (Fig. 9.7), offer an alternative to

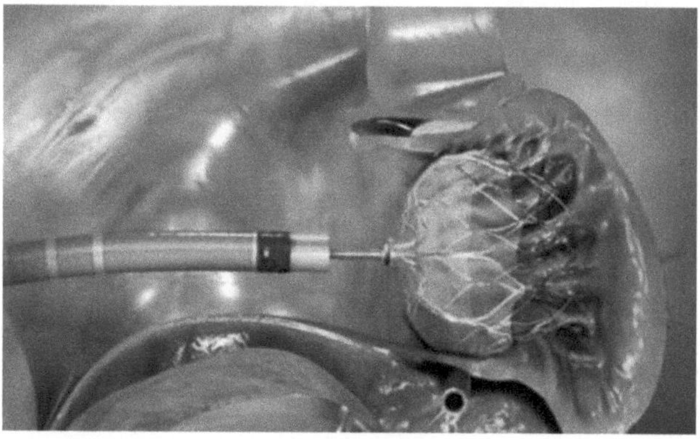

FIGURE 9.7 Boston Scientific Watchman left atrial appendage (LAA) closure device. *Source: Boston Scientific Corporation. Reprinted with permission.*

long-term anti-coagulation and to patients contraindicated for the anti-coagulant therapy. Clinical data suggested that over 90% of the emboli in AF patients derive from thrombus in LAA [3]. These devices are designed to close the LAA and to prevent the clots from migrating into the bloodstream where a stroke or a myocardial infarction (heart attack) event can occur.

9.2.4. Stents and Stentgrafts

A common cardiovascular disease is stenosis, narrowing of the arterial lumen, which causes blockage of blood flow leading to insufficient blood supply (ischaemia) in downstream tissues. Stenosis often results from atherosclerosis, a chronic buildup of plaques due to accumulation of fatty materials and inflammatory response on the arterial wall. Stable plaques are progressive diseases that can remain asymptomatic for a long time. Unstable plaques, on the other hand, can suddenly rupture and form emboli leading to acute ischaemia and tissue death in the affected anatomic site.

In the early years of intervention for the atherosclerotic arterial diseases, a balloon catheter was used to mechanically open up the stenotic artery to restore blood flow. The technique is named 'percutaneous transluminal angioplasty (PTA)'. Although the PTA balloon angioplasty demonstrated the benefit of immediate restoration of blood flow in diseased vessels, the effect is temporary as re-narrowing (restenosis) of the artery often occurs necessitating re-intervention with time. In more recent decades, a stent, which is a small tubular structure made of metallic alloys or polymeric materials, is placed via catheterization into the stenotic artery to hold the vessel open by scaffolding. A balloon angioplasty may be performed prior to stent implantation. Stents are designed to be either balloon expandable or self-expanding. A balloon expandable stent is mounted on a balloon catheter and deployed by inflating the balloon once it is introduced into the diseased site. Balloon expandable stents are made of metallic alloys such as 316L stainless steel or cobalt–chromium alloy of MP35N or L605. L605 contains 15 wt% of tungsten. Although more difficult to process, L605 offers better visibility under fluoroscopy (radiopacity) than does MP35N, a benefit for stents of small strut widths. Self-expanding stents are often made of Nitinol or cobalt–chromium wire mesh. They are mounted inside a distal cover sheath of a delivery catheter and then deployed by simply pulling the cover sheath once properly positioned inside the target lesion. Both balloon expandable and self-expanding designs may use radiopaque markers of gold, platinum, tantalum or tungsten for better visualization and positioning.

Today, balloon angioplasty and stenting are used to treat stenosis in the interventions of coronary, renal, peripheral, carotid and neurovascular arterial diseases. When utilizing stents for supporting a tubular graft structure of polymeric fabric, a stentgraft is used to repair aortic aneurysm or dissection due to trauma.

9.2.4.1. *Coronary Stents*

For patients with coronary artery diseases, the options of treatment include medication, percutaneous coronary interventions (PCI), and coronary artery bypass surgery. When the stenosis is less severe and the blood flow is not compromised, medications and lifestyle change are generally prescribed to control the disease progression. Bypass surgery is recommended for patients with severe multiple lesions and co-morbidities such as diabetes and/or heart failure, who are difficult to treat by PCI procedures [4]. If patients are indicated for PCI, the procedure may include percutaneous transluminal coronary angioplasty (PTCA) and stenting.

9.2.4.1.1. Bare Metal Stents

The first coronary stent approved by the Food and Drug Administration (FDA) was the Palmaz-Schatz stent commercialized by Cordis, a Johnson and Johnson Company, in 1994. The Palmaz-Schatz stent is a balloon expandable stent laser-machined from 316L stainless steel tubing. Since then, coronary stents of various designs and materials have been introduced. A good stent design seeks a delicate balance among flexibility, trackability, radial strength and elastic recoil. Radiopacity and biocompatibility are also desirable with any stent material consideration. A few of the early generation coronary stents were made of coil forms of wire. These include the balloon expandable Wiktor stent (tantalum, Medtronic) as well as self-expanding stents such as Cardiocoil (Nitinol, InStent Medtronic) and Wallstent (wire mesh of cobalt–chromium alloy with platinum core, Schneider/Boston Scientific). Other stents based on the tubular mesh design include Nir stent (316L stainless steel, Medinol), Bx Velocity (316L stainless steel, Cordis/Johnson and Johnson), Multi-Link PENTA (316L stainless steel, Guidant/Abbott Vascular) and Multi-Link Vision (cobalt–chromium alloy, Guidant/Abbott Vascular).

9.2.4.1.2. Drug-Eluting Stents

Although stenting reduces the restenosis rate of PTCA procedure, restenosis of bare metal stents still occurs in about 20% of cases after 6 months of the procedures and remains a clinical issue to be addressed. The primary cause for restenosis is related to the blood vessel injury, which occurs during stenting. In response to this trauma, smooth muscle cells grow during the healing stage forming scar tissue and causing re-narrowing of vessel lumen. This is known as neointimal hyperplasia. The emergence of drug-eluting stents came from the concept that local delivery of drugs, which inhibit smooth muscle cell proliferation, can control neointimal hyperplasia and reduce stent restenosis. The first-generation drug-eluting stents include Cypher (Cordis, Johnson and Johnson) and Taxus (Boston Scientific). The first FDA-approved drug-eluting stent, Cypher (Fig. 9.8a), used an immunosuppressive and anti-proliferative substance, sirolimus, which was mixed in a polymeric carrier spread-coated

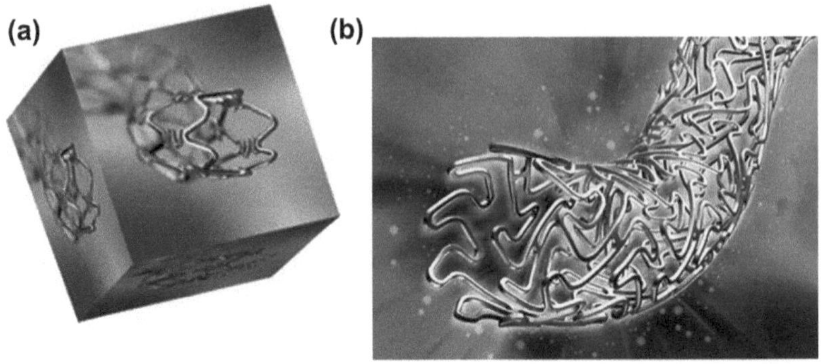

FIGURE 9.8 (a) Cordis Cypher drug-eluting stent. *Source: Cordis Corporation. Reprinted with permission* and (b) Boston Scientific Taxus Liberté drug-eluting stent. *Source: Boston Scientific Corporation. Reprinted with permission.*

onto a balloon expandable Bx Velocity stent of 316L stainless steel. Paclitaxel, an anti-proliferation drug, is used for the Taxus drug-eluting stent (Fig. 9.8b). Again a polymer carrier is used to coat the stent and control the drug-release rate. The metallic substrate was based on the balloon expandable Express stent platform of 316L stainless steel.

There were two problems associated with the first-generation drug-eluting stents, which raised public concern about the safety of these novel devices. The polymer coating had the propensity of inducing inflammation and the cytotoxicity of the drugs caused poor endothelialization leading to a significant increase in late thrombosis. Development of the more recent generation drug-eluting stents adopted bioresorbable polymers. In some recent designs, no polymer was used and the drug was directly applied onto the stent. The selection of drug regimen also moved towards a combined pro-healing and anti-proliferation therapy to promote endothelialization while preventing neo-intimal hyperplasia. The recent generation drug-eluting stents include the Endeavor zotarolimus-eluting stent from Medtronic and the Xience everolimus-eluting stent from Abbott Vascular [5].

9.2.4.1.3. Bioresorbable Stents

Vessel healing after stent placement normally takes place within a few months post the procedure; the need for scaffold is therefore temporary. Stent technology based on a permanent metallic implant presents a number of limitations. The presence of metallic stents in arteries not only limits the vessel vasomotion and positive late lumen remodelling, it can also interfere with future re-intervention and imaging modalities such as computed tomography (CT) and MRI due to image artefacts. In addition, the risk of having uncovered metallic stent struts as a result of incomplete endothelialization poses a long-term concern for late thrombosis. The emerging bioresorbable stent technology intends to address these limitations. Constructed a bioresorbable polymer or a biodegradable metal, a bioresorbable stent provides a temporary radial support to the vessel for a few months and then gradually integrates into the vessel wall with time, hence leaving nothing behind after completion of the bioresorption process. Bioresorbable stents can easily incorporate drug-eluting functions by blending the drug into the resorbable polymer or by coating onto the surface. Multiple drug-eluting regiments are also possible by adopting various combinations of coating, reservoir as well as controlled bioresorption and drug-eluting kinetics.

The first commercialized bioresorbable polymer stent is the balloon expandable ABSORB bioresorbable vascular scaffold (BVS) (Abbott Vascular) (Fig. 9.9a), which received the CE Mark approval in early 2011. The BVS stent is made of poly-L-lactic acid (PLLA) semi-crystalline polymer with amorphous poly-D,L-lactic acid (PDLLA) coating. The PDLLA coating contains and regulates the elution of an anti-proliferative drug of everolimus. These polymers gradually break down in 1–2 years through hydrolysis into lactic acid, CO_2 and H_2O, which are metabolized.

An example of metallic bioresorbable stent was the absorbable metallic stent (AMS) of magnesium alloy (Biotronik) (Fig. 9.9b). The benefit of a magnesium alloy stent is that it provides a much stronger scaffold than do polymeric stents. Early clinical experience suggests that the magnesium stent is quickly endothelialized and then gradually degrades into inorganic salts which are metabolized. Depending on the alloy composition, the magnesium scaffold can be controlled to be completely resorbed within a few months after

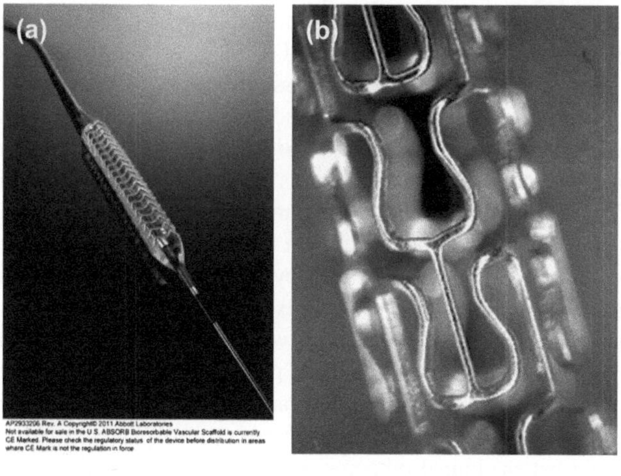

FIGURE 9.9 Photographs of bioresorbable stents: (a) Abbott vascular BVS. *Source: ©2013 Abbott. All Rights Reserved. Reprinted with permission* and (b) Biotronik AMS stents. *Source: Adapted with permission from Temporary scaffolding of coronary arteries with bioabsorbable magnesium stents: a prospective, non-randomised multicentre trial; Raimund Erbel, et al. from the PROGRESS-AMS (Clinical Performance and Angiographic Results of Coronary Stenting with Absorbable Metal Stents) Investigators; The Lancet, Volume 6, Issue 9576, Page 1870.*

implantation. Long-term efficacy of these devices is still unclear, requiring randomized clinical trials and long-term follow-up.

9.2.4.2. Peripheral Stents and Stentgrafts

Peripheral artery diseases (PADs) are arteriosclerotic vascular diseases developed in arteries outside of the coronary circulation. PADs, in a broad sense, include renal, lower extremity, carotid, cerebral and aortic arterial diseases.

9.2.4.2.1. Renal Vascular Stents

Progression of renal artery stenosis can lead to hypertension and renal insufficiency. For patients who fail to respond to medical therapy, PTA and stenting have replaced surgery as the recommended therapy for renal artery stenosis management. Balloon expandable systems such as the RX Herculink Elite Renal stent (CoCr alloy, Abbott Vascular) and Express SD (316L stainless steel, Boston Scientific) are used for this indication. Although renal artery stenting has been widely practiced, the clinical benefit of stenting when compared with the outcome of optimal medical therapy remains controversial [6].

9.2.4.2.2. Lower Extremity Peripheral Stents

Lower extremity PAD affects iliac, femoral, popliteal and infra-popliteal arteries which control blood supply to the leg and foot. Stenotic lesions in these arteries can cause symptoms of claudication (pain and weakness in legs of walking), rest pain and tissue damage due to ischaemia. In severe cases, patients may end up going through limb amputation as the last resort.

For patients with lower extremity PADs, medications are used to relieve symptoms and to control the disease progression. Surgery or interventions maybe needed to treat the disease if symptoms are not easily manageable by medications. For long-segment, diffuse and occlusive lesions, venous bypass surgery remains as the gold standard. For localized lesions in large vessels, iliac and femoral arteries, PTA and stenting provide a less-invasive option of treatment. Primary stenting is not generally recommended in smaller femoral, popliteal or tibial arteries, but as a provisional salvage therapy for failed balloon angioplasty.

Various arterial segments of lower extremity are subjected to complex mechanical deformation with the extreme worst cases occurring near the flexion area of the joint. The anatomic mechanical motion can impose bending, torsion, as well as axial tension and compression on the stent *in vivo*. Because of these severe mechanical conditions, more flexible self-expanding Nitinol stents generally yield better clinical outcome when compared to balloon expandable stents. Even with self-expanding stents, device fracture is not uncommon, especially in long-segment and overlapping stent placements. Severe stent fractures result in poor clinical outcomes of reduced patency rate due to restenosis [7].

Self-expanding Nitinol stents for treating lower extremity PAD include Smart (Cordis) (Fig. 9.10a), Luminexx (C.R. Bard), Absolute (Abbott Vascular)

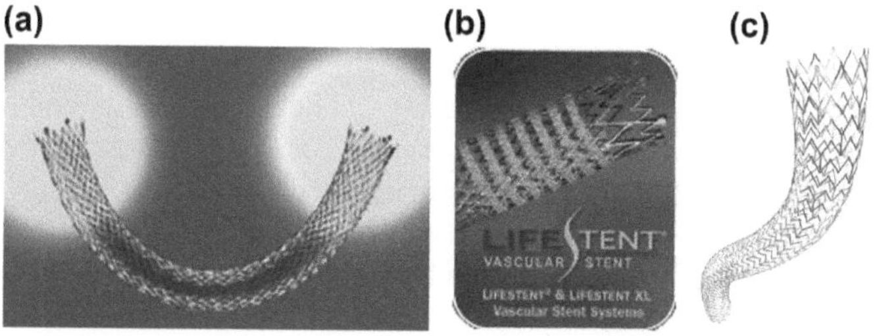

FIGURE 9.10 Photographs of (a) Cordis Smart stent. *Source: Cordis Corporation. Reprinted with permission* (b) Bard LifeStent. *Source: C. R. Bard Inc. Reprinted with permission* and (c) Cook Zilver stent. *Source: Cook Medical. Reprinted with permission.*

and Protégé Everflex (ev3) stents. Designed to improve the structural durability, newer generation stents such as LifeStent NT (Nitinol, Edwards-C.R. Bard) (Fig. 9.10b) adopt a helical cell structure to further enhance the axial and torsional flexibility. Drug-eluting stents are used for these indications and for potential treatment of in-stent restenosis; examples include Everolimus-eluting Dynalink-E (Abbott Vascular) and Paclitaxel eluting Zilver (Cook) (Fig. 9.10c).

9.2.4.2.3. Carotid Stents

Atherosclerosis in carotid arteries can reduce blood supply to the cranial circulation. It can also be a source of emboli, which can break off from the plaque and cause ischaemic stroke by blocking the blood flow in downstream vessels. Medications of anti-hypertension, anti-platelet and cholesterol-control drugs are used to reduce stroke risk. For patients with symptomatic carotid stenosis, surgical endarterectomy as well as carotid angioplasty and stenting offer treatment options. Recent guidance document [8] suggests that both treatment options yield comparable clinical outcomes, hence carotid angioplasty and stenting may be considered as an alternative procedure to surgical endarterectomy.

Because carotid arteries are located in superficial locations, carotid stenting favours self-expanding designs to prevent stent crushing, which can impair blood supply to the cranial circulation. Current self-expanding Nitinol carotid stents include RX Acculink (Abbott Vascular), Precise Pro RX (Cordis, Johnson and Johnson) and Exponent (Medtronic). The other type of self-expanding carotid stent is Wallstent (Elgiloy, Boston Scientific), which has a unique design of braided wire mesh structure.

9.2.4.2.4. Neurovascular Devices

Neurovascular disorders include atherosclerosis and aneurysm. While atherosclerosis increases the risk of ischaemic stroke, cerebral aneurysm posts risk of haemorrhagic stroke. Surgical procedures, such as placing an aneurysm clip to exclude aneurysm from the circulation, are highly invasive. Transcatheter placements of embolic coils and stents provide less-invasive options for treating these diseases.

Ischaemic disorders (atherosclerosis) – balloon angioplasty and stenting are used for treating stenotic lesions. Because arteries in the neurovascular system are smaller and more torturous than other vasculatures, neurovascular stents are design to be smaller in profile with high flexibility. Currently, the self-expanding Wingspan™ (Nitinol, Boston Scientific/Stryker) is the only approved stent for treating intrastenosis under the FDA humanitarian device exemption (HDE) program in the United States.

Haemorrhagic disorders (cerebral aneurysm) – endovascular embolic coils are delivered through a catheter to fill up the aneurysm. The clotting of blood and the healing around the coil create a barrier to exclude the aneurysm from the circulation in preventing rupture. For wide-neck aneurysms, a stent may be used to hold the embolic coil securely inside the aneurysm and to prevent it from dislodging. Current embolic coils in the market include GDC (Boston Scientific/Stryker), Axium (ev3) and Deltapaq (Depuy, Johnson and Johnson). Self-expanding neurovascular stents are generally used for treating cerebral aneurysm. These include Solitaire (ev3), Neuroform EZ (Nitinol, Boston Scientific/Stryker) and Enterprise (Nitinol, Cordis, Johnson and Johnson). A photograph of ev3 Solitaire stent with embolic coil is shown in Fig. 9.11.

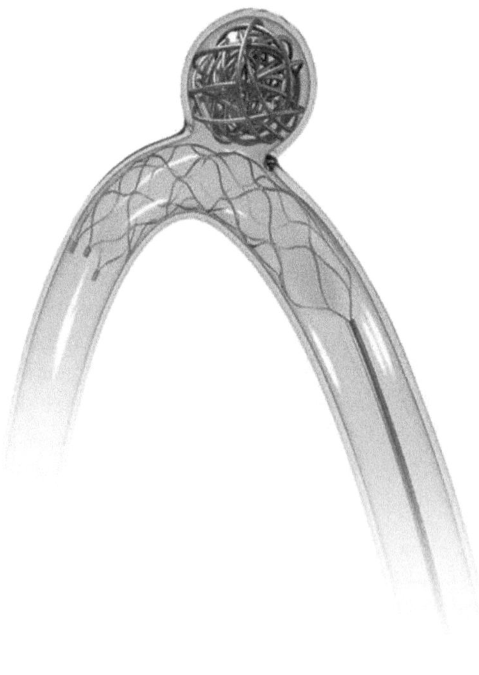

FIGURE 9.11 A photograph of the ev3 Solitaire neurovascular stent with embolic coil. *Source: Solitaire™ is a trademark of a Covidien company. ©Covidien All rights reserved.*

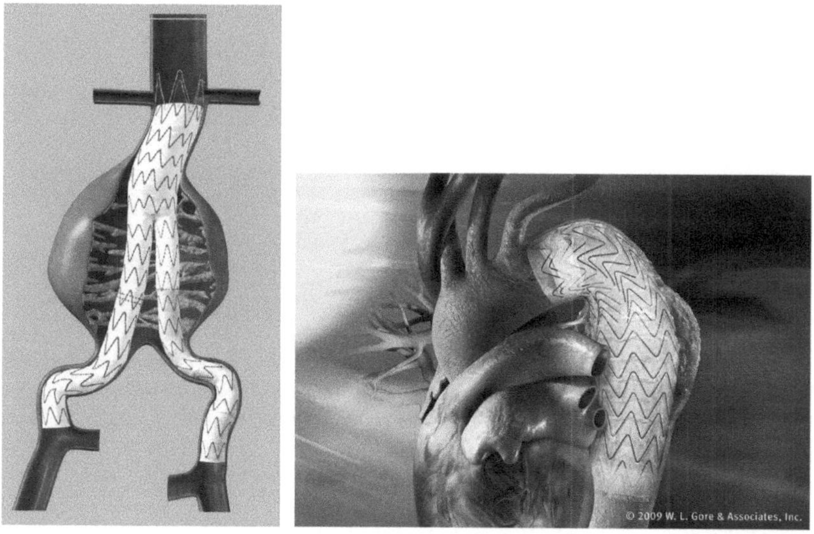

FIGURE 9.12 Endovascular stentgrafting in abdominal (left) *Source: Medtronic Inc. Reprinted with permission* and thoracic (right) aortic aneurysms. *Source: W.L. Gore & Associates, Inc. Reprinted with permission.*

9.2.4.2.5. Aortic Aneurysm Stentgraft

Aortic aneurysm is a disease of local weakening and dilatation, which can develop at various locations of the aorta. The most common location is the abdominal aorta before the iliac bifurcation. Other locations include aortic root and thoracic aorta of both ascending and descending segments. Aortic aneurysm is often asymptomatic but presents a great risk of rupture, which is life-threatening, as it results in mass internal bleeding within a short period of time. When the aneurysm is large, e.g. >5 cm for abdominal aortic aneurysm (AAA), it poses a significant risk of rupture. Open surgery is recommended to replace the diseased section of this major artery. The recent developments of endovascular aortic stentgrafting for AAA and thoracic aortic aneurysm (TAA) offer a less-invasive option for treating this type of disease. In this procedure, a stentgraft of synthetic fabric (polyester or polytetrafluoroethylene) with metallic stent support is delivered through a transcatheter and placed inside the aneurysm to carry the blood flow and isolate the aneurysm from the vascular pressure (Fig. 9.12).

Examples of current AAA stentgraft include Endurant (polyester with Nitinol stents, Medtronic) (Fig. 9.13a), Zenith (polyester with 316L stainless steel self-expanding stent support) (Fig. 9.13b) and Excluder (expanded polytetrafluoroethylene and fluorinated ethylene propylene with Nitinol wire stent support, W.L. Gore) (Fig. 9.13c).

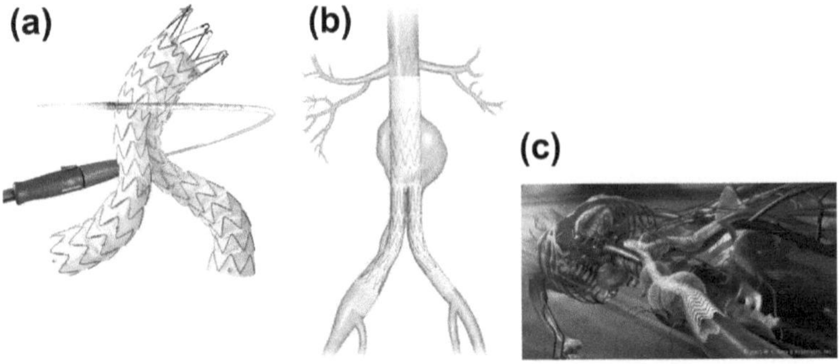

FIGURE 9.13 (a) Medtronic Endurant. *Source: Medtronic Inc. Reprinted with permission* (b) Cook Zenith. *Source: Cook Medical. Reprinted with permission* and (c) W.L. Gore Excluder AAA stentgrafts. *Source: W.L. Gore & Associates, Inc. Reprinted with permission.*

9.2.5. Embolic Protection Devices

One of the major risks of PTA and stenting is distal vessel embolization. The event can occur when emboli are dislodged from plaques during device manipulation and stent deployment. Embolic protection devices are used in conjunction with angioplasty and stenting to protect the patient from these embolic events. There are three types of embolic protection devices as illustrated during carotid stenting in Fig. 9.14. The most common is the distal filter (Fig. 9.14a) which is constructed with a net having pore sizes of about 100 μm. The device allows continuous blood flow while capturing embolic debris during the procedure. Examples of embolic protection filter devices shown in Fig. 9.15 are Emboshield NAV (Abbott Vascular) and SpiderFX (ev3). Another design is the distal occlusion balloon (Fig. 9.14b), which is placed downstream to the stenotic lesion to prevent the debris and blood flow into the distal vessel. The debris captured are then aspired though the catheter.

Unique for carotid angioplasty and stenting, an alternative approach is to use a proximal occlusion balloon as illustrated in Fig. 9.14c. The advantage of this approach is to avoid crossing the lesion under any unprotected procedure. The system first inflates an occlusion balloon in the common carotid artery proximal to the lesion. A second balloon is then inflated to occlude external carotid artery to assure that the antegrade blood flow is completely blocked in the internal carotid artery (ICA). Under the blockage of these two balloons, the Willis circuit imposes a back pressure and prevents any antegrade blood flow in the ICA. After the procedure, the debris captured in the stagnant blood in ICA is aspired and removed.

9.2.6. Cardiac Valve Prosthesis

The four cardiac valves facilitate blood circulation throughout the body in an orderly manner. Replacement of the heart valves may be needed if any of the

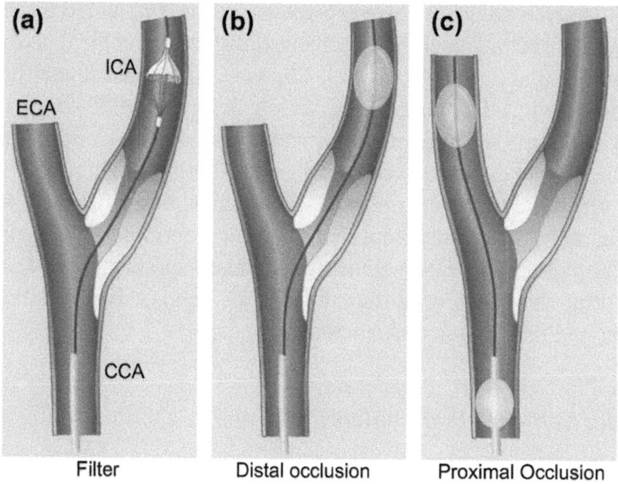

FIGURE 9.14 Three types of embolic protection approaches during carotid stenting: (a) distal filter, (b) distal occlusion and (c) proximal occlusion. *Source: Adapted with permission from Carotid Artery Stenting vs Endarterectomy; Marco Roffi, Debabrata Mukherjee and Daniel G. Clair; European Heart Journal (2009) 30, Page 2696.*

four native valves become dysfunctional due to disease. Heart valves are basically check valves, meaning the flow is unidirectional. Malfunction can manifest in either inadequate opening because of stenosis, or improper closing, causing regurgitation. More than half a century has passed since Starr and Edwards' description of successful prosthetic valve replacement in 1961 [9]. Until recently, replacement of the heart valve is performed by a cardiothoracic surgeon in an open-chest procedure, where excorporeal cardiopulmonary bypass is employed to temporarily support the circulation demand of the patient [10]. There are many different types of prosthetic heart valves classified according to the construction material, design and placement methods. Surgical heart valves

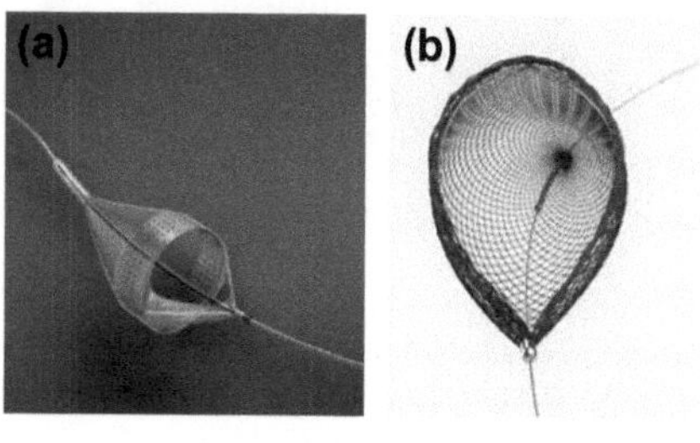

FIGURE 9.15 Photographs of (a) Abbott Vascular Emboshield NAV. *Source: ©2013 Abbott. All Rights Reserved. Reprinted with permission* and (b) ev3 SpiderFX embolic protection filters. *Source: SpiderFX™ is a trademark of a Covidien company. ©Covidien All rights reserved.*

refer to the conventional heart valve prostheses that are implanted by performing an open heart surgery, including mechanical valves, biological tissue valves, stentless xenograft and homograft [11]. The emerging transcatheter heart valve replacement introduced commercially within the years 2000–2010 has transformed the landscape of future management of valvular diseased patient where valve replacement can be performed without open heart surgery [12]. These heart valve prostheses are constructed with biological tissue for valve leaflet and a collapsible metal frame structure that can be crimped into a low profile and delivered through a catheter without open heart surgery. This section will provide a brief overview of different types of heart valve prostheses, their construction and performance characteristics.

9.2.6.1. Mechanical Heart Valve Prosthesis

The first-generation mechanical heart valve consisted of a silastic ball, which seated in an open alloy cage structure with sewing ring for attachment. The elastomeric ball seals the opening when closed and moves forward into the cage when open (Fig. 9.16a). The second-generation mechanical heart valve consisted of tilting or translating disc models where a disc-shaped occluder enabled the opening and closing of the prostheses through the rotation or translation of the disc within an orifice constraint (Fig. 9.16b). These models were soon surpassed by the third-generation bileaflet prosthetic heart valves. Modern mechanical valves are comprised of pyrolytic carbon orifice housing and two pyrolytic carbon leaflets (Fig. 9.16c).

Some earlier mechanical valve models utilized metallic alloys as the housing material but had since been replaced by pyrolytic carbon designs due to its better biocompatibility and superior wear resistance [13]. The bileaflet pyrolytic carbon valves constitute the predominant mechanical

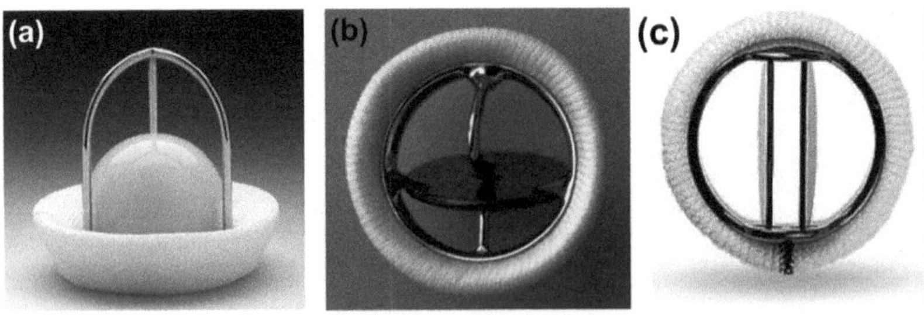

FIGURE 9.16 Typical mechanical heart valve designs: (a) caged ball. *Source: Reprinted with permission from Edwards Lifesciences LLC* (b) tilting disc. *Source: Adapted with permission from Gary L Grunkemeier, Hui-Hua Li, David C Naftel, Albert Starr, Shahbudin H Rahimtoola, Long-term performance of heart valve prostheses, Current Problems in Cardiology, Volume 25, Issue 2, February 2000, Pages 78–154* and (c) bileaflet designs. *Source: Adapted with permission from Hadi Mohammadi, Kibret Mequanint, Prosthetic aortic heart valves: Modeling and design, Medical Engineering & Physics, Volume 33, Issue 2, March 2011, Pages 131–147.*

valve implanted today because of its excellent structural durability [14]. However, all mechanical prostheses require anti-coagulant treatment after the replacement valve implant. The potential advantage of avoiding the hazards of anti-coagulation has led to the search for a valve replacement of suitable biological material, which would not require long-term anti-coagulant therapy.

Mechanical heart valves can exhibit wear and mechanical damage due to impacts of fluid–structure interaction. Fracture of the mechanical heart valve components can be catastrophic should it happen. Therefore, careful design, rigorous testing and extensive non-destructive evaluation are the essential components for the current mechanical heart valve prosthesis [15,16]. The critical structural components are designed with the fracture-mechanics-based damage-tolerant methodology. Accelerated wear testing with flow visualization are usually employed to ensure adequate structural reliability.

9.2.6.2. Biological Heart Valve Prostheses

To mitigate the risk associated with life-long anti-coagulation therapy required for mechanical heart valves, a number of different approaches to the problem of finding a suitable biological valve have been made. An autologous or autogenous valve was hand-crafted from the patient's own tissue such as fascia lata or pericardium intra-operatively. The procedure was technically demanding, the valves had very limited durability, and this approach was abandoned. An autograft valve is one translocated from one position to another—for example, when the patient's own pulmonary valve is used to replace a diseased aortic valve. This procedure is still in limited practice today, primarily for use in the so-called 'Ross Procedure' [17], wherein the patient's diseased aortic valve is replaced with their own pulmonary valve, which is in turn replaced by a homograft or a prosthetic tissue valve. The procedure is of particular value in children as the translocated pulmonary trunk grows with the child. Most late problems have been related to failure of the pulmonary replacement valve. The procedure requires a double-valve replacement operation with increased surgical risk. A homograft (or allograft) valve is one transplanted from a human donor. They are sterilized using an antibiotic solution and either stored in fixative or cryopreserved. A heterograft (or xenograft) valve is one transplanted from another species such as a pig, or manufactured from tissue such as bovine pericardium. Porcine valves (Fig. 9.17a) are treated with glutaraldehyde, which sterilizes the valve tissue and renders it biologically acceptable to the recipient. Most bioprostheses are mounted on stents attached to a sewing ring but more recently, stentless valves, which are sewn in free hand, have become available. Stentless valves (Fig. 9.17b) have a greater effective orifice area compared with stented valves, but are technically more difficult to implant. Porcine heart valves are subject to structural deterioration and calcification and thus may have limited durability.

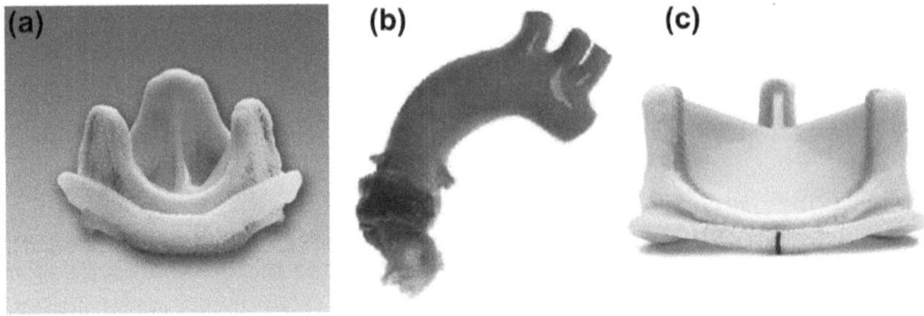

FIGURE 9.17 Representative tissue valve: (a) porcine valve, (b) stentless allograft and (c) bovine pericardial valve. *Source: Reprinted with permission from Edwards Lifesciences LLC.*

Bovine pericardial valves (Fig. 9.17c) are engineered from chemically stabilized bovine pericardium mounted on a metallic structural frame. The market leading pericardial valve is fabricated by anchoring the pericardial tissue behind the alloy stents, which prove to have excellent long-term durability [18]. Companies also applied anti-calcification treatments that further improved the durability of the pericardial valves.

For patients undergoing aortic valve replacement, choice of prosthesis is easier for the elderly patient. Bioprosthetic valves degenerate more slowly in elderly patients than in the young. A bioprosthesis would be the best choice as the patient would avoid the risk of anti-coagulant treatment. The American Heart Association/American College of Cardiology (AHA/ACC) task force recommends bioprostheses for those over 65 undergoing aortic valve replacement [19]. This strategy carries the risk that some patients will outlive their prosthesis and face the need for repeat surgery in their 80s with increased mortality and morbidity at this age. The emerging transcatheter heart valve may be used in the so-called valve-in-valve procedure, potentially opening up a new paradigm for heart valve replacement based on the biological tissue valve technology [20].

9.2.6.3. *Transcatheter Heart Valve Prostheses*

Surgical aortic valve replacement (SAVR) is the gold standard treatment known to improve symptoms and survival in patients with severe, symptomatic aortic stenosis. Peri-operative mortality, however, is higher among patients of old ages and/or with severe co-morbidities for whom SAVR may be indicated. Transcatheter aortic valve implantation (TAVI) is an emerging, catheter-based technology that allows for implantation of a prosthetic valve without open heart surgery.

TAVI allows for delivery of a prosthetic heart valve within the diseased native aortic valve without the need for open heart surgery or cardiopulmonary bypass. Transcatheter aortic valve (TAV) prosthesis is manufactured with bovine or porcine pericardium and mounted within a metal scaffold. The device

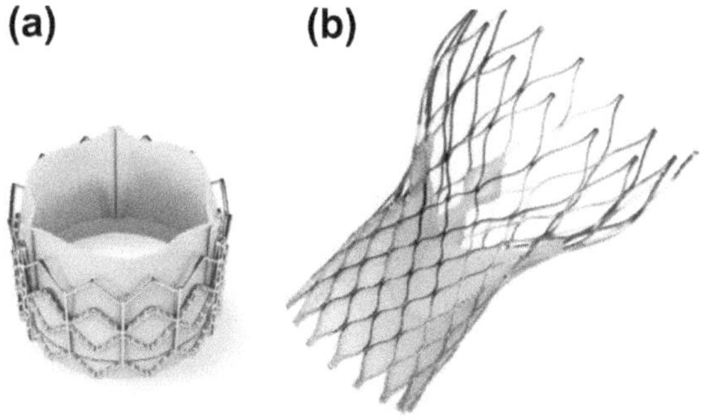

FIGURE 9.18 Examples of the emerging transcatheter heart valves: (a) Edwards' SAPIEN. *Source: Reprinted with permission from Edwards Lifesciences LLC* and (b) Medtronic's CoreValve Revalving System. *Source: Adapted with permission from Pierre M. Sfeir, Antoine B. Abchee, Ziyad Ghazzal, Jean Beresian, Pamela Gellad, Chakib M. Ayoub, Endovascular Transcatheter Aortic Valve Implantation: An Evolving Standard, Journal of Cardiothoracic and Vascular Anesthesia, Available online 22 December 2012.*

is delivered by transcatheter procedure across the stenotic aortic valve either through the femoral artery (transfemoral), sub-clavian artery, axillary artery, or ascending aorta. These are retrograde approaches. The device can also be delivered using an antegrade approach directly through the apex of the heart by means of thoracotomy incision, i.e. transapical procedure. The first successful TAVI procedure in a human was reported in 2002 [21]. Both the SAPIEN of balloon expandable design (316L stainless steel, Edwards Lifesciences) (Fig. 9.18a) and the self-expanding CoreValve (Nitinol, Medtronic) (Fig. 9.18b) have been approved for use in Europe since 2007 for symptomatic patients with severe aortic stenosis at high surgical risk or with other contraindications to open heart surgery. More than 30,000 patients have been treated with TAVI globally to date. In the United States, only Edwards' SAPIEN TAV is currently approved by the FDA for inoperable or high risk patients, as of the date of publication for this book.

9.2.6.4. Repair Devices – Annuloplasty Rings

Not all valvular diseases call for valve replacement. In patients with mitral regurgitation (MR), mitral valve repair should be considered when the patient becomes symptomatic or when left ventricular (LV) systolic dysfunction starts to develop. By repairing the mitral valve instead of replacing it, the sub-valvular apparatus remains intact, helping to preserve the LV function. If the mitral annulus is dilated, the reconstructive procedure may include annuloplasty, often with the use of a prosthetic ring to aid in remodelling the annulus [22]. Photographs of exemplary annuloplasty rings are shown in Fig. 9.19.

Prosthetic annuloplasty rings may be rigid, flexible, or semi-flexible. Annuloplasty rings are available in a range of sizes and designs (e.g. complete

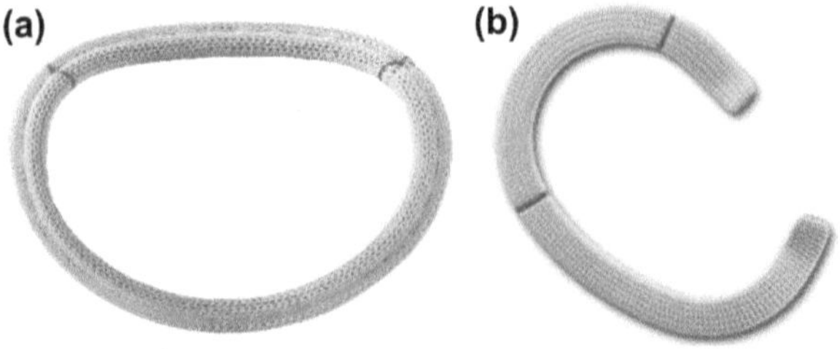

FIGURE 9.19 Example of annuloplasty ring: (a) semi-flexible mitral annuloplasty ring and (b) partially open tricuspid annuloplasty ring. *Source: Reprinted with permission from Edwards Lifesciences LLC.*

or partial; asymmetric, D-shaped, kidney-shaped or saddle-shaped) to provide appropriate modelling of annular configuration and support for valve leaflets as needed. Composition typically includes a metal alloy, silicon, or synthetic polymer core. The prosthesis may be covered with tubing material, such as polyfluorotetraethylene (PTFE), and with or without an outer layer of a knit synthetic fabric.

Severe tricuspid regurgitation (TR) has been treated successfully with tricuspid annuloplasty. Tricuspid valve repair or replacement is indicated in cases of Ebstein anomaly that results in symptomatic TR.

9.2.6.5. Cardiac Rhythm Management Devices

Heart rhythm disorders, or arrhythmias, are caused by irregular electrical impulses that disturb heart contractions, leading to inefficient pumping of blood to the body. In general, a normal resting heart rate ranges from 60 to 100 beats/min. Different types of arrhythmias may be classified into four main categories according to the heart rate and affected regions:

(1) **Ventricular arrhythmias**, which begin in the heart's lower chambers (ventricles).
(2) **Supraventricular arrhythmias**, which begin above the ventricles, in the atria, or in the atrioventricular node.
(3) **Bradyarrhythmias**, which are slow heart rhythms caused by the slow generation of electrical impulses or the blocking of impulse propagation. These typically lead to heart rhythms slower than 60 beats/min.
(4) **Tachycardias,** which are rapid heart rhythms that exceed normal resting heart rate, typically more than 100 beats/min.

Treatments for heart rhythm disorders usually include medication management, cardioversion, catheter ablation and the implantation of a cardiac defibrillator or pacemaker [23]. Only permanent implantable devices such as implantable cardioverter-defibrillators (ICDs) and pacemakers are covered within the scope of this chapter.

Implantable pacemakers are devices for treating hearts that beat too slowly. New models also incorporate diagnostic monitor to help physicians determine unexplained abnormal cardiac events.

ICDs are devices for monitoring and correcting heart rates that are dangerously out of rhythm. It works by emitting a high-energy pulse to restore the normal cardiac rhythm. Latest development in these devices also incorporated features for fluid monitoring and trending data, so that physicians can adjust medical treatment for patients at risk of acute heart failure. Development of telemetry allows patient data to be transferred to physicians through the Internet without patient visiting the doctor. This remote data management has a great potential to significantly transform patient cares.

Risks associated with an implantable defibrillator system implant include infection at the surgical site and/or sensitivity to the device material, failure to deliver therapy when needed, or receiving therapy when it is not needed.

Both devices consist of a signal generator and a lead. The signal generator is usually implanted under the skin in the upper chest with a lead tunnelling through the vein into the cardiac chambers while another electric lead is anchored into the endocardium. A major challenge to the overall system reliability is the fatigue fracture of the lead body [24]. Design considerations and verification testing will be discussed in a later session of this chapter.

9.3. *IN VITRO* CHARACTERIZATION OF CARDIOVASCULAR IMPLANTABLE DEVICES

U.S. and international regulatory requirements for medical devices and drugs provide specific guidance for bench-top testing, animal studies, clinical research, manufacturing, sterilization, and post-market surveillance for various device categories. All implantable cardiovascular devices fall within the Class II and III categories. Commercialization of these products in the United States requires FDA's Premarket Notifications (510(k)), Investigational Device Exemptions (IDE), PMAs, or Exemptions [25–27]. Recent regulatory environment has lead to significant overhaul of the 510(k) process. The regulation is still under update. It is important to recognize that regulatory agencies, such as the FDA, do not approve biomaterials. Instead, the agency regulates medical devices manufactured from biomaterials. Exceptions are devices that take the final form *in situ* where the final form cannot be evaluated ex vivo.

Rigorous device testing is an integral part of implantable medical device product development process in transforming innovative concepts into reliable commercial medical devices. It is also an important component in design control, risk management, and quality assurance. Failure modes and effects criticality analyses (FMECA) and regulatory requirements are generally considered when formulating *in vitro* testing strategies and protocols. These testings are performed either by in-house testing engineers or by ISO certified laboratory that has an extensive inventory of testing equipment and the expertise to perform a wide variety of testing per FDA, CE, ASTM and ISO medical testing requirements. *In vitro* testing ensures anticipated clinical risks, such as safety and long-term durability, associated with the devices are reasonably addressed, which is a mandatory measure before human clinical studies.

For the present discussion, only implantable devices are considered. Delivery systems and accessories are excluded from the present scope.

9.3.1. *In Vitro* Characterization of Exemplary Devices

9.3.1.1. *Intravascular Stents*

In vitro characterization of intravascular stents generally follows the FDA Guidance Document on Non-Clinical Engineering Tests and Recommended Labeling for Intravascular Stents and Associated Delivery System [28]. The basic testing and documentation requirement include the following:

(1) Material definition
- (a) Chemical composition
- (b) Transformation temperature (for self-expanding stents)
- (c) Corrosion resistance and biocompatibility

(2) Dimensional definition
- (a) Strut dimensions
- (b) Stent dimensions in both crimped and deployed conditions
- (c) Stent coverage area
- (d) Foreshortening during stent deployment
- (e) Recoil (for balloon expandable stents)

(3) Integrity and mechanical attributes
- (a) Imperfections and defects
- (b) Mechanical properties of raw material as well as deployed finished stents
- (c) Radial stiffness and strength
- (d) Radial outward force (for self-expanding stents)
- (e) Crush and kink resistance (for self-expanding stents)
- (f) Stress–strain computer simulation analysis under *in vivo* loading conditions
- (g) Fatigue analysis and accelerated durability testing

(4) Imaging
- (a) MRI safety and compatibility
- (b) Radiopacity

9.3.1.2. *Surgical Heart Valves*

Based on the risk analysis, pre-clinical *in vitro* testing on the replacement heart valve should follow the methods described in ISO 5840 [29] or equivalent methods. Unless scientific rational can be provided for omitting some, the following tests are recommended by the FDA and ISO guidelines:

(1) Material property testing
(2) Biological safety
(3) Hydrodynamic performance
(4) Structural performance
(5) Device durability
(6) Component fatigue assessments
(7) Device-specific testing

For each test, the test report should include the protocol, acceptance criteria and results. The test protocol should include sample size, environmental conditions, test parameters, and test duration. In addition, design controls apply to all heart valves product development process. These procedures must include design validation procedures. Design validation must ensure that devices conform to defined user needs and intended uses and must include testing of production units under actual or simulated use conditions.

In vitro characterization of surgical heart valves should follow the widely recognized standard testing requirements such the ISO 5840 and/or FDA replacement heart valve draft guidance. While some of these standards are in draft format, they still provide important guidelines for the proper test methodology and protocol designs. Since this is an emerging field, new standards and guidance are being updated, and practitioners should follow the most recent update. Some regulatory agencies, such as U.S. FDA, are available to discuss testing requirements through pre-IDE communications.

As discussed in the introductory section, the main function of a heart valve is to provide an efficient unidirectional flow. The flow characteristics are represented by the forward flow pressure drop and regurgitation flow during the closing phase. Ideal valve prosthesis should have a flexible leaflet that opens easily with a very small pressure gradient and have a large effective orifice area to maintain low gradient during peak flow. It should also close quickly to minimize closing volume and completely to eliminate regurgitation flow. Bench-top hydrodynamic testing can be combined with animal *in vivo* evaluation to assess the complete haemodynamic performance.

9.3.1.3. Transcatheter Heart Valves

Since transcatheter heart valves include metal scaffold and surgical heart valve technologies, it is evident that they need to meet the performance criteria set forth for both technologies with exceptions of certain device-specific requirements such as scaffold coverage area. In addition, due to the unique design and delivery methods, the transcatheter heart valve should also be subjected to additional engineering evaluation for its intended application. Testing specific to the transcatheter heart valve includes assessment of the perivalvular leak, valve migration, and irregularly deployed valve configurations. In addition, special tests are required to evaluate the delivery system. The new ISO 5840 is being revised to incorporate testing requirements for minimally invasive and transcatheter heart valve prostheses [30].

9.3.2. *In vitro* Test Methods

9.3.2.1. *Transformation Temperature Testing for Self-Expanding Structures*

Transformation temperatures are used for specifying Nitinol raw material as well as finished Nitinol self-expanding medical devices.

For Nitinol raw materials, the transformation temperatures are highly sensitive to the nickel and titanium composition; 1 wt% change in nickel or titanium content results in more than 100 °C shift in transformation temperatures. Therefore, conventional analytical methods such as DC plasma (DCP) spectroscopy, inductively coupled plasma spectroscopy and atomic absorption do not have the sufficient precision to ensure consistency in transformation temperatures. Instead, transformation temperatures are directly measured for the specification and control of Nitinol raw materials after full annealing at 650–950 °C. The full annealing treatment is intended to eliminate all processing effects which can affect the transformation temperature. The test methods of differential scanning calorimetry (DSC), bend and free recovery (BFR), electrical resistivity measurement and constant-load dilatometry are viable techniques for measuring Nitinol transformation temperatures. The industry has, however, adopted DSC as the standard technique for measuring Nitniol raw material transformation temperatures (also referred to as 'ingot transformation temperatures'). One of the martensite to austenite transformation temperatures, A_s, A_p or A_f, is commonly selected for raw material specification [31,32].

For finished self-expanding Nitinol devices, specifying the transformation temperatures and the crimping process define the mode of action, i.e. if the device expands via superelasticity or shape memory mechanism. Among the various transformation temperatures, the austenitic transformation finish (A_f) temperature, which designates the lowest temperature of superelasticity, is commonly used in the industry for Nitinol device specification. Although DSC is used for device A_f testing, the BFR test following ASTM F 2082, *Standard Test method for Determination of Transformation Temperature of Nickel–Titanium Shape Memory Alloys by Bend and Free Recovery*, is generally prescribed [33]. The BFR A_f temperature is also referred to as 'functional A_f temperature' or 'active A_f temperature'.

9.3.2.1.1. Differential Scanning Calorimetry

DSC is a thermal analytical technique commonly used for characterizing phase transitions and chemical reactions. As illustrated in Fig. 9.20, in

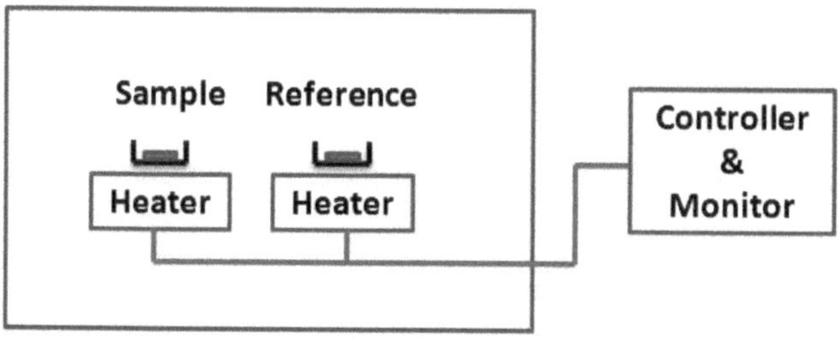

FIGURE 9.20 A schematic illustration of a DSC instrument.

a typical DSC instrument are two thermal chambers where the specimen and a reference of inert material are placed under cover gas protection. Either nitrogen or helium is used for cover gas. During a temperature scan, the energy consumed for maintaining an equal temperature of the specimen and the reference is monitored. Because this energy is compensated by changes in heat capacity and heat flows in (endothermic) and out (exothermic) of the test specimen, both changes in heat capacity, ΔC_p, and phase transition enthalpy, ΔH, can be determined at various temperature increments under a constant pressure condition.

For example, glass transition in polymers is a second-order endothermic transition where a polymer changes from its brittle, amorphous glassy state to a flexible rubbery state on heating. As illustrated in Fig. 9.21, glass transition is observed of a step change in heat capacity and the inflection point of this transition curve is commonly designated as 'glass transition temperature (T_g)'. When a polymer is heated above its T_g, molecules arrange themselves into an orderly crystalline form. This crystallization reaction is a first-order exothermic

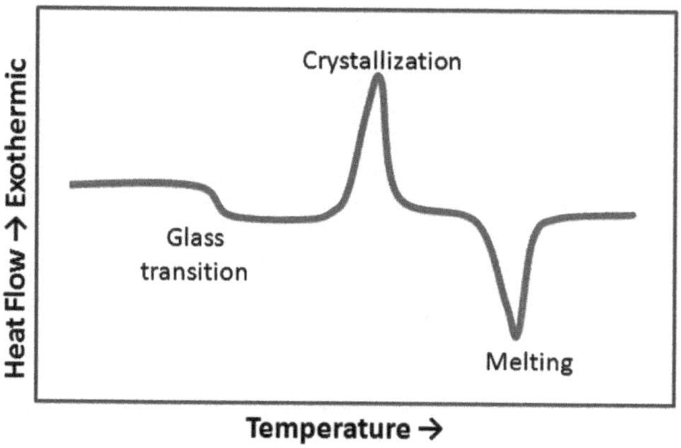

FIGURE 9.21 A schematic of a DSC thermogram illustrating various reactions in polymeric materials.

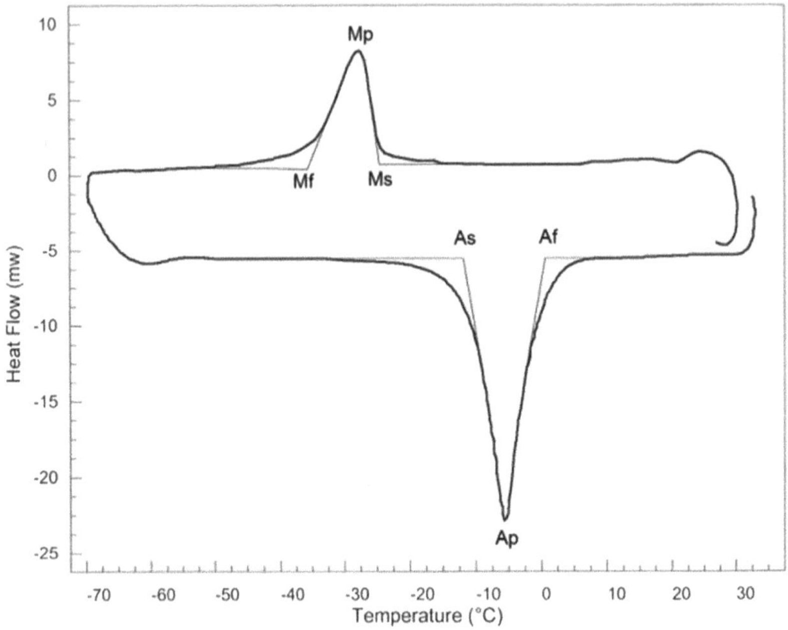

FIGURE 9.22 A DSC thermograph of a fully annealed Nitinol specimen exhibiting a single-stage transformation.

transformation and the DSC thermogram exhibits an exothermic peak. Continuing heating to its melting temperature, T_m, the material absorbs a latent heat during melting as a result of the first-order phase transition of melting from solid to liquid. The DSC thermograph exhibits an endothermic peak associated with this enthalpy of melting.

Both the R-phase and the martensitic transformations in Nitinol are exothermic while the reverse transformations are endothermic first-order phase transformations. The R-phase transformation has a second-order component, of which the lattice distortion continues to increase with decreasing temperature. However, this second-order lattice distortion influences only the baseline of the DSC thermograph and does not affect the determination of the transformation temperatures. An exemplified DSC thermograph of a fully annealed nitiniol specimen is shown in Fig. 9.22. The fully annealed specimen exhibits a single-stage transformation with no thermal activities associated with the R-phase transformation present in this DSC thermograph. The transformation temperatures are determined by a slope-intercept technique according to ASTM F2004 [32].

Another example of DSC thermograph shown in Fig. 9.23 is for a heat-treated specimen such as those sampled from finished medical devices. Here, the specimen exhibits a two-stage transformation where both R-phase and

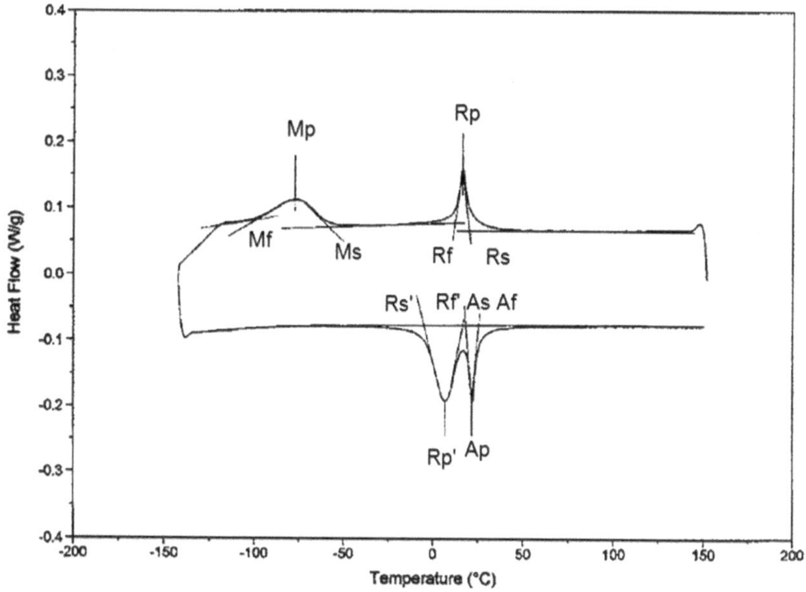

FIGURE 9.23 A DSC thermograph of a heat-treated Nitinol specimen exhibiting two-stage transformation. *Source: Reprinted with permission from ASTM F2005-05, copyright ASTM International, 100 Barr Harbor Drive, West Conshohocken, PA 19428.*

martensitic transformations are detected upon cooling as well as heating. The transformation temperatures are determined by the slope-intercept technique. Definitions of Nitinol transformation temperatures are listed in Table 9.1 following *ASTM F2005, Standard Terminology for Nickel-Titanium Shape memory Alloys* [34].

9.3.2.1.2. BFR Test

In BFR testing, the temperatures of the reverse transformation back to austenite of a Nitinol sample are determined by measuring the recovery of deformation during heating. The test specimen can be a piece of sample material or a finished Nitinol device. Figure 9.24a illustrates a test setup where a specimen, such as a straight piece of wire or a small-diameter tube, is first deformed into a curved shape in its martensitic state. Ideally, the deformation of specimen should occur in the test bath, so that there is no risk of premature shape recovery due to potential temperature excursion during specimen transfer from the deformation fixture to the test bath. Cooling is necessary if the martensitic transformation temperature is below the ambient temperature and an alcohol or a water bath is used depending on the range of test temperatures. To avoid over-deformation which can affect the reverse transformation temperatures, ASTM F2082 (*Standard Test Method for Determination of Transformation Temperature of Nickel-Titanium Shape Memory Alloys by bend and Free Recovery*) [33]

TABLE 9.1 Nitinol Transformation Temperatures

	Definition	
Transformation Temperature	**One-Stage Transformation**	**Two-Stage Transformation**
R_s	–	The temperature at which the transformation from austenite to R-phase begins on cooling.
R_p	–	The temperature of the exothermic peak position on the DSC curve (maximum transformation rate) upon cooling for the austenite to R-phase transformation.
R_f	–	The temperature at which the transformation from austenite to R-phase is completed on cooling.
M_s	The temperature at which the transformation from austenite to martensite begins on cooling.	The temperature at which the transformation from R-phase to martensite begins on cooling.
M_p	The temperature of the exothermic peak position on the DSC curve (maximum transformation rate) upon cooling for the austenite to martensite transformation.	The temperature of the exothermic peak position on the DSC curve (maximum transformation rate) upon cooling for the R-phase to martensite transformation.
M_f	The temperature at which the transformation from austenite to martensite is completed on cooling.	The temperature at which the transformation from R-phase to martensite is completed on cooling.
R_s'	–	The temperature at which the transformation from martensite to R-phase begins on heating.
R_p'	–	The temperature of the endothermic peak position on the DSC curve (maximum transformation rate) upon heating for the martensite to R-phase transformation.

TABLE 9.1 Nitinol Transformation Temperatures—cont'd

	Definition	
Transformation Temperature	One-Stage Transformation	Two-Stage Transformation
R_f'	–	The temperature at which the transformation from martensite to R-phase is completed on heating.
A_s	The temperature at which the transformation from martensite to austenite begins on heating.	The temperature at which the transformation from R-phase to austenite begins on heating.
A_p	The temperature of the endothermic peak position on the DSC curve (maximum transformation rate) upon heating for the martensite to austenite transformation.	The temperature of the endothermic peak position on the DSC curve (maximum transformation rate) upon heating for the R-phase to austenite transformation.
A_f	The temperature at which the transformation from martensite to austenite is completed on heating.	The temperature at which the transformation from R-phase to austenite is completed on heating.

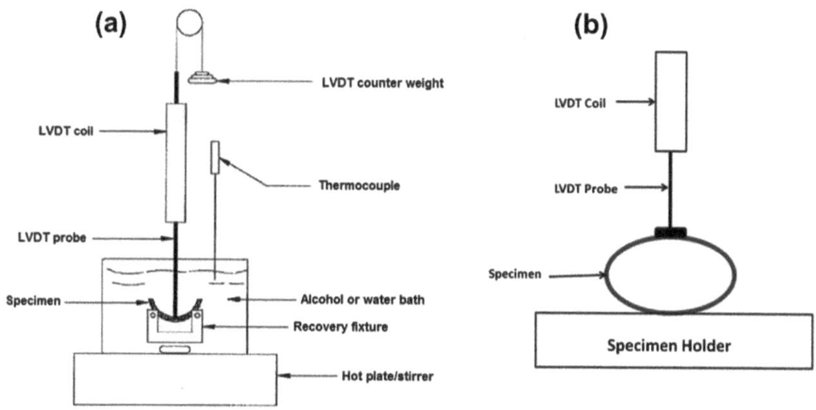

FIGURE 9.24 Illustrations of bend and free recovery (BFR) test setup for (a) material specimens *Source: Reprinted with permission from ASTM F2082-06, copyright ASTM International, 100 Barr Harbor Drive, West Conshohocken, PA 19428* and (b) large-diameter tubes or finished devices, respectively.

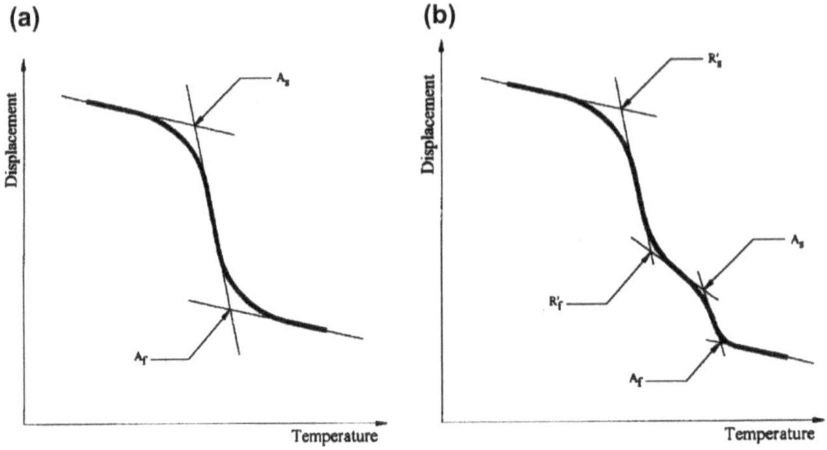

FIGURE 9.25 BFR displacement–temperature curves and the reverse transformation temperatures for (a) one-stage transformation and (b) two-stage transformation. *Source: Reprinted with permission from ASTM F2082-06, copyright ASTM International, 100 Barr Harbor Drive, West Conshohocken, PA 19428.*

specifies the strain of deformation to be within 2.0–2.5%. For large-diameter tubing or finished devices which cannot be easily bent along the longitudinal direction, a section of the tube or the whole device can be deformed by crushing as shown in Fig. 9.24b as long as the maximum local strain is properly controlled within the ASTM F2082 specification.

Upon heating through the reverse transformation temperature range, the displacement of the linear variable differential transducer probe registers the recovery of the deformation against the bath temperature recorded with the thermocouple. Uniformity of the test bath temperature and heating rate are important factors that can influence test results. Magnetic stir is generally required to maintain the temperature uniformity of the bath and the heating rate is controlled to be no more than 4 °C/min.

The shape recovery curves for one-stage and two-stage transformations are shown in Fig. 9.25a and b, respectively. The reverse transformation temperatures are determined by tangent-intersect technique as illustrated in the figures.

9.3.2.2. Structural Integrity

9.3.2.2.1. Radial Stiffness and Radial Strength

Deployment of a balloon expandable device in an arterial vessel exerts radial loading on the device (Fig. 9.26) due to vessel elasticity, cardiac pulse pressure and muscle-skeletal motions. Radial stiffness, defined as the change in radial pressure/force as a function of device diameter, is an important device

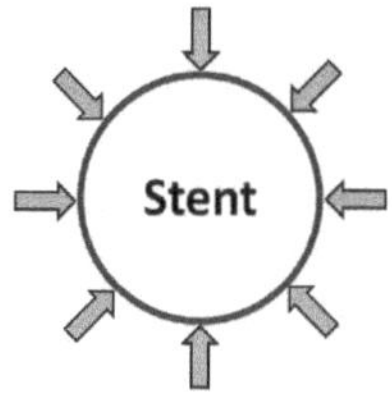

FIGURE 9.26 A graphic illustration of radial loading on a stent.

characteristic which determines the equilibrium diameter of the device *in vivo*. Radial strength, on the other hand, is defined as the pressure/force at which the device experiences irrecoverable deformation. It defines the dimensional stability of the device in resisting collapse under radial loading. The clinical significance of radial strength can influence the stability of vascular patency as well as the risk of strut malapposition and device migration for a stent. For a transcatheter heart valve, radial strength can affect valve diameter, haemodynamic function as well as the risk of device migration.

One method for testing radial stiffness and radial strength is to use a segmented head radial force tester. A photograph of such a tester and an illustration of its detailed construction are shown in Fig. 9.27a and b, respectively. The tester consists of a group of wedge-shaped force elements that are rotatable around the axis of the pivotal hinge. The distal ends of the force elements form a central aperture, where the test specimen is placed during testing, while the proximal ends of the force elements are coupled to a rotatable force collector. A drive mechanism is used to actuate and control the rotation of the force collector, which in turn rotates the force elements to either close or open the aperture. A force transducer is coupled to the drive mechanism to

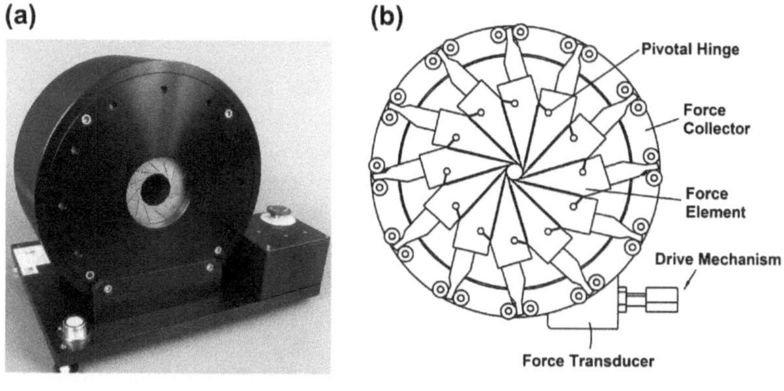

FIGURE 9.27 (a) Photograph and (b) construction of a segmented head radial force tester. *Source: United States Patent and Trademark Office; US7,069,794 B2.*

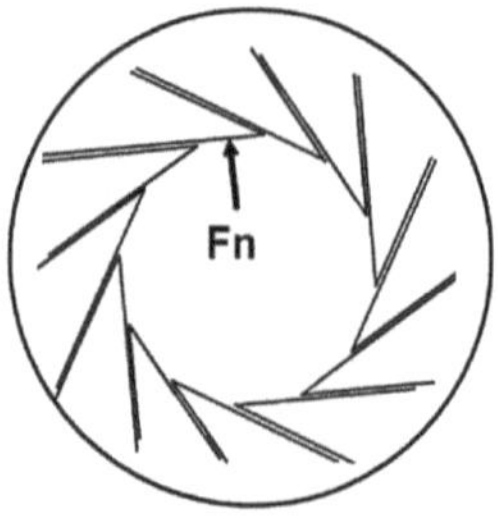

FIGURE 9.28 Radial force component, F_n, acting on each of the force elements. *Source: United States Patent and Trademark Office; US7,069,794 B2.*

measure the force associated with the rotation of the force collector [35]. Friction of the system is an important factor that can affect the measurement. When friction is significant, the baseline friction should be monitored by running the tester without a specimen and subtracted from the test data.

After the effect of friction has been properly addressed, the radial force component of the specimen acting on each of the force element, F_n, (Fig. 9.28) can be derived from the geometrical construction of the force elements, force collector and the force transducer. The radial force, F_r, of the test specimen is then determined by summing up the force acting on each of the force elements,

$$F_r = n * F_n \tag{9.1}$$

Where F_n is the force acting on each of the force element and n is the total number of the force elements. The radial pressure, P, can be converted from the radial force, F_r, according to the following relationship:

$$P = F_r/(\pi * D * L) \tag{9.2}$$

Where D is the diameter and L is the length of the specimen. The hoof force, H_{f}, is thus derived using the equation,

$$H_{\mathrm{f}} = (P * D * L)/2 = F_r/(2\pi) \tag{9.3}$$

Another test method for radial force is to use a sling-type of radial force tester as shown in Fig. 9.29. In this test setup, a sling is wrapped around the test article and slide around a low-friction pins before going through a slit of a fixture plate [36]. The closing and opening of the loop of sling is controlled by pulling and releasing the end of the sling, which is mounted in cross-head jaw of a tensile tester for monitoring force and motion. Ideally, the sling material must have low friction and be stiff in tension but flexible in bending. Again, the system friction needs to be monitored and subtracted from the test data.

The radial to linear force and distension relationships are described in the following equations:

$$F_r = \pi * F_{\mathrm{L}} \tag{9.4}$$

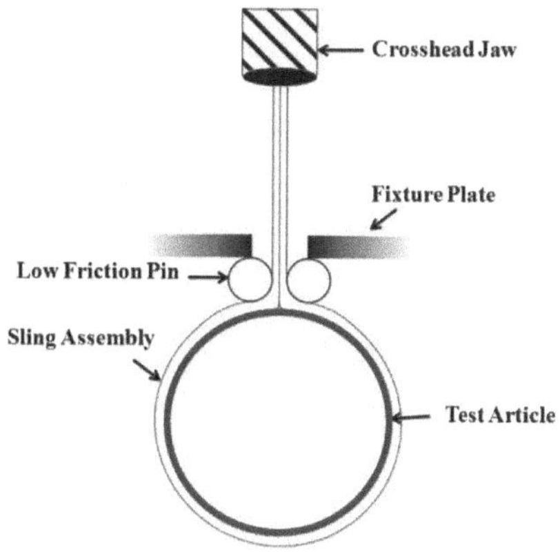

FIGURE 9.29 A sling-type of radial force testing fixture.

$$\Delta D = \frac{2\Delta x}{\pi}\left(1 + \frac{F_L}{K}\right) \tag{9.5}$$

$$\text{and } K = \frac{2EA}{L_0} \tag{9.6}$$

where

F_r: radial force
F_L: linear force measured by the tensile machine
ΔD: change in the outer diameter of sling and test article
Δx: linear displacement of the cross-head of tensile machine
E: elastic modulus of sling material
A: cross-sectional area of sling material
L_0: relaxed length of sling material between grip locations

A typical radial force test result for a balloon expandable device is shown in Fig. 9.30. Here, the radial load can be expressed in Newton (N) for the whole device or N/mm when normalized to unit length. Alternatively, the radial load can be presented as pressure in Pascal. When a device goes through crimping, it first goes through the initial loading of elastic deformation (stage 1). As the radial loading continues, the plasticity gradually increases until the final crimped diameter, in this case, 2 mm, is reached (stage 2). The unloading cycle then goes through elastic recoil (stage 3) before the specimen aperture or the sling loop diameter returns to its original setting (stage 4). Because a deployed balloon expandable device must go through plastic deformation during crimping and balloon expansion, the resulting structural stiffness *in vivo* is more representative by the slope associated with the unloading curve B–B′. For this reason, the unloading slope of B–B′ is used to define the radial stiffness.

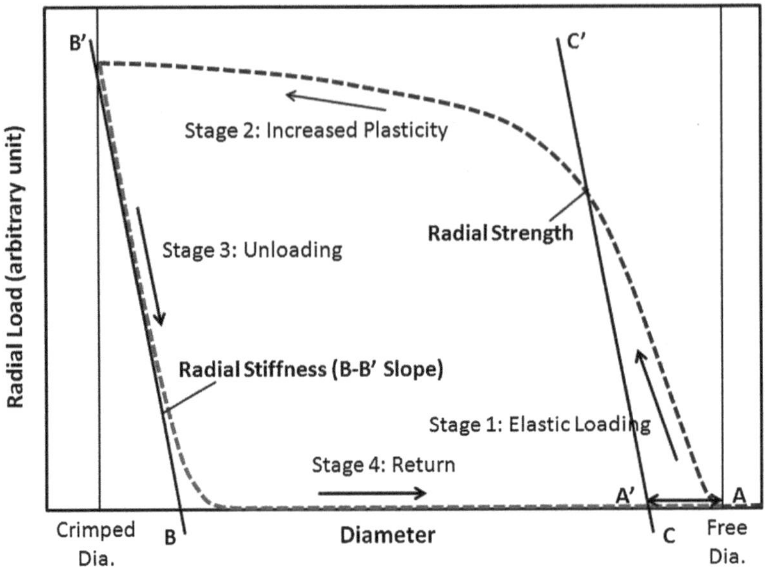

FIGURE 9.30 Typical radial force test data of a balloon expandable device.

To determine the radial strength, a line C–C′ parallel to line B–B′ is drawn with an offset of A–A′ from the original free diameter. Here, the offset A–A′ represents a clinically relevant amount of potential radial deformation on the device and the intersection of C–C′ with the initial loading curve defines the radial strength [36].

Alternatively, the radial force curve can be measured from a device after crimping and balloon expansion such that the mechanical characteristics are representative of the actual device *in vivo*. In this case, the radial stiffness can be calculated from the slope of the initial elastic loading curve and the radial strength by the intercept point between the initial loading curve and a parallel line to the loading curve with an offset of relevant clinical risk of radial deformation. However, the geometrical and mechanical consistencies of a deployed device after crimping and deployment are much more difficult to control that the results are less reproducible.

9.3.2.2.2. Radial Resistive Force and Chronic Outward Force of Self-Expanding Structures

Radial resistive force (RRF) and chronic outward force (COF) are relevant to Nitinol self-expanding devices only. For the purpose of definition, RRF is defined as the radial force of the device in resisting radial compression while COF is the continuing radial expansion force of the device acting on the vessel wall. While RRF is important in maintaining the vessel lumen stability against

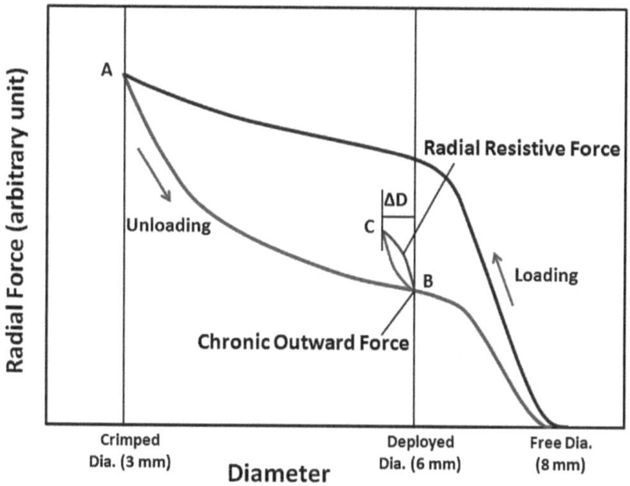

FIGURE 9.31 A radial force test result of a Nitinol self-expanding device.

potential anatomic compression, well-designed COF is required to achieve proper lumen opening and to avoid vessel wall injury due to over-stress associated with excessively high COF.

Radial force of a Nitinol self-expanding device can be measured using either a segmented head tester or a sling-type radial force test setup. Because the mechanical property of a Nitinol device is temperature sensitive, the test must be carried out at 37 °C to represent the *in vivo* environment. A typical radial force test result for a Nitinol self-expanding device is shown in Fig. 9.31. Here, the device is radially compressed from its free diameter of 8 mm and the radial force increases following the loading curve until it reaches the minimum crimped diameter of 3 mm at point A. The device is then unloaded and the radial force decreases following the unloading curve through point B and back to zero at its free state. Following this test, the COF is determined by the radial force along the unloading curve at a specific diameter of equilibrium between the device expansion and the vessel constriction. For example, the COF at an equilibrium diameter of 6 mm is defined by the radial force at point B.

The interaction between device and anatomic loading is dynamic and often complex. The resistance of a Nitinol self-expanding device against the anatomic compression depends closely on various modes of this interaction. For simplicity, RRF can be measured by an excursion test from the equilibrium device diameter. In this test, the device is radially compressed inside the segmented head tester aperture or the loop of a sling-type tester by a clinically relevant amount of compression in device diameter, ΔD. As illustrated in Fig. 9.31, compressing the device from its equilibrium diameter at point B by an amount of ΔD causes the radial force to increase following the blue line from point B to point C. The radial force following this blue line determines the RRF at every excursion points during the compression while the radial force at point C is the maximum RRF at the maximum compressive excursion. After reaching the maximum compression excursion at point C, the device is then unloaded and the radial force returns following the red line back to the value at its equilibrium diameter of point B [36].

9.3.2.2.3. Crush Resistance

In addition to radial strength, the ability of a self-expanding or non-self-expanding device to resist permanent deformation against non-radial localized compression (crush deformation) can be evaluated by crush test using parallel plates as illustrated in Fig. 9.32. In the figure, a round device (red dotted circle) is crushed between the two crush plates into an oval shape (blue solid oval). Such a fixture can be setup in a tensile testing machine to monitor loading force and displacement. The change in device dimension after unloading from simulated deformation provides an indication of device's ability in resisting permanent deformation due to non-radial compression. Crush resistance is important for devices implanted in superficial anatomic locations. For example, carotid stents may experience external, non-cardiac compression and require good spring-back characteristic to maintain vessel patency after crush deformations.

For self-expanding devices, the slope of the load–displacement curve generated by crush test is defined as 'pinch stiffness'. Pinch stiffness can be used to gauge the ability of the device in achieving circular deployment. For transcatheter heart valves, the higher the pinch stiffness, the more likely the valve will achieve better circularity.

9.3.2.3. Structural Durability

For cardiovascular applications including stent and prosthetic heart valves, the device is often subjected to cyclic loading associated with pulsatile flow

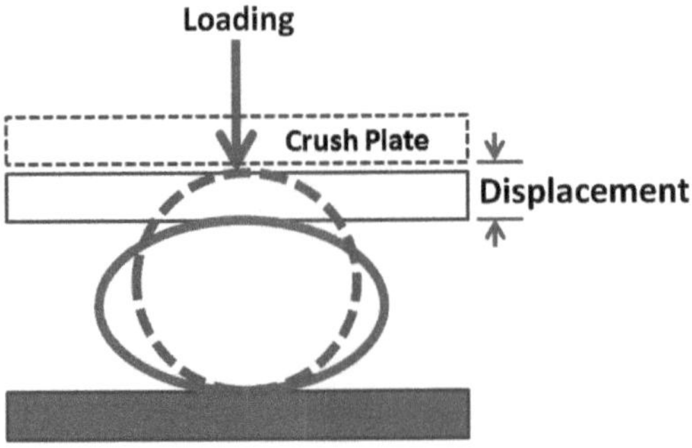

FIGURE 9.32 An illustration of parallel plate crush test fixture.

and pressure waveform due to the ventricular contraction and relaxation during a cardiac cycle. Musculoskeletal-motions-associated respiration and physical activities also contribute to this cyclic deformation. Depending on the device and implantation location, one or more of these cycling modes may be the dominant dynamic loading mode for structural durability considerations. Structural durability and reliability assessment is a critical aspect of design for structural components. While the test methodology for medical devices can utilize conventional fatigue testing and assessment technology, there are important differences from other engineering designs. Most notably, medical devices implanted into humans are subjected to high safety expectations with respect to very high-cycle fatigue durability. Furthermore, the inability to perform periodic inspections and/or repairs once implanted means that structural reliability for these devices must be established and verified prior to clinical uses. Additionally, the typical cardiovascular implantable devices contain structural components of small physical dimension on the order of sub-millimetres or microns. The small physical dimension has an important implication to the materials fatigue resistance due to the well-known small-crack phenomenon [37] and the effects of micro-structural inhomogeneity and inclusions. This section provides a general framework for establishing structural durability and reliability of safety-critical structural components commonly seen in cardiovascular implanted devices. The specific requirements should confirm to the regulatory guidance for the relevant devices.

Broadly speaking, a comprehensive structural durability and reliability assessment, or life analysis, may include the following components: (i) assessment of *in vivo* loading; (ii) materials fatigue resistance characterization; (iii) stress–strain analysis; (iv) fatigue life analysis; (v) structural reliability validation testing; and (vi) dynamic failure mode.

9.3.2.3.1. Assessment of *In Vivo* Loading

Evaluation of *in vivo* loading is concerned with the interaction of device with patient anatomy, including structural and haemodynamic interactions. The loading modes differ depending on the specific devices under investigation. For example, endovascular stents interacts with the blood vessel that is subjected to pulsatile pressure loading. Additional non-pulsatile loading such as flexion, torsion, axial tension and compression may exist depending on the location of the implant. As illustrated in Fig. 9.33 for tissue heart valves, the valve frame provides the structural support to the tissue leaflet that is subjected to the haemodynamic loading including the difference between systolic pressure (P_{sys}) and diastolic pressure (P_{dias}), pressure on native leaflets (P_{LF}), force on commissures, as well as the structural interactions with the annulus of the patient anatomy, and/or the native valve leaflets (in the case of TAVI), P_{Ann}. For pacemaker leads, the variety of tension, bending, and twisting

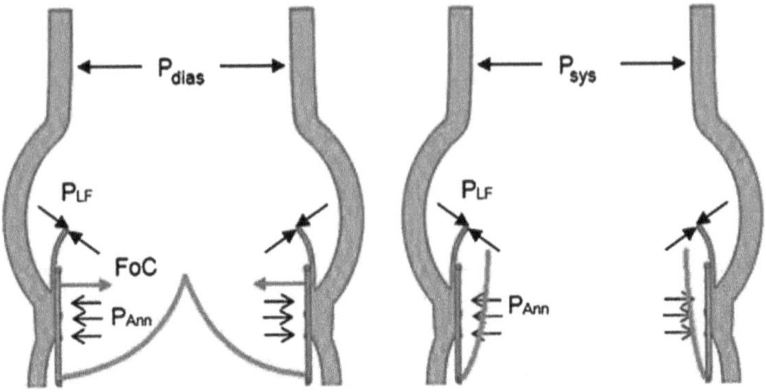

FIGURE 9.33 Typical loading modes on surgical heart valve under in-service conditions.

modes of loading may exist due to the interaction of the leads with the internal organs. Characterization of these conditions may be obtained from a combination of bench-top simulation, animal study, and clinical imaging investigation.

9.3.2.3.2. Material Fatigue Resistance Characterization

Material fatigue resistance characterization can be performed following general fatigue methods. Since medical device is usually small in physical dimensions, the test specimen should mimic the general dimensional characteristics of the device to ensure the applicability of the fatigue resistance data. Many specimen geometries have been devised to address the special requirements of the medical device. For example, the diamond-shaped stent-cell specimen was proposed for characterizing the fatigue resistance of Nitinol stents. Rotating wire bending specimen has been proposed to characterize the fatigue resistance of thin wires. Due to the very high cycle fatigue condition imposed on these devices, the fatigue data usually exhibit substantial statistical fluctuation in the fatigue life as illustrated in Fig. 9.34, where the survival reliability is shown at different stress levels. Consequently, a statistical analysis is necessary to establish the fatigue life distribution for these material fatigue data.

Because the fatigue life has been observed to obey weak-link statistics, a common statistical treatment of fatigue data makes use of Weibull analysis [38]. Specifically, the cumulative fatigue life-distribution function, $\Phi(t)$, represents the population fraction fractured by time (t), given by the parametric function as:

$$\Phi(t) = 1 - \exp\left[- \left(\frac{t}{\alpha}\right)^{\beta}\right] \tag{9.7}$$

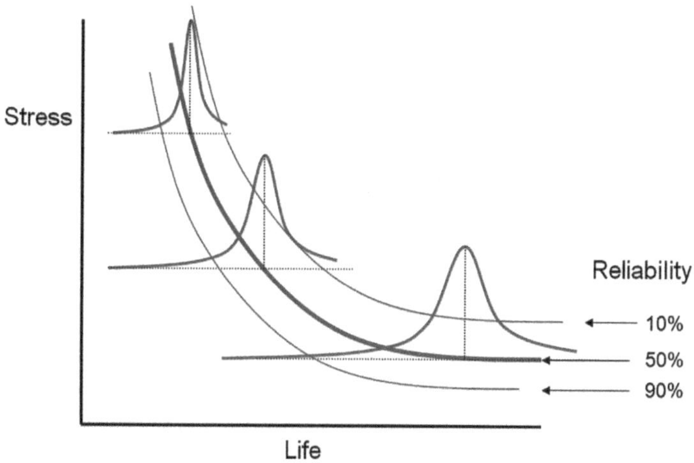

FIGURE 9.34 A schematic example of fatigue statistical behavior.

In Eqn (9.7), β is called the shape parameter, α is called the scale parameter, and time (t) is measured by the number of fatigue cycles. The scale parameter (α) in the Weibull distribution, also known as the characteristic life, is a constant for a single stress testing condition. The characteristic life is therefore a function of stress. The relationship between life and stress in high cycle fatigue is usually expressed using a power law equation [39].

$$\alpha(\sigma) = \frac{\tau}{\sigma^n} \tag{9.8}$$

In Eqn (9.8), τ and n are characteristic constants, and σ is the effective fatigue test stress, which incorporates the effects of both the cyclic stress and the mean stress.

In addition to cyclic stress, it has been well documented that the mean stress plays an important role in the fatigue of metals. In order to include the effects of both cyclic stress and mean stress, it is a common practice to combine these two types of stress into a single parameter of 'effective fatigue stress' to correlate with fatigue life. The Basquin model of effective stress is based on the equation for a linear constant-life line, and is recommended in the military design handbook [40].

$$\sigma_{\mathrm{eff}} = \sigma_{\mathrm{a}} + \xi \sigma_{\mathrm{m}} \tag{9.9}$$

In Eqn (9.9), ξ is a proportional constant that incorporates the effect of mean stress, and σ_a and σ_m are the fatigue stress amplitude and the mean stress, respectively. It should be noted that other forms of non-linear functional dependence may be used in place of Eqn (9.9), depending on the material behavior. The general method described in this session is still applicable for

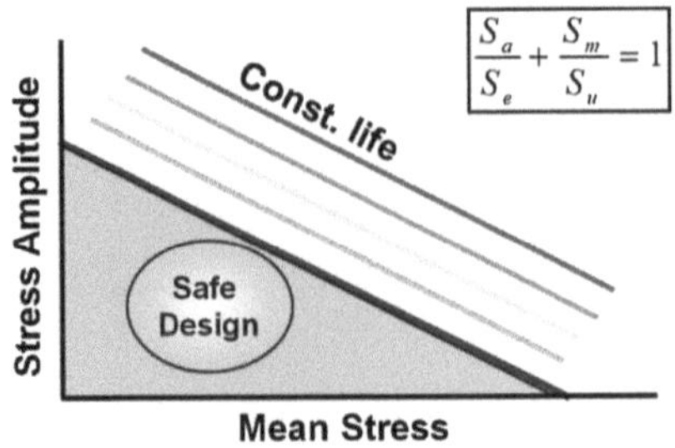

FIGURE 9.35 Modified Goodman reliability diagram.

more generic effective stress/strain curves. Eqn (9.9) has been found to describe the fatigue relations for metals under very high cycle life fatigue conditions.

Finally, the complete life-distribution function can be obtained by combining Eqns (9.7), (9.8) and (9.9), where the cumulative distribution function (cdf) is expressed in terms of four parameters: τ, β, n, and ξ:

$$\Phi_s(t;\tau,\beta,n,\xi) = 1 - \exp\left[-\left(\frac{t(\sigma_a + \xi\sigma_m)^n}{\tau}\right)^{\beta}\right] \tag{9.10}$$

Maximum likelihood estimation (MLE) was used for the determination of parameters, α, β, ξ, and ρ in Eqn (9.10) because of the multi-dimensional nature of the problem and because MLE is well suited for accommodating censored data. Acceptability of the cdf can be assessed by the chi-squared goodness-of-fit test. Confidence bounds on the estimated parameters can be obtained by using the Fisher information matrix and standard likelihood ratio methods [41].

The fatigue relationship in the mean-stress and stress amplitude domain is often expressed in terms of a Goodman diagram. By incorporating the statistical fatigue reliability, a modified Goodman reliability diagram can be represented in a typical graph shown in Fig. 9.35.

9.3.2.3.3. Stress–Strain Analysis

Stress analysis, combined with fatigue analysis and accelerated durability testing, provides an indication of device structural reliability. Stress analysis is usually performed using finite element analysis (FEA) on a high-performance computer system. FEA is a numerical method to compute the maximum stress and strain in the device subject to the prescribed boundary and loading conditions through the device manufacturing, delivery and service history. The constitutive properties of the materials must be determined using materials and processing condition similar to those of the finished devices. In addition, verification and validation is an important aspect of the FEA to demonstrate the accuracy and creditworthiness of the stress analysis results. The general guideline for FEA verification and validation published by the ASME standards committee is a good reference to consult in the design of a device FEA verification and validation plan [42].

Once the principal stresses are obtained from the output database, the mean stresses and cyclic stress amplitude can be computed according to the standard definitions of mean stress and stress amplitude as provided by ASTM E1823, *Standard Terminology Relating to Fatigue and Fracture Testing* [43]. The mean stresses, σ_{m}, and stress amplitude, σ_{a}, were computed from the maximum principal stresses with the following equations:

$$\begin{aligned}\sigma_{\mathrm{m}} &= \frac{\sigma_{\mathrm{p3,diastole}} + \sigma_{\mathrm{p3,systole}}}{2} \\ \sigma_{\mathrm{a}} &= \frac{\left|\sigma_{\mathrm{p3,diastole}} - \sigma_{\mathrm{p3,systole}}\right|}{2}\end{aligned} \tag{9.11}$$

The mean stress (σ_{m}) is equal to the average of the maximum principal stress between the end of systole ($\sigma_{\mathrm{p3,systole}}$) and the end of diastole ($\sigma_{\mathrm{p3,diastole}}$). The stress amplitude (σ_{a}) is half of the absolute difference in the maximum principal stress between the end of diastole and the end of systole. The extracted pairs of (σ_{m}, σ_{a}) can be plotted on a constant-life diagram for comparison with the material fatigue data. It should be noted that Eqn (9.11) is applicable to predominantly proportional loading where the principal stress and strain vector orientations at the minimum and maximum loading states are maintained. More sophisticated equations are needed when significant stress and strain vector rotation occurs between the minimum and maximum loading states. To evaluate the fatigue resistance, the mean stress and stress amplitude can be combined into an effective fatigue stress parameter, σ_{EFS}, using Eqn (9.10).

9.3.2.3.4. Fatigue Life Analysis

Fatigue analysis can be performed by comparing the fatigue stress under physiological loading condition with the intrinsic fatigue resistance of the material to define the fatigue safety factor. Following the standard engineering convention, the traditional fatigue safety factor is defined by the average stress amplitude limit and the average effective fatigue stress, i.e.

$$S_F = \frac{\sigma_{\mathrm{E}}(N, 50\%, 95\%)}{\sigma_{\mathrm{EFS}}} \tag{9.12}$$

where $\sigma_{\mathrm{E}}(N,\ 50\%,\ 95\%)$ is the y-intercept of the mean (50% reliability) constant-life line for $N = 600$ million cycles, and σ_{EFS} is the effective fatigue stress value for the nominal-case stress condition under typical physiological conditions. For critical structural components, a fatigue performance index can be computed by comparing the effective fatigue stress under worst case conditions (σ_{EFS}) and the statistically defined minimum fatigue strength with adequate reliability and confidence level (σ_{E}), which is a function of the

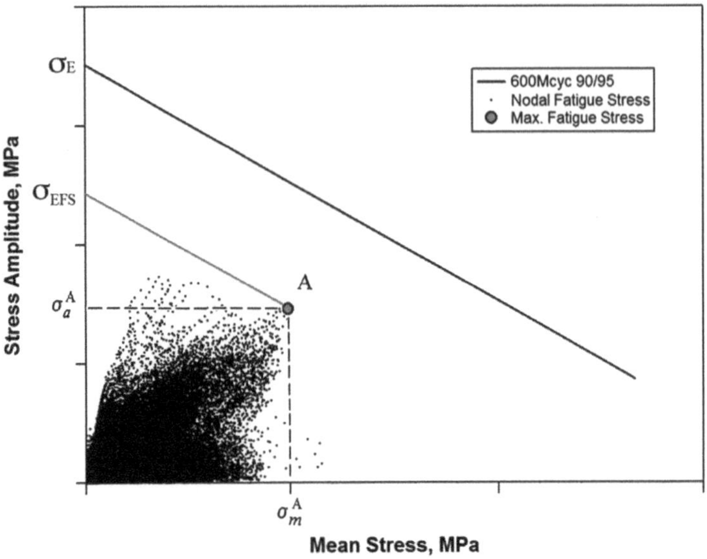

FIGURE 9.36 Graphical representation of calculating a traditional fatigue safety factor based on material fatigue strength and the FEA fatigue stress analysis.

appropriate lifetime (N), reliability (R) and confidence (α), i.e. $\sigma_{\mathrm{E}}(N, R, \alpha)$. Explicitly, the fatigue performance index can be expressed as

$$I_{\mathrm{F}} = \frac{\sigma_E(N, 90\%, 95\%)}{\sigma_{\mathrm{EFS}}} \tag{9.13}$$

For some devices, the commonly used Goodman analysis may be adequate to demonstrate the durability by calculating the fatigue safety factor or fatigue performance index. A typical fatigue Goodman diagram analysis is shown in Fig. 9.36. Adequate fatigue safety factor and fatigue performance index should be evaluated with respect to the conservative assumptions made throughout the analysis.

It should be noted that the fatigue performance index may be small even though adequate safety factor can be obtained. This often results from stacking up too many conservative assumptions on top of each other during the numerical stress analysis. For those cases, a more comprehensive reliability analysis can be performed using Monte Carlo simulation to assess the device system reliability under the in-service condition [44]. This technique makes use of the stress distribution under a variety of service and manufacturing conditions as well as the fatigue life distribution of the material to predict the overall failure probability of the device. The risk is assessed within the intersecting area under the stress and life distribution curves, as illustrated in Fig. 9.37.

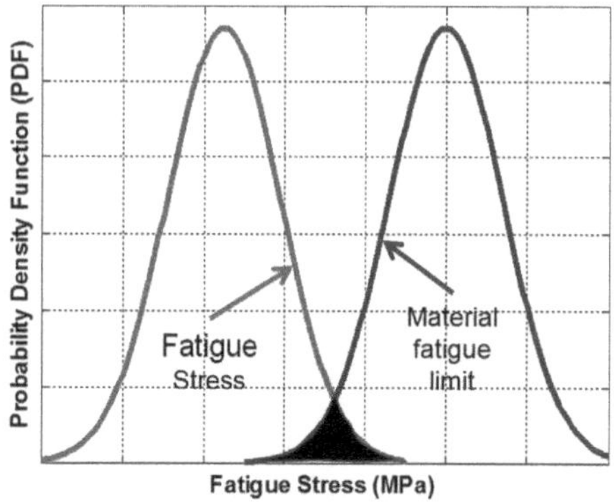

FIGURE 9.37 Total fatigue risk assessed within the intersecting area under the stress- and life-distribution curves.

9.3.2.3.5. Structural Reliability Validation Testing

For highly critical devices, the regulatory agency generally requires additional durability testing to confirm the reliability analysis performed for the device. Such tests usually involve subjecting the actual manufactured devices or subcomponents to the conservative simulated use conditions with the worst-case test parameters. For instance, pulsatile fatigue is often performed for coronary stent device where the stents are deployed into an appropriate compliant mock artery vessel that is subjected to pulsatile dilation simulating the blood vessel pulsation under cardiac cycles. A photograph of such a tester is shown in Fig. 9.38. Heart valve frame fatigue test may be performed under simulated

FIGURE 9.38 Accelerated fatigue testing of stents in a pulsatile mock artery test system. *Source: Reprinted with permission from Bose Corporation.*

leaflet loading in a simulated physiological fluid. Pacemaker leads can be tested by subjecting the lead body to the worst-case cyclic motion. Depending on the regulatory agency and the types of devices, up to 10 years of service life durability, i.e. 400 million cardiac cycles, may be required. Some device may require even higher number of cycles to ensure long-term fatigue reliability.

9.3.2.3.6. Dynamic Failure Mode

Alternatively, a test-to-fracture strategy has been proposed in order to more efficiently assess the fatigue resistance of medical devices. This testing strategy makes use of the simulated accelerated fatigue testing by imposing progressively higher fatigue stresses on the device for a shorter number of cycles until fracture is induced. This test is usually referred to as the dynamic failure test. In this test, the location of fracture can be compared with those predicted by fatigue life analysis to enhance the trustworthiness of the overall structural durability assessment program. The fatigue life can be related to the test-loading parameter to obtain a device stress–life characteristic curve. This strategy is often used in the design-development stage as well as design-verification stage.

9.3.2.4. Device Functional and Durability Testing: Heart Valve Prosthesis

9.3.2.4.1. Haemodynamic Performance Testing

Hydrodynamic performance testing shall be performed to provide information on fluid mechanical performance of the heart valve prosthesis in terms of the load to the heart and potential for blood stasis and damage. Since heart valve functions essentially as a check valve, the hydrodynamic performance should include forward flow efficiency during opening phase and back flow regurgitation during closing phase. It has been proposed that the total energy efficiency is the optimal performance standard for heart valves. In practice, the forward flow gradient and back flow regurgitation can be characterized independently under steady flow conditions. The actual performance can be evaluated more comprehensively using a mock circulation loop under simulated pulsatile flow conditions with a variety of pressure and total cardiac output conditions.

For surgically implantable valves, the device should be placed into the appropriate compliant annulus-like fixture to simulate the intended implant destination. For transcatheter heart valves, the implant shall be deployed into the test fixture using the loading and deployment steps in accordance with the instructions for use. The test chamber shall be representative of the critical aspects of the target implant site (e.g. compliance, geometry, etc.) for the target patient population. The design details of the test chamber along with the measurement accuracy and repeatability of the test system shall be evaluated and documented. A pulsatile mock circulation test system is often called the pulse duplicator as depicted in Fig. 9.39.

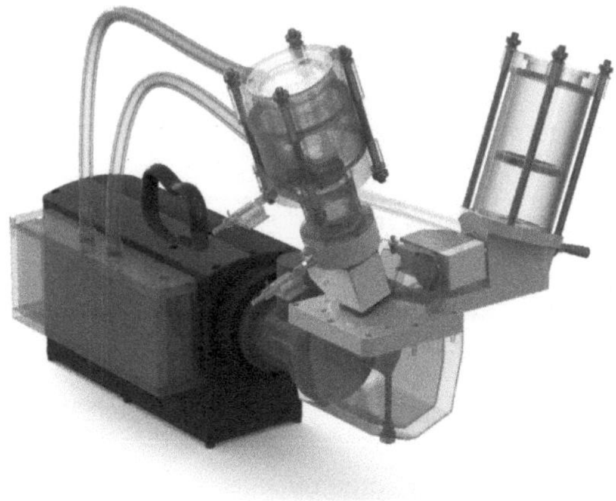

FIGURE 9.39 Mock-loop pulsatile test system for prosthetic heart valve fluid dynamics performance characterization. *Source: Reprinted with permission from BDC Laboratory.*

The pulse duplicator is a hydraulic model of the left human circulatory system. In order to obtain similarity in the pulsatile flow process, the physiological input and output impedances of the left human circulatory system are simulated by means of tuned hydraulic elements such as adjustable compliances and resistances. Ventricular contraction can be simulated directly either by positive displacement pump or by means of a flexible ventricular model coupled with a volume displacement drive unit moved by pneumatic, electromechanical or hydraulic actuators. The computer-controlled drive system facilitates simulation of the natural ventricular dynamics. Different circulatory conditions are adjustable by variations in frequency and actuation amplitude. For pressure determination, the volume, energy loss pressure and flow signals were measured at several positions in the pulse duplicator. Pressure tracings were recorded in the apex region of the ventricle and in the ascending aorta with high fidelity pressure transducers. The aortic pressure transducer is located four diameters downstream of the aortic valve since pressure recovery evolves up to this distance. Cardiac output is measured using a flow-metre placed downstream of the ascending aorta.

A guideline for the performing and reporting of hydrodynamic tests can be found in the regulatory guidance document. The minimum performance requirements as defined by ISO 5840 are provided in Table 9.2 [29,30]. These parameters assume a circular deployed valve diameter. The anticipated out-of-round variations in deployed shapes shall be evaluated using equivalent diameter approach. The pulse duplicator is used for routine measurements with prosthetic heart valves according to FDA standards. Typically, cardiac outputs of 2, 4, 5, and 7 L/min, and heart rates of 70 and 110 beats/min, respectively, are used.

When assessing retrograde flow, both the transvalvular regurgitant volume and the combined transvalvular and paravalvular regurgitant volume should be

TABLE 9.2 Minimum Performance Requirements

Position	Aortic							Mitral			
Valve size (TAD, mm)	19	21	23	25	27	29	31	25	27	29	31
A_{EO} (cm^2)	≥0.70	≥0.85	≥1.00	≥1.20	≥1.40	≥1.60	≥1.80	≥1.20	≥1.40	≥1.60	≥1.80
Regurgitant fraction (%)	≤10	≤10	≤10	≤15	≤15	≤20	≤20	≤15	≤15	≤20	≤20

evaluated independently for comparison against the corresponding values listed in Table 9.2.

The minimum performance requirements correspond to the following pulsatile flow conditions: beat rate = 70 cycles/min, simulated cardiac output = 5.0 L/min, mean aortic pressure = 100 mm Hg, and systolic duration = 35%. These pulsatile flow conditions are based on a healthy normal adult and might not be applicable for paediatric device evaluation. The minimum performance requirements are based on values in the published scientific literature [45,46].

The total regurgitant fraction shall include closing volume and both transvalvular and paravalvular leakage volumes (Fig. 9.40), as

$$A_{EO} = \frac{q v_{RMS}}{51{,}6 \times \sqrt{\dfrac{\Delta p}{\rho}}} \tag{9.14}$$

where EOA is the effective orifice area (cm^2); q_V is the root mean square forward flow (mL/s) during the positive differential pressure period; Δp is the mean pressure difference (measured during the positive differential pressure period); and ρ is the density of the test fluid (g/cm^3).

Recent development of transcatheter heart valves has placed increased attention on paravalvular leakage volumes. Consequently, developers may need to construct simulated deployment environment during the hydrodynamic performance testing to explicitly characterize the individual contributions from the transvalvular and paravalvular leakage volumes.

9.3.2.4.2. Accelerated Wear and Durability Testing

To demonstrate *in vitro* device durability, the prosthetic heart valves should be subjected to simulated device function under appropriate loads in an appropriate fluid environment to a specified number of cycles. To achieve the required long-term durability over 5–15 years, accelerated testing methodology is employed. According to ISO 5840, the pressure measurement system used to measure the transvalvular pressure difference should have a natural frequency of at least 1000 Hz and a minimum measurement accuracy of 5 mm Hg.

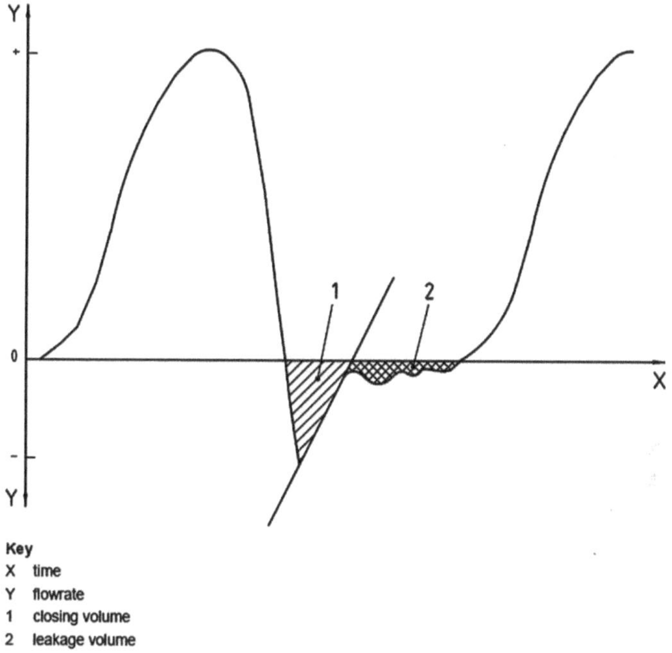

FIGURE 9.40 Forward flows versus retrograde flow through heart valve. *Source: This excerpt is taken from ISO 5840:2005, Figure 3, with the permission of ANSI on behalf of ISO. © ISO [2013] – All rights reserved.*

Durability testing for prosthetic heart valves can be conducted using accelerated wear tester. Some minor damage is expected on valves after completing durability testing. Failures, however, are characterized by excessive structural damage and/or functional impairment. Examples of structural deterioration include holes, tears, gross delamination, fraying, incomplete coaptation, fracture, excessive deformation, failure of any individual component as well as other mechanical breakdown and/or wear. Examples of functional impairment include excessive regurgitation and/or excessive transvalvular forward flow pressure difference. Usually, a comparison of the durability results between the test and reference valves should be performed.

Valve mounting is an important consideration to ensure adequate testing conditions. For example, a rigid mount may induce high transient stress in the device and pressure spike in the fluid due to water–hammer effect. The design of the mounting apparatus should consider the proper clinical application environment to simulate the intended application. A silicon gasket with proper durameter shown in Fig. 9.41 has been widely used in surgical heart valve testing.

9.3.2.5. *Corrosion Assessment*

Corrosion of implantable medical devices results in loss of materials through electrochemical processes. Device corrosion raises clinical concerns including

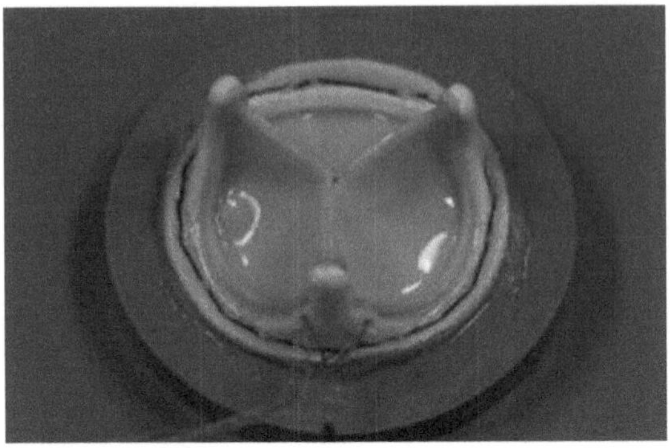

FIGURE 9.41 Mounting apparatus for accelerated durability testing of heart valves. *Source: Reprinted with permission from Edwards Lifesciences LLC.*

premature structural failure as a result of material loss and metallic ion release as corrosion by-products, which can cause adverse local tissue reactions and systemic toxicity. Although implantable medical devices are generally constructed of passive alloys where the passivity of surface oxides protects the materials from severe corrosion attacks, design, processing and the device function *in vivo* can very often influence the stability of the passive oxide layer and pose risks of corrosion.

Depending on the device design and application, potential risks of following corrosion mechanisms need to be evaluated.

9.3.2.5.1. Pitting and Crevice Corrosion Resistance

The presence of aggressive ionic species, such as chloride ions, reactive proteins and local change in pH may degrade the passive protective oxide layer of implantable medical devices and induce localized corrosion of pitting that can rapidly penetrate into the depth of material. Crevice corrosion can develop in areas where mass transport is limited as oxygen depletion, change in pH value and concentration of aggressive ions in such an environment can exacerbate the risk of pitting corrosion and impede the ability of the material to repassivate.

The resistance to pitting and crevice corrosion can be evaluated by following ASTM F746 [47], Standard Test Method for Pitting or Crevice Corrosion of metallic Surgical Implant Materials, and ASTM F2129 [48], Standard Test Method for Conducting Cyclic Potentiodynamic Polarization Measurements to Determine the Corrosion Susceptibility of Small Implant Devices. While F746 is fundamentally a potentiostatic test method intended for material screening, F2129 is most commonly used for assessing pitting and crevice corrosion susceptibility using cyclic potentiodynamic polarization.

Figure 9.42 illustrates the corrosion test cell used in F746 and F2129. In this test setup, the electrochemical activity of the test specimen (referred to as the working electrode) is controlled and monitored by a potentiostat through

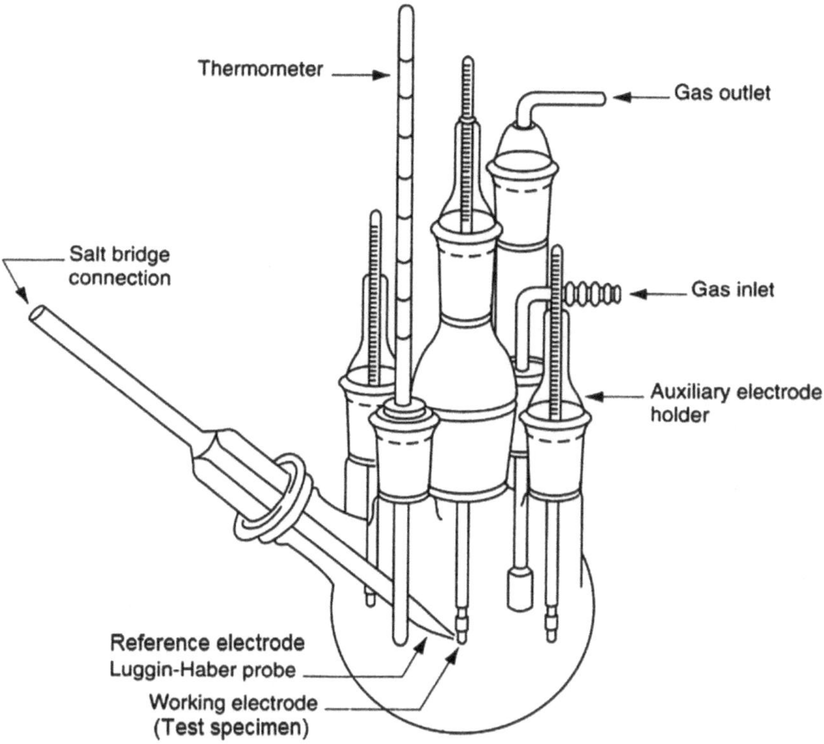

FIGURE 9.42 ASTM F746 and F2129 corrosion test cell. *Source: Reprinted with permission from ASTM G5-94, copyright ASTM International, 100 Barr Harbor Drive, West Conshohocken, PA 19428.*

a reference electrode of saturated calomel electrode, SCE, placed before the Luggin–Haber probe, and an auxiliary electrode (platinum or high purity graphite) in a simulated physiologic solution. Although other electrolytes such as Hank's solution and Ringer's solution can be used, phosphate buffered saline is specified as the standard test solution in these two ASTM test methods. The auxiliary electrode inputs current and controls the differential electrochemical potential between the specimen and the reference electrode. At the same time, it measures the current flow between the specimen and the auxiliary electrode.

Before running the test, the potentiostat and corrosion cell are first calibrated following ASTM G5, Standard Reference Test method for making Potentiostatic and Potentiodynamic Anodic Polarization Measurements [49]. In assessing the accuracy of the instrument, a standard sample of ferrite type 430 stainless steel (UNS S43000) is tested according to the standard procedure in a 1.0 N H_2SO_4 solution. The results are compared to the acceptable range of the reference polarization plot provided with the standard specimen. The accuracy of the potentiostat and corrosion cell is considered acceptable if the test result of the standard specimen falls within the range of the reference polarization plot.

During the F2129 testing, the test solution is first heated to 37 °C and then deaerated by purging with high purity nitrogen gas. Deaeration lowers the oxygen content dissolved in the simulated physiologic solution and should continue during the entire test cycle. Specimen representative of the final device condition after deployment *in vivo* is subsequently immersed in the test cell. The rest potential (E_r), also referred to as 'open-circuit potential', is monitored for 1 h or until the value stabilizes before the start of polarization scan. During the scan, the voltage is reversed if the current density increases beyond a threshold current density (i_t) from its passive state. In general, it is defined as a current density two decades higher from that recorded at the breakdown potential (E_b) while E_b is identified as the potential where the current density starts to increase due to pitting. Alternatively, the scan is reversed at a minimum reversing potential or vertex potential (E_v) of 800 mV where oxygen evolution due to water breakdown can dominate the reaction. During the reverse scan, the polarization potential is gradually brought down back to E_r if repassivation is not observed (Fig. 9.43b). Alternatively, the scan can be stopped when a protection potential (E_p) is observed as the current density decreases below the passive current density (Fig. 9.43a). The scan may also be stopped if no hysteresis is observed during the reverse scan, indicating current density increase due to oxygen evolution with no stable pitting (Fig. 9.43c). In all cases, final potential (E_f) of the test cycle is recorded.

To interpret the test results obtained from F2129 testing, higher breakdown potential (E_b) implies better resistance to pitting corrosion while the protection potential (E_p) gauges the ability of the specimen to repassivate after pitting corrosion. E_p provides an indication on the resistance to crevice corrosion. Because the F2129 test method evaluates the pitting susceptibility of a device in a deaerated simulated physiological solution, it does not take into account the effects of cells, proteins, physiological oxygen content and other biological factors that can potentially influence the corrosion behavior *in vivo*, the F2129 corrosion testing is thus intended as a comparative test method against a predicated device for similar indications, which has demonstrated safe clinical corrosion resistance. It is however a significant challenge to assess the clinical risk of corrosion for small cardiovascular implants if the occurrence is low as extensive explanting is not possible. With recent progress in medical device processing, most commercial devices today do exhibit fairly high E_b indicative of good resistance to pitting corrosion. It may be difficult and clinically unnecessary to demonstrate equivalency or superiority against such predicated devices as test controls. Hence, there have been numerous discussions to establish acceptance criterion.

One proposal is to use the difference of $E_b - E_r$ as a safety margin against pitting corrosion [50]. The basis for this approach is that pitting corrosion occurs only when E_r approaches E_b and the larger the gap ($E_b - E_r$), the greater is the safety margin against any biological factors that can potentially perturb the electrochemical state of the device *in vivo*. The counter argument is that both E_b and E_r measured using the F2129 test method may not represent the true values *in vivo* after long-term implantation.

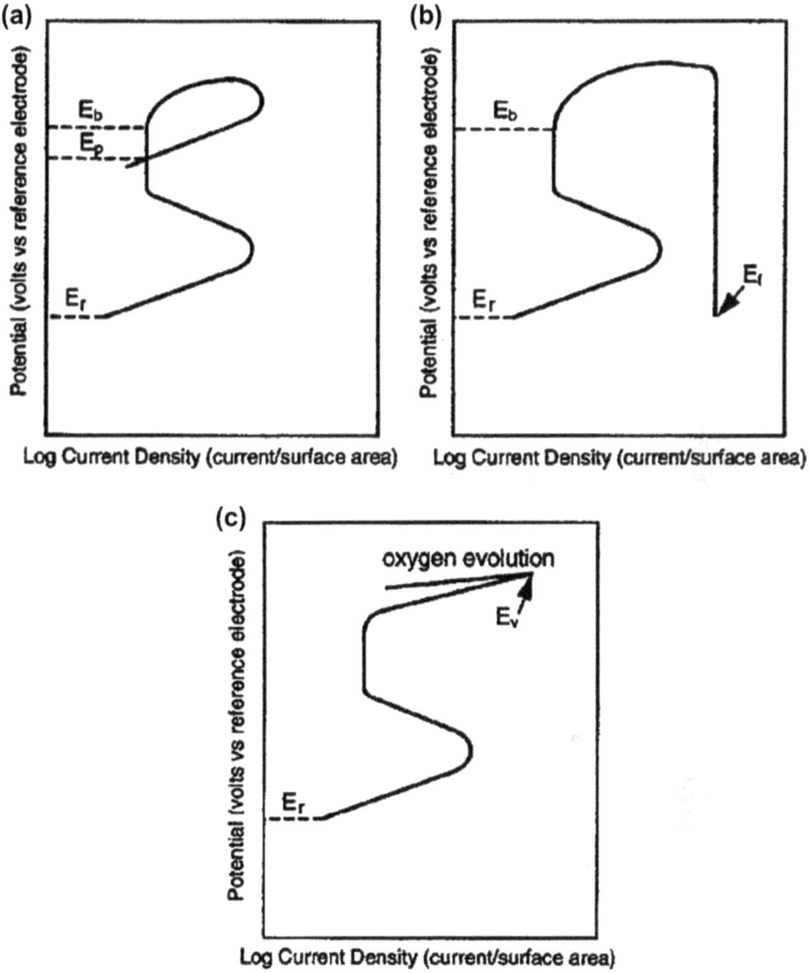

FIGURE 9.43 Schematic of cyclic potentiodynamic curves of (a) specimen exhibiting passivity breakdown at E_b and repassivation at E_p, (b) specimen exhibiting passivity breakdown at E_b but no protection potential and (c) specimen exhibiting oxygen evolution. *Source: Reprinted with permission from ASTM F2129, copyright ASTM International, 100 Barr Harbor Drive, West Conshohocken, PA 19428.*

Another proposed criterion is to judge the risk of pitting corrosion based on breakdown potential (E_b) [51]. In reviewing the published *in vivo* data, the authors concluded that for most biomedical metallic alloys including 316L stainless steel, cobalt–chromium alloys, titanium alloys and Nitinol, the rest potentials were found to be below 300 mV SCE. Hence, devices exhibiting breakdown potential >600 mV SCE are considered safe and those exhibiting breakdown potential <300 mV SCE would be considered questionable in resisting pitting corrosion. Devices having breakdown potentials below 600 mV SCE but above 300 mV SCE are considered marginal and required additional assessments to determine the suitability for implantation.

9.3.2.5.2. Galvanic Corrosion Risk Assessment

The risk of galvanic corrosion needs to be considered when dissimilar metals are likely to be in contact to setup a galvanic cell. Examples of potential galvanic corrosion risks include stents with radiopaque markers and overlapping stents of different metallic materials. In a galvanic cell, the more noble metal is protected from corrosion attack at the expense of the less noble counterpart, which can experience enhanced corrosion damage.

ASTM G71, Standard Guide for Conducting and Evaluating Galvanic Corrosion Tests in Electrolytes [52], provides guidance for conducting and evaluating galvanic corrosion tests. The corrosion test cell is set up according to Fig. 9.39 and calibrated according to G5 reference test method. The potentiostat should be a high-impedance instrument (e.g. $>10^{12}\,\Omega$) and configured as a zero-resistance ammeter. Before the test, the specimens are evaluated to determine the relative electrochemical potentials. If the relative electrochemical potentials between the two specimens of galvanic couple is uncertain, the open circuit potential can be measured for each specimen. The less noble specimen is the anode and is placed in the cell as the working electrode. The more noble specimen is the cathode and is connected to a holder as auxiliary electrode. In contrast to the F2129 test method, galvanic corrosion should be evaluated in an aerated solution at 37 °C to approximate the oxidizing environment in arterial blood. To aerate the test cell, it is suggested that air is continuously bubbling in the electrolytic solution throughout the test. The galvanic current, or current density when normalized by the surface area, between the anode and the cathode is then recorded for a specific period of time or until the galvanic current stabilizes. The corrosion rate (material loss per unit surface area per unit time) of the corroding specimen (anode) due to galvanic coupling can thus be calculated from the measured galvanic current based on Faraday's law.

Alternatively, the test can be designed as a comparative test against the self corrosion rate of the less noble material. In this case, the self-corrosion current density, I_{corr}, of the anode is measured by Tafel extrapolation using the F2129 dynamic polarization scan. The scan, however, needs to start from a potential more cathodic to the rest potential, so that the cathodic polarization segment of the curve can be captured for Tafel analysis. The assessment of the galvanic effect is then expressed as the galvanic factor, i.e. the galvanic corrosion rate divided by the self-corrosion rate.

9.3.2.5.3. Fretting Corrosion Consideration

When an implantable device has sliding surfaces between components or is placed in contact with another device, such as overlapping stents, where the contact surface can experience oscillatory motion, the risk of fretting corrosion needs to be evaluated. Fretting motion can destroy the passive film and change the local corrosion behavior of the device. The ability and kinetics for the material to repassivate become crucial to protect the device from corrosion attack. The release of debris generated by fretting motion also presents risk for toxicity and adverse tissue reactions.

The evaluation for fretting corrosion is generally incorporated into the device fatigue durability testing in simulated physiological solutions. The fretting surface is examined under an optical or scanning electron microscope for signs of corrosion after fatigue durability testing.

9.3.2.6. Biocompatibility and Haemocompatibility Testing

Biocompatibility testing refers to a battery of testing used to evaluate the biological reaction of the human body to the implanted device. There is a well-recognized ISO standard that governs the specific test protocol requirements (ISO 10993) [53]. Safety evaluation studies (*in vitro* and *in vivo*) are conducted on a variety of biomaterials, medical devices, and related products to identify the presence of toxins or any other potentially harmful effects. Biocompatibility testing can be used from the initial screening of new materials to product release testing, periodic audit testing, and non-clinical or premarket safety evaluations to meet current FDA and international standards.

Biocompatibility testing assays must be designed to be compatible with the intended use of the component material or finished product. These biocompatibility tests challenge various biological models with the test material or a suitable extract. Specific safety programs must follow FDA guidance and ISO 10993 standards. These programs must be selected with regulatory consultation and meet the requirements of the guidance for specific devices. Primary test categories used to determine biological effect include the following:

(1) Cytotoxicity
(2) Haemocompatibility
(3) Sensitization
(4) Intracutaneous irritation
(5) Systemic toxicity
(6) Sub-chronic toxicity
(7) Genotoxicity
(8) Implantation
(9) Haemocompatibility
  (a) Haemolysis
  (b) Complement activation
  (c) *In vivo* thrombogenicity
(10) Pyrogenicity

(11) Chronic toxicity
(12) Carcinogenicity
(13) Biodegradation
(14) Reproductive or developmental toxicity

Cytotoxicity testing – ISO 10993-5, USP 87 (*in vitro*) Cytotoxicity is a rapid, standardized test that is very sensitive and inexpensive for determining if the implant material contains significant quantities of harmful extractables and their effect on cellular components.

Haemocompatibility testing – ISO 10993-4 (*in vitro* and *in vivo*), this group of tests determines the compatibility of devices designed for direct or indirect contact with circulating blood.

Sensitization testing – ISO 10993-10 (*in vivo*) Sensitization or hypersensitivity tests for adverse reactions in animals by exposing the animal to the material. Sensitization reactions are noted by observing redness and swelling as it interacts with the body's immune system.

Irritation testing – ISO 10993-10 (*in vivo*) Irritation tests the reaction to a single, repeated or continual exposure from device materials that may produce irritation. Different from sensitization in that it is without the involvement of an immunological mechanism.

Systemic toxicity testing – ISO 10993-11 (*in vivo*) Systemic toxicity (acute) evaluates the potential adverse effects of medical devices on the body's organs and tissues that are remote from the site of contact. There are four categories: acute (24 h), sub-acute (14–28 days), sub-chronic (90 days or 10% of an animal's life span), and chronic (anything longer).

Genotoxicity testing – ISO 10993-3 (*in vitro* and *in vivo*) Genotoxicity tests use bacterial or mammalian cells to determine gene mutations, changes in chromosome structure and number, or other gene toxicities caused by medical devices, material or their extracts.

Implantation testing – ISO 10993-6 (*in vivo*) Implantation determines the local pathological effects on living tissue surrounding the implanted device. A sample is implanted directly into the animal, appropriate for use of the device and is evaluated at the gross level and the microscopic level.

All these tests can often be performed by certified contract testing service laboratories. But choosing the correct testing program for a particular device requires interpretation of the guidance based on an understanding of the intended clinical use of the device, with assays designed to mimic the actual clinical application. Examples of such test laboratories include WuXi AppTec, NAMSA, Charles River, and Nelson Lab.

9.3.2.7. Imaging

Imaging is one of the most important considerations in the design of cardiovascular devices. For interventional devices delivered through transcatheter means, intraoperative imaging is a key factor affecting the procedure success and accuracy. For both the surgical implants and transcatheter devices, post-procedure diagnostic imaging using X-ray, ultrasound and magnetic resonance play a critical role in patient management. Interaction of the implant device and delivery system with the imaging environment, therefore, should be evaluated in the design and development stage.

9.3.2.7.1. Radiopacity

Medical devices often contain a radiopacifier to enhance visualization during implantation for temporary implantation devices, such as catheters or guide-wires, or for monitoring the position of permanently implanted medical devices, such as stents, joint implants, and heart valves. Metal implants usually have sufficient radiocontrast that additional radiopacifier may not be necessary. Polymer-based devices, however, usually incorporate materials with high electron density contrast compared to the surrounding tissue. When testing a new medical device for regulatory submission, device manufacturers will usually evaluate the radiocontrast according to ASTM F640 'Standard Test Methods for Determining Radiopacity for Medical Use' [54].

For temporary implantation devices or delivery system, radiopaque markers can be seen under typical fluoroscopic methods. For permanent implantable devices, a qualitative or quantitative indication of the visibility of the implant on real-time and plane film X-ray is recommended. Additionally, images of animal implants, *in vitro* phantoms, or equivalent models, may be used for such assessment.

ASTM standard test methods cover the determination of the radiopacity of materials and products utilizing X-ray based techniques, including fluoroscopy, angiography, CT and dual energy X-ray absorptiometry (DEXA), also known as DXA. The results of these measurements are an indication of the likelihood of locating the product within the human body.

According to ASTM standards, three different test methods may be used to determine radiopacity of a medical device.

Method A – Radiopacity is (1) qualitatively determined by viewing image(s) of a test sample and the image background, with or without the use of a body mimic, or (2) quantitatively determined as a specific difference in optical density or pixel intensity between the image of a test sample and the image background, with or without the use of a body mimic.

Method B – Radiopacity is determined by (1) qualitatively comparing image(s) of a test sample and a user-defined standard without the use of a body mimic, or (2) quantitatively determining the specific difference in optical density or pixel intensity between the image of a test sample and the image of a user-defined standard without the use of a body mimic.

Method C – Radiopacity is determined by (1) qualitatively comparing image(s) of a test sample and a user-defined standard with the use of body mimic or (2) quantitatively determining the specific difference in optical density or pixel intensity between the image of a test sample and the image of a user-defined standard with the use of a body mimic.

Radiopacity is one of the key considerations in the design of various devices such as guidewires or stents that are used during radiological intervention. The radiopacity of a given endovascular device is important since it allows the device to be tracked during the interventional procedure. The two main factors contributing to a material's radiopacity are density and atomic number. Examples of commonly used radiocontrast materials include titanium, tantalum, platinum, tungsten, barium sulphate, and zirconium oxide. The high-radiocontrast elements can be alloyed in metallic implants or combined with devices made of lower density metal through mechanical attachment, lamination, co-extrusion or coating. The high-radiocontrast elements and compounds may be dispersed into lower density matrix by other physical or chemical means. There are commercial sources that place radiopaque markers on catheters and other transcatheter delivery system components.

9.3.2.7.2. MRI Safety and Compatibility

Magnetic resonance imaging (MRI) safety and compatibility are important issues for all medical implants since they may be exposed to an MRI environment during subsequent diagnosis procedures after the implantation. MRI testing is required for of medical device approval by regulatory agencies, such as the FDA and the European Union (EU) Notified Bodies. Currently, ASTM F2503-05, which specifies the marking practice for items brought into the MRI environment, requires MRI safety labelling using the categories and icons MRI safe, MRI conditional, or MRI unsafe [54].

The main issues affecting the safety and compatibility of passive implants in the MRI environment concern magnetically induced displacement force and torque, radiofrequency (RF) heating, and image artefacts. The MRI static field induces displacement forces and torques on magnetic materials. RF heating in the body is created by currents induced by the RF excitation pulses applied during MRI scanning. Projectile effect and the rotations produced by magnetically induced force and torque, and RF heating have resulted in severe patient injury and death during an MRI scan. The presence of an implant may produce an image artefact that may appear as a void region or as a geometric distortion of the true image. If the image artefact is near the area of interest, the artefact may make the MRI scan uninformative or may lead to an inaccurate clinical diagnosis, potentially resulting in inappropriate medical action.

To comply with the MRI safety labelling requirement, medical device manufacturers may provide a scientific rationale or the MRI safety testing data to support the claim. Specifically, testing should encompass the range of sizes of the device. If not all sizes of the device are tested, the sample should include a size or combination of sizes that represent the worst-case scenario for each test.

A science-based rationale rather than test data may be sufficient to support identifying an implant as 'MRI Safe', for example, a non-conducting or a non-magnetic item, such as a plastic Petri dish, poses no known hazards in all MRI environments. Similarly, a science-based rationale rather than test data may be sufficient to support identifying an item as 'MRI Unsafe'.

If a device was to be labelled as 'MRI Conditional', appropriate experimental data should be provided to support identifying the device as 'MRI Safe' or 'MRI Unsafe'. In each case, it is recommended to follow the test methods described in the standards below or equivalent methods.

(1) Magnetically Induced Displacement Force (ASTM F2052) [56]

A basic requirement is the measurement of magnetically induced displacement force, which can be measured indirectly via a deflection angle between the magnetically induced force of a device or part and its force due to gravity, thus demonstrating whether or not the magnetically induced force is lower than the gravitational force of the device. In general, the safe area in this worst-case consideration is indicated by a deflection angle below 45°. The angle is determined by suspending the device or component from a string in the static spatial gradient magnetic field, which is created by the decay of the static magnetic field at the magnet-bore entrance.

(2) Magnetically Induced Torque (ASTM F2213) [57]

The static magnetic torque is another basic interaction that can affect a device for MRI safety concerns. Forces and torques can be static or dynamic. The magnitude of the static magnetic induced torque depends on the device dimensions and the magnetic saturation of the material, assuming that a device is completely magnetically saturated. Additionally, it depends on the strength of the static magnetic field if the device is not completely saturated. The maximum static torque will be experienced by the device at the isocentre of the magnet, where the static magnetic field is homogeneous and maximum. The static magnetic torque can be measured, for example, by using a non-magnetic measurement apparatus with torsion springs. Other setups that use force gauges and a lever distance for calculating the torque are feasible. In general, the magnetically induced torque shall not exceed the worst-case torque (largest dimension × gravity force of the device).

(3) Heating by RF Fields (ASTM F2182) [58]

RF pulses in the megahertz-frequency range are the primary source of heating energy in comparison to the switched-gradient magnetic fields in the kilohertz range. RF heating is complex and is dependent on multiple parameters. To achieve worst-case heating conditions in the phantom, you should pay attention to the local electric and magnetic field distribution near the implant inside the phantom. These fields need to be similar to the local electric and magnetic field distribution near the implant inside the patient, so that the heating of the implant in the phantom is comparable to the heating inside the patient. Anatomical positioning of the implant in the phantom does not reliably predict the implant heating in the patient. Therefore, the field conditions and system geometry should be chosen to be comparable to worst-case clinical conditions.

(4) Image Artefact (ASTM F2119) [59]

In MRI examinations of patients with metallic implants like stents, various artefacts can occur in MRI images, because of magnetic and electro-magnetic interactions of the implant with the MRI environment. The standard test method F2119-07 is generally used for evaluation of susceptibility as well as RF artefact appearance and size of implants. Measurements according to this standard use an *in vitro* model filled with copper sulphate solution ($CuSO_4$). However, $CuSO_4$ is a conductive liquid, and standing waves may be detected in MRI with a high magnetic field strength ($\geq$3 T) and wavelength in dimension or smaller than the phantom used. Standing waves inhibit precise artefact measurements due to strong inhomogeneities of the background signal. This may be overcome by, for example, changing the transmitter voltage, the phantom dimensions or using a multi-transmit coil. Finally, it should be recognized that regulatory agency provide the most up-to-date information testing and compliance requirements. For US FDA submissions to fulfil 510(k), IDE or PMA requirements, the FDA website [60] is a useful resource.

REFERENCES

[1] Despopoulos A, Silbernagl S. Color atlas of physiology. 5th ed. Stuttgart and New York: Thieme; 2003.

[2] Christoffels VM, Moorman AFM. Development of the cardiac conducting system. Circ Arrhythm Electrophysiol 2009;2:195–207.

[3] Blackshear JL, Odell JA. Appendage obliteration to reduce stroke in cardiac surgical patients with atrial fibrillation. Ann Thorac Surg 1996;61:755–9.

[4] Michaels AD, Chatterjee K. Angioplasty versus bypass surgery for coronary artery disease. Circulation 2002;106:e187–90.

[5] Doostzadeh J, Clark LN, Bezenek S, Pierson W, Sood PR, Sudhir K. Recent progress in percutaneous coronary intervention: evlolution of the drug-eluting stents, focus on the XIENCE V drug-eluting stent. Coron Artery Dis 2010;1:46–56.

[6] Rabbia C, Pini R. Evidence-based medicine in renal artery stenting. J Cardiovasc Surg 2010;51(5):755–63.

[7] Scheinert D, Scheinert S, Dax J, Piorkowski C, Braunlich S, Ulrich M, et al. Prevalence and clinical impact of stent fractures after femoropopliteal stenting. J Am Coll Cardiol 2005;45:312–5.

[8] ACCF/AHA Writing Committee. Guideline on the management of patients with extracranial carotid and vertebral artery disease. Circulation 2011:123.

[9] Starr A, Edwards ML. Mitral replacement clinical experience with a ball-valve prosthesis. Ann Surg 1961;154:726–40.

[10] MedlinePlus: http://www.nlm.nih.gov/medlineplus/ency/article/002954.htm.

[11] Schoen FJ, Levy RJ. Tissue heart valves: current challenges and future research perspectives. J Biomed Mater Res 1999;47(4):439–65.

[12] Généreux P, Head SJ, Wood DA, Kodali SK, Williams MR, Paradis JM, et al. Transcatheter aortic valve implantation 10-year anniversary: review of current evidence and clinical implications. Eur Heart J July 31, 2012. ehs220 first published online.

[13] Bokros JC, Gott VL, LaGrange LD, Fadall AM, Vos KD, Ramos MD. Correlations between blood compatibility and heparin adsorptivity for an impermeable isotropic pyrolytic carbon. J Biomed Mater Res 1969;3(3):497–528.

[14] Cao HC. Mechanical performance of pyrolytic carbon in prosthetic heart valve applications. J Heart Valve Dis 1996;5:S32–49.

[15] Ryder JK, Cao HC. Structural integrity assessment of heart valve prostheses. A damage tolerance analysis of the CarboMedics prosthetic heart valve. J Heart Valve Dis 1996;5:S86–96.

[16] Ritchie RO. Fatigue and fracture of pyrolytic carbon: a damage-tolerant approach to structural integrity and life prediction in "ceramic" heart valve prostheses. J Heart Valve Dis 1996;5(Suppl. 1):S9–S31.

[17] Ross DN. Replacement of aortic and mitral valves with a pulmonary autograft. Lancet 1967;2(7523):956–8.

[18] Welke KF, Wu Y, Grunkemeier GL, Ahmad A, Starr A. Long-term results after Carpentier-Edwards pericardial aortic valve implantation, with attention to the impact of age. Heart Surg Forum 2011;14(3):E160–5.

[19] Bonow RO, Carabello BA, Chatterjee K, de Leon AC Jr, Faxon DP, Freed MD, et al. 2008 Focused update incorporated into the ACC/AHA 2006 guidelines for the management of patients with valvular heart disease: a report of the American College of Cardiology/American Heart Association Task Force on Practice Guidelines (Writing Committee to Revise the 1998 Guidelines for the Management of Patients with Valvular Heart Disease): endorsed by the Society of Cardiovascular Anesthesiologists, Society for Cardiovascular Angiography and Interventions, and Society of Thoracic Surgeons. Circulation 2008;118:e523.

[20] Webb JG, Wood DA, Ye J, Gurvitch R, Masson JB, Rodés-Cabau J, et al. Transcatheter valve-in-valve implantation for failed bioprosthetic heart valves. Circulation 2010;121:1848–57.

[21] Cribier A, Eltchaninoff H, Bash A, Borenstein N, Tron C, Bauer F, et al. Percutaneous transcatheter implantation of an aortic valve prosthesis for calcific aortic stenosis: first human case description. Circulation 2002;106:3006–8.

[22] Filsoufi F, Carpentier AC. Principles of reconstructive surgery in degenerative mitral valve repair. Semin Thorac Cardiovasc Surg 2007;19:103–10.

[23] Epstein AE, DiMarco JP, Ellenbogen KA, Estes 3rd NA, Freedman RA, Gettes LS, et al. ACC/AHA/HRS 2008 guidelines for device-based therapy of cardiac rhythm abnormalities: a report of the American College of Cardiology/American heart Association task force on practice guidelines (Writing Committee to Revise the ACC/AHA/NASPE 2002 Guideline Update for Implantation of Cardiac Pacemakers and Antiarrhythmia Devices): developed in collaboration with the American Association for Thoracic Surgery and Society of Thoracic Surgeons. Circulation 2008;117:e350–408.

[24] Borleffs CJ, Erven L, Bommel RJ, Velde ET, Wall EE, Bax JJ, et al. Risk of failure of transvenous implantable cardioverter-defibrillator leads. Circ Arrhythm Electrophysiol 2009;2:411–6. http://dx.doi.org/10.1161/CIRCEP.108.834093.

[25] FDA IDEs (21 CFR Part 812): [available on line] http://www.fda.gov/MedicalDevices/DeviceRegulationandGuidance/HowtoMarketYourDevice/InvestigationalDeviceExemptionIDE/ucm046164.htm.

[26] FDA PMAs (21 CFR Part 814): [available on line] http://www.fda.gov/MedicalDevices/DeviceRegulationandGuidance/HowtoMarketYourDevice/PremarketSubmissions/PremarketApprovalPMA/default.htm.
[27] FDA PDPs (21 CFR Part 814.19): [available on line] http://www.fda.gov/MedicalDevices/DeviceRegulationandGuidance/HowtoMarketYourDevice/PremarketSubmissions/PremarketApprovalPMA/ucm048168.htm#pdp.
[28] Guidance for Industry and FDA Staff – Non-Clinical Engineering Tests and Recommended Labelling for Intravascular Stents and Associated Delivery Systems, [available on line] http://www.fda.gov/medicaldevices/deviceregulationandguidance/guidancedocuments/ucm071863.htm.
[29] BS EN ISO 5840–2009, Cardiovascular implants – Cardiac valve prostheses.
[30] ISO 5840-2009, Part 3, Cardiovascular implants – Cardiac valve prostheses.
[31] Marquez J, Slater T, Sczerzenie F. Determining the transformation temperatures of NiTi alloys using differential scanning calorimetry. Proceedings of the second international conference on shape memory and superelastic technologies. Pacific Grove (California); 1997. p. 13–18.
[32] ASTM F2004, Standard test method for transformation temperature of Nickel–Titanium alloy by thermal analysis.
[33] ASTM F 2082, Standard test method for determination of transformation temperature of nickel–titanium shape memory alloys by bend and free recovery.
[34] ASTM F2005, Standard terminology for nickel–titanium shape memory alloys.
[35] Motsenbocker T, Goff E. US patent 7,069,794 B2, Radial expansion force measurement technology, 2006.
[36] ASTM Working Document WK15227, Standard Guide for Measuring Radial Loading of Balloon-Expandable and Self-Expandable Stents, 2011.
[37] Ritchie RO, Peters JO. Small fatigue cracks:mechanics, mechanisms and engineering applications, materials transactions. In: Higashida K, Narita N, editors. Recent progress in understanding of materials fracture; Feb 2001.
[38] Weibull W. Fatigue testing and analysis of results. New York: Pergamon Press; 1961.
[39] Basquin OH. The exponential law of endurance tests. ASTM Proc, vol. 10; 1910.
[40] Metallic materials and elements for aerospace vehicle structures, MIL-HDBK-5E, vol. 1; 1987.
[41] Meeker WQ, Escobar LA. Statistical methods for reliability data. New York: John Wiley and Sons; 1998.
[42] ASME V&V 10. Guide for verification and validation in computational solid mechanics. New York, NY; 2006.
[43] ASTM E1823, Standard terminology relating to fatigue and fracture testing.
[44] Marrey RV, Burgermeister R, Krever M, Sleeper A. Probabilistic analysis methods for medical device design. MPMD Proc ASM Int 2007.
[45] American College of Cardiology/American Heart Association 2006 guidelines from the management of patients with valvular disease.
[46] Yoganathan AP, Travis B. Fluid dynamics of prosthetic valves, The practice of Clinical Echocardiography. 2nd ed. Philadelphia, PA: Otto. WB Saunders; 2002.
[47] ASTM F746. Standard test method for pitting or crevice corrosion of metallic surgical implant materials.
[48] ASTM F2129-08. Standard test method for conducting cyclic potentiodynamic polarization measurements to determine the corrosion susceptibility of small implant devices.

[49] ASTM G5. Standard reference test method for making potentiostatic and potentiodynamic anodic polarization measurements.

[50] Pound BG. Susceptibility of nitinol to localized corrosion. J Biomed Mater Res A 2006; 77(1):185–91.

[51] Rosenbloom SN, Corbett RA. An assessment of ASTM F2129 electrochemical testing of small medical implants – lessons learned, paper no. 07674, NACE 2007. Houston, TX; 2007.

[52] ASTM G71. Standard guide for conducting and evaluating galvanic corrosion tests in electrolytes.

[53] ISO-10993 set (part 1 through 20) entails a series of standards for evaluating the biocompatibility of a medical device prior to a clinical study.

[54] ASTM F640-07. Standard test methods for determining radiopacity for medical use.

[55] ASTM F2503-05. Standard practice for marking medical devices and other items for safety in the Magnetic Resonance environment; 2005.

[56] ASTM 20522–F2106e1. Standard test method for measurement of magnetically induced displacement force on medical devices in the magnetic resonance environment; 2006.

[57] ASTM F2213-06. Standard test method for measurement of magnetically induced torque on passive implants in the magnetic resonance environment; 2006.

[58] ASTM 21822–F2202a. Standard test method for measurement of radio frequency induced heating near passive implants during magnetic resonance imaging; 2002.

[59] ASTM F2119-01. Standard test method for evaluation of Magnetic Resonance image artifacts from passive implants; 2001.

[60] FDA Magnetic Resonance safety and compatibility guidance website: http://www.fda.gov/MedicalDevices/DeviceRegulationandGuidance/GuidanceDocuments/ucm107705.htm.

Index

Note: Page numbers followed by "f" and "t" indicate figures and tables respectively

B

D

G

H

I

J

K

L

M

N

O

T

U